The
Encyclopedia
of World
Geography

BRG

BROWN REFERENCE GROUP

THE ENCYCLOPEDIA OF
WORLD GEOGRAPHY

Consultant Editor	Graham Bateman
Project Editor	Victoria Egan
Advisory Editor	Susan Kennedy
Text Editors	Lauren Bourque, Rita Demetriou, Peter Lewis
Art Editor	Steve McCurdy
Designers	Chris Munday, Martin Anderson
Cartographic Manager	Richard Watts
Cartographic Editor	Tim Williams
Picture Research Manager	Claire Turner
Picture Research	Alison Floyd
Jacket Design	Ayala Kingsley
Typesetting	Brian Blackmore
Production	Clive Sparling, Nicolette Colborne

This Brown Reference Group edition
published in 2008

The Brown Reference Group plc.
First Floor
9-17 St. Albans Place
London N1 ONX
www.brownreference.com

© 2004 The Brown Reference Group plc.

ISBN 978-1-84044-251-9

Printed in Thailand

08 09 10 11 10 9 8 7 6 5 4 3 2 1

Introductory photographs
Previous page: Fall scene at Half Dome,
Yosemite National Park, central California
(A. Gin/Powerstock Zefa)
This page: Mount Paektu, North Korea
(Magnum/H. Kubota)

CONTENTS

INTRODUCTION
6

CANADA AND THE ARCTIC
10

Canada 12

Dependencies in the region
Greenland 22
St Pierre and Miquelon 23

THE UNITED STATES
24

United States 26

THE STATES OF THE UNION
48

Alaska 50
Washington 50
Oregon 50
California 51
Hawaii . . . . : 52
Idaho 53
Nevada 53
Montana 53
Wyoming 54
Utah 55
Colorado 55
Arizona 56
New Mexico 56
Oklahoma 56
Texas 57
North Dakota 58
South Dakota 58
Kansas 58
Minnesota 59
Iowa 59
Nebraska 59
Missouri 60
Wisconsin 60
Illinois 60
Michigan 61
Indiana 61
Ohio . . . , ... 62
Maine 62
New Hampshire 63
Vermont 63
Massachusetts 63
Connecticut 64
Rhode Island 65
New York 65
Pennsylvania 66
New Jersey 66
Delaware 67
Maryland 67
District of Columbia 68
Arkansas 68
Louisiana 68
Kentucky 69
Tennessee 69
Mississippi 70
Alabama 70
West Virginia 71
Virginia 71
North Carolina 72

South Carolina 72
Georgia 72
Florida 73

MEXICO, CENTRAL AMERICA
AND THE CARIBBEAN
74

Mexico 76
Guatemala 80
Belize 81
El Salvador 82
Honduras 83
Nicaragua 84
Costa Rica 85
Panama 86
Cuba 87
Bahamas 88
Haiti 88
Jamaica 89
Dominican Republic 90
Saint Kitts-Nevis 90
Antigua and Barbuda 91
Dominica 92
Saint Lucia 92
Saint Vincent and the Grenadines 93
Barbados 94
Grenada 95
Trinidad and Tobago 96

Dependencies in the region
Turks and Caicos Islands 97
Bermuda 97
Cayman Islands 97
Puerto Rico 97
Virgin Islands of the United States 98
British Virgin Islands 99
Anguilla 99
Montserrat 99
Aruba 99
The Netherlands Antilles 99
Guadeloupe 100
Martinique 101

SOUTH AMERICA
102

Colombia 104
Venezuela 106
Guyana 108
Suriname 109
Bolivia 110
Ecuador 111
Peru 112
Brazil 116
Paraguay 120
Chile 121
Argentina 124
Uruguay 127

Dependencies in the region
Falkland Islands 128
French Guiana 128

NORTHERN AFRICA
130

Morocco 132
Algeria 134
Tunisia 136
Libya 138
Egypt 140
Mauritania 144
Mali 145
Niger 146
Chad 147
Sudan 148
Ethiopia 150
Eritrea 151
Djibouti 152
Somalia 153

CENTRAL AFRICA
154

Senegal 156
Gambia 157
Guinea 158
Guinea-Bissau 159
Sierra Leone 160
Liberia 161
Côte d'Ivoire (Ivory Coast) 162
Burkina Faso 163
Ghana 164
Togo 166
Benin 167
Nigeria 168
Cameroon 170
Central African Republic 171
Equatorial Guinea 172
Gabon 173
Congo (Republic of) 174
Congo (Democratic Republic of) 174
Burundi 176
Rwanda 177
Uganda 178
Kenya 179
Tanzania 180
Seychelles 181
São Tomé & Príncipe 182
Cape Verde 183

SOUTHERN AFRICA
184

Angola 186
Zambia 188
Malawi 190
Mozambique 192
Namibia 194
Botswana 195
Zimbabwe 196
South Africa 198
Lesotho 202
Swaziland 203
Madagascar 204
Mauritius 206
Comoros 207

Dependencies in the region
Mayotte 208
Réunion 208
St Helena 208

THE NORDIC COUNTRIES
210

Norway 212
Sweden 216
Finland 222
Denmark 226
Iceland 230

THE BRITISH ISLES
232

United Kingdom 234
Ireland 244

FRANCE AND ITS NEIGHBORS
248

France 250
Andorra 260
Monaco 261

THE LOW COUNTRIES
262

Netherlands 264
Belgium 270
Luxembourg 274

SPAIN AND PORTUGAL
276

Spain 278
Portugal 286

Dependencies in the region
Gibraltar 289

ITALY AND GREECE
290

Italy 292
Vatican City 300
San Marino 301
Greece 302
Malta 306
Cyprus 307

CENTRAL EUROPE
308

Germany 310
Switzerland 318
Liechtenstein 321
Austria 322

EASTERN EUROPE
326

Poland 328
The Czech Republic 332
Slovakia 334
Hungary 336
Romania 338
Bulgaria 340
Yugoslavia (Serbia and Montenegro) . 342
Slovenia 347
Croatia 348
Bosnia Herzegovina 349
Macedonia (FYROM) 350
Albania 351

RUSSIA AND NORTHERN EURASIA
352

Estonia 354
Latvia 355
Lithuania 356
Moldova 357
Belarus 358
Ukraine 360
Russia 362
Georgia 370
Armenia 372
Azerbaijan 374
Turkmenistan 376
Uzbekistan 377
Tajikistan 378
Kyrgyzstan 379
Kazakhstan 380
Mongolia 381

THE MIDDLE EAST
382

Turkey 384
Syria 386
Lebanon 388
Israel 390
Jordan 392
Iraq 394
Kuwait 396
Iran 397
Yemen 399
Afghanistan 400
Saudi Arabia 402
Bahrain 404
Qatar 405
United Arab Emirates 406
Oman 407

THE INDIAN SUBCONTINENT
408

India 410
Bhutan 415
Pakistan 416
Bangladesh 418
Nepal 420
Maldives 421
Sri Lanka 422

Dependencies in the region
British Indian Ocean Territory 423

CHINA AND ITS NEIGHBORS
424

China 426
Taiwan 434

Special Administrative Regions
Hong Kong 436
Macao 437

SOUTHEAST ASIA
438

Myanmar (formerly Burma) 440
Laos 443
Thailand 444
Cambodia 446
Vietnam 448
Malaysia 450
Singapore 452
Brunei 453
Indonesia 454
East Timor 457
Philippines 458

JAPAN AND KOREA
460

Japan 462
North Korea 470
South Korea 472

AUSTRALASIA, THE PACIFIC ISLANDS AND ANTARCTICA
474

Australia 476
New Zealand 482
Papua New Guinea 486
Solomon Islands 487
Tuvalu 488
Nauru 488
Kiribati 489
Fiji 490
Vanuatu 490
Tonga 491
Samoa 492
Palau 493
Marshall Islands 493
Micronesia 494

Dependencies in the region
Northern Mariana Islands 494
Guam 494
Wake Island 494
Johnston Atoll 494
Midway Islands 494
American Samoa 495
New Caledonia 495
Wallis and Futuna Islands 495
French Polynesia 495
Antarctica 496

GLOSSARY
498

FURTHER READING AND ACKNOWLEDGMENTS
502

INDEX
504

INTRODUCTION

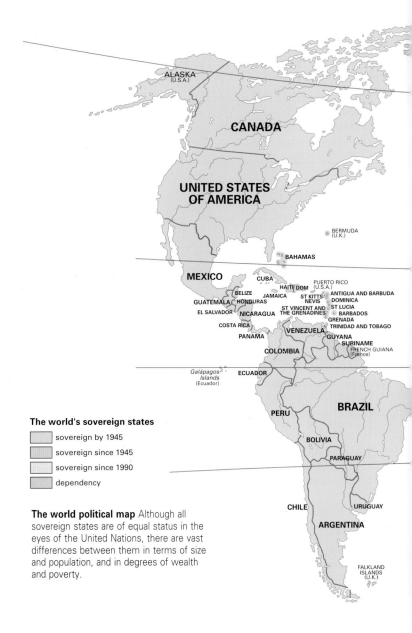

The world's sovereign states

- sovereign by 1945
- sovereign since 1945
- sovereign since 1990
- dependency

The world political map Although all sovereign states are of equal status in the eyes of the United Nations, there are vast differences between them in terms of size and population, and in degrees of wealth and poverty.

THE WORLD IN WHICH WE LIVE IS CHANGING MORE RAPIDLY than at any time in human history. Politically, the collapse of communism in the former Soviet Union and Eastern Europe has ended the Cold War and altered the way that many countries are governed. Borders have been redrawn, new nations have emerged and old countries have been reborn. Notably there are now 15 republics in place of the Soviet Union, Yugoslavia has fragmented, Czecho-slovakia has split in two and Germany has reunited. Although most of these changes have been comparatively peaceful, political realignments and economic reorganizations have created new tensions. In some instances, particularly in the Balkans and some of the former states of the Soviet Union, age-old hostilities between ethnic groups have re-emerged in the form of outright civil war. Environmentally, too, the world is at a crossroads; the speed and degree of our exploitation of global resources have increased dramatically in recent decades, and threaten the very future of the planet. In some cases these abuses have already caused irreversible damage, the ultimate consequences of which can only be surmised.

At the same time, the world is growing smaller; modern communication means that images of an event – whether an earthquake or a revolution – in one country are instantly flashed onto television screens on the other side of the world. The same communications network has inextricably linked the world's financial markets and created a truly global economy where no one currency or stock exchange is truly independent.

Despite the wealth of superficial information we are bombarded with on a daily basis, educators have become increasingly aware of the fact that most of us really know very little about the world beyond our immediate geographical boundaries. While better forms of communication allow us to see other cultures more easily, they do little to help us gain insight into and understanding of the different customs, religions, languages and the wealth and lifestyles of the many nations of the world. All these have been influenced by the physical environment, the actions of other countries and the deeds of particular individuals – soldiers and politicians, religious leaders, artists, scientists and industrialists – who have contributed to the development of each nation. The aim of the *Encyclopedia of World Geography* is to provide, in a single volume, a complete portrait of each of the countries in the world by looking in detail at the

climate and physical features of each, its vegetation and animal life, the major events of its history, the lifestyles of its people, and the components of each country's industry and economy. The result provides the reader with a better view of the world than can be gained from any atlas or almanac and contains more information on each country than any general encyclopedia.

STRUCTURE OF THE ENCYCLOPEDIA

The Encyclopedia of World Geography contains detailed individual country profiles for all 193 nation-states of the world, as well as many dependencies. At the primary level the book is divided into 22 regional sections. These regions have been defined either in geographic terms (South America, Northern Africa, Central Africa and Southern Africa, for example) or according to political criteria (United States, British Isles, Italy and Greece, France etc). Dependencies are given special coverage in one of two ways. If a dependency or island territory is located in the same region as the country to which it belongs, it is treated as a special feature within that country

GREENLAND (Denmark)

SVALBARD (Norway)

Arctic Circle

ICELAND

NORWAY FINLAND
SWEDEN
ESTONIA
LATVIA
DEN LITH
R

UNITED KINGDOM
IRELAND
NETH
BEL GER POLAND
LUX
FRANCE
CZ
SL
AU
CR B
YU
HUN
UKRAINE
BELARUS
MOLDOVA
ROMANIA

RUSSIA

KAZAKHSTAN

MONGOLIA

NORTH KOREA
SOUTH KOREA

JAPAN

PORTUGAL SPAIN
ITALY AL
M
BULGARIA
GEORGIA
GREECE TURKEY
AR AZ
UZBEKISTAN
KYRGYZSTAN
TAJIKISTAN
TURKMENISTAN

CHINA

MOROCCO
Canary Is (Spain)
TUNISIA
MALTA
CYP
ISRAEL L SYRIA
JORDAN
IRAQ
KUWAIT
IRAN
AFGHANISTAN

WESTERN SAHARA (Morocco)
ALGERIA LIBYA EGYPT
BAHRAIN QATAR
UAE
SAUDI ARABIA
OMAN
PAKISTAN
NEPAL BHUTAN
BANG

TAIWAN

Tropic of Cancer

CAPE VERDE
MAURITANIA MALI NIGER CHAD SUDAN
ERITREA YEMEN
INDIA
MYANMAR
LAOS
THAILAND VIETNAM
PHILIPPINES

NORTHERN MARIANA ISLANDS (U.S.A.)

SENEGAL
GAMBIA
GUINEA-BISSAU GUINEA
SIERRA LEONE
BURKINA FASO
BE
CÔTE D'IVOIRE G T
NIGERIA
CENTRAL AFRICAN REPUBLIC
DJIBOUTI
ETHIOPIA
Socotra (Yemen)
SRI LANKA
CAMBODIA
BRUNEI

GUAM (U.S.A.)
MARSHALL ISLANDS

LIBERIA
CAMEROON
SAO TOME & PRINCIPE
EQ
GABON
REPUBLIC OF CONGO
RW
CABINDA (Angola)
DEMOCRATIC REPUBLIC OF CONGO
BU
UGANDA
KENYA
TANZANIA
SOMALIA
MALDIVES
MALAYSIA
SINGAPORE
PALAU
FEDERATED STATES OF MICRONESIA
NAURU KIRIBATI Equator

SEYCHELLES

INDONESIA
EAST TIMOR
PAPUA NEW GUINEA
SOLOMON ISLANDS
TUVALU

COMOROS
ANGOLA
ZAMBIA MALAWI
MOZAMBIQUE
MADAGASCAR
MAURITIUS

SAMOA

NAMIBIA
ZIMBABWE
REUNION (France)
VANUATU
FIJI

BOTSWANA
SWAZILAND
SOUTH AFRICA LESOTHO
NEW CALEDONIA (France)
TONGA
Tropic of Capricorn

AUSTRALIA

NEW ZEALAND

A	ANDORRA
AL	ALBANIA
AR	ARMENIA
AU	AUSTRIA
AZ	AZERBAIJAN
B	BOSNIA HERZEGOVINA
BANG	BANGLADESH
BE	BENIN
BEL	BELGIUM
BU	BURUNDI
CR	CROATIA
CYP	CYPRUS
CZ	CZECH REPUBLIC
DEN	DENMARK
DOM	DOMINICAN REPUBLIC
EQ	EQUATORIAL GUINEA
G	GHANA
GER	GERMANY
HUN	HUNGARY
L	LEBANON
LITH	LITHUANIA
LUX	LUXEMBOURG
M	MACEDONIA (Former Yugoslav Republic of)
NETH	NETHERLANDS
R	RUSSIA
RW	RWANDA
S	SLOVENIA
SL	SLOVAKIA
SW	SWITZERLAND
T	TOGO
UAE	UNITED ARAB EMIRATES
YU	YUGOSLAVIA

entry. For example, Corsica is treated as a special feature within the France entry. However, if the dependency is located in a different region, it is featured at the end of its regional section. For example, French Guiana is featured at the end of the regional section on South America, not France.

Each region is introduced by a high-quality topographic map showing key physical features (rivers, mountains, lakes, etc), national boundaries and country names, plus major cities and towns. These maps give the fundamental physical background to the country profiles that follow.

COUNTRY PROFILES
The core of the Encyclopedia comprises individual entries on every country of the world, which vary in length from one page (or less for small island states) to 50 pages for the United States. The United States' section has been divided into two parts: an extensive national profile followed by a section covering each of the States separately. Whatever the length of the entry, the text has been prepared to present information in a consistent and logical fashion that enables the reader easily to make comparisons between countries. Each profile is divided into three major parts – GEOGRAPHY, SOCIETY and ECONOMY.

The GEOGRAPHY section describes the physical landscape including main areas of mountains, deserts and plains, and the main water features – rivers and lakes. The characteristic climatic conditions are outlined and native plants and animals described.

The second major section, SOCIETY, puts the people into the context of their environment. First the history of the country is told chronologically from earliest known times up to the present. This historical perspective is crucial to our understanding of the country today; subsequent subsections describe the present-day system of government, the ethnic groups that make up the population, their religious beliefs and other cultural distinctions.

The final section – ECONOMY – describes how the people have utilized the land and its resources to build an economy that funds (or fails to fund) their nation's requirements. It lists the key features of agriculture and industry, explains how local trade and commerce operate, and describes the transportation and communications systems. The last subsection looks at health, welfare and education.

△ Opening pages to one of 22 regions with topographic map and locator within its continent.

▷ Typical two-page entry on a country.

1 The country's popular name followed by its official name in smaller type.

2 Map including major cities, rivers, lakes and seas, plus a locator map and national flag.

3 Authoritative text organized under three major headings: Geography, Society and Economy.

△ Typical pages from a longer country entry.

4 Diagram explaining the system of government of selected countries.

5 Fact panel with current statistics on population, geography, economy and government.

6 Special features focusing on key people, notable historical events and island territories.

FEATURE PANELS
Within many entries special features have been included to give a deeper appreciation of the country. These self-contained, illustrated articles focus on elements of history, culture or geography that are integral to that nation's "personality". They include discoveries such as Peary's race for the North Pole (Canada), monuments of culture such as Florence (Italy), lost civilizations, for instance, the Incas (Peru), technological achievements like the Space program (US) and biographies of the individuals who are forever linked to the history or heritage of their country – whether Lincoln (United States), Shakespeare

(United Kingdom) or Erik the Red (Canada). Features on geography include farflung territories such as Easter Island (Chile), unique agricultural regions, for instance, Madeira (Portugal) and pivotal sites like Egypt's Suez Canal.

FACTS AND STATISTICS
Reference data is incorporated throughout the Encyclopedia in text, maps and artwork. In addition, special fact panels have been compiled for every country. The structure of these panels broadly mirrors the sequence of information in each country profile. Summary details and the latest statistics are given for land area, climate, major physical features, population, form of government, composition of armed forces, largest cities (including the capital), official language, ethnic

ARCTIC OCEAN
Franz Josef Land
Severnaya Zemlya
New Siberian Islands
Laptev Sea
East Siberian Sea
Greenland
Svalbard
Kara Sea
Arctic Circle
Novaya Zemlya
Barents Sea
Iceland
Bering Sea
Lena
Central Siberian Plateau
Kamchatka Peninsula

Physical zones
ice and snow
tundra
mountains/barren land
forest
grassland
semidesert
desert

▲ mountain peak (meters)

Ob
Yenisei
West Siberian Plain
Sea of Okhotsk
British Isles
North Sea
Baltic Sea
URAL MTS
Irtysh
L. Baikal
Amur
Sakhalin
North European Plain
Carpathian Mts
Volga
Altai Mts
Kuril Islands
Hokkaido
Azores
Mt Blanc 4807 ALPS
Danube
Aral Sea
Syr Darya
L. Balkhash
Tien Shan
Gobi Desert
Sea of Japan
Pyrenees
Caucasus Mts
Caspian Sea
Amu Darya
Kunlun Shan
Honshu
Black Sea
Huang
Atlas Mts
Mediterranean Sea
HIMALAYAS
Yangtze
East China Sea
Tropic of Cancer
Canary Islands
SAHARA
Libyan Desert
Euphrates
Tigris
Indus
Mt Everest 8850
Thar Desert
Ganges
Taiwan
PACIFIC OCEAN
Ahaggar Plateau
Tibesti
Nile
Arabian Peninsula
Persian Gulf
Deccan Plateau
Salween
Verde Islands
SAHEL
Red Sea
Arabian Sea
Bay of Bengal
Hainan
South China Sea
Mariana Islands
Marshall Islands
Niger
Socotra
Mekong
Philippine Islands
Caroline Islands
Ethiopian Highlands
Maldives
Andaman Islands
Nicobar Islands
Sri Lanka
Equator
Gilbert Islands
Mt Cameroon 4070
Congo
L Turkana
Borneo
Gulf of Guinea
L Victoria
Sumatra
Sulawesi
Bismarck Archipelago
Ascension
Kilimanjaro 5895
Seychelles
Java
New Guinea
Solomon Islands
Ellice Islands
Rift Valley
L Tanganyika
INDIAN OCEAN
Timor
Samoa Islands
St Helena
Comoros
L Malawi
Coral Sea
New Hebrides
Fiji Islands
Tonga Islands
Zambezi
Madagascar
Mauritius
Réunion
New Caledonia
Tropic of Capricorn
Limpopo
Mozambique Channel
Kalahari Desert
Great Sandy Desert
Macdonnell Ranges
L Eyre
Great Dividing Range
Darling
Drakensberg Mts
Cape of Good Hope
Tristan da Cunha
Mt Kosciusko 2230
Tasmania
Tasman Sea
New Zealand
uth Georgia
Mt Cook 3764

Weddell Sea
Enderby Land
Wilkes Land
Antarctic Circle
Queen Maud Land
Victoria Land
Antarctica
Coats Land

The landscapes of the world are a reflection of relief, climate and vegetation. Vegetation is strongly affected by latitude. As precipitation decreases forests give way to grasslands and deserts.

composition and religious affiliations, currency, gross domestic product for the country and gross national product per person, average life expectancy of men and women and the major resources that contribute to the economy, natural and agricultural. Information for these panels has been researched from numerous sources providing a unique data resource for student and researcher alike.

MAPS AND OTHER ILLUSTRATIONS

No book on geography would be complete without maps to depict the location of features described. *The Encyclopedia of World Geography* is no exception. In the early pages there are maps showing all the nation-states of the world and the major global physical features. In addition to the topographic map for each regional section, each country entry

has a small map showing major cities (including the capital), rivers and lakes, and adjacent countries, oceans and seas. The position of the country within the continent appears on an inset locator map. The national flag of each country is also depicted. Other maps show major federal divisions of certain countries, where relevant.

While this Encyclopedia is bulging with reference information, it is also a visual celebration of the diversity of the lands and peoples that occupy this planet. Hundreds of color photographs have been specially selected to present a stunning panorama that displays scenic beauty and evokes local life. These are supplemented with detailed captions. Schematic diagrams showing how the systems of election and government work in selected countries are also included.

Canada and the
Arctic

COUNTRIES IN THE REGION

CANADA

DEPENDENCIES IN THE REGION

GREENLAND · ST PIERRE AND MIQUELON

The maple leaf, Canada's national symbol, emblazoned on garage doors in Saskatchewan. The maple leaf's official status was only confirmed in 1965; it became the central motif of the new national flag introduced in that year. Syrup is made from the sweet sap of the sugar maples that are native to the deciduous forests of the eastern part of the country, and whose leaves in fall turn a brilliant red.

ARCTIC OCEAN

Ellesmere Island

GREENLAND
(Denmark)

Thule
(Qaanaaq)

North Magnetic Pole
1985

Queen Elizabeth Islands

Devon Island

Banks Island

Resolute

Somerset Island

Baffin Bay

Prince of Wales Island

Baffin Island

Inuvik

Victoria Island

Boothia Peninsula

Godthåb
(Nuuk)

Cape Farewell

Great Bear Lake

Melville Peninsula

Davis Strait

Mt Logan
5951

Selwyn Mts

Mackenzie Mountains

Mackenzie

Back

Iqaluit

ATLANTIC OCEAN

Whitehorse

ROCKY MOUNTAINS

Cassiar Mts

Southampton Island

Hudson Strait

Great Slave Lake

Yellowknife

PACIFIC OCEAN

Coast Mountains

Hay River

Lake Athabasca

Ungava Peninsula

Labrador

Peace

Hudson Bay

Churchill

Goose Bay

Queen Charlotte Islands

Williston Lake

Great Plain

Athabasca

Reindeer Lake

Churchill

Nelson

Canadian Shield

Newfoundland

Mt Robson
3954

Fraser

Edmonton

Saskatchewan

Laurentian Highlands

St John's

Mt Waddington
3994

Selkirk Mts

Saskatoon

Lake Winnipegosis

Lake Winnipeg

Fort Albany

CANADA

Albany

Sept-Iles

Gulf of St Lawrence

Cape Race

St-Pierre & Miquelon
(France)

Vancouver Island

Vancouver

Calgary

Regina

Lake Manitoba

Winnipeg

Quebec

St Lawrence

Appalachian Mts

Prince Edward I

Cape Breton Island

Thunder Bay

Sault Ste Marie

Montreal

Ottawa

Nova Scotia

Halifax

Lake Superior

L Huron

Toronto

Lake Ontario

Cape Sable

Niagara Falls

Lake Michigan

Lake Erie

- capital city
- major town

height of land (meters)

3000
2000
1000
500
200

permanent ice

▲ mountain peak

Canada

CANADA

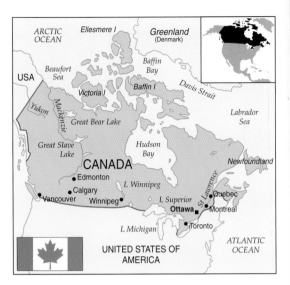

ARCTIC OCEAN
Ellesmere I
Greenland (Denmark)
USA
Beaufort Sea
Baffin Bay
Victoria I
Baffin I
Davis Strait
Yukon
Mackenzie
Great Bear Lake
Labrador Sea
Great Slave Lake
Hudson Bay
Newfoundland
CANADA
Edmonton
Calgary
L Winnipeg
Vancouver
Winnipeg
L Superior
St Lawrence
Quebec
Ottawa
Montreal
L Michigan
Toronto
ATLANTIC OCEAN
UNITED STATES OF AMERICA

CANADA OCCUPIES MOST OF THE NORTHERN half of the North American continent. Its economy and culture show many similarities with the United States – so much so that some outsiders tend to think of it as no more than a satellite. However, nothing could be further from the truth: for Canada has its own distinct character and rich ethnic mix. Its unique bilingual and bicultural society can be traced back to the nation's French and British origins.

NATIONAL DATA – CANADA

Land area	9,093,507 sq km (3,511,023 sq mi)			
Climate		Temperatures		Annual
	Altitude m (ft)	January °C(°F)	July °C(°F)	precipitation mm (in)
Resolute	64 (200)	-32 (-26)	4 (40)	138 (5.4)
Vancouver	0 (0)	3 (38)	17 (63)	1,199 (47.2)
Winnipeg	248 (813)	-18 (0)	20 (67)	514 (20.2)
Montreal	30 (98)	-12 (10)	20 (67)	1,065 (41.9)
Halifax	30 (98)	-6 (21)	19 (65)	1,542 (57.1)

Major physical features highest point: Mount Logan 5,951 m (19,524 ft); longest river: Mackenzie 4,240 km (2,635 mi); largest lake: Lake Superior (part) 83,270 sq km (32,150 sq mi)

Population (2006 est.) 33,098,932

Form of government federal multiparty parliamentary monarchy with two legislative houses

Armed forces army 33,000; navy 12,000; air force 17,100

Largest cities Toronto (4,726,194); Montreal (3,290,804); Vancouver (1,840,441); Calgary (1,013,675); Ottawa (capital - 896,048); Edmonton (841,202); Hamilton (699,795); Quebec (650,026); Winnipeg (633,107); London (350,340)

Official languages English, French

Ethnic composition British 28%; French 23%; other European 15%; Amerindian 2%; other mainly Asian, African, Arab 6%; mixed background 26%

Religious affiliations Roman Catholic 42.6%; Protestant 23.3%; other Christian 4.4%; Muslim 1.9%; other and unspecified 11.8%; none 16%

Currency 1 Canadian dollar (CAD) = 100 cents

Gross domestic product (2006 est.) $1.165 trillion

Gross domestic product per capita (2006 est.) $35,200

Life expectancy at birth male 76.86 yr; female 83.74 yr

Major resources coal, natural gas, petroleum, hydropower, iron ore, nickel, zinc, copper, gold, lead, molybdenum, potash, diamonds, uranium, rock salt, fisheries, tourism, cereals, dairy and meat products, food stuffs, timber, tobacco

GEOGRAPHY

Canada covers a much greater expanse than the United States. This vast and often mountainous country is second only in land area to Russia, though its population is very much smaller.

The land
Canada can be divided into several geographical areas, of which the largest is the so-called Canadian Shield, occupying almost half the country. This bowl-shaped area centered on Hudson Bay consists of ancient rocks that were once raised into mountainous forms. It extends as far east as the Gulf of St Lawrence, and west as far as the lowlands around Great Bear Lake and Great Slave Lake in the Northwest Territories.

Southeast of the Shield lie the lowlands of the Great Lakes–St Lawrence river area, occupying southern Ontario and southwestern Quebec. This much smaller area is the most densely populated part of Canada. The gently rolling country of the Bruce Peninsula is bounded by Lake Huron and Georgian Bay, Lake Erie and Lake Ontario. It is traversed by the Niagara Escarpment over which the Niagara Falls cascade. The Ontario Lowlands to the northeast of the escarpment are an ancient glacial plain covered in

The Kicking Horse river (*above*) A canoe follows the path of early explorers through the Rocky Mountains in British Columbia. Today five of Canada's national parks lie within the spectacular scenery of the Rockies.

The Klondike highway (*left*) cuts through a forested landscape of snow-clad mountains that captures the romantic image of Canada's far north. In the two years between 1897 and 1899, some 30,000 people came this way in search of gold.

drumlins – low ridges of rocky debris left by the retreating glaciers. Farther northeast lie the lowlands of the Ottawa Valley and southern Quebec.

East of Quebec is the region known as Appalachian Canada, which includes the Atlantic provinces of New Brunswick, Prince Edward Island, Nova Scotia and the island of Newfoundland. Here ancient rocks have been folded and reshaped over time into craggy, mountainous country, now much eroded by weather and glaciers. Among the highest of the mountain chains are the Nôtre Dame Mountains that run northeast from Quebec, and the Long Range Mountains of Newfoundland. Southwest of the Shield and Hudson

Bay is the Great Plains area of Canada – a vast triangular expanse of more level country covering about one-fifth of the land area. The Manitoba Lowlands are seamed with a network of lakes, most of them remnants of one immense lake that covered the whole area during the last glacial retreat. The deep sediments left behind provide rich farming country.

Westward, the land rises over the dissected edge of the Manitoba Escarpment, beyond which are found the rolling plateaus of the Saskatchewan Plain. Here, torrential meltwaters from the ice-age glaciers have carved wide, steep-banked river channels through the countryside. Westward through Alberta the land con-

tinues to rise. It has been severely eroded since the glaciers last retreated to form a rough plateau area.

West of the Great Plains is the Canadian Cordillera – the mountainous area that borders Canada's Pacific coast. A great mountain belt some 800 km (500 mi) wide extends from Alaska through the Yukon Territory, western Alberta and British Columbia. The western spine of the Cordillera is formed by the Coast Mountains. The range then turns south, interrupted by plateaus, to form the Mackenzie Range and eventually the Rocky Mountains that form the eastern spine of the Cordillera. These mighty peaks of relatively young rock rise to heights of more than 3,000 m (10,000 ft).

Along the western seaboard lies a parallel and even higher chain. The St Elias Mountains to the north include Mount Logan, which at nearly 6,000 m (20,000 ft) is Canada's highest peak. The Coast Mountains run southward along the coast, cut by deep fjords or inlets of the sea. Offshore, Vancouver Island and the Queen Charlotte Islands are the peaks of yet another parallel mountain range, while north of the Shield lies the Arctic.

To the northwest the broad archipelago

of the Arctic islands border the permanent ice of the Arctic Ocean. Much of the ocean's surface is covered throughout the year by a layer of ice at least 3 m (10 ft) thick, yet less than half the surrounding land is ice-covered. The islands rise toward the northeast, becoming mountainous in Ellesmere Island and along the east coast of Baffin Island. Their barren, rocky terrain is largely covered with permanent or semipermanent fields of snow and ice, and around the mountain peaks there is extensive glaciation.

Climate
More than half of Canada has a subarctic climate, with cool summers and extremely cold winters – often colder than in the Arctic itself, where great extremes of temperature are moderated by the waters of the Arctic Ocean just below the ice. In the polar regions winter begins in August. Temperatures drop rapidly until December, then remain low until March. The more equable waters of the Atlantic Ocean also moderate the climate where they flow into the Arctic, producing warmer winters, colder summers and higher precipitation. Summer temperatures are usually more uniform throughout the Arctic, though some continental areas may enjoy short periods of almost Mediterranean warmth.

Farther south the climate is typically continental, with hot summers and cold winters. Southern Ontario, among the Great Lakes, has the warmest summers. The southeast coast and especially the west coast enjoy the mildest winters. Onshore winds bring heavy rain and snow to the Coast Mountains, but the air over the interior is dry, especially in winter. In the summer, the famous prairie thunderstorms help to redistribute moisture across the Great Plains. Farther east the Great Lakes provide some moisture, while moist Atlantic air supplies ample rainfall to the Atlantic seaboard, where the winter snowfall is particularly heavy.

Vegetation
The treeline in northern Canada arches northward from Labrador on Canada's east coast across the Ungava Peninsula, then down the east side of Hudson Bay. Across this bay it sweeps sinuously northwest to the lower Mackenzie Valley and Alaska. North of it there is little or no fertile soil, and most of the country, known as the Barren Grounds, is tundra.

Farthest from the pole the tundra vegetation includes low shrubs, grasses and sedges. In the most northerly lands, the cushion plants of the polar desert cover less than one-tenth of the ground surface.

South of the treeline, the boreal region forms one of the world's largest coniferous forests, stretching from Alaska to Newfoundland.

The eastern forests, from the Great Lakes to the coasts, are mostly mixed, with sugar maple, beech, birch, pine and hemlock. But the lowland plains of the far south are covered with broadleaf woodland: hickory, oak and elm flourish here, along with chestnut, maple and walnut. In the western mountains spruce, Douglas fir and lodgepole pine are the commonest, with aspen and yellow pine in

The icy grip of winter blankets Montreal in snow. Although lying in the same latitude as Florence in Italy, Montreal has bitter winters with temperatures falling below 0°C (32°F) for more than 150 days each year.

plateau areas. On the Pacific coast, with its high rainfall, are some of the world's most imposing forests – dense, towering ranks of Douglas fir, western red cedar and hemlock.

The prairie country is too dry for more than a scattering of tree cover and was originally wide rolling grassland. Today little of this remains, the rich black soils having been plowed up and turned into the famous wheat-growing country.

Wildlife
Canada's Arctic waters support whales, walruses and seals, along with the semi-aquatic Polar bear. On the tundra there are Musk oxen, caribou, wolves, Arctic foxes and lemmings; many migratory birds also spend the summer here, including auks, sea ducks, gulls, terns and

shorebirds. Caribou flourish in the northern forests, along with moose (known as elk in Europe), lynx, and Black and Brown bears. Beavers, martens, muskrats, mink and other fur animals were once, and to a degree remain, the basis of the fur trade.

Farther south the White-tailed deer is common, while the more settled country harbors mostly smaller mammals such as Gray and Red squirrels, chipmunks, weasels and otters. Among the varied and noisy bird life are cardinals, wood warblers, Baltimore orioles and catbirds.

The prairies support smaller creatures such as jackrabbits, gophers and Sharp-tailed grouse as well as some surviving bison and Pronghorn antelope. The western mountains support highly adapted species such as the Bighorn sheep and the Mountain goat.

Robert Peary and the race for the North Pole

As early as 1607 Henry Hudson (c.1565–1611) set out by ship in search of a route across the North Pole to China. He started by sailing northward near the east coast of Greenland, but turned east when he encountered the Arctic pack ice. For 300 years the North Pole remained unconquered. Many attempts were made to reach it but they all failed. In 1895–96 the Norwegian explorer Fridtjof Nansen (1861–1930) in a specially constructed

Robert Peary's 1905 expedition (*above*) This contemporary artist's impression shows the team that penetrated as far north as 87° 6′ N. Another three years were to pass before he reached his goal.

Nansen's ship, the *Fram* (*left*), as shown on a Wills' cigarette card. The *Fram* was specially constructed to withstand the pressure of Arctic ice.

vessel, the *Fram*, proved for the first time that the polar seas were entirely covered in ice; but his attempt to reach the pole by sledge did not succeed.

By this time an American naval officer named Robert Edwin Peary (1856–1920) had begun his own preparations for a sledge journey to the pole. He used a "pyramid" approach whereby, beginning with a large group, he would set up a chain of depots, sending back some of the support sledges each time. After several close approaches, in 1908 he

finally achieved his goal. His expedition took him from Ellesmere Island's Cape Columbia all the way to the North Pole. There were five people on the final run: Peary himself, three Inuit, and his dog-driver Matthew Henson (1866–1955).

On his return in 1909, Peary learned that his former companion Frederick Albert Cook (1865–1940) was claiming to have reached the pole before him. Cook's proof was unsatisfactory, but the evidence is hard to judge, and the debate continues.

SOCIETY

Canada's democratic government derives from that of Britain, of which it was once a colony. It has also retained elements of its French heritage. It is a country that has come increasingly to accept its obligations toward its indigenous peoples. After centuries of immigration, its population has great diversity.

History

The first peoples to enter the Americas almost certainly came from Siberia via the

Bering Strait and Alaska during the last ice age between 30,000 and 10,000 years ago and gradually spread southward through the rich, uninhabited land over thousands of years into the heart of the continent. The ancestors of today's Inuit were the last people to move into North America from Asia about 4000 BC, reaching Greenland about 2500 BC. In the western Arctic the traditional Inuit culture, based on hunting for food in small family groups, still survives, but old patterns are changing: snowmobiles have replaced dog teams, while guns and whaling bombs have replaced harpoons, and today's Inuit boats are driven by outboard motors. Their summer tents and winter igloos have been replaced by modern buildings, and many have relocated to large settlements.

Europe's first definite contact with Canada came with the arrival of Viking adventurers such as the Norse explorer Leif Eriksson (originally from Greenland) and the Icelander Thorfinn Karlsefni in about AD 1000. The Vikings did not stay, however, perhaps because of the hostility of the local Native Indian peoples.

In 1497 an Italian navigator called John Cabot (1450–98) led an English expedition to discover unknown lands, and arrived at Newfoundland. In the 16th century explorers began to penetrate the interior. Among them were Henry Hudson (c. 1550–1611) from England, Jacques Cartier (1491–1557) who is supposed to have given Canada its name, and his fellow-Frenchman Samuel de Champlain (1567–1635), who is known as the Father of New France for his work in founding the colony that grew along the St Lawrence river from Quebec to Montreal.

French influence and colonization also invaded the Maritime Provinces, and from its St Lawrence core spread with the fur trade across the Great Lakes to the Mississippi and Saskatchewan river watersheds. However, French northern and western fur interests came increasingly to be rivaled by those of the British Hudson's Bay Company, founded in 1670. During the next century, growing friction between England and France culminated in the capture of Quebec and Montreal by the British, led by General James Wolfe (1727–59) in 1759. This was followed by the Treaty of Paris in 1763, by which New France was ceded to Britain as the colony of Quebec, with its Maritime territory.

During and after the American War of Independence (1775–83), when the North American colonies were in revolt against British rule, several thousand British loyalists fled to Canada, especially to Nova Scotia and Quebec. In 1791 the separate colonies of Upper Canada (Ontario) and Lower Canada (Quebec) were created by the Constitutional Act of 1791. In 1837 dissatisfaction in both Upper and Lower Canada with British economic policy, along with the continuing tight constraints on self-government, brought open revolt. The uprising failed, but it did bring about political reform and the reunification of the two Canadas. In 1849 the tiny colony of Victoria was created on Vancouver Island in the far southwest of the region. Following the subsequent gold rush in the Fraser Valley in the southwest, the colony of British Columbia was established in 1858.

In 1867 the British North America Act created the Dominion of Canada, comprising the provinces of Quebec, Ontario, Nova Scotia and New Brunswick. The Dominion Government bought the Hudson's Bay Company lands, which became the Northwest Territories. In 1870 Manitoba joined the Dominion, followed in 1871 by British Columbia. Prince Edward Island followed in 1873, the Yukon Territory in 1898 and Saskatchewan and Alberta in 1905. Newfoundland remained nominally independent until 1949. Canada today is composed of 10 provinces and three national territories, including the new territory of Nunavut, which came into existence in 1999 as a response to Inuit land claims in the Northwest Territories.

Meanwhile, Canada's prosperity had risen, swelled by the enormous grain production across the recently settled prairies, by the exploitation of northern forests for lumber and pulp, and also by the development of other natural resources. The discovery of gold deposits along Klondike Creek in the Yukon in 1896 precipitated the famous Gold Rush.

All these developments disrupted the way of life of the indigenous population far more than the withdrawal of the fur trade or the destruction of the bison. There followed an armed uprising among the *Métis* – a rural people of mixed French and Native Indian descent – led by Louis Riel (1844–85). Riel was hanged in 1885.

After Canada's vigorous support of the Allies (Britain, France, Italy, Russia and the United States) during World War I, Britain gave the Dominion greater autonomy by the Statute of Westminster of 1931. Not until 1982 did Canada gain full control of its own constitution.

Since the 1960s French-Canadian nationalists in Quebec have campaigned vigorously for independence from Canada, and in 1976 the Parti Québecois won control of the provincial government. In 1980 Quebec voters decisively rejected independence in a referendum and in 1995 a similar referendum was defeated. Nationalists have continued to press for independence, but in 1998 the Supreme Court ruled that without a majority vote, Quebec had no right to unilaterally secede from Canada.

Government
Essentially Canada still remains the federation of states set up by the 1867 British North America Act. As a constitutional monarchy, Canada is headed by a sovereign, now specifically called the sovereign of Canada and not the sovereign of the British Isles.

In practice the sovereign's power is exercised by a governor-general, usually appointed for a six-year term on the advice of the Canadian prime minister. The governor-general gives the royal assent to new legislation, without which it cannot become law. To receive assent, legislation must first pass both houses of the federal parliament: the senate (upper house) and the house of commons. Both houses can introduce legislation, but only the commons can introduce new taxation or expenditure. The federal parliament has authority over all international affairs and all major domestic ones.

Each of Canada's 10 provinces has a governmental structure mirroring that of the nation. A lieutenant-governor appointed by the governor-general summons a single-chamber elected legislative assembly, in which the majority leader becomes premier and appoints a cabinet. Provincial governments are responsible for local concerns such as civil law, local taxation, land management, local trade

The Canadian provinces (*below*) Quebec was first settled by the French, from 1534, but in 1763 the whole country was ceded to Britain. Names like Saskatchewan preserve Canada's Native Indian inheritance.

Provinces of Canada
■ national capital
● provincial capital

YUKON TERRITORY

NORTHWEST TERRITORIES

NUNAVUT

Iqaluit ●

Whitehorse ●

Yellowknife ●

BRITISH COLUMBIA

ALBERTA

CANADA

MANITOBA

NEWFOUNDLAND

St. John's ●

Edmonton ●

SASKATCHEWAN

ONTARIO

QUEBEC

St.-Pierre & Miquelon (France)

Victoria ●

Regina ●

Winnipeg ●

PRINCE EDWARD I.
Charlottetown ●
NEW BRUNSWICK
Quebec ● Fredericton ●
Halifax ●
NOVA SCOTIA

Ottawa ■

Toronto ●

Francis McClintock's Arctic expedition (*below*) in 1859 cuts roads for sledges through heavy snow hummocks in Queen's Channel in Canada's far northwest. McClintock's expedition found the logbook of an earlier British explorer, John Franklin, detailing the illnesses that fatally overwhelmed his expedition in 1845.

and commerce, hospitals, health, welfare and education. The huge but very sparsely populated Yukon, Nunavut, and Northwest Territories are administered directly by the federal government.

People

Canadians of British descent form the largest ethnic group, accounting for somewhat less than half the total. French Canadians form about one-quarter of the

Canada's system of government has grafted the federalism of the neighboring United States onto a parliamentary system modeled on that of the British. The two-chamber federal parliament consists of the senate, whose 104 members are appointed for life, and the 282-member house of commons, elected for a maximum of five years. Legislation must be passed by both houses to become law.

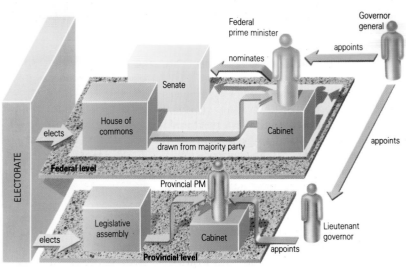

national population and some four-fifths of Quebec's. A further one-tenth of Canadians are of mixed origin, the largest group being the *Métis*. Most of the rest are descended from European immigrants, especially German, Italian, Ukrainian and Dutch. Native Indians and Inuit, the original inhabitants, now account for less than one-fiftieth of the population.

Most Canadians speak English as their first language. The exception is in Quebec, which has a French majority – rigorous laws here protect and encourage French language and culture. Roman Catholics slightly outnumber Protestants in Canada's current Christian population. However, heavy immigration from Asia in recent decades has greatly increased the variety of religions.

ECONOMY

Canada is rich in resources, with massive mineral deposits, enormous areas of commercial-grade timber and vast areas under cultivation. The resulting prosperity has created one of the highest standards of living in the world.

Agriculture and forestry

Agriculture is one of Canada's major industries, though today it employs only one-twentieth of the labor force on one-twentieth of the total land area. Most agricultural land lies in the south; in more northerly latitudes the growing season is too short.

British Columbia's extensively irrigated southern plateau and Fraser delta lands are almost entirely given over to fruit growing, while cattle and sheep ranching prevail farther north in parts of the interior. Almost four-fifths of Canada's farmland lies between the Rocky Mountains and the Great Lakes. In the prairie provinces hard wheat remains the dominant crop, though there is increased competition from other grains.

In Alberta, cattle, pigs and sheep are raised, principally in dryer areas and in the foothills. Similar livestock emphasis is characteristic of the more marginal parts of Manitoba. Southwestern Ontario is Canada's corn and soybean belt, but it also produces most of the country's tobacco and much of its fruit and vegetables. Apart from wheat, other agricultural export crops include feed grains, oilseeds, apples, potatoes and maple syrup.

Canada is the world's leading exporter of wood products. Sawn timber, plywood and pulp for newsprint account for more than one-tenth of exports, and there is potential for even higher production levels. The most productive area is British Columbia, the majority of which is covered by forests that are largely exploited by major multinational companies. Quebec's massive forests are not yet fully exploited and most of the Atlantic provinces have a well-developed woodland industry.

Fishing and the fur trade

Historically, the Canadian fishing industry centered on Newfoundland. Today's catches are taken principally by large trawlers fishing the banks out to the extended 400 km (250 mi) limit set by the Canadian government in 1977. Ashore,

Downtown Toronto (*left*), and the CN tower. Modern buildings in the lower part of the city have replaced those that were destroyed by fire in 1904.

A baseball player slides into base (*right*) Baseball is one of the less controversial imports from the United States, whose economic domination is still sometimes seen as a threat in Canada.

French influence (*below*) in Montreal, the main commercial center in Quebec province, where French language and culture are protected and encouraged by law.

fish are frozen or canned for export. These limits and subsequent quota regulations have brought sharp disputes with France over fishing rights around the islands of St Pierre and Miquelon, but have helped to revive the ailing Pacific fisheries, famous for their salmon. Canada's extensive freshwater lakes are also exploited for both commercial and sport fishing.

Furs remain an important export, notably from Quebec, Saskatchewan, the Northwest Territories and the Yukon.

Industry and power

Canada's mineral wealth is very considerable and makes a major contribution to its exports. The western provinces, particularly Alberta, have the greatest part of

Canada's known fossil fuel deposits, including oil, gas and coal. Gas and oil have also attracted much recent attention in the Mackenzie Valley and under the Beaufort Sea. Northern Ontario and Quebec have rich iron mines exploited in the past, and they contain some of the largest untapped low-grade iron-ore deposits in the world. Other important minerals include zinc (notably in New Brunswick), nickel, copper, lead, uranium (in the north of Quebec, Ontario and Manitoba), potash (in Saskatchewan) and molybdenum (in British Columbia).

Manufacturing accounts for nearly one-fifth of Canada's gross national product and about one-sixth of the workforce. The most important industries are petroleum refining and automobile production. Apart from pulp and paper production, other significant activities include meat-packing and iron and steel production. Ontario is the country's most important industrial area. The prairie provinces produce food, beverages, petroleum and chemicals. British Columbia produces machinery and chemicals.

Hydroelectricity remains the most important source of power, but there are also thermal and nuclear power stations.

Trade and commerce
Canada's economy is heavily dependent on trade, and some three-quarters of all Canada's trade is with the United States. The economy achieved an average real growth rate of over 2 percent GDP in the late 1990s. Canada remains a rich country with prospects of solid economic growth.

Transportation and communications
Many northern communities are accessible only to all-terrain vehicles and aircraft where many so-called bush pilots operate. Air Canada and Canadian Airlines provide both domestic and international flights.

There are two transcontinental railroad systems: the privately owned Canadian Pacific Railway and the Canadian National Railway. The British Columbia Railway, which is controlled by the provincial government, provides a vital link to the province's central and northern areas. A similar provincial railway links Moosonee on James Bay at the southern tip of Hudson Bay to southern Ontario. The hub of the whole system is Montreal, which connects not only with the Canadian West, but also with the American Midwest, New York, the seaboard cities and the ports of the Atlantic provinces.

In the Atlantic provinces ferries still

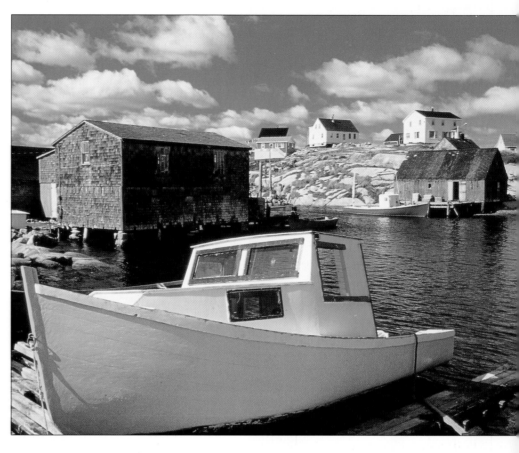

The fishing village of Peggy's Cove on the coast of eastern Canada. It was to exploit the yearlong coastal fishing, notably around Newfoundland, that Europeans first settled permanently in Canada.

provide essential road, rail and passenger links. The ports of Halifax and St John in Nova Scotia and Newfoundland respectively carry much of Canada's external Atlantic trade, especially in winter, while Vancouver and Prince Rupert on the west coast are Canada's year-round Pacific outlets. The St Lawrence Seaway connects the Atlantic Ocean lanes directly with Great Lake ports such as Chicago and Thunder Bay in the heart of the continent.

In British Columbia one of the largest ferry fleets in the world plies between Vancouver Island and the mainland, where Vancouver itself serves the northwestern coastal ports with freight and passengers. In the Northwest Territories the Mackenzie water system connects the Arctic Ocean with the Mackenzie river, the Great Slave Lake and Lake Athabasca; the Great Slave Lake Railway and the Mackenzie Highway connect with the south.

About half Canada's 885,000 km (550,000 mi) of roads are in the prairie provinces. The Trans-Canada Highway links the capitals of the four Atlantic provinces to Toronto, Montreal, the prairies and Vancouver, while superhighways serve the Windsor–Quebec City corridor and major western centers.

The media
Canada has over 100 daily newspapers, mostly privately owned by a number of multimedia conglomerates. Broadcasting is controlled by the Canadian Radio-Television and Telecommunications Commission, an independent authority established by parliament in 1968. Two national TV networks are operated by the publicly owned Canadian Broadcasting Corporation, one broadcasts in English and one in French. There are three private television networks and over 2,000 cable TV systems. There are also hundreds of independent radio stations. Advances in telecommunications technology have been of benefit to isolated northern and Arctic communities.

Health and welfare
Federal government provides pensions, unemployment insurance and family allowances, while the provinces provide everything else. Most people belong to the federal medical insurance program, and are covered by provincial hospital insurance schemes.

Each province has its own education system, reflecting local needs. Children usually attend pre-school classes, then elementary and secondary education leading to college or one of 60 or so universities. Most education is free and compulsory from the age of six to secondary level.

An engineering triumph

The great Saint Lawrence river has been a trade route since the earliest days of settlement. Today ocean-going cargo vessels ply the waters where once the fur-trappers paddled their canoes. Flowing 1,200 km (750 mi) from Lake Ontario to the sea, the Saint Lawrence travels through the attractive Thousand Islands area before narrowing as it passes over a series of turbulent rapids above Montreal. It widens beyond Montreal to form Lake Saint Francis and Lake Saint Louis, and flows into Lake Saint Pierre. The final section follows a gently meandering course, eventually broadening out at the Gaspé Peninsula to enter the Atlantic via the Gulf of Saint Lawrence.

Until the 1950s the rapids denied passage from the Great Lakes to the sea to all but smaller river craft. In 1954 the United States and Canada began an ambitious joint project to open the river to deep-sea vessels, at a total cost of over $340 million to the Canadians and over $130 million to the Americans. The project involved building a total of 16 locks and a complex of canals and deepened waterways. Far to the northwest of the Saint Lawrence itself, Lake Huron and Lake Superior are linked by the Sault Sainte Marie Canal with its four locks. The Welland Ship Canal between Lake Ontario and Lake Erie has eight locks, and bypasses the Niagara Falls. The main seaway has seven locks between Lake Ontario and Montreal, and was opened in 1959.

Tolls are levied on vessels passing through the system. These are intended to cover the building costs (which must be met by the year 2009), interest on the funds raised for construction, and the normal operating and maintenance costs of the system. The seaway can accommodate most ocean-going vessels apart from the very largest. This makes the whole stretch from Montreal to the Great Lakes accessible to about four-fifths of the world's merchant shipping.

The seaway is closed by ice from mid-December to early April – about 115 days a year – but at other times it provides passage to the Atlantic from numerous inland cities in both Canada and the United States. Altogether 66

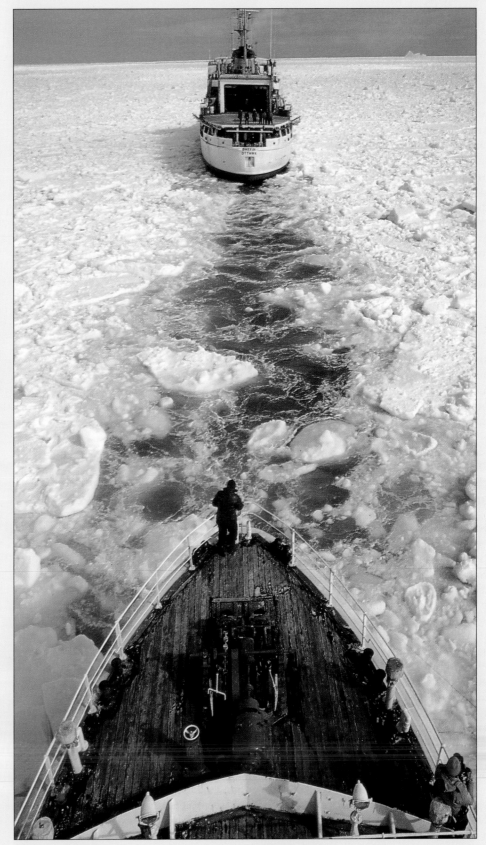

American and 12 Canadian ports have access to the seaway. The routes from these ports to western Europe are shorter than the transatlantic routes from the United States' eastern seaboard. About nine-tenths of the cargoes carried on the seaway are bulk commodities such as

Crashing through frozen water, an icebreaker in the St Lawrence Seaway clears a passage for a merchant ship in early spring. The seaway is closed by ice for about four months from mid-December.

grain, iron ore and coal, but other vessels also use it to transport a great variety of general cargoes.

Dependencies in the region

GREENLAND
DENMARK

Greenland, the world's largest island, lies largely within the Arctic Circle. Some 85 percent of the land area is covered by the northern hemisphere's biggest ice cap. Near its center it is over 3,000 m (10,000 ft) thick and its base is about 300 m (1,000 ft) below sea level. At its edges, along the coast, glaciers have created a complex landscape of deep fjords and offshore islands. The ice sheet is partly enclosed by mountains to the west and east, where they rise to Greenland's highest peak, Mount Gunnbjørn.

The climate is arctic, but precipitation is noticeably higher in the south, which is warmed by the North Atlantic Drift. The east coast is cooled by the Labrador Current, which flows from the Arctic Ocean to the Atlantic. Wildlife and vegetation in the ice-free areas are characteristic of the Arctic tundra. There are salmon and trout in rivers and streams, while cod, salmon, halibut, flounder and capelin are found offshore.

A series of Inuit migrations probably reached Greenland between 4000 BC and 1000 AD, and Inuit have lived here ever since. In the 10th century Viking settlers from Norway established two colonies, but contact with Scandinavia was lost some 500 years later. Colonization was only resumed in 1721, when Greenland became a Danish colony. Hans Egede, a missionary from Norway, founded a trading post – with Danish permission – near Nuuk (Godthåb) (now the capital). In 1776 the Danes assumed a trade monopoly (free trade was not reestablished until 1951). During World War II, while Denmark was under Nazi German occupation (1940–45), the United States protected Danish interests in Greenland. After the war Denmark met local demands for extensive reform, and in 1953 Greenland became part of the Kingdom of Denmark, no longer a colony.

Home rule was instituted in 1979. However, Denmark retains control of foreign affairs, and Greenlanders are Danish citizens. The 31-member Landsting (parliament) is elected by all adults aged 18 or over, and is chaired by the leader of the majority party. Greenland also elects two members of the Danish parliament. Today more than four-fifths of the population are native Greenlanders, mostly descended from Inuit with an admixture of early European settlers. The remainder are largely Danes. The official languages are Greenlandic (Eskimo) and Danish, and most people belong to the Danish Lutheran church.

Settlement is effectively limited to the coastal fringe. Sheep and reindeer are raised in the extreme south, where hay is grown for fodder. Vegetables are cultivated in the south and along the warmer west coast. Fishing has displaced seal-hunting as the leading industry. It is largely state-financed, as are the onshore canning and freezing plants. Hunting in the north produces pelts for export.

Greenland's mineral resources are large but difficult to exploit. Mining of the world's largest cryolite deposits was abandoned as uneconomic in 1963. Exploration has revealed valuable metal deposits, of which zinc and lead are now being mined. Electricity, from thermal

The colorful corrugated houses of Ilulissat (Jakobshavn) (*above*) on the west coast of Greenland. Fishing is the main occupation, and there is a large processing and canning plant beside the bay.

The frozen sea (*below*) at the Taasiilaq (Ammassalik) trading post on Greenland's eastern coast, just south of the Arctic Circle, prevents the movement of cargo.

has its own hospital. Nine years of elementary education are free and compulsory; four years of secondary education are optional. Both vocational and university education are available for Greenlanders in Denmark.

ST PIERRE AND MIQUELON
FRANCE

This tiny archipelago lies off the south coast of Newfoundland on Canada's eastern seaboard. The largest island, Miquelon, consists of two islands linked by a narrow isthmus. St Pierre to the southeast is smaller, but accommodates most of the islands' small population. The scenery is barren, with few trees and extensive peat bogs. The tall cliffs and rocky islets are home to many sea birds.

The islands were first settled by fishermen from western France in the 17th century, and the population has remained staunchly French ever since. The islands became a French overseas *département* in 1976, but in 1985 they were promoted to the status of a *collectivité territoriale*. The people vote both in local and in French national elections; the governor is appointed from France.

There is little agriculture apart from the growing of a few vegetables for local consumption, and the economy depends primarily on fishing from the rich waters nearby. Most food is imported from Canada and France, with which good transportation links are maintained. The islanders enjoy the benefits of French citizenship, including free education and good health and welfare provisions.

stations, relies on imported fuels. Other major imports include machinery and transportation equipment, manufactured goods and food.

Greenland has no railroads. The small road network is well used, but dog sleds remain the chief form of surface transportation. Most freight arrives by sea, but there is a well-developed internal and external air service. A sophisticated telecommunications network is available to most of the population. Radio and television broadcasts, like Godthåb's one newspaper, *Atuagagdliutit*, are in both Greenlandic and Danish.

A full range of welfare services is funded by the Danish government. Free health care is available to everyone, and each of the three administrative districts

Erik the Red – Viking explorer

In about 982 AD a man called Erik the Red set out from Iceland on a dangerous voyage into the Atlantic. According to a 13th-century saga Erik was a violent man who had been sentenced to three years' exile for "some killings" – but he was also a brave man. His target was a little-known island far to the west of Iceland – a place that had brought disaster and death to a group of Icelandic colonists only a few years before.

Sailing due west, Erik made landfall near a massive glacier. However, recalling an earlier ill-fated expedition by another explorer, he decided to sail southward. After rounding the cape, he discovered fertile valleys hidden among the glaciers on the western coast, and called the place Greenland – believing "people would be much more interested in going there if it had a pleasant name".

Some four years after his first voyage he set sail again from Iceland with 300 prospective colonists in 25 ships. Just 14 of these vessels reached the fjord that Erik had named for himself – present-day Tunugdliarfik. On a steep, grassy hillside he founded his Eastern Settlement, where the site of his house can still be seen. A breakaway group then founded a Western Settlement farther up the coast.

Greenland in the 10th century was appreciably warmer than it is today. The Vikings raised cattle and sheep in the south, but also ventured north to hunt caribou, whales, seals and bears; Polar bears were much prized as curiosities at European courts. For a while the colony flourished. About 280 farms are recorded in written sources, and archaeologists believe there were many more. Christianity reached Greenland in about 1000 AD, brought by the Norse explorer Leif Eriksson on his return from the recently Christianized Norway, and over the next 200 years nearly 20 churches were built, along with a convent and a monastery. The remains of a 12th-century cathedral at Gardar still survive.

In 1261 Greenland, like Iceland, came under Norwegian rule, but within a century the Western Settlement had been abandoned, and the Eastern Settlement went into a long decline. No one knows exactly why this happened, but it is known that in the late Middle Ages the climate changed for the worse, and this may have had a serious impact on the settlers. Inuit stories also suggest battles between their ancestors and Viking settlers. Animals and people would certainly have died in the bitter cold of the Arctic winter. The last recorded contact was in 1406. After that, the last Viking settlement on Greenland vanished into the cold and darkness of the Arctic night.

The United States

COUNTRIES IN THE REGION

THE UNITED STATES OF AMERICA

The torch of freedom "Give me your tired, your poor, your huddled masses yearning to breathe free" run the famous lines on the Statue of Liberty in New York Harbor. Erected in 1886, and partly paid for by the French people to symbolize their friendship with the United States, it was the first sight that greeted generations of European immigrants, signaling their safe arrival in the land of opportunity.

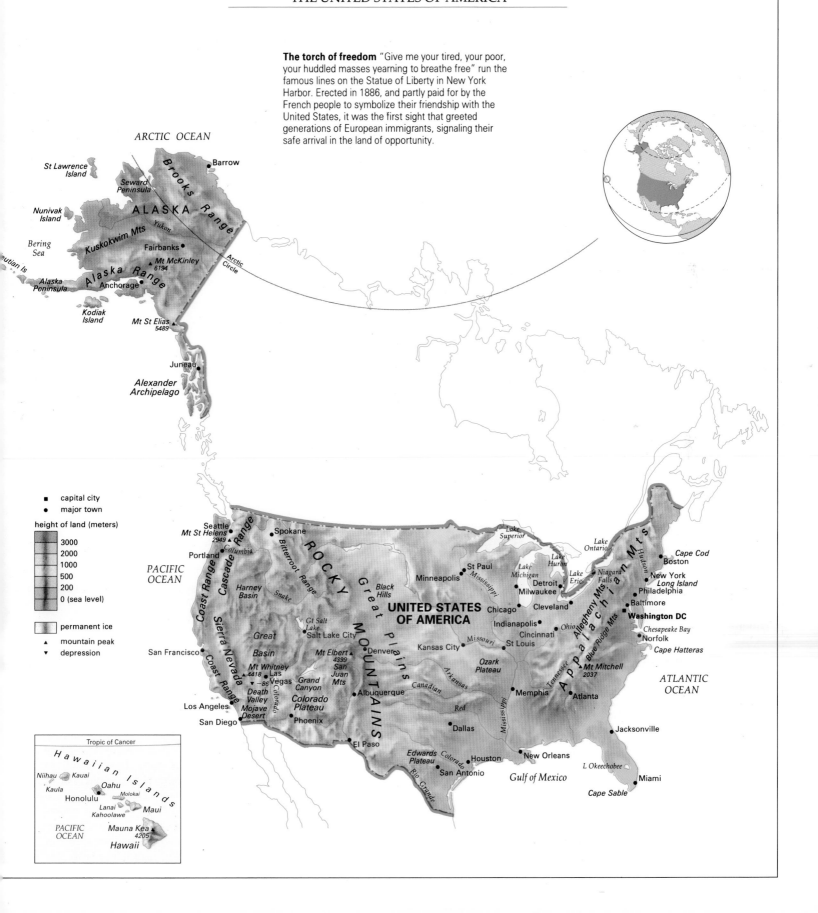

United States

THE UNITED STATES OF AMERICA

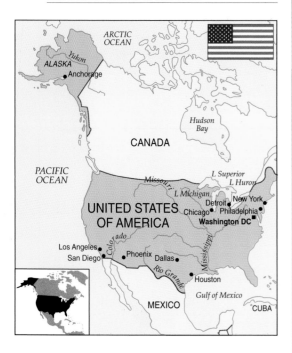

THE UNITED STATES OF AMERICA OCCUPIES part of North America, bordered by Canada to the north and Mexico to the south, and washed by the Atlantic Ocean to the east, the Gulf of Mexico to the south and the Pacific Ocean to the west. This area contains 48 of the 50 American states, and is known as the coterminous United States. The other two states are Alaska at the northwestern tip of North America, and the island group of Hawaii in the central Pacific. Other outlying territories include Puerto Rico and the Virgin Islands of the United States in the Caribbean, and Guam, American Samoa and the Trust Territory of the Pacific Islands.

The United States has the third largest population after China and India; in geographical extent it is the fourth largest country in the world, after Russia, Canada and China. But in terms of economic, political and cultural influence, it is the leading nation of the world. It owes its success to its plentiful natural resources, a rich cultural mix, and a strong sense of national identity.

GEOGRAPHY

Great diversity is, not surprisingly, a dominant characteristic of the landscape of the United States. The towering peaks of the west give way suddenly to the grasslands of the central lowlands, which in turn merge gradually into the more varied landscapes of the east. The climate is similarly diverse, ranging from Arctic in northern Alaska to subtropical in the southeast; the warmest areas include both the arid heat of the Arizona deserts and the balmy everglades of Florida.

The land
There are five main topographic areas within the coterminous United States. To the east and southeast, along the Atlantic and Gulf coasts, a broad coastal plain rises inland to the Appalachian Mountains. West of the Appalachians is a vast lowland area that rises steeply in the west to the mountains of the cordillera – a continuation of Canada's Rocky Mountains. The eastern branch of the cordillera flanks the central lowlands; its western branch runs parallel to the Pacific coast, and between these two branches lies a variegated landscape of plateaus, basins and mountain ranges.

The eastern and southeastern coastal plain is broad and generally low-lying, with many areas of swamp. The northern portion is largely submerged beneath the Atlantic; almost all that remains is a series of islands, most notably Long Island in New York State, and the peninsula of Cape Cod in Massachusetts. South of Long Island the lowlands become wider, reaching their greatest extent in the southern coastal states. Offshore, a line of barrier islands and sandbars runs for

Big Sur, California (*above*), on the Monterey Peninsula. The Santa Lucia range is separated from the Coast Range inland by a narrow valley and the main railroad linking San Francisco and Los Angeles.

Vermont in the fall (*left*) Vast tracts of New England forest were cleared by early settlers, and by the 1850s some 50 percent of the area was farmland. Today only 12 percent is farmed.

nearly 1,700 km (over 1,000 mi) parallel to the east coast. Inshore, the lower reaches of many river valleys have been flooded by the sea, thus creating large estuaries such as Chesapeake Bay on the coast of Virginia. This is in stark contrast to the Mississippi delta on the south coast around New Orleans, where masses of river-borne silt have collected to create new land. In Florida part of the offshore continental shelf has been raised to form a long, flat, swampy peninsula.

Beyond the Atlantic coastal plain, and parallel to it, the Appalachians extend southwestward from Canada to northeastern Alabama. They form a generally low but almost unbroken mountain chain largely covered with deciduous forests. Toward the southeast, the Blue Ridge Mountains rise in a steep escarpment above the rolling landscape of the Piedmont, once a highly productive agricultural area. The Blue Ridge Mountains reach their highest point at Mount Mitchell in North Carolina. Northwest of the geologically older Blue Ridge, and parallel to it, is a series of valleys separated by mountain ridges. The largest of the valleys is the Cumberland Valley, which is part of a long trench system that runs across the main Appalachian ridge to meet the Shenandoah Valley in Virginia, continuing through Pennsylvania and New York State to the Valley of Vermont in New England.

Westward, beyond the Appalachian chain, is a vast continental basin up to 2,500 km (1,500 mi) wide. It extends south from the Canadian border and the Great Lakes to meet the coastal plains that

NATIONAL DATA – UNITED STATES OF AMERICA

Land area 9,161,923 sq km (3,537,438sq mi)

Climate	Altitude m (ft)	Temperatures		Annual precipitation mm (in)
		January °C(°F)	July °C(°F)	
Barrow	42 (13)	-27 (17)	5 (41)	107 (4.2)
San Francisco	5 (16)	9 (49)	17 (63)	500 (48.4)
New Orleans	9 (30)	11 (51)	28 (82)	1,572 (61.8)
Chicago	190 (623)	-6 (21)	23 (73)	910 (35.8)
Washington, DC	22 (72)	-1 (35)	27 (80)	981 (38.6)

Major physical features highest point: Mount McKinley 6,194 m (20,320 ft); longest river: Mississippi-Missouri 6,020 km (3,740 mi); largest lake: Lake Superior (part) 83,270 sq km (32,150 sq mi)

Population (2006 est.) 298,444,215

Form of government federal multiparty republic with two legislative houses

Armed forces army 595,946; navy 376,750; air force 347,400

Largest cities New York (8,141,241); Los Angeles (3,945,097); Chicago (2,818,628); Houston (2,059,430); Phoenix (1,474,184); Philadelphia (1,426,259); San Diego (1,311,612); San Antonio (1,299,646); Dallas (1,222,053); San Jose (898,624); Detroit (858,803); Jacksonville (822,117); Indianapolis (770,634)

Official language English

Ethnic composition White 81.7%; Black 12.9%; Asian 4.2%; Amerindian and Alaska native 1%; Native Hawaiian and other Pacific islander 0.2%

Religious affiliations Protestant 52%; Roman Catholic 24%; Mormon 2%; Jewish 1%; Muslim 1%; other 10%; none 10%

Currency 1 United States dollar (USD) = 100 cents

Gross domestic product (2006 est.) U.S. $12.98 trillion

Gross domestic product per capita (2006 est.) U.S. $43,500

Life expectancy at birth male 75.02 yr; female 80.82 yr

Major resources coal, copper, lead, molybdenum, phosphates, uranium, bauxite, gold, iron, mercury, nickel, potash, silver, tungsten, zinc, petroleum, natural gas, timber, fisheries, tourism, cereals, citrus fruits, cotton, fish, livestock, oilseeds, potatoes, soybeans, sugar beet, timber, tobacco, vegetables border the Gulf of Mexico. The whole of this enormous area is drained by the great Mississippi–Missouri and Ohio river systems. In the north there are upland extensions of the Canadian Shield, such as the hills bordering Lake Superior. The eastern side of the basin consists of the Central Lowlands, whose southern limits are marked by the low plateaus of Kentucky and Tennessee, and farther west by the isolated uplands of the Ozark Plateau. The western side of the basin is formed by the semiarid Great Plains, which rise gradually from east to west to meet the western cordillera.

The western cordillera covers almost one-third of the North American continent. Within the coterminous United States it forms two main mountain systems that run south from the Canadian border. On the eastern side the Rocky Mountains face out across the Great Plains. In the north they are really an extension of the Canadian Rockies. Farther south, the Wyoming Basin provides the main corridor through the Rockies, used by pioneer trails and interstate highways alike. The nearby Grand Teton and Yellowstone National Parks include some spectacular scenery. South again, in Colorado and New Mexico, the Rockies rise steeply from the Great Plains, providing some breathtaking views.

West of the Rockies in the northwest lies the Columbia Basin, where the rocks are still being formed by a continuing upflow of lava that has buried many older landscape features. The rocks of the basin have been cut by deep river valleys and by meltwaters from ice-age glaciers. Farther south the landscape is dominated by low mountain ranges interspersed with desert basins floored with alluvium. One such basin has been filled by Utah's Great Salt Lake. East and south of this area the Colorado Plateau extends south into Arizona, with a remarkable landscape of mesas, buttes and canyons. The Grand Canyon of the Colorado river is just one of many national parks in this region.

The western branch of the western cordillera has two main strands. The Coast Ranges rise close to the shoreline, while the eastern line consists of the Cascade Range and the Sierra Nevada. Between them lies part of a discontinuous trench that appears all along the Pacific coast of North and South America. It includes Puget Sound in Washington, Oregon's Willamette Valley and the Central Valley of California.

The Cascade Range extends from Washington through Oregon to Lassen

The Bald eagle (*above*), America's national bird, is fully protected and has been reintroduced into New York State from captive-bred stock.

Snowcapped Mount Tom (*left*), in California's Sierra Nevada range, has the rough, jagged profile of relatively young mountains.

The stark beauty of Utah (*below*), where stratified rock formations surround Lake Powell. The buttes and mesas are familiar from many Western movies.

Peak in California, and includes a chain of high volcanoes. Among them is Mount Saint Helens, which erupted with devastating force in 1980. California's Sierra Nevada range includes Mount Whitney, the highest mountain in the coterminous United States, and is cut by spectacular glacial valleys. The eastern side forms a steep escarpment, whereas the western side slopes more gradually toward the Central Valley of California.

The northern outliers of the Coast Ranges are the glacier-clad Olympic Mountains of western Washington. A much lower section in western Oregon climbs to dramatic outcrops on the Californian state line. From here to Point Conception on the coast of southern California the ranges become low and linear, with major fault lines running parallel to them. An earthquake on the San Andreas Fault brought devastation to San Francisco in 1906, and a rather smaller earthquake in 1989 provided a grim reminder of this. In southern California the Coast Ranges become higher again but more fragmented.

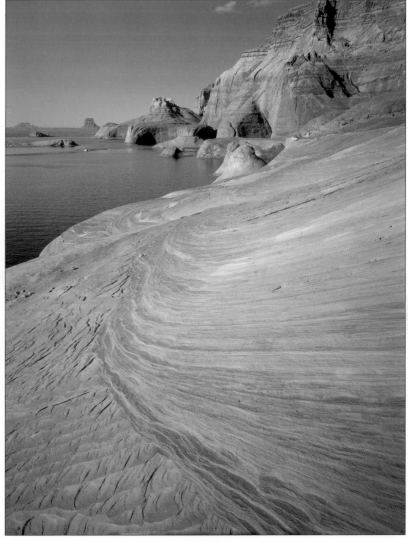

Climate

The coterminous United States lies in a region of prevailing westerly winds, with the landmass of Canada to the north, a warm, shallow sea to the south, and broad oceans to both east and west. In general, however, the sheer size of the North American continent produces a continental climate throughout most of the country, marked by cold winters, warm or hot summers and a broad diurnal range of temperatures.

The west coast benefits from its maritime position, which brings much milder winters than elsewhere. Cool, moist winds from the Pacific shed much of their precipitation as they rise over the Coast Ranges, so that Oregon and Washington have the heaviest rainfall in the country, especially during the fall. Farther south, in California, the summers tend to be hot and very dry. Rainfall is consequently much lower, especially in the Central Valley; agriculture here is dependent on irrigation. In the far south there is little rain at all. Farther inland much of the remaining moisture falls as rain or snow over the Cascade Range and the Sierra Nevada. The westerlies then become warm and dry as they descend the eastern slopes, creating the deserts of Nevada and Arizona.

East of the Rockies the Great Plains are semiarid, but farther east the Central Lowlands receive more rain in summer from hot, humid air flowing north from the Gulf of Mexico. Less beneficial, though, are the spectacular weather conditions produced throughout the central United States when humid air from the south meets colder air from the western cordillera. The result may be anything from thunderstorms and tornadoes to hailstorms and blizzards. Not surprisingly, this American heartland suffers the most violent extremes of temperature, from season to season and sometimes even from one hour to the next. Spring and fall are pleasant but very short, and winters are longer toward the north.

Areas near to the Gulf of Mexico have much shorter winters, but are liable to hurricanes in the late summer and early fall. Equally vulnerable are the states in the southeast adjoining Florida, where warm, humid conditions prevail all year round. North of here, the Appalachians receive plentiful rain throughout the year. The northeast, however, has a largely continental climate, only slightly modified by the nearby ocean. This is because the prevailing westerly winds blow offshore. Low winter temperatures combined with unstable air conditions lead to some spectacular snowfalls, especially over the mountains and along the coast.

Vegetation

The first European settlers in the eastern United States found a land covered in rich deciduous forest, including hardwoods such as hickory, oak and walnut. Relatively few of these trees now remain; most of the forest cover in the east is secondary growth, though a few elms, maples and beeches can still be seen. Conifers grow in the colder northern areas and on high ground.

About halfway across the central basin the tree cover gradually gives way to

The gold rushes

A pioneering prospector from Minnesota, George Northrup, poses for a daguerreotype – an early type of photograph – in about 1860, with the tools and pistols of his trade, and its reward: a bag of gold dust worth the then amazing sum of $90,000.

It is largely thanks to the Hollywood film industry that the term gold rush evokes such vivid images – whiskery prospectors tramping through northern snows, panning the sand of icy Alaskan rivers, starving in lonely cabins, or "playing hard" in rough boom towns of hastily erected clapboard houses. But in fact gold rushes have happened all over the world, and the pattern is generally much the same. Their importance for the United States, both in its real history and in its mythology, lies in the fact that there were several gold rushes within a very short time, all of them along the pioneer routes of the West.

The first of them – the so-called Fortynine – occurred beyond the mountains of the Sierra Nevada in central California, where John Augustus Sutter (1803–80), a bankrupt Swiss emigrant, had founded the trading colony of Nueva Helvetia (New Switzerland; later Sacramento). In 1848, one of his workmen found particles of gold in the tailrace of his sawmill. Sutter tried to keep this discovery secret, but failed, and in the next year at least 80,000 prospectors swarmed over the area now known as Sutter's Mill in a frantic search for the "Mother Lode" – the rich gold seam from which other seams were thought to branch.

The land could not possibly support such hoards of settlers; many Fortyniners lived in hardship and died in squalor. Sutter's trade was shattered, his land devastated, his property and live-stock stolen. He became bankrupt yet again, and returned east in 1851. When he died many years later, he was still fruitlessly seeking government compensation. He had at least a moral right to it, for the gold rush swelled settlement in the area, creating the prosperous city of Sacramento.

Other gold rushes followed in Colorado (1858) and South Dakota (1876). Hastily built townships were abandoned when local deposits ran out, leaving many ghost towns scattered across the countryside. These later became famous, and some have been restored as tourist attractions, such as Deadwood below the Black Hills of western South Dakota.

It was the Alaskan Rush, however, that really caught the world's imagination. It began in 1886, when gold was struck at Fortymile Creek on the Yukon river, and was fueled by further discoveries in 1896, both in the Yukon and in the far west of Alaska on the site of present-day Nome. Pioneering was by now in decline, with less land to be settled and fewer fortunes to be made. Alaska promised a last chance, both for the rootless men of the west and for discontented migrants from the east.

Few of them could afford to make

the hazardous sea passage to Alaska around South America, so most took the landward route. It was a path strewn with dangers that were hardly mentioned in the attractive though largely fictional manuals produced by opportunistic publishers in the east. Thousands set off, but many perished on the snow-covered mountain crossings, and those who arrived found conditions little better. Many of them starved, and some considered themselves lucky to be able to revert to their former occupations. One such person was bank clerk Robert William Service (1874–1958), whose once popular ballads captured much of the atmosphere created by the men who joined the Alaskan Rush.

By 1900 some 22 million dollars' worth of gold was being panned annually, but at a high cost. Few real fortunes were made, and as early as 1911 yields were declining. However, while the rush lasted, it spurred the settlement of inhospitable Alaska as almost nothing else could, making possible today's exploitation of Alaska's many other valuable resources. But for the gold rushes, the Last Frontier might have remained unconquered for a much longer period.

grassland, although much of the area concerned is now agricultural land. As rainfall decreases to the west, the long prairie grassland is replaced by shorter, thinner steppe grasses.

Alpine vegetation covers the western mountain ranges, with desert vegetation in the arid areas in between. Plant life here is surprisingly rich, with a wide variety of cacti and succulents. Westward again, on the coasts of Oregon and California, are Douglas firs and the last surviving stands of redwood, the world's tallest tree. Redwoods can reach as high as 90 m (300 ft), and some living specimens are thousands of years old.

Central and southern California are characterized by chaparral vegetation, with its dense thickets of shrubby plants that are resistant to the long summer droughts. Palm trees are native to southern California and Florida, though they are also grown elsewhere.

Wildlife

Animal life in the United States shows the dramatic effects of human settlement. The first European colonists brought horses, cattle, sheep and other European species such as sparrows and starlings. Settlers moving west in the 19th century all but eliminated the huge herds of buffalo and pronghorn that once roamed the Great Plains. Very few of these now protected species survive, and today even the Bald

Magnificent buffalo in Yellowstone National Park, Wyoming. These animals once dominated the landscape in herds that reached from horizon to horizon, but few survived. Today they live in small, isolated populations where inbreeding is a danger.

eagle, symbol of the United States, is in danger of extinction, as are the American bighorn sheep and the cougar.

Mammals with a wide distribution include White-tailed deer, American black bears, bobcats, raccoons, skunks, opossums, beavers and muskrats. Moose, Red foxes, otters and wolverines inhabit the northern coniferous forests. The forest rivers of the east and southeast support an abundance of fish species, and there are seven different species of salamander in the Appalachians. The deserts of southern California, Nevada and Arizona boast many reptile species, including the poisonous Gila monster. The Great Plains are the home of prairie dogs (gregarious rodents), while the Rockies harbor marmots, Mountain goats and pikas.

Among marine mammals, seals are found on both coasts, but sea lions only in the Pacific, and the Florida manatee only in the larger rivers of the southeast.

Conservation and management

The landscapes of the United States show startling contrasts between areas of manmade urban and rural wilderness and vast tracts of natural wilderness carefully conserved for future generations.

Armed for attack (*above*) Pronghorn bucks use their sharp, hooked antlers for sparring with rival bucks during the breeding season. Some 35 million once roamed the plains, but – like the buffalo – they were all but exterminated by European settlers.

A manatee (*above*) browses in a Florida stream. These huge, slow-moving creatures are in danger of extinction from sewage pollution of their freshwater habitats, and from injuries caused by speedboat propellers.

Zigzagging to the horizon (*right*) the Florida Keys are a chain of offshore islands linked by the Overseas Highway. Completed in 1938, this runs southwest from the mainland across Card Sound to the principal islands of Key Largo and Key West.

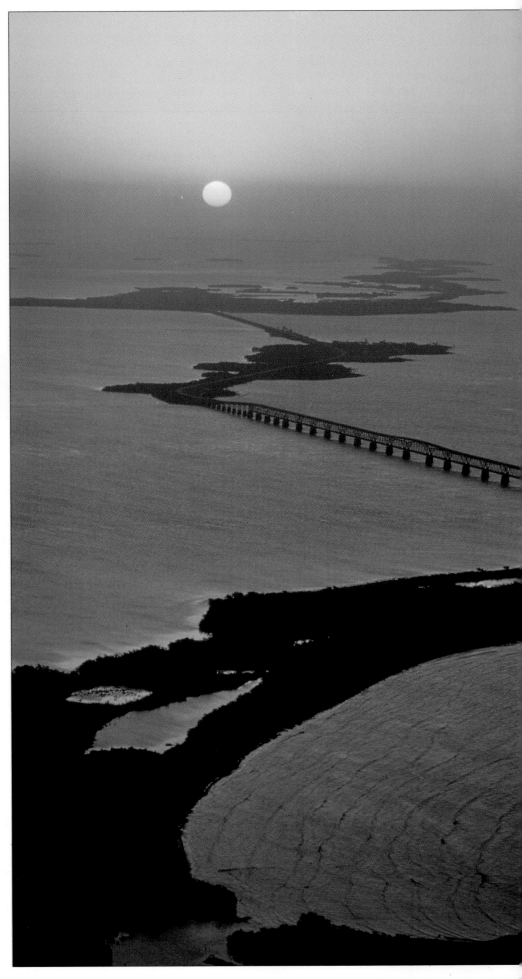

The first national parks, both of them in California, were established in 1890. The Sequoia National Park, which has over 300 lakes and some of the highest peaks of the Sierra Nevada, contains a fine stand of giant sequoias. The Yosemite National Park includes Half Dome, a peak that has been sliced in half by a glacier. The other national parks include Mount Rainier in Washington, Yellowstone in northwest Wyoming, and also the Great Smoky Mountains National Park, which straddles the Tennessee–North Carolina border in the southern Appalachians.

Many other areas enjoy similar levels of protection, but many are also suffering increasing pressure from millions of visitors. This pressure, combined with atmospheric pollution and the effects of fertilizers and pesticides, means that even wilderness areas are now under threat. The population of redwoods, for instance, is declining. Pollution controls such as those initiated in Oregon may yet be able to turn the tide, but the fact remains that even a national park cannot be completely protected from the highly industrialized society that surrounds it.

Alaska

The state of Alaska is located in the far northwest of North America, with Canada to the east and the Bering Strait, which separates it from the Soviet Union, to the west.

The southernmost part of Alaska is called the Panhandle – a narrow strip of land that runs south along the northern Pacific coast bordering Canada inland. It is mountainous and flanked by offshore islands. The coast along the Gulf of Alaska is lined by an almost continuous mountain chain characterized by glaciers, volcanoes and severe earthquakes. It runs from the Boundary Range in the Panhandle via the Wrangell Mountains and the Alaska Range to the southwestward line of the Aleutian chain.

North of the Alaska Range is the broad central plateau, whose tablelands are greatly fragmented by the Yukon river and its tributaries. North of here, the Brooks Range runs clean across the state from the Canadian border to the Bering Strait. Northward again, there is a broad coastal plain, covered in tundra, bordering the Arctic Ocean.

The climate is generally cold, but varies greatly with altitude and latitude. The greatest extremes of temperature are to be found in the drier areas of the interior and north. Southern coastal areas are less cold in winter, but are generally subject to heavy rain and snow.

Hawaii

The Hawaiian Islands are the peaks of a submerged volcanic chain that stretches in a wide arc across the central Pacific Ocean. The largest of the eight main islands are Hawaii and Maui, and there are some 124 smaller islands. The climate is warm and humid, and rainfall varies from sparse in some areas to extremely heavy on slopes facing the northeasterly trade winds, where it supports a rich vegetation. There are no native reptiles or mammals, but the bird life is remarkable and varied; most resident native species are unique to these islands.

SOCIETY

In the course of the 20th century, the United States has achieved worldwide cultural dominance. Spread by the mass media, its popular culture enjoys worldwide appeal, helping to make English the most international of languages. Its standard of living is among the highest in the world, and its ethos of individual liberty is widely admired.

Nevertheless the country has considerable social and economic problems. Organized crime, political corruption and the wasteful use of resources characterize American society as much as the skill and enterprise that produced the Moon landings, medical research, the Philadelphia Orchestra, Hollywood or Mickey Mouse.

History

Archaeological evidence suggests that America's first settlers probably crossed from Siberia to Alaska via the Bering Strait during the ice ages between 30,000 and 10,000 years ago, and spread rapidly across the continent. Their descendants, the Native-Americans, developed a wide variety of societies and cultures. Their homes, for example, ranged from the cliff villages of the Pueblos in the southwest to the bark lodges of the Iroquois in the northeast and the huge decorated wooden halls of the sophisticated Northwest Coast Indians.

The question as to which European peoples reached the Americas first has caused much controversy. There is con-

siderable evidence pointing to a Viking seafarer from Greenland, Leif Eriksson (c. 1000 AD), who with others settled briefly in a land they called Vinland; this has been variously located along the northeast coast from Labrador (Canada) to New Jersey. The Italian navigator and explorer Christopher Columbus (1451–1506), who arrived almost 500 years later in 1492, was attempting to find a westward route to India; he christened the Caribbean islands the West Indies and their inhabitants Indians. It soon became clear, however, that these lands were not Asia but a new continent; it received its name from Amerigo Vespucci (1454–1512), one of several navigators who followed Columbus west. Vespucci published a widely read account of his explorations.

The earliest settlements were Spanish and Portuguese. The first English settlement (following a number of abortive efforts such as that of Sir Walter Raleigh (1552–1618) at Roanoke, Virginia), was at Jamestown, Virginia, established by the Virginia Company in 1607. The first of the New England colonies, Plymouth, followed in 1620. Its founders, the so-called Pilgrim Fathers, were strict religious dissenters hoping to found an ideal community free of political harassment. The huge Massachusetts Bay colony, which eventually incorporated Plymouth, was founded by even more authoritarian Puritans. Maine and New Hampshire began as its offshoots, while the New Haven and Rhode Island settlements were established by opponents of their strict regime. Maryland, by contrast, was

Christopher Columbus (*above*) explorer and visionary. The wreck of his flagship, Santa Maria, which sank off the island of Hispaniola in 1492, was located in 1968.

A painting of William Penn (*left*), making a treaty with the Native Americans. Penn is renowned for his fair and friendly dealings with them.

founded in the 1630s by Roman Catholics.

Meanwhile, in 1624 the Dutch had established the prosperous colony of New Amsterdam, centered on Manhattan Island. In 1664 it was captured by the British and renamed New York; New Jersey then separated from it in 1665. Pennsylvania was in many ways the most successful of the early colonies. It was founded in 1682 by a moderate Quaker from England, William Penn (1644–1718). Pennsylvania was unusual in maintaining fairly good relations with the Native Americans, though this was eventually undone by Penn's son, who swindled the Delaware Indians out of a large area of their land in a deal known as the Walking Purchase. Part of Pennsylvania later became the colony of Delaware. North and South Carolina were established in 1663 by English investors, and Georgia in 1732 as a philanthropic alternative to the debtors' prison.

During the 18th century these colonies remained dependent on Britain in order to survive and expand. There was a mutual need for defense both against the French and the Native-Americans, upon whose territories the colonists were encroaching. By 1763, when the French had been finally defeated, the British colonies were close to being self-governing republics, and they began to resent paying British taxes. The British government's ban on expansion into Native-American lands beyond the Appalachians, together with economic decline, fueled the growing unrest, and criticism at home led the British authorities to impose even sterner controls. This only lent support to the radical separatist minority. Incidents such as the so-called Boston Massacre of 1770, in which five people were killed by British troops, became propaganda showpieces. In 1776 the Second Continental Congress duly declared the 13 colonies independent, and proudly proclaimed their confederation as a new nation – the United States.

In the War of Independence that followed, the American Continental Army, backed by state militia, faced British garrisons, later reinforced by German mercenaries. The ill-equipped American forces were often rescued from defeat only by capable generalship, especially that of George Washington (1732–99). The turning point came when revolutionary France entered the war, sending military and financial aid to the rebels, harrying transatlantic supply lines and threatening the British mainland. The surrender of the forces under the command of the British military leader Lord Cornwallis (1738–1805) at Yorktown in 1781 marked the effective end of Britain's hold on the United States, although the war did not end until 1783. The American victory was marked by savage reprisals against the loyalists, many of whom retreated north to British Canada.

Even during the later stages of the war America's great expansion had begun. During the next few years, Vermont, Kentucky, Tennessee and Ohio joined the Union. In 1803 the Louisiana Purchase

George Washington, first president of the United States. His characteristically stern, set mouth in this and other portraits was due to his swollen gums, caused by primitive false teeth. His sets, carved in wood or ivory, have been preserved.

added extensive but sparsely settled French territories in the west and south, which came to form the heartland of modern America. A year later Thomas Jefferson (1743–1826), third president of the United States, despatched the American explorers Meriwether Lewis (1774–1809) and William Clark (1770–1838) on an expedition to explore the new territory. They followed the Missouri river to its source, crossed the Rocky Mountains, and were the first Europeans to make the overland journey to the Pacific coast. In 1806 they returned to Saint Louis on the Mississippi, and their account of the journey helped to establish the strong

Wharfside laborers, Virginia (*above*) At the end of the American Civil War in 1865, the 13th, 14th and 15th amendments to the US constitution abolished slavery altogether and guaranteed citizenship to former slaves.

The Battle of Kennesaw Mountain, Georgia (*left*), June 1864; a major Civil War confrontation that cost the Union some 3,000 men, and the Confederates 600.

Abraham Lincoln – patron saint of common sense

"In the present civil war it is quite possible that God's purpose is something different from the purpose of either party – and yet the human instrumentalities, working just as they do, are of the best adaptation to effect His purpose." So wrote Abraham Lincoln (1809–65) in 1862, during the second year of his presidency – and his words suggest the true character of the man known to history as the Great Emancipator.

Abraham Lincoln's life story is the stuff of legend, but the legend often obscures the known facts about his life. Born in a backwoods cabin in Kentucky on 12 February 1809, he was six when his family moved to Indiana; three years later his mother died and his father remarried. In 1830 the family moved to Illinois, where he worked in a variety of occupations. In 1832 Lincoln volunteered for the Black Hawk War, in which the Sauk and Fox Indians came into conflict with the Illinois militia over contested lands. In 1834, after the first of his four elections to the state assembly, he began to

Abraham Lincoln, seen here in 1860, belonged to the first generation of historical figures whose features were accurately recorded for posterity in photographs rather than paintings.

study law; he passed his bar examination two years later. Lincoln soon built a prosperous legal practice and a reputation for straightforwardness and common sense. He married in 1842, and had four children.

During his term in Congress (1847–49), Lincoln proposed a limited form of emancipation for slaves, but nothing came of his suggestions. Then, in 1854, Stephen Arnold Douglas (1813–61) steered the Kansas–Nebraska Act through Congress, which gave Kansas and Nebraska settlers the option to permit slavery in their territories. Lincoln and his fellow Republicans thought the territories should be free soil where anyone, black or white, could better themselves. In 1858 Lincoln contested Douglas' seat in the Senate in a series of sharp public debates. He failed to win the seat, but in May 1860 he won the presidential nomination, and on 6 November he was elected. Shortly after, South Carolina left the Union.

Lincoln held to his principles and to the constitution, resisting all efforts to compromise on slavery in the territories. Before his inauguration on 4 March 1861, six more states had seceded, forming, with South Carolina, the Confederate States of America. On 12 April Confederate guns fired on Fort Sumter in Charleston harbor on the coast of South Carolina. The American Civil War had begun.

Lincoln applied himself to the war with the same common sense that had served him as a lawyer, and his careful management of the high command brought him eventual victory. His chief object was always to save the Union, but in 1863 he declared freedom for all slaves in rebel territories. Lincoln always insisted that this Emancipation Declaration was a war measure, but it opened the way to the abolition of slavery. In 1864 Lincoln was reelected on the promise of a constitutional amendment that would outlaw slavery for ever. State after state ratified the amendment, but before it could be adopted Lincoln was assassinated. He died on 15 April 1865, and the Thirteenth Amendment was adopted later the same year.

Lincoln undoubtedly hated slavery – but only such an honest and clear-thinking statesman could have turned that hatred into such a lasting and positive achievement.

sense of "Manifest Destiny" that was to fuel westward settlement.

In 1812 anti-British resentment led the United States Congress to declare war again. Despite American military success, the war brought the country close to fragmentation in 1814; but Britain was unwilling to fight another long war, and agreed to equitable peace terms. The United States settled down to a period of consolidation in the east, with the growth of great cities such as New York, Boston (Massachusetts), Philadelphia (Pennsylvania) and Charleston (South Carolina). Meanwhile, exploration and expansion continued westward. Texas and other territory was acquired from Mexico in 1848, followed by further gains in the Gadsden Purchase of 1853; Alaska was bought from Russia in 1867.

In the meantime, however, conflict over state rights was increasing between the northern and southern states, with the south resisting the centralizing authority of the north. The plantation economy of the southern states depended on slave labor, while the religious idealists of the more industrialized northern states found slavery repugnant. Matters at last came to a head in 1860 under the presidency of Abraham Lincoln (1809–65). The southern states seceded and formed the Confederacy. The Civil War of 1861–65 devastated the south, which suffered further privations after Lincoln's assassination in 1865. Slavery had been abolished, but crushing poverty led to interracial resentment that endured for more than a century.

The expansion westward continued, helped by the spread of the great railroad networks. A flood of immigration from Europe was encouraged in order to open up the new lands to farming. The westward movement was hastened by the apparent freedom and opportunity of frontier life, and by the prospect of gold on the west coast. The reality was often cruelly disillusioning, but the potent legends of the Wild West were born. The treatment of the Native-American population at this time was insensitive and often brutal, leading to atrocities on both sides. Meanwhile, industrial expansion in the northeast produced the first great industrial tycoons – selfmade men such as John D. Rockefeller (1839–1937) and Andrew Carnegie (1835–1919). It also created serious social problems.

The United States established itself as the dominant power in the region, following the declaration of the Monroe Doctrine in 1823, which denied Europe

Driving the Golden Spike, a painting by H. Charles McBarron commemorating the ceremonial linking of the Union Pacific and Central Pacific railroads at Promontory, Utah, in May 1869.

any right to intervene in the New World states. The United States encouraged revolutions in Latin American colonies, including one in Cuba, which was supported by a US invasion force. This led to war with Spain in 1898, which America won. As a result they acquired other Spanish territories, notably Puerto Rico and the Philippines. The islands of Hawaii were also annexed during this period.

As World War I raged throughout Europe, President Woodrow Wilson (1856–1924) committed the United States to neutrality, but in 1917 the German U-boat campaign, which hit American shipping, forced him to declare war. A strong isolationist reaction followed once peace was won, causing the United States to reject membership of the League of Nations – an international organization created to preserve the territorial integrity of member countries.

At home, the passing of the Volstead Act in 1920 banned the sale of alcohol. This proved unenforceable, encouraging the unprecedented growth of organized crime, and creating another American legend – the Prohibition era. The apparent prosperity of these years ended abruptly with the stock-market crash of 1929. This precipitated the Great Depression – a spiral of decline that was worsened by the European financial crisis of 1931. When Franklin D. Roosevelt (1882–1945) was elected president in 1932, he introduced emergency measures that effectively halted the decline. However, his New Deal policies had limited success in restor-

ing growth until the forthcoming war effort regenerated production.

The political unrest in Europe strengthened isolationist sentiment. When World War II broke out, the United States did no more than supply arms to Britain and France. Controversy continued at home until the end of 1941, when an attack on Pearl Harbor, Hawaii, by Japan, Germany's ally, brought the United States into the war. American industrial might was instrumental in the final victory, culminating in the use of the first atomic bombs to bring about Japan's surrender.

In the postwar period the United States established itself as the leading nation of the West in opposition to the Soviet Union. Financial aid programs such as the Marshall Plan reconstructed Western Europe and Japan as buffers against communism. As the Cold War tensions between the Soviet Union and the United States increased, so the spread of communism became the dominant fear in American society. This led to the political purges of communists of the early 1950s associated with Senator Joseph McCarthy (1909–57). Under United Nations auspices, the United States took the lead in the Korean War, repelling communist forces invading South Korea.

A period of prosperity followed, paving the way for the liberal aspirations of President John F. Kennedy (1917–63). In

foreign policy, however, Kennedy approved an abortive invasion of Cuba and faced down a Soviet attempt to establish missile sites there. He also committed the United States to protecting South Vietnam against communist invasion from the north. After Kennedy's assassination in 1963, this escalated into the Vietnam War, which eventually ended in political defeat for the US and withdrawal in 1974.

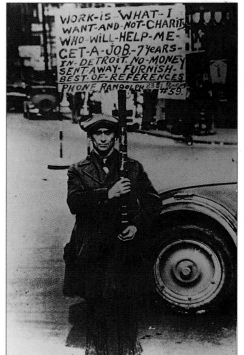

A victim of the Great Depression (*above*) with a handwritten placard proclaiming his need of a job. When this photograph was taken in 1932 the Depression was at its height, and one out of every four workers in the United States was unemployed.

Enforcing the law (*above*) – bootleg whiskey goes into the sewer. The blackmarket trade in illicit liquor was the inevitable result of Prohibition, and gave rise to a wave of organized crime and gang warfare.

City neighborhood (*left*) Iron fire escapes decorate the tenement buildings on Delancey Street on New York's Lower East Side, a dilapidated neighborhood of bars, hotels and small, oldfashioned businesses, reflecting the ethnic mix of urban society.

The subsequent moral crisis that overtook the United States owed much to the corruption that had been exposed by the Watergate scandal, and the resignation in 1974 of President Richard Nixon (1913–94). Throughout this time, however, the United States continued to lead the world in science and technology, as witnessed by its space program and the first Moon landing in 1969. Meanwhile, civil rights leaders such as Martin Luther King Jr. (1929–68) campaigned to remove social and political inequalities from the United States' black population.

During the 1970s America's international standing suffered as it entered an economic recession, partly as a result of an oil crisis in the Middle East. In 1990 the United States became involved in military action in the Middle East after Iraq seized Kuwait. On September 11 2001 a recession-struck US suffered great loss when suicide terrorists used commercial passenger jets to attack New York City and Washington, D.C., annihilating the World Trade Center, and partially destroying the Pentagon. The US declared total war on terrorism and on Afghanistan's Taliban government for failing to deliver the suspects, Osama bin Laden and his al-Qaeda terrorist network.

Government

The governmental system of the United States was based on that of 18th-century Britain, in the sense that its creators were attempting to reproduce its democratic strengths while correcting its flaws. It differs in that it is a federal system, in which certain powers are delegated to the national government and all other powers fall to the legislators of the individual states. The use of names such as Capitol and Senate for its institutions sprang from their intention to embody idealized notions of Roman and Greek democracy.

A written constitution was chosen in preference to the half-statutory, half-traditional British model as a better guarantor of rights. Included in the constitution are provisions that allow it to be amended. Amendments may be proposed by a two-thirds vote of both houses and ratified by three-quarters of the state legislatures – the method used for all 26 amendments adopted so far. A constitutional convention to consider amendments may be called if two-thirds of the states wish it.

The first 10 amendments are known collectively as the Bill of Rights and were adopted in 1791. They explicitly guarantee freedom of speech, religion, and the press and the right to a speedy and fair trial. The Tenth Amendment reserves all governmental powers not explicitly mentioned in the constitution to the states or to the people.

Three amendments – the Thirteenth, Fourteenth and Fifteenth – were adopted after the Civil War. These extended political rights to black Americans by abolishing slavery, establishing the citizenship rights of native-born and naturalized Americans and recognizing the rights of blacks to vote. The Fourteenth Amendment guarantees "equal protection under the laws" to all Americans.

Later amendments have continued to enlarge the political rights of ordinary Americans and to clarify archaic and confusing clauses in the original constitution. The Nineteenth Amendment (passed in 1919) extended the franchise to women in federal elections (some states had done so earlier). The Twentieth streamlined the period of transition between administrations after a presidential election, and the Twenty-fifth clarified the succession process following the death or resignation of a president in mid-term.

The drafters of the constitution tried to limit the power of the federal government and retain that of the states. Central to

A system of checks (*above*) balances the legislative (Congress), the executive (the president) and the judiciary (the Supreme Court). The president's veto can be overridden by Congress; the Supreme Court is able to declare presidential and congressional decisions invalid.

their effort was the concept of separation of powers, by which lawmaking, administration and enforcement are kept separate. Instead of a monarch as head of state, a president was to be elected. This was to be done via an electoral college, whose elected representatives would choose the president, for it was feared that a simple contest for public favor might easily lead to corruption and the election of demagogues. The electoral college soon became a formality, and it has been suggested that the worst fears of the constitution-makers have come true. The prolonged dispute in the 2000 election over the victory of George W. Bush – who lost the popular vote but won the electoral college vote – led to calls to abolish the electoral system.

The president must be at least 35 years old, and must be a native-born United States citizen who has been resident there for at least 14 years. In practice he must normally possess considerable wealth and political support. In theory he heads only the executive branch of government, but in practice his authority has increased to cover foreign policy and (as party leader) the legislative program in Congress. Because of the separation of powers, however, he often lacks direct authority to enforce his decisions, especially when his party is in the minority in Congress. In such cases, his limited powers of veto provide opportunities for mutual obstruction, leading hopefully to compromise but sometimes stalemate.

Congress alone can make laws, levy taxes, declare war and determine its own operations. In extreme cases it can impeach the president. It consists of two houses: the House of Representatives and the Senate. Representatives are elected for a two-year term from each state, the number being determined by each state's

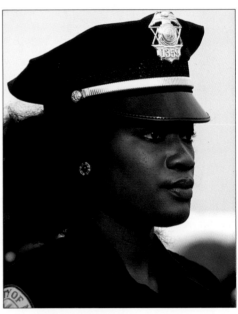

Miami cop (*above*) Most law enforcement in the United States is entrusted to city, town or state police authorities, and is paid for by local taxes. Television movies give the job a glamorous image, and recruits are attracted from every social group.

population, up to an overall total of 435. They must be at least 25 years old, state residents and United States citizens of seven years' standing. Senators must be at least 30, state residents and United States citizens of at least nine years' standing. Each state elects two senators to serve six-year terms, and one-third of the Senate is reelected every two years. The Senate possesses greater power and status than the lower house, and its 16 standing committees have now become powerful organs of government.

At the head of the judiciary is the United States Supreme Court. This is the final authority on the interpretation of laws and of the constitution, and the ultimate appeal court for decisions of the

Native American culture The Pueblos of New Mexico live in the same type of adobe-built dwellings that their pre-Columbian ancestors did. They preserve much of their traditional lifestyle, but have adopted some features of modern life, such as television.

Life in the sun (*above*) A volleyball tournament on the beach at Santa Barbara, California, attracts a large crowd of young people. The relaxed and affluent way of life that can be found on America's west coast continues to attract immigrants, most of them young, to the state.

Fast food (*right*) A New York street vendor caters for the lunchtime needs of city workers. Foods such as pretzels, hot dogs and knish were introduced by German or East European Jewish immigrants, but are now regarded as typically American.

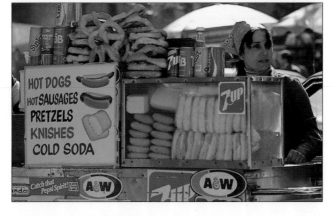

various specialized federal courts and state supreme courts. The nine justices are appointed for life by the president with the consent of the Senate. A majority decision is conclusive, but a tie will sustain the decision of a lower court. Political considerations almost always affect the appointment of judges.

Below the federal government are the individual state governments, whose structures generally mirror those of the federal government. Each has its own constitution, with a governor at the head and (apart from Nebraska) a two-chamber legislature. The governor is elected, as are most state officials, including judges. Municipal governments in the larger cities are most often headed by an elected

mayor and council, but other local municipalities favor an elected commission or a professional manager appointed by an elected body.

There are only two significant political parties, because the enormous electorate makes it hard for any third party to achieve a foothold nationally. Broadly speaking, Republicans are conservative and Democrats liberal. However, both parties reflect a wide spectrum of opinions, while local differences may modify or even reverse this balance.

People

The United States, with its millions of settlers from all over the world, is remarkable for its ethnic diversity – so much so

that it has been characterized as a melting pot of peoples and cultures. But this is in some ways misleading. The years of heavy immigration were marked by a strong movement toward assimilation, which was seen as the best way to find employment and fulfill the "American dream". But as society has become more uniform, the process has reversed; ethnic pride is now a legitimate means of self-assertion, especially for marginalized minority groups.

There was considerable linguistic and cultural diversity among the original Native American inhabitants, but the impact of the European invasion was such that only a few have retained their distinctive cultural identity. The first Old World settlers came largely from northwestern Europe, especially Great Britain, Ireland, the Netherlands, Germany, France and Spain, and from Africa (chiefly the west coast regions) brought to work on plantations as slaves.

After the Civil War, however, a deliberate program of mass immigration brought in large numbers from Italy, Scandinavia and the troubled countries of Eastern Europe, including Russia; many of these were Jewish immigrants fleeing poverty and pogroms. Settlement patterns varied: many Italians and Jews settled in the big cities, founding their own distinctive neighborhoods such as Little Italy in New York; Scandinavians and some Balkan peoples took farmland in areas such as Minnesota. The degree of cultural acceptance and assimilation also varied very widely.

The economic collapse of the south following the Civil War swelled the class of so-called "poor whites", who lived in much the same conditions as the newly liberated black slaves, providing a potent breeding ground for racial hatred. Both in the agrarian south and in the more industrial north, a large, generally poor black community grew up, undereducated and exploited for cheap labor; in the south they were often denied basic civil rights. A gradual change in attitudes, aided by legislation, especially following World War II, has removed some of these distinctions; in many areas of society a degree of "positive discrimination" toward black (or African) Americans has been favored. But thousands of African-Americans are still effectively segregated in inner-city neighborhoods, or ghettos.

Later in the 19th century, an influx of indentured Chinese workers, especially to the west coast, created a strong Asian community that has spread far beyond areas such as Chinatown in San Francisco, California, and has been added to in recent years by many immigrants from Southeast Asia, especially the Philippines and Korea. Immigration from Central and South America has created a large Hispanic population, especially in the big cities of both coasts. Much of this immigration is illegal, creating serious social problems. Immigration is now limited by law, though the United States still has the highest level of immigration anywhere in the world. It welcomes refugees, and large influxes of people from countries such as Cuba and Vietnam.

All of these very diverse people are united by one major factor: the English language. It is the first language for most people, although the growth in immigration from Hispanic countries has made Spanish a second official language in some states. Nevertheless, a command of English opens the door not only to work and advancement, but to the enjoyment of popular culture – sports, music, movies and television.

Religious freedom is guaranteed under the Constitution, and there is remarkable religious diversity. The largest group are the Protestant Christians. Denominations such as Baptists, Methodists, Lutherans and Episcopalians are well represented, with a higher proportion of worshipping (as opposed to merely nominal) Christians than in much of Protestant Europe.

Roman Catholics form the next-largest group; they are largely of Irish, Italian or Hispanic descent. Jews account for only a small minority of the population as a whole, but form a very much higher proportion of the urban population of cities, particularly New York. All the other major world faiths are found here, and there are innumerable sects and cults, from Black Muslims to Scientologists and the American Indian Church. Some of these are identified with the particular areas where they have settled, especially the Mormons of Utah.

Tele-evangelist Jimmy Swaggart (*above*), at his Family Worship Center Church, is one of several popular preachers who have in recent times publicly confessed to misconduct in their private lives. But "born again" Christianity continues to attract huge numbers of followers.

Rice-Eccles Stadium, Salt Lake City, Utah (*below*) Fireworks herald the start of the 19th Winter Olympic Games in February 2002 and the end of three long years of preparations for the two-week event.

The United States possesses the largest economy in the world, and its people enjoy almost the highest standard of living, exceeded only by Japan. Nevertheless, not all Americans are in a position to benefit from its vast wealth.

The country's economy is dominated by the industrial and business sectors, and its agriculture, though vast, plays only a small part in wealth creation. Recent years have seen an increase in service industries at the expense of the traditional manufacturing sector. A consequent decline in the industrial northeastern states has been offset by the rise of new industries and technologies in the south and southwest.

The US has immense natural resources, yet even these cannot supply all the energy and raw materials demanded by the massive industrial output and adventurous foreign policy. As a result, recent years have been characterized by increasing trade and domestic deficits while in 2001 the economy spiralled into a recession.

Agriculture

About half the land area of the United States is used for agriculture. Production is well above domestic demand, and the country is a leading food exporter. The use of machinery and scientific farming methods means that only a small part of the labor force works on the land.

Arable farming accounts for less than half the agricultural area, but the vast tracts of land available have made the country a major world producer. Crop yields have risen, but so has the use of fertilizers, pesticides and herbicides. There are plans to reverse this trend by modifying existing government subsidies and incentives. However, measures to deal with the droughts of the late 1980s delayed the implementation of this policy. Soil erosion is a major problem in some areas of the country.

Crops vary greatly across the country on account of the vast differences in climate and topography. The main arable area is the great central basin between the Appalachians and the Rockies. This includes the Wheat Belt of the western prairies and the Corn Belt of the Central Lowlands, where maize – generally called corn in the United States – is still the principal crop. There is a wide variety of other crops, including soybeans, sunflowers and sugar beet. Cotton is grown in the Mississippi delta area, and farther

Vast orange groves California's Mediterranean climate makes it a center for truck farming: fruit, vegetables and salad crops, grown under irrigation almost all the year round, are transported by air and road to urban destinations all over the country.

west with the help of irrigation. Other crops in the south include groundnuts (peanuts), sugar cane and tobacco.

Fruit and vegetables are grown all over the country, but California is the leading producer, especially in the winter. The state's mild, dry climate is ideal for an expanding wine industry.

Livestock production varies similarly depending on the area. Dairying is the prime occupation in the north and in the mainly industrial northeast, though poultry is important here too. Farther south is a mixed farming area, but dairying is

again important in the west. Ranches and irrigated farms dominate an area extending from southern Nevada to the Rockies. In recent years the United States has found itself embroiled in trade wars with the European Union over agricultural produce. Disputes have ranged from tariff wars on bananas to European attempts to ban imports of US beef and genetically modified (GM) foods on health grounds.

Forestry and fishing

About one-third of the land area is forested, and of this area about two-thirds are commercially exploited or available for exploitation. Commercial forests are mainly in private ownership, but a significant proportion is under local, state or federal control. Between

Drilling for oil (*above*) The oil industry in the United States has led the world in developing new technology for discovering and exploiting petroleum and natural gas reserves.

The mighty Hoover Dam (*below*), also known as the Boulder Dam, is built across the Colorado river on the Arizona–Nevada border, and provides power for the three states of Arizona, Nevada and California

1990 and 1995 the United States planted more trees than any other country. It is the world's largest producer and exporter of roundwood (uncut timber).

The United States is among the five leading fishing nations of the world. Its catch is largely dominated by marine fish from fleets in the Pacific, Atlantic and Gulf of Mexico. Stocks are beginning to recover from the overfishing of the 1970s, but pollution remains a major threat. In 1989 a vast oil spill from the tanker *Exxon Valdez* devastated part of the shoreline of southern Alaska.

Industry and resources
The United States is rich in mineral deposits, and is a leading producer of valuable metals such as copper, lead, silver, gold and zinc. There are large accessible reserves of iron ore in the Great Lakes area, and especially around Lake Superior. There are also large copper-mining operations in several of the western mountain states. Lead mining is mainly

concentrated in southern Missouri, whereas zinc is extracted in a variety of scattered locations. Gold and silver production is limited, but many other minerals are extracted in significant quantities, notably uranium.

The United States has about one-third of the world's recoverable high-quality coal and one-eighth of its lignite. Coal deposits are largely concentrated in the Appalachian states of West Virginia, Kentucky and Pennsylvania, but there are significant fields in Illinois and Ohio. The country also possesses vast resources of natural gas and petroleum. The main petroleum and gas fields are in northern Alaska, Texas and Louisiana, but there are additional reserves off California and in some states bordering the Rockies.

Even these resources are not enough to meet demand within the United States, although the majority of the country's electricity comes from thermal power plants. The choice of alternatives is very limited. Virtually all hydroelectric

The space program

"I believe that this nation should commit itself to achieving the goal, before this decade is out, of landing a man on the moon and returning him safely to earth. No single space project in this period will be more exciting, or more impressive to mankind, or more important for the long-range exploration of space; and none will be so difficult or expensive to accomplish."

This dramatic announcement by President John Fitzgerald Kennedy (1917–63) in May 1961 marked the official beginning of the "space race" – but the starting pistol had already been fired in 1957, when the Soviet Union had launched Sputnik 1. The world's first artificial satellite had been little more than a spherical case broadcasting a simple electronically generated signal. Yet the United States Moon-landing program was partly driven by the fear that the Soviet Union might otherwise gain superiority in space. Big budgets, and the massive Saturn V launch vehicle, took the program rapidly through Earth-orbit and Moon-approach tests to the dramatic Apollo 11 mission of 1969. In July of that year, television viewers all over the world watched transfixed as Neil Alden Armstrong (b. 1930) put the first human footprints on the Moon.

Apollo 11 was the first of seven similar flights, but of these only the near-disastrous Apollo 13 mission captured the public imagination. Lunar landings were expensive, moreover, and their scientific and military value was limited. The Soviet Union, meanwhile, had made no apparent attempt to reach the Moon; instead, it was building and researching earth-orbit space stations. Skylab, the only American space station, was launched in May 1973, partly with equipment left over from canceled Moon-landing missions. It was never used as intensively as the Soviet stations, and in 1979 it fell back to Earth after five years of neglect.

The focus of the American manned space program had by now shifted to reusable vehicles. Budgets, however, were dwindling, and progress was slow. The first Space Shuttle was not launched until 12 April 1981. After four successful test flights the Shuttle came into its own as a satellite launch and repair platform, an experimental station and a military spacecraft. But on 28 January 1986 the orbital vehicle Challenger exploded just 75 seconds after the 25th Shuttle lift-off. There were no survivors, and further Shuttle flights did not resume until September 1988.

Ironically, some of the most impressive achievements of the American space program have been the least expensive ones. The unmanned Ranger Moon probes helped Apollo mission planners to select suitable landing sites. Today's communications and survey satellites support global television networks, and provide up-to-the-minute information on a range of subjects, from weather patterns to land and water resources, mineral deposits, agriculture, forestry and military maneuvers. Satellite data are also used in town planning, earthquake prediction and measurement, mapping and exploration, astronomical observation and many other widely varied applications.

Meanwhile, out in the wider solar system, Mariner spacecraft also made studies of the inner planets, and the Viking 1 and 2 missions put landing craft on the surface of Mars; there was no convincing sign of life, but the missions provided new and important information about the geology and geography of the planet. The Voyager missions discovered rings around Jupiter and Uranus, explored the rings of Saturn, and revealed a wealth of new information about the moons orbiting the outer planets.

In the 1990s highlights of the space program included the spacecraft Galileo reaching Jupiter in 1995 and Pathfinder landing on Mars in 1997. Construction began in 1998 on an international space station which will support long-term stays by US and Russian crews conducting research.

The dream comes alive The Space Shuttle, one of the most dramatic achievements yet in the United States' space program, rises in a plume of exhaust from the launch pad at Cape Canaveral, Florida.

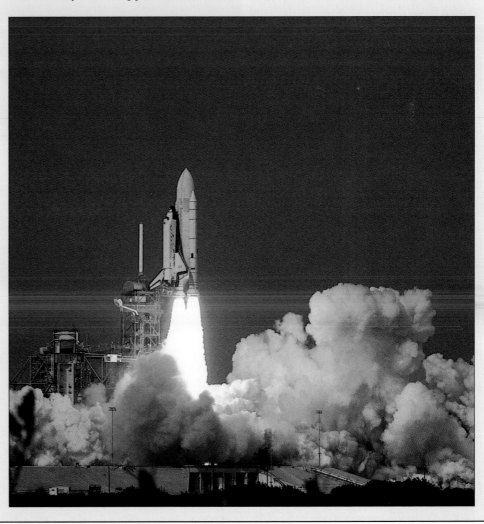

potential has already been harnessed. Nuclear power has become increasingly unpopular since a serious accident in 1979 at Three Mile Island (Pennsylvania). Ten years later, California voters became the first to bring about the closure of a nuclear plant. The disposal of radioactive waste also presents significant problems. Aside from solar, tidal and wind power, geothermal energy is the only resource that is still largely untapped.

Manufacturing industry accounts for a further significant portion of gross domestic product. In recent decades the traditional heavy industries – aluminum, iron and steel products, machinery, cement and building materials – have been strongly challenged by the rise of many new electronics-based industries. With the so-called microchip revolution, electronics and telecommunications became the country's fastest-growing industry in the 1990s, making California's "Silicon Valley", south of San Francisco, world-famous as a high-tech industrial center. By 2000, three of the world's top five companies were American and their size related directly to the exponential growth of personal computers and the Internet. Between them, Cisco Systems, Intel and Microsoft have a market value of over $1.1 trillion.

Aerospace and associated industries were temporarily affected by a slowdown in the early 1990s. The backbone of manufacturing industry remains transportation equipment, especially automobiles, trucks and buses. Today most of this equipment – along with refined petroleum, paper, cigarettes and pharmaceuticals – comes from only a few major corporate producers. Other important manufactures include rubber and plastic products, textiles and clothing. Construction also remains a major industry, vital for providing and maintaining the vast interstate road network.

Trade and commerce

The United States handles a higher volume of trade than any other country in the world. Since the signing of the controversial North American Free Trade Agreement (NAFTA) in 1992, Canada has become the country's largest trading partner by far, overtaking the previous leader, Japan. Restrictions on United States imports by Japan, however, caused political friction between the two countries throughout the 1990s. Other major trading partners are Mexico (the third member of NAFTA), the European Union, and, increasingly, China.

The tangled skein of a freeway intersection (*above*) More than any other country in the world, the United States depends on the automobile – and is suffering for it in terms of congestion, pollution and cost.

Twin banners (*right*) Wall Street, heart of the United States' financial community, signifies its commercial might as clearly as the Stars and Stripes represents its national identity.

The United States is a major producer of fuel and raw materials, yet imports are substantially greater than exports. This is mainly due to unusually high industrial production and consumer demand. Machinery, transportation and telecommunications equipment dominate both exports and imports.

That the US was able to sustain growing current-account deficit through the 1990s indicates the pace of technology-driven expansion that the country experienced and the high level of foreign investment.

Business and finance provide the mainstay of the American economy. The country is responsible for around 15 per-

Desert storage After the terrorist attacks on the US in 2001, the downturn in the airline industry forced airlines to cut costs and store many older planes in airfields such as this one at Victorville in the Mojave Desert.

cent of world trade. In this area above all the United States exercises its influence over the rest of the world as the leading source of investment capital. Canada and Central and South America in particular are dependent on US investment.

Tourism brings with it significant revenue earnings. Visitors are attracted to coastal and mountain resorts, to major cities, and to national parks and monuments. Other major attractions include Disney-World, Disneyland and Hollywood. Service industries are growing in importance, and cities such as Washington DC and Denver, Colorado, earn most of their revenue in this way. Suburban superstores and shopping malls have changed the pattern of consumer spending, accelerating the flight from city centers to suburbs.

Transportation
The United States possesses the best-resourced transportation network in the

world. The country depends for its prosperity on the efficient movement of passengers and freight across vast distances, and an emphasis on the development of its infrastructure has been a significant factor in its economic success.

The national road network has the longest mileage of any in the world. Most cities and towns are linked by a carefully planned network of interstate highways. The country is dominated by the automobile, and the majority of households own at least one vehicle. Most people travel to work in automobiles, cramming the commuter highways with rush-hour traffic from ever more distant suburbs. Pollution from exhaust fumes has forced many states to follow Oregon's lead and to introduce legal limits. Yet as automobile ownership increases, some public and city transit systems are declining.

The railroad network, too, is the largest in the world, but since the 1920s it has seen a marked decline both in mileage and in passenger traffic. Yet the railroads still carry a significant portion of the nation's freight traffic, including bulk freight, containers and "piggyback" loads (highway trailers carried on flatcars). Despite these general trends, railroads and subway systems still handle much commuter traffic in long-established cities such as Chicago, Illinois, and New York. As towns and cities face increasing congestion, some advances have been made in creating practical alternatives to the automobile. Examples include San Francisco's BART (Bay Area Rapid Transit) system and the "people movers"

of cities such as Pittsburgh, Pennsylvania.

Long-distance passenger traffic within the United States is largely controlled by the airlines, though a comprehensive network of bus routes provides a cheap and easily accessible alternative. Railroads are scarcely contesting this market, as transcontinental passenger trains are a thing of the past. The twin hubs of the air transportation network are the airports at Atlanta, Georgia, and Chicago, but there are more than 800 other airports receiving scheduled flights. Thousands more serve local traffic, including private airlines and company-owned air fleets. Recent years have seen increasing financial pressure on the airlines, but this seems unlikely to diminish their importance.

The country's extensive network of rivers, canals and seaways serves more than 50 ports, and carries a significant portion of freight traffic. The busiest inland waterway is the Mississippi river, while New York and New Orleans, Louisiana, are the main ports. The Great Lakes in the north are connected by canal to New York, and are accessible to ocean-going vessels through Canada's Saint Lawrence Seaway.

Media and communications
All media in the United States are free of government control, and on controversial issues they have frequently challenged government policy. Media coverage of the Vietnam War helped bring American involvement to an end, and media investigation of the Watergate affair toppled the Nixon administration.

Television is the dominant medium, and almost every home has a television receiver. There are five major networks – NBC, ABC, CBS, Fox and PSBS – and over

1,500 commercial stations as well as 9,000 cable networks. Satellite broadcasting enables transmission to a worldwide audience. In 1997 there were 575 million radios, vastly more than in any other country, and around 12,000 commercial radio stations. Legislation of 1996 deregulated the media and telecommunications market, although there were restrictions on maximum network coverage.

There are more than 1,500 daily newspapers with a combined circulation of 56.7 million, the second highest in the world behind Japan. Newspapers have traditionally served regional or local readerships with only a few national dailies.

Prior to its dissolution in 1995, the American Telephone and Telegraph Company (AT&T) carried out most manufacturing, technology and research into communications. It has been replaced by regional private companies who operate telephone, telegraph and electronic transmissions. Some companies now provide both cable TV and long-distance telephone services, and the entire network has been expanded to include transmission by fiber-optic cable, microwave radio relay, coaxial cable and domestic satellites. There were also 86 million cellular phone subscribers in 1999, the world's highest national average.

By 2000, over 50 percent of American homes had a personal computer (PC) and 40 percent had access to the Internet. The United States led the world in the development of the technology which created the World Wide Web in the 1990s. By the end of the century the 300 million pages estimated to exist on the Web underlined the significance of the technology that had revolutionized commerce, education, and leisure equally. The possibility of interconnection between television, telephones and computers via the Internet is part of a dynamic, still-expanding field.

Health and welfare

Poverty and racial issues still affect many aspects of welfare provision, and around one-tenth of the population live below the poverty line. Welfare benefits cover unemployment and work injury, sickness and disability, old age, maternity and widowhood. They are funded both by the private sector and by government, and include cash payments and help to poor families. However, the amount of help available varies between states. Anyone who has been employed is entitled to a federal pension, but the scheme is proving difficult to fund and its future is in some doubt. Employers often provide extra

benefits, as well as health and life insurance, though employees normally contribute to such schemes.

Medical facilities are good in most areas, and people's health is generally improving. The federal government provides some subsidies for medical care, but for most people it is a commodity that has to be bought, and this produces inequalities. The mortality rate of black infants, for example, is twice as high as for white infants, and many people cannot afford health insurance of any kind. Medical facilities are good in most areas but are being strained by the growing problem of drug abuse and the incidence of AIDS/HIV infection, which is the highest in any country outside Africa. Health insurance is often denied to those infected with AIDS/HIV.

Education

Education is largely the responsibility of local and state authorities, though some schools are run by private or religious institutions. In most states it is free and compulsory from the ages of 6 to 16, and students generally stay on to complete their secondary education at 17 or 18. The United States has many world-renowned universities and institutes of higher education. These, too, are mainly state-supported, though the federal government underwrites student loans to cover course and other fees. It also provides funds for teaching materials, school meals, research grants, and education programs for Native-American children.

Federal laws have had a profound effect on many schools. For instance, the insistence on integrated schools in the 1964 Civil Rights Act led to the controversial practice of "bussing in" black children over long distances.

American football had its origins as an amateur intercollegiate sport. Now it is the United States' major professional sport. The campus football stars of today still rise to become the most valued players of the professional league tomorrow.

The worlds of Walt Disney

"I don't want the public to see the world they live in ... I want them to feel they're in another world."

The dawn of 17 July 1955 saw an 11-km (7-mi) traffic hold-up on the Santa Ana Freeway at Los Angeles in southern California, and there was similar congestion on other roads for miles around. This temporary chaos resulted from the vision and determination of one man – Walter Elias Disney (1901–66) – and its cause was the opening of Disney's newest venture, Disneyland.

In the early 1950s Disney was known as the man who had turned the animated cartoon into a highly profitable art form. For some people that would have been enough, but Disney wanted much more. He began to visit places such as amusement parks, zoos, circuses, museums and exhibitions, collecting ideas for what he called "a new concept in entertainment". Financial backers were hard to come by, but a Disney television series provided the necessary income.

When the first invited guests streamed into Disneyland in 1955, they found all that had been promised – and more. Adventureland offered a boat trip through a jungle landscape filled with mechanical animals. Frontierland took them back to the days of the Wild West, with scenes made famous by Disney's three films about the folk hero Davy Crockett (1786–1836), who died defending fort Alamo during the war for Texan independence. Dominating the whole scene was Sleeping Beauty's castle, the heart of Fantasyland, while Tomorrowland featured a rocket ride to the moon.

Despite a series of dispiriting problems on opening day, the world's first theme park was an unqualified success, and Disney's ambitious ideas for it continued to proliferate. The following year saw the arrival of Storybook Land and Tom Sawyer's Island. America's first daily monorail opened there in 1959, along with a scale replica of the Matterhorn in Switzerland and an imaginative undersea ride that owed something to the book *Twenty-thousand Leagues Under the Sea* (1864) by Jules Verne (1828–1905).

Walt Disney (*above*) is acclaimed by Mickey Mouse, his first and most famous creation, now known and loved by children the world over.

Fairytale magic (*right*) The floodlit towers of Cinderella's castle are a beacon for visitors escaping to the romantic fictions of Disneyland.

Disneyland, however, was only the beginning. When Walt Disney died in 1966, his plans for a new project in Florida were already well advanced. Walt Disney World finally opened near Orlando in central Florida in 1971, and the first phase was fully completed in 1975. It featured a host of new rides and attractions, including a fascinating audio-animatronic gallery of American presidents.

Then 1982 saw the opening of the Experimental Prototype Community of Tomorrow, better known as EPCOT, at nearby Lake Buena Vista. Disney had envisaged "a community of tomorrow that will never be completed ... a living blueprint of the future, where people actually live a life they can't find anywhere else today." What finally emerged was a marriage between entertainment, education and technology. Future World included a series of themed pavilions, including a time journey to the age of dinosaurs and a history of transportation. Meanwhile, the World Showcase featured pavilions in different national styles, and Spaceship Earth provided a ride through the history of communication.

Other highly successful theme parks have opened in Tokyo and near Paris, France, while Hong Kong Disneyland is expected to open in 2005. The company earned $25 billion in 2000 and is a major producer of home videos in addition to its successful films and TV shows. Future expansion is likely to include e-commerce and delivery of pay-per-view movies via the Internet.

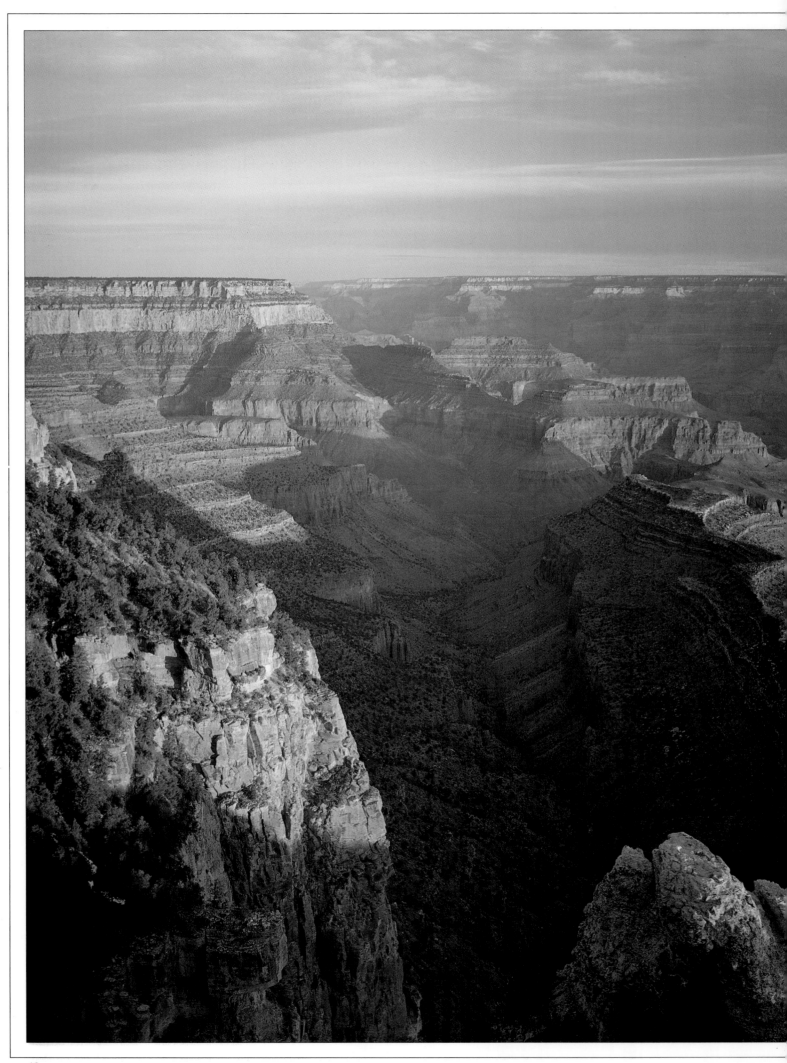

The States of the
Union

THE STATES OF THE UNITED STATES OF AMERICA

Pacific Coast
Alaska
Washington
Oregon
California
Hawaii

Mountain Region
Idaho
Nevada
Montana
Wyoming
Utah
Colorado

The Southwest
Arizona
New Mexico
Oklahoma
Texas

The Middle West
North Dakota
South Dakota
Nebraska
Kansas
Minnesota
Iowa
Missouri
Wisconsin
Illinois
Michigan
Indiana
Ohio

New England
Maine
New Hampshire
Vermont
Massachusetts
Connecticut
Rhode Island

Middle Atlantic Region
New York
Pennsylvania
New Jersey
Delaware
Maryland
District of Colombia

The South
Arkansas
Louisiana
Kentucky
Tennessee
Mississippi
Alabama
West Virginia
Virginia
North Carolina
South Carolina
Georgia
Florida

Sunset on the Grand Canyon, the spectacular gorge cut by the Colorado river through the mountains of northwestern Arizona. Its majestic beauty attracts millions of visitors each year.

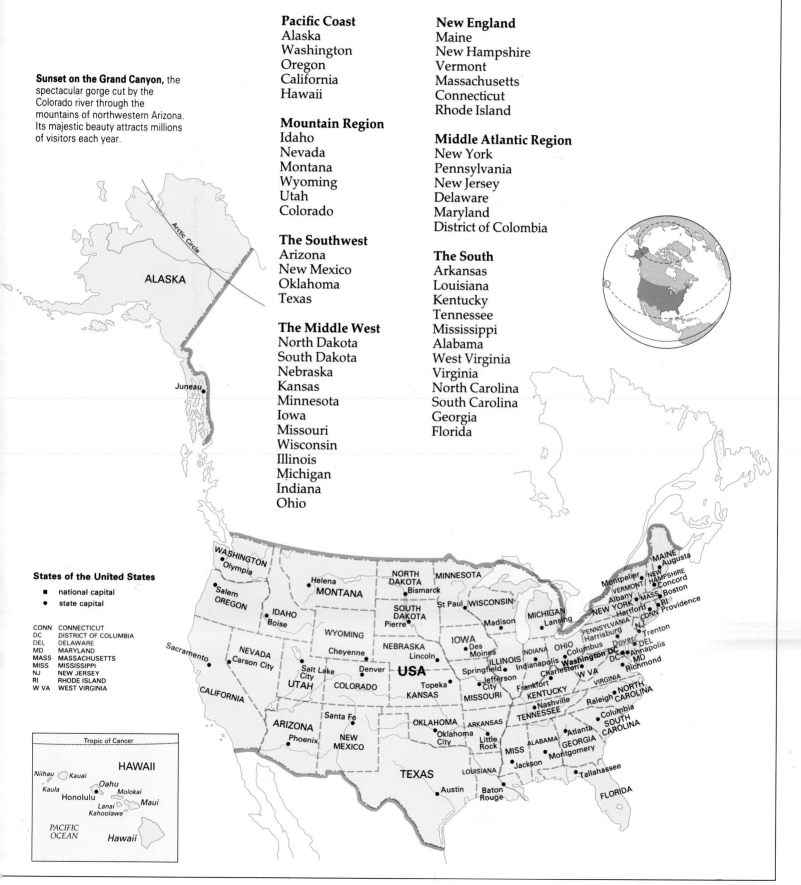

States of the United States

■ national capital
● state capital

CONN   CONNECTICUT
DC   DISTRICT OF COLUMBIA
DEL   DELAWARE
MD   MARYLAND
MASS   MASSACHUSETTS
MISS   MISSISSIPPI
NJ   NEW JERSEY
RI   RHODE ISLAND
W VA   WEST VIRGINIA

ALASKA

Nickname	The Last Frontier
Joined the Union	1959
Capital	Juneau
Area	1,530,693 sq km (591,004 sq mi)
Population	670,053
Outstanding features	Mount McKinley, Glacier Bay National Park
Principal industries	gas and petroleum mining, tourism, fishing

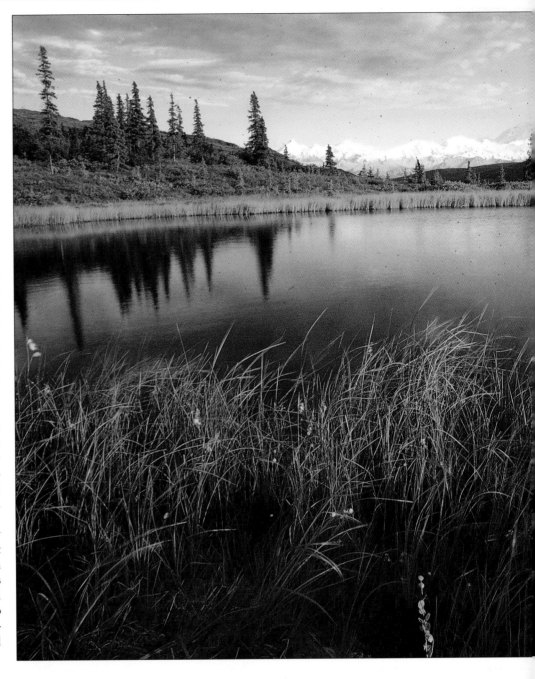

Alaska is America's largest state, and the newest apart from Hawaii. It is a wild and often mountainous country, lying at the northwestern tip of the continent, facing Russia across the Bering Strait.

It was across this Strait that the first peoples are thought to have entered the Americas some 30,000 years ago. Their descendants, the Inuit and Aleut peoples, still inhabit the remoter parts of Alaska. In the 18th century the first European settlers also arrived from Russia, and Alaska was controlled by the Russian-American Company until United States Secretary of State William Henry Seward (1801–72) negotiated its purchase from Russia in 1867.

Since then there has been a steady stream of immigrants, attracted by its vast mineral wealth and by huge projects such as the Trans-Alaska Oil Pipeline. Towns such as Anchorage in the south and Fairbanks in the center have expanded to become considerable conurbations. However, most of the state remains a vast and magnificent wilderness.

WASHINGTON

Nickname	Evergreen State
Joined the Union	1889
Capital	Olympia
Area	176,479 sq km (68,139 sq mi)
Population	6,395,798
Outstanding features	Mount Rainier (Cascade Range), Puget Sound
Principal industries	aircraft, wood and paper products, foodstuffs, information technology

During the 16th century European navigators were searching for a Northwest Passage to Asia. Their search was fruitless – but they did discover the existence of America's Pacific Northwest.

At first the area that is now Washington state was part of the Oregon Country – the whole of the Pacific Northwest – whose northern boundary was set in 1846 along the present Canadian frontier.

From the 1830s to the 1850s, white settlers moved into the area from the Midwest along the Oregon Trail. Washington finally became a separate territory in 1853, benefiting from the demand for construction materials created by the California gold rush. Railroad construction in the 1880s brought rapid growth, and during the 1890s Washington was once again a base for miners and prospectors, this time heading for Alaska and the Klondike. The 20th century saw the damming of the Columbia river to provide power, irrigation and improved navigation.

Washington's economy has been largely dependent on natural resources, such as forestry. However, in recent years, Seattle has become a major economic center, being the home of successful high-tech and software companies. The state is also a center for the aerospace industry, and there are numerous military establishments along the shores of Puget Sound.

OREGON

Nickname	Sunset State
Joined the Union	1859
Capital	Salem
Area	251,418 sq km (97,073 sq mi)
Population	3,700,758
Outstanding features	Mount Hood (Cascade Range), Crater Lake National Park, Columbia River Gorge
Principal industries	timber products, machinery, foodstuffs

Oregon has a dramatic landscape of high, densely forested mountain ranges, with plateaus and plains cut by deep valleys and gorges. The forests of the west and northeast provide much of its wealth, while in the east are arid deserts.

In the 16th century the area was claimed by both Britain and Spain. However, the large and diverse Native-American population that inhabited the area remained largely undisturbed until 1792,

Mighty mountains (*above*) The jagged flanks of the Cascade Range in the state of Washington were raised by movements of the Earth's crust, which also cause earthquakes or volcanoes in all the states along the Pacific coast.

Northern wilderness (*left*) Alaska's lakes and wetlands, havens for wildlife, are among Earth's last natural wildernesses. Seen here in the background, Mount McKinley, the tallest peak in North America, lies within Denali National Park, which has been declared a Biosphere Reserve.

CALIFORNIA

Nickname	Golden State
Joined the Union	1850
Capital	Sacramento
Area	411,047 sq km (158,706 sq mi)
Population	36,457,549
Outstanding features	Yosemite National Park (Sierra Nevada), Death Valley, Lake Tahoe
Principal industries	foodstuffs, aircraft, tourism, microelectronics, information technology

when Robert Gray (1755–1806) entered from the west via the Columbia river. Fur traders soon arrived from the east, and in 1811 John Jacob Astor (1763–1848) paved the way for white settlement when he established the trading post of Astoria. Subsequent decades saw a growing tide of settlement as pioneers arrived along the Oregon Trail; most settled in the Williamette Valley between the Cascade and Coast Ranges. In 1842 Britain abandoned its claim to Oregon, which later became the 33rd state of the Union.

Oregon depends largely on farmland and forestry for its prosperity, and was among the first states to introduce stringent control of vehicle exhaust emissions. In 1980 ash from the volcanic eruption of Mount Saint Helens in the neighboring state of Washington caused massive environmental damage. The state economy, however, has become more diversified in recent decades, and suffered less than might otherwise have been expected. Tourism is an increasingly important activity in the state.

Portland, Oregon, is the state's largest city, principal port and also its cultural and educational center. It grew up in the mid 19th century, boosted by its position as the main supply port to the west coast gold rushes.

California is the most populous of the United States, and its economy, though facing severe difficulties throughout 1994, can stand comparison with that of the world's largest nations. It has the world's heaviest concentration of motor vehicles, its most extensive freeway system and a consequent pollution problem – yet until the late 18th century it remained an unspoiled wilderness.

In 1542 the Spanish navigator Juan Rodríguez Cabrillo (d.1543) became the first European to see the coast of California. The first official Spanish expedition arrived in 1769. In the very same year Junípero Serra (1713–84), a Spanish priest, set up a Franciscan mission at Monterey on the rugged coast of central California.

In 1821 Mexico won its independence from Spain, and within 20 years Mexican settlers had taken over the mission ranches, expelling and exploiting the Native-American population. However, the area had also begun to attract white American settlers. In 1848, after a brief war with Mexico, the territory was ceded to the United States. Only a year later the discovery of gold to the east of Sacramento started a further massive influx of prospectors and settlers.

The state remained loyal to the Union during the Civil War. Waves of immigration from the east brought farmers to the inland valleys and industries to the growing cities. Hollywood, near Los Angeles, became the film and television capital of the world. California's population, which doubled on average once every two decades between 1860 and 1960, is ethnically very mixed, with large populations of Spanish-speakers, Chinese and Vietnamese, Native-Americans and African-Americans. More than 90 percent are urban dwellers.

Today California is not only the prime agricultural state; it also has enormous reserves of petroleum, a thriving industrial sector and a massive income from tourism. Silicon Valley, situated south of San Francisco Bay, has become famous in the field of information technology. The San Andreas Fault runs through to the north of San Francisco, making the area vulnerable to earthquakes. Movement along the fault caused the disastrous San Francisco earthquake of 1906 and that in southern California in January 1994.

HAWAII

Nickname	Aloha State
Joined the Union	1959
Capital	Honolulu
Area	16,760 sq km (6,471 sq mi)
Population	1,285,498
Outstanding features	Mauna Kea (the world's highest island peak), Mauna Loa
Principal industries	tourism, foodstuffs, defense industries

Hawaii is America's most far-flung state, and perhaps its most unusual. It comprises a chain of steep volcanic islands in the Pacific some 3,800 km (2,400 mi) southwest of the Californian coast. The state is named after the largest island.

The islands were first settled about 1,500 years ago by Polynesian peoples. The first European visitor was the British explorer James Cook (1728–79) in 1778, who called them the Sandwich Islands. In about 1795 they were united for the first time under the rule of the formidable warrior monarch Kamehameha I (1758–1819). However, European settlement progressed, and in 1851 his successor Kamehameha III (1814–54) placed the islands under United States' protection. The native population declined considerably owing to introduced diseases among other causes; many new settlers

San Francisco's Golden Gate Bridge (*top*) frames the leisurely sailing craft in the bay, and the city's glittering skyline – a reminder of the fierce commercial drive behind California's easy-going lifestyle.

Icons of Hollywood culture (*above*) A movie star's hand and footprints left in the famous stretch of sidewalk outside Graumann's Chinese Theater, Los Angeles.

Hawaii's golden beaches (*right*) attract a multitude of vacationers to the island, to swim, surf or sunbathe on the beaches. Inland, other attractions include spectacular lava formations and tropical bird life.

arrived, especially from eastern Asia. Today there are more Hawaiians of Japanese than of Polynesian descent.

American influence grew until in 1900 the islands were finally annexed by the United States. A huge naval base was established at Pearl Harbor, and in 1941 this became the target of the Japanese attack that brought the United States into World War II. Hawaii was by now so firmly American in character that incorporation as the 50th state was inevitable.

Hawaii supports several industries, notably sugar refining; but its greatest asset is its natural beauty. The climate combines tropical sunlight with mild, temperate weather. Rainfall can be fierce, but it supports the lush forests that cover the slopes of the spectacular volcanic mountains, many of which are still active. Clear air and cloudless skies make their upper slopes an ideal site for astronomical observatories. Not surprisingly, the islands are a great tourist attraction, giving millions of mainland Americans each year a taste of Polynesia.

IDAHO

Nickname	Gem State
Joined the Union	1890
Capital	Boise
Area	216,430 sq km (83,564 sq mi)
Population	1,466,465
Outstanding features	Hell's Canyon, Craters of the Moon National Monument, Shoshone Falls
Principal industries	foodstuffs, timber processing, chemicals

Idaho lies in the northwest of the United States, its short northern border touching the Canadian border. Its center is dominated by several of the Rocky Mountain ranges, and the south by the strange lava landscapes of the Snake River Plain.

Native-American peoples such as the Shoshoni have lived here as hunter–gatherers for at least 10,000 years, and their descendants still inhabit the area. Idaho was originally part of Oregon Country, and was obtained by treaty from Great Britain in 1846. It was first settled by gold prospectors, and later by Confederate refugees after the Civil War. Many Mormons also settled here, fleeing persecution for their religion in the east; they did much to build up the state, and still form the majority in the southeast.

Idaho has a small but growing manufacturing base, but remains a strongly agrarian area, famous for its vast potato crop, its wheatfields and its lumber.

NEVADA

Nickname	Silver State
Joined the Union	1864
Capital	Carson City
Area	286,352 sq km (110,561 sq mi)
Population	2,495,529
Outstanding features	Boundary Peak, Carson Sink, Lake Mead (Boulder Canyon)
Principal industries	tourism, gold and mineral extraction, electronics, chemicals

Nevada lies east of the Sierra Nevada mountains (after which it is named). It is a state of surprising contrasts. The land is largely arid, unpromising and sparsely populated, and yet Nevada is one of the most prosperous states – for reasons that are often controversial.

Native-American peoples have lived here for more than 20,000 years. The first European settlers were Spanish missionaries and Canadian fur traders in the 18th and early 19th centuries. They were followed by American explorers such as John Charles Frémont (1813–56), who opened up the west to settlers and prospectors. In 1848 the area was ceded by Mexico to the United States. The Comstock Lode discovery in 1859 caused a silver rush, attracting settlers in great numbers. However, monetary reforms in the 1870s produced a dramatic slump in the silver industry, and the cattle ranching that replaced it declined shortly afterward, leaving large numbers of empty ghost towns. Mining and ranching revived early in the 20th century, only to be wiped out again during the Great Depression of 1929.

In the early 1930s, however, the state government passed the legislation that

A gambler's paradise Las Vegas began as a Mormon settlement in the 1850s, later becoming a small agricultural town in the desert. Now, as a result of the state's liberal gaming laws, it attracts millions of visitors each year to its casinos.

legalized gambling and prostitution, and minimized the statutory requirements for marriage and divorce. This led to an influx of investment from elsewhere in the United States; major resort centers grew up in the scenic areas around Las Vegas in the south and Reno in the west, offering luxurious hotels, night life and casinos. The resulting boom in tourism brought immediate prosperity to the area. Meanwhile, more orthodox industry was boosted by cheap hydroelectricity from the Hoover (Boulder) Dam on the Colorado river, and by the discovery of further mineral deposits. Most of the population remains crowded into the Las Vegas and Reno areas, leaving vast open spaces that have proved ideal for military and civil defense installations, including nuclear testing sites.

MONTANA

Nickname	Treasure State
Joined the Union	1889
Capital	Helena
Area	380,847 sq km (147,046 sq mi)
Population	944,632
Outstanding features	Granite Peak (Rocky Mountains), Glacier National Park
Principal industries	foodstuffs, copper and petroleum extraction, timber processing

In 1862 gold was discovered at Bannack in the northern Rockies, attracting a wave of prospectors and opportunists. Two years

Proud of the past (*above*) This reconstruction of a Wild West frontier town in Montana preserves a sense of the self-sufficiency of these isolated pioneer communities, where the bar and the bath-house stand side by side.

Reflected glory (*right*) Dark pines and golden aspens reach toward the snow-capped peaks of the majestic Maroon Bells, which tower above Maroon Lake near the famous mountain ski resort of Aspen, Colorado.

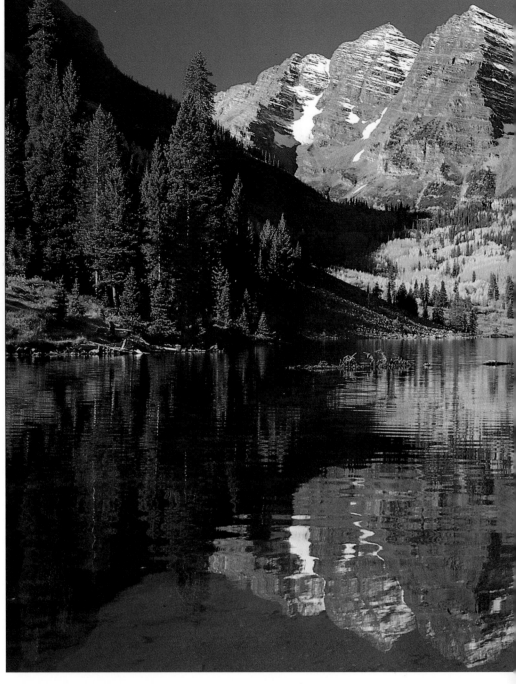

later the Montana Territory was created, and the first cattle ranchers moved in to graze their huge herds on the vast expanses of gently rolling land in the east of the territory.

The Native-Americans fought hard to defend their hunting grounds. Names such as Sioux and Cheyenne have become legendary, as have the battlefields where they won their last victories – Little Bighorn (1876) in southeastern Montana, and the Bighole Basin (1877) in the southwest. But shortly afterward the Native-Americans were defeated by United States troops, and this opened the way for white settlement.

The United States had acquired the territory as part of the Louisiana Purchase in 1803, but until the gold rush of 1862 little had been done to exploit it. With the Native-American population very greatly reduced, settlers turned their attention from gold to the massive local copper deposits. Offers of land brought a wave of immigrant homesteaders from northern Europe from 1909 onward, but the severe droughts of the 1920s ruined the livelihoods of many of the new farmers.

Agriculture and mining are among the mainstays of the economy. The chief crops are wheat, barley and sugar beet, but numerous minerals are extracted, apart from copper – notably lead, silver and phosphates. A growing number of tourists are attracted to the state's dramatic scenery and historic battle sites.

WYOMING

Nickname	Equality State
Joined the Union	1890
Capital	Cheyenne
Area	253,324 sq km (97,809 sq mi)
Population	515,004
Outstanding features	Yellowstone and Grand Teton National Parks
Principal industries	petroleum, gas and coal extraction, tourism, foodstuff

When the United States bought this area from France as part of the 1803 Louisiana Purchase, it was still a mountainous wilderness inhabited by the Arapaho and Shoshoni peoples. For many years the arid prairies and mountains discouraged settlement, and Wyoming was little more than a staging point for people heading farther westward along the Oregon and Overland Trails. The arrival of the telegraph and railroad gave rise to retaliatory raids by the Arapaho and Shoshoni, whose lands were taken and hunting grounds disturbed. By 1869, however, most of the Native-Americans had been confined to the Wind River Reservation on the eastern slopes of the Rockies.

Cattle ranchers were quick to move in once the railroads had arrived, and large areas of the state are still pastureland. However, mineral fuel production is the dominant industry, and easily available uranium and hydroelectricity have made Wyoming a leading producer of power. Wyoming was the first state to create national parks, and tourism has benefited correspondingly. It was also the first state to give women the vote – hence its nickname Equality State.

were resettled on reservations. By the 1890s the Mormons had established a solid, stable community, based largely on agriculture and small-scale cooperative businesses. A disapproving United States Congress refused successive statehood applications until 1896, when polygamy was officially renounced and the Mormon Church relinquished its political status. Its social dominance, however, remained, and its traditions of the work ethic and mutual assistance were a major factor that encouraged the state's industrial and commercial growth.

Today the Mormon influence endures, liberalized by pragmatic "revelations" such as on the admission of black people into the church. Utah's popular tourist centers, and the prosperity of its capital, Salt Lake City – a major financial and commercial center – are living testimonies to its worldly success.

COLORADO

Nickname	Centennial State
Joined the Union	1876
Capital	Denver
Area	269,594 sq km (104,091 sq mi)
Population	4,753,377
Outstanding features	Mount Elbert (Rocky Mountains), Black Canyon, Garden of the Gods
Principal industries	microelectronics, foodstuffs, machinery

Colorado, lying astride the Rockies, is the archetypal Western state. It was home to various Native-American peoples, ranging from the Mesa Verde cliff-dwellers to Plains peoples such as the Cheyenne and Arapaho. These were later overrun by the Spanish, who named the area Colorado ("red" or "ruddy") after its reddish rocks.

The eastern side, annexed by France, was sold in the Louisiana Purchase in 1803; the west passed from Mexico to the United States in 1848. The 1859 gold rush brought an immense influx of settlers, creating numerous frontier towns. Their need for food stimulated the development of irrigated agriculture, and cattle and sheep ranching replaced the exterminated buffalo. In the late 19th century discoveries of rich mineral reserves made the new state a major industrial center.

In recent years, Colorado's older industries have suffered setbacks, but have to a certain extent been replaced by modern computer and aerospace industries. The awesome scenic beauty of the Rockies, and the ideal skiing conditions at resorts such as Aspen, support a fast-expanding tourist industry.

UTAH

Nickname	Beehive State
Joined the Union	1896
Capital	Salt Lake City
Area	219,887 sq km (84,889 sq mi)
Population	2,550,063
Outstanding features	Great Salt Lake, Canyonlands National Park, Lake Powell
Principal industries	armaments, electronics, foodstuffs, tourism

The landscape of Utah is generally barren, arid and inhospitable, but it has nevertheless nurtured some hardy peoples and distinctive cultures. Among the earliest of these, about 400 AD, were the Anasazi or Pueblo peoples. Their society decayed long before the arrival of Spanish explorers in the 18th century, but their awesome cliff villages survive, along with pottery and rock paintings. Their place was taken by the Navajo and Ute peoples, many of whom still live in the state.

In 1848 the United States annexed the area from Mexico. At about this time, members of the Church of Jesus Christ of Latter-day Saints, or Mormons, were forced to flee from persecution, sparked off by their unconventional practices, in particular that of polygamy. In 1847 Brigham Young (1801–77), claiming a divine mission to transform the desert into a paradise, led wagon trains of Mormons into the Great Salt Lake Valley. With hard labor and missionary zeal, vast tracts of land were irrigated, and Young's vision was accomplished in less than 15 years.

The Native-Americans were seldom directly persecuted, but their food supplies suffered; clashes followed, and they

ARIZONA

Nickname	Grand Canyon State
Joined the Union	1912
Capital	Phoenix
Area	295,259 sq km (114,000 sq mi)
Population	6,166,318
Outstanding features	Grand Canyon, Painted Desert, Petrified Forest National Park
Principal industries	electronics, tourism, copper mining

Arizona is one of America's youngest states, but one of the oldest inhabited. Native-American peoples such as the Anasazi and Hohokam flourished in this arid but beautiful area for at least 25,000 years. They were succeeded by the Apache and Navajo not long before the Spanish came. The area remained Spanish until 1848, when it was ceded to the United States; further territory was added by the Gadsden Purchase of 1853. It is in Arizona that many of the legends of the Wild West are rooted; Tombstone, for example, was the site of the notorious gunfight at the OK Corral.

During the 1970s Arizona's population was nearly doubled by immigrants from other states, drawn by the sunny climate and employment opportunities in cities such as Tucson and Phoenix. Most areas, however, remain sparsely populated. Native-Americans are more numerous than in most other states, but are heavily outnumbered by Spanish-speakers.

Arizona's spectacular scenery, with its great mountains rising abruptly above the plains, is a major asset, and its many national parks form the basis for a thriving tourist industry.

NEW MEXICO

Nickname	Land of Enchantment
Joined the Union	1912
Capital	Santa Fe
Area	314,924 sq km (121,593 sq mi)
Population	1,954,599
Outstanding features	Wheeler Peak (Rocky Mountains), Carlsbad Caverns
Principal industries	mining, tourism, foodstuffs

New Mexico has a long history of habitation. Ruined cliff dwellings give evidence of the Pueblos' agricultural civilization; the nomadic Navaho and Apache peoples probably entered the area only a few centuries before the Spanish. In 1540 Francisco Vásquez de Coronado (1510–54) came here in search of the legendary "seven cities of gold", but left empty-

handed. Early in the 17th century, Juan de Oñate (1550–1630) built the first European settlements, including Santa Fe in 1610. The colony was later destroyed by Native-Americans defending their way of life from the white settlers. In the 18th century Spain regained control, and in 1821 the area became part of Mexico.

The eastern section was annexed by the United States in 1845. During the subsequent Mexican War the whole area was occupied by the Army of the West, before Mexico formally ceded it to the United States in 1848. Conflict with the Navaho and Apache peoples fueled the legends of the Wild West. But the arrival of the railroad in 1880 brought a massive wave of immigration by white settlers.

During World War II the atomic bomb was developed at Los Alamos to the northeast of Santa Fe, and scientific research remains an important activity. Agriculture is limited by the low rainfall, but cattle ranching is still important. More than one-third of the population are of Spanish origin, about one-half of them Mexican-Americans, and Spanish is the state's second official language. The state attracts many tourists, and there is a thriving market for Native-American and Mexican craft souvenirs, such as pottery, blankets and silver jewelry.

OKLAHOMA

Nickname	Sooner State
Joined the Union	1907
Capital	Oklahoma City
Area	181,185 sq km (69,956 sq mi)
Population	3,579,212
Outstanding features	Black Mesa (Panhandle), Wichita Mountains
Principal industries	machinery, petroleum extraction, metals

The area of Oklahoma made up part of the Louisiana Purchase, sold by France to the United States in 1803. In 1828 Congress created an Indian Territory here, reserved

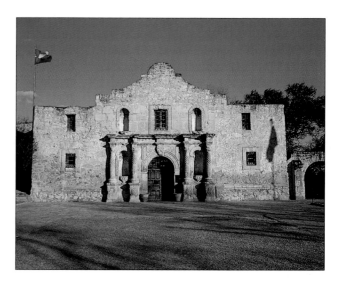

Monument Valley, Arizona (*left*), its buttes and mesas sculpted by nature from red sandstone. This stark but beautiful land is the ancestral home and present-day reservation of the Navajo people.

The Alamo fortress (*right*) at San Antonio, Texas, a small shrine to an immense legend. It was here, in 1836, that a tiny band of volunteers, notably Davy Crockett (1786–1836), sacrificed their lives to defend Texan independence.

A florid facade (*below left*) belonging to the missionary church of San Xavier del Bac at Tucson, Arizona. Established by Spanish Jesuits in 1700, this mission has served the Pagago Native-Americans down to the present day.

exclusively for Native-Americans and consisting of five republics or nations. It became known as Oklahoma, from the Choctaw words meaning "red people". In the Civil War the Native-Americans, unhappy with the federal government, sided with the Confederate army, and suffered accordingly. In 1889 Congress started to allow homesteaders to open up the Indian Territory land for farming, and Oklahoma was swamped by a series of "land runs". Today only a minority of the population are Native-Americans. In 1995 the terrorist bombing of a federal building in Oklahoma City took 168 lives; two local men were later convicted of the bombing.

A mining boom brought other settlers, but farming remained the most important occupation of the new state. However, many farmers lost their land and their livelihood through drought in the 1930s, but new prosperity came with the discovery of massive petroleum reserves. Modern Oklahoma remains primarily industrial, though less dependent purely on its extensive mineral resources.

TEXAS

Nickname Lone Star State	
Joined the Union 1845	
Capital Austin	
Area 691,027 sq km (266,807 sq mi)	
Population 23,507,783	
Outstanding features Mount Livermore, Guadalupe Peak, Big Bend National Park	
Principal industries petroleum, machinery, foodstuffs, clothing	

Texas is the largest of the 48 coterminous states. Its sheer size and its distinctive character have done much to shape international perceptions of the United States as a whole. The wide open alluvial plains extend inland from the Gulf of Mexico, rising to meet the Great Plains to the north. The southwestern border with Mexico is formed by the waters of the Rio Grande, which flows down through the arid hill country of the west.

The first Spanish explorers found the area only sparsely inhabited by Native-American peoples. It was claimed by the United States following the Louisiana Purchase of 1803, but remained under Spanish rule. A coup in Mexico in 1833 drove the Texans, led by Anglo-American settlers, to stage a rebellion and declare independence. In 1836, following the famous siege of the Alamo at San Antonio in central-southern Texas, the Texans finally routed the Mexicans at San Jacinto in the east. Nine years later the Republic of Texas was annexed by the United States, and Mexican opposition was finally crushed in 1847.

Texas seceded from the Union in 1861, and came under martial law during the Reconstruction. In the 1870s and 1880s the Native-American population was forcibly expelled; railroad construction opened up the state to swift settlement, and a boom in cattle ranching created an image of Texas as the archetypal "Wild West" state.

The discovery of petroleum in 1901 began an even greater boom, in which Houston and Dallas in the east became two of the most affluent cities in the world. A wide range of other industries followed, making Texas a leading manufacturing center. Meanwhile, agriculture has retained its importance, especially in the areas of cotton and livestock production. The boom has slackened during the various periods of economic depression, but the economy has always revived. Recent years have seen a growth in the important electronics and consumer-product sectors.

leading to the eventual ratification of the Federal Constitution. During the 19th century the state became increasingly industrialized, and immigrants, especially from Ireland, swelled the labor force. The Abolitionist movement also became strong here, and the state sided firmly with the Union in the Civil War.

Massachusetts today maintains many legacies of its past, whether in the wealth and influence of older families (the "Proper Bostonians"), its Irish heritage, or the many historical sites and monuments. However, immigration from many parts of the world has created a highly cosmopolitan society. The traditional industrial base in textiles has declined in recent decades, but the gap has been filled by many new "high-tech" industries, often stimulated by the presence of internationally famous academic institutions such as Harvard University, Boston University and the Massachusetts Institute of Technology (MIT).

The Old State House, Boston, Massachusetts
(*below*) stands at the heart of this crowded city of skyscrapers – a reminder of its colonial past. Today the city is host to sizable Irish, Italian, African-American and Puerto Rican communities.

CONNECTICUT

Nickname	Constitution State
Ratified the Constitution	1788
Capital	Hartford
Area	12,997 sq km (5,018 sq mi)
Population	3,504,809
Outstanding features	Connecticut River, Long Island Sound
Principal industries	military equipment, insurance, electrical equipment

Connecticut is one of the United States' oldest and proudest communities, lying on the New England coast between the great cities of New York and Boston.

The colony was founded during the 1630s by Puritans – originally from Britain – who had left the Massachusetts Bay Colony. The colonists enjoyed extremely good relations with the local Algonkin peoples. They settled in two main areas: the Connecticut River Valley (1633–35), and the coastal strip between Saybrook and New Haven (1635–38). In 1662 a royal charter provided effective self-government, and Connecticut citizens continued to value their independence. In 1776 more than half of the army under the command of George Washington (1732–99), based at New York, was drawn from Connecticut; and later that year Connecticut citizens adopted the original 1662 charter as their

state constitution. In 1814–15 Connecticut joined four other New England states in the Hartford Convention – an attempt to revise the federal constitution. During the Civil War, however, the state strongly supported the Union.

The 19th century saw radical changes. A stream of Irish immigrants began in the 1840s, and this was swelled later in the century by an influx of French Canadians and immigrants from Europe. Textile factories were built, and industry soon displaced agriculture as the main source of local employment.

Today farming plays only a small and dwindling part in the state economy. Some of the original 19th-century industries still thrive, notably gun- and clock-making, silverware and the manufacture of sewing machines. However, textile industries have largely disappeared as the economy has become more service-based. The ancient forest that once covered most of Connecticut has long been felled, and today's mixed woodland is a later growth. The original towns and villages have gradually merged along the main traffic arteries, and most people are urban-dwelling. However, there are no large cities. Local autonomy is fiercely prized, and local history is vehemently preserved in the form of historic buildings, sites and monuments.

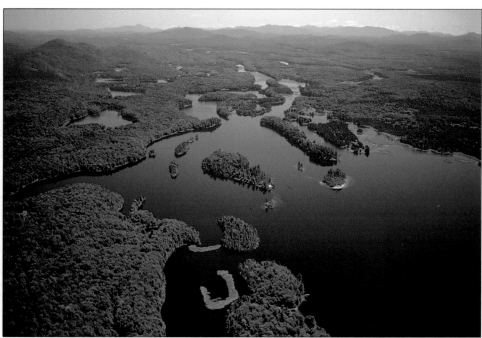

Unspoilt beauty (*above*) in upstate New York, seen at Lows Lake in the Adirondack Forest Preserve.

South Manhattan, New York City (*left*) Two days after the terrorist attacks of September 11 2001, smoke and dust still rise from the site of the World Trade Center. The cost of the damage was an estimated $105 billion.

RHODE ISLAND

Nickname	Little Rhody
Ratified the Constitution	1790
Capital	Providence
Area	3,139 sq km (1,212 sq mi)
Population	1,067,610
Outstanding feature	Narragansett Bay
Principal industries	jewelry, machinery, textiles, electronics

The smallest state in the Union is Rhode Island on New England's Atlantic seaboard; it takes its name from one of several islands situated in Narragansett Bay. The first European settlement was the mainland town of Providence, founded in 1836 by the Puritan minister Roger Williams (1603–83) after his expulsion from Massachusetts. Williams was a strict disciplinarian, but maintained a spirit of religious tolerance and dealt honestly with the local Native-Americans. This highly independent tradition made the colony swift to rebel but slow to join the new nation. Rhode Island was the last of the original 13 colonies to ratify the Constitution, and the last to introduce universal adult suffrage – the subject of Dorr's Rebellion in 1842.

Economically, Rhode Island has always been progressive. It was at the forefront in the Industrial Revolution, and today its economy relies mainly on sophisticated light industries and the service sector. The state retains strong historical roots, however, particularly in the old seaports of Providence and Newport, where whole districts are carefully preserved, including some of the oldest buildings in the United States.

NEW YORK

Nickname	Empire State
Ratified the Constitution	1788
Capital	Albany
Area	136,583 sq km (52,735 sq mi)
Population	19,306,183
Outstanding features	Mount Marcy (Adirondack Mountains), Long Island, Niagara Falls
Principal industries	publishing, clothing, pharmaceuticals, machinery, finance

New York was one of the 13 original American colonies, and the state, like the city of New York, reflects the diversity of its many immigrant settlers. The largest groups came originally from Italy, Germany, the Soviet Union, Poland, Ireland and Puerto Rico.

The first European to arrive was the Englishman Henry Hudson (1565–1611), who explored the river that was later named after him. The first settlers, however, were the Dutch, who founded the colony of New Amsterdam on Manhattan Island. In 1664 the British took the colony by force, renaming it New York. A troubled period followed, during which the conflicting claims of France and Britain were further complicated by rivalries between Native-American groups; the situation was largely resolved with the Treaty of Paris in 1763. During the War of Independence, much of the colony remained loyal to Britain. New York City remained in British hands until the war ended, but then became the capital of the new nation from 1789 to 1790.

The state's already cosmopolitan population was constantly swelled by an uninterrupted stream of immigrants. Immigration continued unabated after the Civil War, doubling the population of New York City between 1880 and 1890. The city today is among the world's most populous cities, and a national center for finance and the arts. It is also the nation's largest seaport, and there are three international airports located within the metropolitan area. So dominant is New York City that the rest of the state is often forgotten. Upstate New York contains a variety of landscapes, from the Great Lakes and the Niagara Falls in the west, via the mountains and lakes of the Appalachians, to the Hudson Valley in the east.

On September 11 2001 New York City was the target of a brutal terrorist attack which completely destroyed the twin towers of the World Trade Center – a potent symbol of the financial world and a great landmark recognized the world over. Over 3,000 people were killed.

PENNSYLVANIA

Nickname	Keystone State
Ratified the Constitution	1787
Capital	Harrisburg
Area	119,251 sq km (46,043 sq mi)
Population	12,440,621
Outstanding features	Mount Davis (Allegheny Mountains), Susquehanna River
Principal industries	steel and other metals, foodstuffs, machinery

Pennsylvania is one of the oldest and most populous states, extending westward from the Delaware river as far as the shores of Lake Erie. Its official name, the Commonwealth of Pennsylvania, reflects the idealism of the state's founder, William Penn (1644–1718).

In 1681 Penn was granted a large expanse of American land as payment of a debt owed to his father by the British monarch. The following year he founded a colony based on Quaker principles of tolerance and democracy on the site of present-day Philadelphia by the Delaware river. Penn was notably honest in his dealings with the Native-American inhabitants, and the Quaker policy of tolerance encouraged many religious minorities to settle in the colony. A significant proportion were Germans, known today as Pennsylvania Dutch, and included communities such as the Mennonites and the Amish. Pennsylvania played a central role in the independence struggle; the Declaration of Independence was signed at Philadelphia, and the city became the first capital and seat of government. The state was also a Civil War battleground, and saw a major Union victory at Gettysburg in 1863.

Pennsylvania's fortunes were founded on its rich agricultural land, but huge deposits of coal and other minerals helped to make it the United States' first really great industrial and manufacturing center, especially around Pittsburgh, where "steel kings" such as Andrew Carnegie (1835–1919) established gigantic works. These traditional industries have since declined, but Pennsylvania remains a primarily industrial state. Even today, though, much of the land is still cultivated, and there are beautiful stretches of farm and woodland. The city of Philadelphia, with its great art collections, orchestras and museums, is a world-famous cultural center, while Pittsburgh, its industrial environment dramatically transformed by a clean-up campaign, is rapidly achieving the same status.

NEW JERSEY

Nickname	Garden State
Ratified the Constitution	1787
Capital	Trenton
Area	20,168 sq km (7,787 sq mi)
Population	8,724,560
Outstanding features	Kittatinny Mountains, Pine Barrens (heath, dunes, and woodland)
Principal industries	chemicals, electrical machinery, scientific research

Elfreth's Alley (*above*), a historic enclave in downtown Philadelphia – very British in architectural style.

Rural tranquility in Maryland (*right*) The pace of life here is relatively slow, the climate mild, and on the undulating hills dairy and poultry farming are common.

New Jersey lies on the Atlantic seaboard, sandwiched between New York and Pennsylvania. It calls itself the Garden State, but to travelers on its highways it may seem crowded and industrialized, plagued by pollution, organized crime and urban decay. In fact both aspects are true; as one of the smallest and most densely populated states, New Jersey harbors many contradictions.

European explorers found the area inhabited by the Lenni Lenape peoples. The Swedes and the Dutch established the first settlements, but the English took over in 1664. New Jersey, created in 1702 from East and West Jersey, became a separate colony in 1738. The War of Independence sharply divided the colony, and some of the fiercest battles were fought here, notably at Trenton (1776) and at nearby Princetown (1777).

New Jersey was in the forefront of the Industrial Revolution, site of the United States' first factory town (albeit unsuccessful) in 1791, and a center of canal, steamship and later railroad building. North-central New Jersey became the main transportation hub of the eastern seaboard, whereas the south developed into a major truck-farming area, growing salads and vegetables for the city markets – hence the state's nickname. After the Civil War, liberal commerce laws encouraged massive corporations, leading to runaway industrialization, which was only partly curbed by Federal legislation in 1913. The United States relied heavily on the state's enormous industrial capacity in both World Wars.

Northeastern New Jersey rapidly became an overspill area for neighboring New York City, so that today a majority of the population lives within a 30-mile radius of that city, along highways that are among the world's busiest. Farmland is rapidly diminishing in this area, but elsewhere – both in the heavily cultivated south and in the lush pastures of the northwest – New Jersey can still lay claim to its nickname. The long stretches of good beaches along the Atlantic Ocean are a major attraction for the thousands of visitors to the area, and Atlantic City has become a world-famous resort.

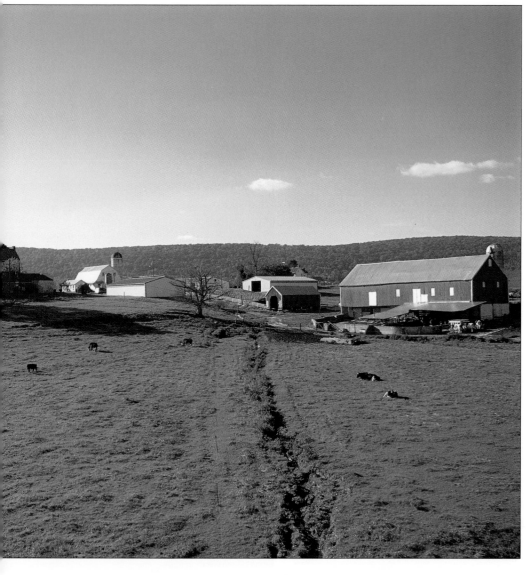

MARYLAND

Nickname	Old Line State
Ratified the Constitution	1788
Capital	Annapolis
Area	27,091 sq km (10,460 sq mi)
Population	5,615,727
Outstanding features	Chesapeake Bay, Backbone Mountain
Principal industries	electrical and electronic equipment, foodstuffs, chemicals

In 1632 a royal charter from Charles I of England (1600–49) granted land in the Chesapeake Bay area to Cecil Calvert, Lord Baltimore (1605–75). He named the new colony Maryland after Henrietta Maria (1609–69), the king's wife, who was a Roman Catholic and made it a haven for Roman Catholics suffering religious persecution in England. However, the Calverts imposed no corresponding restrictions on non-Catholics, welcoming all Christians who believed in the Trinity.

Calvert's brother Leonard (1606–47) founded Saint Mary's City, the first capital, in 1634. However, the rapid growth of tobacco plantations shifted the main centers of population northwestward, and in 1694 the capital was moved to Annapolis; 1729 saw the founding of the great city of Baltimore. Although many plantations used African slaves, the slave trade was penally taxed from 1783; even so, full abolition did not follow until 1864.

Disputes between the Calverts and the Penns were settled in the 1760s by the creation of the Mason and Dixon Line, marking the boundary between Maryland and Pennsylvania. Maryland was active in the War of Independence; the Treaty of Paris, which acknowledged the independence of the colonies, was ratified in Annapolis in 1783. In 1814 a British attack on Baltimore was decisively repulsed. The next decades saw the construction of the first major road and rail routes across the Appalachian Mountains – both of them on Maryland soil. The loyalties of Marylanders were divided in the Civil War, but the state remained in the Union. Maryland never disenfranchised its black population, and this encouraged an influx of black people from the less tolerant states of the South.

The port of Baltimore remains one of the nation's busiest, supported by a strong industrial base that includes many new industries. Service industries are increasing in importance, and the state's many historical sites form the basis for a thriving tourist industry.

DELAWARE

Nickname	First State
Ratified the Constitution	1787
Capital	Dover
Area	5,294 sq km (2,044 sq mi)
Population	853,476
Outstanding features	Delaware River, Delaware Bay
Principal industries	chemicals, textiles, clothing

Delaware today is heavily industrialized and densely populated. The north of the state forms part of the Atlantic megapolis – the long urban corridor that runs from Washington to Boston. Nothing could be more different from the primitive log cabins built by the Swedish settlers who established Fort Christina – modern Wilmington – in 1638.

An earlier Dutch colony had been destroyed by local Native-Americans, but in 1655 Dutch from New Amsterdam (later New York) reclaimed the area, only to lose it to the British in 1664. In 1682 the area was ceded to William Penn (1644–1718), providing his inland colony of Pennsylvania with a corridor to the coast. Delaware gained its own assembly in 1704, and its present name in 1776. The origins of that name went back to early colonial days. Delaware Bay had been so named in 1610 after Thomas West (1577–1617), 12th Baron De la Warr, upon his appointment as governor of Virginia.

During the War of Independence, Delaware was invaded by a British army, and its coastline was constantly under threat from the British navy. Following independence, Delaware's prompt acceptance of the Constitution on 7 December 1787 earned it the title of First State. Although Delaware was strictly speaking a slave state, many slaves had already been granted their freedom before the outbreak of the Civil War, and the state therefore remained a slightly reluctant part of the Union.

The development of industry has had a marked effect on the demographic balance of the state. The older rural communities around Dover and Georgetown in the mainly agricultural south have now declined, and extensive suburban areas have grown up in the industrial north, especially around Wilmington, which is now a major manufacturing center for artificial fibers.

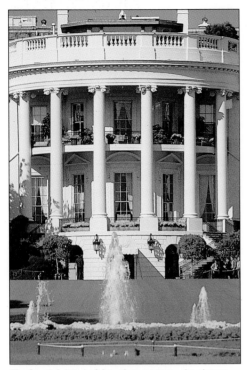

The fountainhead of American power – the elegant balconies of the White House in Washington, D.C., residence of the United States president. The original building was first painted white to hide the scars of its burning by the British in 1814.

DISTRICT OF COLUMBIA

Federal territory authorized	1790
Coterminous with the United States capital	Washington, DC
Area	179 sq km (69 sq mi)
Population	581,530
Outstanding feature	Potomac River
Principal industries	federal administration, tourism

Columbia is the district that was chosen by the first president, George Washington (1732–99), to house a purpose-built capital city – the first in any Western nation.

The French military engineer Pierre-Charles L'Enfant (1754–1825) laid out the city's distinctive plan, based on French principles of formal landscaping, with avenues radiating out from the Capitol and the presidential mansion (the latter became known as the White House following its renovation after the War of 1812). Washington, D.C. also houses the headquarters of America's military, the Pentagon. It was this building which was partially destroyed in the terrorist attacks of September 11 2001, targeting both the capital and New York City. The attack on the Pentagon killed about 200 people.

The District now enjoys most rights of statehood, but is only part of a larger metropolitan area, with all the associated problems. The imposing embassies and government officials' homes almost rub shoulders with slums, which saw serious race riots in the 1960s; a few areas have been cleared in urban renewal programs.

ARKANSAS

Nickname	Land of Opportunity
Joined the Union	1836
Capital	Little Rock
Area	137,754 sq km (53,187 sq mi)
Population	2,810,872
Outstanding features	Boston Mountains, Hot Springs National Park
Principal industries	foodstuffs, chemicals, electronic equipment
Principal industries	electronics, tourism, copper mining

Arkansas lies on the west bank of the great Mississippi river, extending westward to the forested uplands of the Ozark Plateau and Ouachita Mountains.

There were well-established Native-American cultures in the area as early as 500 AD. The first European settlers to arrive were the French, in the 17th century. They then sold the territory to the United States in the Louisiana Purchase of 1803. The lower Mississippi plains became the site of great cotton plantations worked by black slaves transported from Africa. From 1820 until 1836, when statehood was granted, Arkansaw Territory was the area south of the Missouri Compromise line, which separated slave and free states.

Arkansas did not secede from the Union until 1861, when it joined the Confederacy. The Civil War left it impoverished and overly dependent on cotton, and with racial-segregation policies that were maintained until the 1960s. The state today is much more prosperous, but maintains its largely rural character. More varied agriculture and new industries have replaced cotton, and tourism is growing fast in the uplands, which are becoming a popular retirement area.

LOUISIANA

Nickname	Pelican State
Joined the Union	1812
Capital	Baton Rouge
Area	123,677 sq km (47,752 sq mi)
Population	4,287,768
Outstanding features	Mississippi River and Delta
Principal industries	petroleum extraction, foodstuffs, chemicals

When, in 1682, René-Robert Cavelier, sieur de La Salle (1643–87), sailed down the Mississippi and claimed the entire basin of one of the world's largest rivers for King Louis XIV of France (1638–1715), Native-American peoples had been living in the area for some 16,000 years.

In 1718 Jean-Baptiste Le Moyne, sieur de Bienville (1680–1768), founded New Orleans near the Mississippi Delta, and in 1731 the crown colony of Louisiana was born. Many new settlers arrived to swell the colony. They included the Cajuns – French-speakers expelled from Nova Scotia by the British. In 1762 Louisiana fell into Spanish hands and remained Spanish until 1800, when Napoléon Bonaparte (1769–1821) bought it back for France. The effect of this French and Spanish heritage can be seen in the state's unique legal code, and in the architecture and culture of New Orleans.

The second French tenure was shorter even than the first. In 1803 the United States acquired Louisiana as part of the Louisiana Purchase. This very remarkable real-estate deal added more than 2 million sq km (over 800,000 sq mi) of land to United States territory for less than 3 cents per acre.

In Louisiana, as in other southern states, the cotton and sugar cane planters depended for their wealth on the use of large numbers of black slaves. Conflicts of interest soon occurred between wealthy

Mardi Gras (*above*) A reveler in New Orleans' famous parade peers from beneath his gaudy tiger mask. Like jazz, this Caribbean-style carnival is a highlight of Louisiana's African-American culture.

Shrimp boats (*left*) jostle along the wharves at the small fishing port of Delcambre, Louisiana. The shape of the coastline is constantly being changed by deposits of silt, brought down by the Mississippi river.

Competing in the Kentucky Derby (*below*) This world-famous race takes place every May at the Churchill Downs course in Louisville. Together with the Preakness and Belmont stakes it forms the coveted Triple Crown.

KENTUCKY

Nickname	Bluegrass State
Joined the Union	1792
Capital	Frankfort
Area	104,659 sq km (40,409 sq mi)
Population	4,206,074
Outstanding features	Black Mountain, Land of 10,000 Sinks
Principal industries	coal mining, machinery, foodstuffs

The Commonwealth of Kentucky unites the traditions of the frontier and the old South with a modern industrial economy.

The area was once the hunting ground of Native-American peoples such as the Iroquois and the Cherokee. After 1767, it was explored by the frontiersman Daniel Boone (1734–1820), who paved the way for the westward migration of settlers. A substantial black slave population was imported, and the state government opposed abolition. However, Kentucky did not join the Confederacy, and most of its soldiers fought for the Union in the Civil War. In the upland "hillbilly" country – poor and fiercely clannish – Civil War feuds continued into the 20th century.

Modern Kentucky has remained agricultural, famous for its tobacco, horse breeding and bourbon whiskey. However, since the late 19th century massive coal reserves have been exploited, and large areas laid waste by strip mining – controlled since 1966. Kentucky is also well known for its cultural heritage, ranging from the famous Kentucky Derby to folk music, including the celebrated Bluegrass style.

TENNESSEE

Nickname	Volunteer State
Joined the Union	1796
Capital	Nashville
Area	109,152 sq km (42,144 sq mi)
Population	6,038,803
Outstanding features	Clingmans Dome (Smoky Mountains), Great Appalachian Valley
Principal industries	chemicals, foodstuffs, machinery

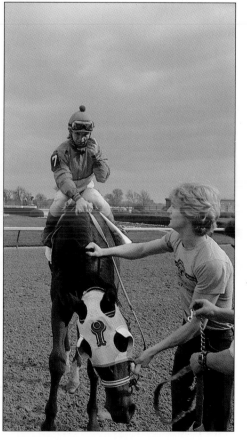

planters and farmers on issues of land use and political suffrage. Louisiana seceded from the Union in 1861, but there was considerable opposition to its secession, and less than a year passed before Union troops reoccupied New Orleans. In 1868 Louisiana was readmitted to the Union, but a legacy of bitterness remained. The constitution of 1898 denied voting rights to most blacks – a situation that was to remain unchanged until the civil rights campaigns of the 1960s.

Modern Louisiana remains primarily agricultural, but cotton is no longer the main crop, having given way to soybeans, rice and other food crops. Beef cattle are also important. Along the coast the distinctive bayous, or creeks, of the Mississippi delta experience a subtropical climate. New Orleans and Baton Rouge are both major ports, and tourists flock to the French Quarter of New Orleans, with its fascinating colonial architecture and long cultural traditions; it is renowned as the birthplace of jazz, and is equally famous for its extremely colorful annual carnival parade or festival of Mardi Gras (Shrove Tuesday).

In 1784 North Carolina offered to cede its western territories to Congress. Angry settlers reacted by creating their own state of Franklin, but its life was short. In 1789 the territory was formally separated from North Carolina, and seven years later it became the state of Tennessee.

Many Tennesseans achieved fame in the War of 1812, including future president Andrew Jackson (1767–1845). Before

Country and Western music in full swing at the Opryland Park Country Music Show in Nashville, Tennessee, where the style first began. The Grand Ole Opry was the music show of the 1920s and 1930s where many legendary singers first appeared.

the Civil War there was strong loyalty to the Union, but deep divisions appeared on the eve of the conflict, and Tennessee became a battleground second only to Virginia. The state escaped the worst excesses of reconstruction. Even so, the white supremacist Ku Klux Klan was born at Pulaski (south of Nashville) in 1865, and blacks were effectively segregated from the 1870s until the 1960s.

From the 1940s onward, power from the massive Tennessee Valley hydroelectric scheme fueled rapid industrial development, ending the state's reliance on agriculture. Nowadays, however, there is a move toward the trade and services sector, and tourism has also become important. Many come to visit Nashville, which is world famous for being the heart of Country and Western music.

MISSISSIPPI

Nickname	Magnolia State
Joined the Union	1817
Capital	Jackson
Area	123,514 sq km (47,689 sq mi)
Population	2,910,540
Outstanding feature	Mississippi River
Principal industries	clothing, foodstuffs, timber processing

Mississippi lies at the very heart of the American cotton belt, a fact that is reflected in its history. The first colonists were French Canadians, who built Fort Maurepas on the coast in 1699. Other French settlements followed, but in 1763 the British took control. Spain occupied the south during the War of Independence, and the area remained in dispute until 1795, when it came to the United States in the Treaty of San Lorenzo.

The Mississippi Territory extended from Tennessee in the north to the Gulf coast in the south, and in 1817 the western part became the state of Mississippi. The Native-Americans were forcibly removed, and Mississippi became a prosperous cotton state. Entrenched attitudes to slavery led to its secession from the Union in 1861. It was devastated during the Civil War, and in 1890 its black population was virtually disenfranchised.

Agriculture suffered in the Depression of 1929, but revived strongly after World War II, and a long, slow process of racial integration and acceptance began. However, the state remains relatively backward economically, and has benefited less from modern industrial development than some neighboring states.

ALABAMA

Nickname	Cotton State
Joined the Union	1819
Capital	Montgomery
Area	133,915 sq km (51,705 sq mi)
Population	4,599,030
Outstanding features	Cheaha Mountain, Mobile Bay
Principal industries	pulp and paper, chemicals, electronics

Alabama is a state with a troubled history, torn by settlement disputes between European colonists and the Native-Americans, by issues of social reform in the 19th century and by civil rights protests in the 1960s.

The first European exploration was made by the Spanish, led by Hernando de Soto (1500–42), who came in search of precious metals. Although they failed to find any, they succeeded in opening up the whole area to European settlement, killing thousands of Native-Americans in the process. For the next three centuries the area was contested by Spain, France and Britain. A French expedition to the Mississippi area established the first permanent settlement on Dauphin Island at the mouth of the Alabama river in 1702. Seventeen years later the first shipload of black slaves from Africa disembarked at Port Dauphin.

In 1763 the British gained control over most of the area, and the Spanish foothold on the coast became increasingly tenuous. The Alabama Territory was created in 1817, achieving statehood two years later. Meanwhile the remaining Native-American populations had been subdued and transported west to Oklahoma, opening the way for another wave of European settlement.

The result was a wealthy cotton-growing state dependent on black slave labor. By 1860 half the population of

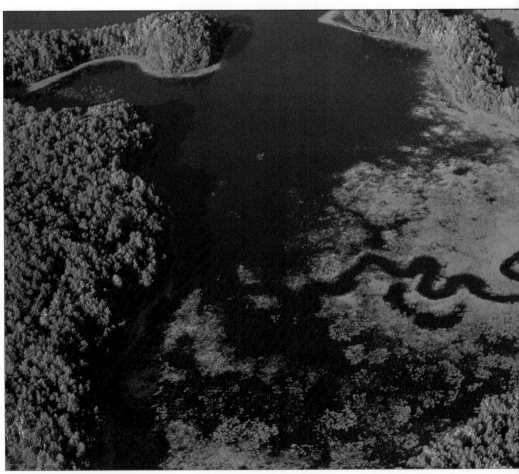

Alabama were black slaves, and secession from the Union in 1861 was all but inevitable; Alabama joined the Confederate States of America. During the Civil War the state suffered appalling casualties, and from 1867 to 1868, following the collapse of the Confederacy, it was placed under military rule. The period of reconstruction (1868–74) merely deepened the gulf between blacks and whites, with increasing support for the white supremacist creed of the Ku Klux Klan. Here, as in other Southern states, blacks had to wait until the 1960s to achieve basic civil rights, and even then progress was slow in the face of discrimination against black voters. Alabama was the scene of many civil rights protests in the 1950s and 1960s, including the march of Montgomery, led by Martin Luther King (1929–68) in 1965.

Agriculture remains important today, with cotton and groundnuts (peanuts) among the chief crops. Manufacturing has diversified from iron and steel and agriculture-based industries to include textiles and modern electronics.

The mighty Mississippi river seen from the air, snaking through dense forest along the western border of the state that bears its name. Its sluggish meanders become ever more tortuous until they meet, creating islands and oxbow lakes.

WEST VIRGINIA

Nickname Mountain State
Joined the Union 1863
Capital Charleston
Area 62,758 sq km (24,231 sq mi)
Population 1,804,200
Outstanding features Spruce Knob (Allegheny Mountains), Ohio river
Principal industries mining, chemicals, machinery, quarrying

West Virginia lies between the eastern and the midwestern states, across the Appalachian Mountains to the north and west of the state of Virginia. Whereas Virginia belongs firmly to the South, West Virginia has always had more in common with the northern states.

Most of West Virginia is mountainous, carved by swift rivers and covered with dense forest. This wild land was home to various Native-American peoples, notably the Hopewell, whose prehistoric earthworks can still be found in the Charleston area. Their successors – the peoples of the Iroquois League – put up strong resistance to the first European settlers when they arrived from Virginia across the Appalachian Mountains in the early part of the 18th century.

These new colonists established a very different order from the slave-worked plantations of Virginia nearer the coast. Cultivation in the mountains was largely limited to small farms, and slavery was virtually nonexistent. The mountain-dwellers soon became dissatisfied with the ruling authorities in the east, and their antipathy to slavery led in 1859 to the seizure of a Federal arsenal at Harper's Ferry (now the easternmost point of West Virginia). This was one of the incidents that precipitated the Civil War. When Virginia seceded from the Union in 1861, the people of the northwest refused to follow them and join the Confederacy; thus the state of West Virginia was born.

After the Civil War, the expansion of the railroad network brought industrial development to the mountains. The rich local resources of salt, limestone and especially coal have been exploited ever since, providing plentiful energy and raw materials for neighboring states – and bringing accompanying social and environmental problems. Agriculture and mining have both declined now, but petroleum and natural gas have since been added to the list of resources; tourism is also expanding.

VIRGINIA

Nickname The Old Dominion
Ratified the Constitution 1788
Capital Richmond
Area 105,586 sq km (40,767 sq mi)
Population 7,642,884
Outstanding features Mount Rogers (Blue Ridge Mountains), Chesapeake Bay
Principal industries textiles, transportation equipment, electronics

Virginia, lying on the Atlantic seaboard between Delaware and North Carolina, has one of the longest histories of any state. With strong links to both south and north, it has often played a pivotal role in the history of the United States.

Virginia, named after Queen Elizabeth I (1533–1603) of England (the "virgin queen"), was the first British colony in America, established in 1607 on the site of present-day Jamestown (southeast of Richmond). The settlers barely survived, plagued by hunger, disease and frequent raids by the Native-Americans who were defending their land from encroachment. However, the colonists' tobacco plantations soon brought increasing prosperity, enabling them to establish a tradition of culture and education, with prestigious institutions such as William and Mary College in Williamsburg (across the river from Jamestown).

The colony was chafing at British rule long before the War of Independence, in which several Virginians played a leading part, notably George Washington (1732–99) and Thomas Jefferson (1743–1826).

Colonial Williamsburg, Virginia, which was founded in 1632, has been the subject of a huge restoration project. More than 400 buildings have been restored, and 18th-century crafts such as spinning yarn have been revived and are demonstrated to the public.

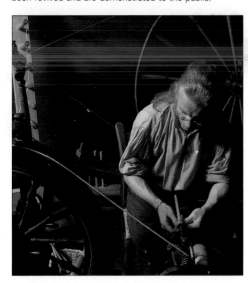

Virginia hosted the first Continental Congress in 1774, and witnessed the decisive British surrender at Yorktown (east of Williamsburg) in 1781.

The importation of slaves from Africa was abolished in 1778, but slavery itself was maintained. In 1861 Virginia led the Confederacy, whose capital was at Richmond. As a result, the northwestern part of the state split away, to become West Virginia. The Civil War ended in 1865, soon after Confederate troops had surrendered at Appomattox (west of Richmond). Many years passed before the state's economy recovered. However, by World War I the old reliance on agriculture, especially tobacco, had been superseded by the growth of various manufacturing industries.

World War II brought very considerable prosperity, with the expansion of military bases and shipbuilding concerns along the coast; the federal government is still a major employer. But Virginia's proudest possession is its history – beautifully preserved at sites such as Williamsburg and Mount Vernon (to the south of Washington, D.C.), which attract a thriving tourist trade.

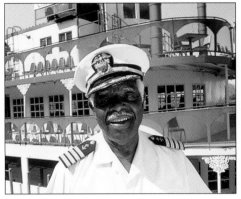

Captain Sam (*above*) in front of the oldstyle sternwheel riverboat in which he runs harbor cruises at Savannah, Georgia.

Classical elegance (*right*) in Charleston, South Carolina, where many plantation houses still stand. It was here that the popular back-kicking dance of the 1920s originated, and became a national, and international, craze.

NORTH CAROLINA

Nickname	Tar Heel State
Ratified the Constitution	1789
Capital	Raleigh
Area	136,412 sq km (52,669 sq mi)
Population	8,856,505
Outstanding features	Mount Mitchell (Blue Ridge Mountains), Cape Hatteras, Cape Lookout
Principal industries	textiles, tobacco, electronics

In 1663 King Charles II of England (1630–85) granted the area between the 31st and 36th parallels to eight "Lords Proprietors" (an earlier colony on Roanoke Island had vanished without trace). The new colony was named Carolina in the king's honor, and was extended northward in 1665 to include areas that had belonged to Virginia. In 1712 North Carolina became a separate colony with its own governor. In 1729 it returned to royal control – which ended in 1781 at the Battle of Guilford Court House.

During the 1830s most of the native Cherokee population were forcibly removed to the west of the Mississippi river. North Carolina seceded from the Union in 1861, and paid dearly for doing so. The constitution of 1868 was more liberal than some, but black people were effectively denied civil rights from 1900 until the 1960s. Today manufacturing has largely replaced agriculture as the chief source of the state's income, but North Carolina remains the center of the tobacco industry in the United States.

SOUTH CAROLINA

Nickname	Palmetto State
Ratified the Constitution	1788
Capital	Columbia
Area	80,582 sq km (31,113 sq mi)
Population	4,321,249
Outstanding features	Sassafras Mountain (Blue Ridge Mountains), Sea Islands
Principal industries	tourism, textiles, chemicals, machinery

In 1670 British colonists built Charles Town near the mouth of the Ashley river. Named after King Charles II of England (1630–85), it was the first lasting European settlement of South Carolina. Ten years later, Charleston was rebuilt across the river, and grew wealthy from its trade in rice, pelts and indigo. Crown control of the colony's finances caused deep resentment, and South Carolina was firm in its support for the American Revolution. In 1780 a British fleet captured Charleston, and many engagements were fought on South Carolina soil.

Cotton plantations brought new wealth inland, but dependence on slavery led to secession from the Union in 1861, when an engagement with Federal troops at Fort Sumter in Charleston Harbor triggered the Civil War. The reconstruction period created lasting racial tensions that were reflected in the 1895 constitution, and black civil rights were not restored until the 1960s.

Modern South Carolina is industrial rather than agricultural, though it is less dominated by textiles than it once was. The warm climate, and the old-world atmosphere of its coastal towns, encourage a thriving tourist industry.

GEORGIA

Nickname	Empire State of the South
Ratified the Constitution	1788
Capital	Atlanta
Area	152,576 sq km (58,910 sq mi)
Population	9,363,941
Outstanding features	Brasstown Bald (Blue Ridge Mountains), Okefenokee Swamp
Principal industries	foodstuffs, textiles, clothing, electrical equipment

In 1732 James Edward Oglethorpe (1696–1785) received a charter from King George II of England (1683–1760) to settle a new colony in North America. He named it Georgia in honor of the king. This, the last of the 13 original colonies, was founded to give Britain's poor greater opportunities in a new land. However, its attempts to produce silk, wine and spices had only limited success; in 1752 the colony reverted to crown control.

Other settlers began to make their homes along the banks of the Savannah river (the eastern border) before the War of Independence. Later on they came in such great numbers that the original inhabitants, the various Native-American peoples, were forcibly displaced beyond the Mississippi river. Black slaves were later brought in to work the state's rapidly developing cotton plantations.

The issue of slavery caused much heart-searching in Georgia, but in January 1861 the state convention voted overwhelmingly for secession from the Union. In 1864, after a series of blockades and invasions, the Union army under General William Tecumseh Sherman (1820–91) destroyed Atlanta and advanced to the coast, leaving a trail of devastation. In 1865 the president of the Confederacy, Jefferson Davis (1808–89), was captured on Georgian soil.

The war was over, but even during the reconstruction period that followed, Georgia's politicians resisted black enfranchisement – an attitude that was to remain entrenched for another century. Integration of black and white pupils in schools was a hotly debated issue as late as 1967. Since then, however, African-Americans have taken a far more active role in the community: the state capital Atlanta was the first major city in the South to elect a black mayor, in 1973.

Miami, Florida, is host to a diverse cultural mix: vacationers from all over the world come to enjoy the beach life; elderly people come here to retire; and thousands of immigrants from Cuba, Haiti, Mexico and South America come to begin a new life.

Cotton is no longer the main crop, having been overtaken by food crops such as groundnuts (peanuts) and maize. Textiles, however, remain important, together with many new industries such as electronics. Federal military installations supplement income from industry. Atlanta is the national headquarters of Coca-Cola, and its rapid growth in recent years makes it not only the commercial and financial center of the southeast, but also a leading national city. Atlanta hosted the 1996 Olympic Games.

FLORIDA

Nickname	Sunshine State
Joined the Union	1845
Capital	Tallahassee
Area	151,939 sq km (58,664 sq mi)
Population	18,089,888
Outstanding features	the Everglades, Florida Keys
Principal industries	tourism, electronics, electrical machinery, foodstuffs

Today tourists flock to the far southeastern state of Florida in search of sun-

shine, relaxation and entertainment, and large numbers of elderly people retire there from northern cities. In Easter 1513 the Spanish soldier and explorer Juan Ponce de León (1460–1521) came in search of the Fountain of Youth, but instead he began a 250-year cycle of conflict and exploitation that almost wiped out the original Native-American population.

In 1565 another Spanish expedition under the leadership of Pedro Menéndez de Avilés (1519–74), utterly destroyed a Huguenot colony on the Saint Johns river, and Menéndez de Avilés founded his own colony of Saint Augustine nearby. For 200 years, control of Florida was disputed between Great Britain, France and Spain; however, the British finally took control in 1763, at the end of the French and Indian Wars. During the War of Independence, Florida remained loyal to Britain, but in 1781 it was seized by the Spanish.

During the war of 1812 an American force seized Fernandina, hoping to prevent the British from using Florida as a military base. The force withdrew in the face of Spanish protests – and the British duly moved in, directing a guerrilla war against American settlements. In 1818 Andrew Jackson (1767–1845) captured Pensacola for the United States, and the following year Spain ceded the whole of Florida. Relations with the remaining Native-Americans deteriorated rapidly, and most were expelled in 1842. After the Civil War new railroads brought settlers and tourists alike, beginning a period of rapid expansion in population that has continued throughout the 20th century. After the Cuban revolution in 1959 many Cuban refugees settled in the state, especially in the southern city of Miami.

Much of the land is covered by forests, lakes and military reserves; there are large areas of farmland, including significant citrus fruit plantations, but most of the population is urban. Tourism is the leading industry, with many visitors to the Everglades National Park in the south and to Disney World near Orlando in central Florida. The state also hosts many major sporting events, including football at the Miami Orange Bowl Stadium, golf tournaments, and auto racing at Daytona Beach. The Kennedy Space Center at Cape Canaveral also makes a significant contribution to the economy, as does the production of electrical machinery.

Florida became the focus of world attention during the presidential elections of 2000, when a dispute over vote counting became crucial to the outcome.

Mexico, Central America and the Caribbean

COUNTRIES IN THE REGION

MEXICO · GUATEMALA · BELIZE · EL SALVADOR

HONDURAS · NICARAGUA · COSTA RICA · PANAMA

CUBA · BAHAMAS · HAITI · JAMAICA

DOMINICAN REPUBLIC · SAINT KITTS-NEVIS

ANTIGUA AND BARBUDA · DOMINICA · SAINT LUCIA

SAINT VINCENT AND THE GRENADINES · BARBADOS

GRENADA · TRINIDAD AND TOBAGO

DEPENDENCIES IN THE REGION

TURKS AND CAICOS ISLANDS · BERMUDA

CAYMAN ISLANDS · PUERTO RICO

UNITED STATES VIRGIN ISLANDS

BRITISH VIRGIN ISLANDS · ANGUILLA

MONTSERRAT · ARUBA · THE NETHERLANDS

ANTILLES · GUADELOUPE · MARTINIQUE

Casting his net A fisherman in Haiti. This impoverished state occupies the western third of Hispaniola, the Caribbean island that was discovered by Christopher Columbus in 1492 and was the site of the first European landing in the Americas.

capital city

major town

height of land (meters)

3000
2000
1000
500
200
0

▲ mountain peak

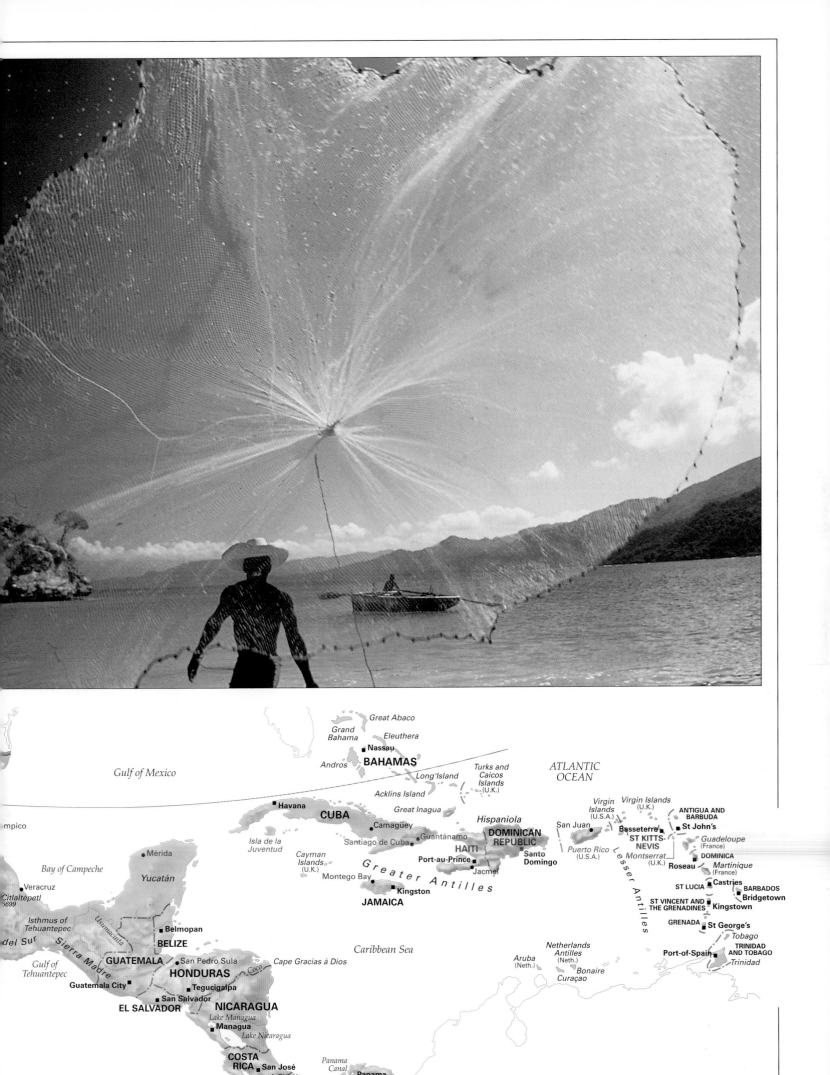

Great Abaco

Grand
Bahama

Eleuthera

Nassau

Andros **BAHAMAS**

Long Island

Turks and
Caicos
Islands
(U.K.)

Acklins Island

Gulf of Mexico

*ATLANTIC
OCEAN*

Havana

CUBA

Great Inagua

Hispaniola

*Virgin
Islands
(U.K.)*

*Virgin
Islands
(U.K.)*

**ANTIGUA AND
BARBUDA**

Camagüey

**DOMINICAN
REPUBLIC**

San Juan

Basseterre

**ST KITTS-
NEVIS**

St John's

*Isla de la
Juventud*

Santiago de Cuba

Guantánamo

HAITI

*Puerto Rico
(U.S.A.)*

Santo
Domingo

*Guadeloupe
(France)*

mpico

*Cayman
Islands
(U.K.)*

Port-au-Prince

Jacmel

*Montserrat
(U.K.)*

DOMINICA

Roseau

*Martinique
(France)*

Mérida

Montego Bay

Greater Antilles

Kingston

ST LUCIA

Castries

BARBADOS

Bay of Campeche

Yucatán

JAMAICA

**ST VINCENT AND
THE GRENADINES**

Bridgetown

Kingstown

Veracruz

*Citlaltépetl
5699*

Belmopan

GRENADA

St George's

Tobago

*Isthmus of
Tehuantepec*

Usumacinta

BELIZE

Caribbean Sea

*Netherlands
Antilles
(Neth.)*

del Sur

Sierra Madre

GUATEMALA

San Pedro Sula

Cape Gracias á Dios

*Aruba
(Neth.)*

Port-of-Spain

**TRINIDAD
AND TOBAGO**

*Gulf of
Tehuantepec*

Guatemala City

HONDURAS

Coco

Bonaire

Curaçao

Trinidad

Tegucigalpa

San Salvador

EL SALVADOR

NICARAGUA

Lake Managua

Managua

Lake Nicaragua

**COSTA
RICA**

San José

*Chirripó
3820*

*Panama
Canal*

**Panama
City**

PANAMA

Mexico

MEXICO IS ONE OF THE LARGEST OF THE Latin American states, possessing a colorful diversity of cultures. However, rich natural resources and swift economic growth have been largely offset by the rapid rise in population.

GEOGRAPHY

Mexico forms a large tongue of mountainous land between the Pacific Ocean and the Gulf of Mexico. The southern half of the country, which lies within the tropics, has a wide variety of landscapes.

The land

Mexico lies in a geologically unstable part of the world. Many areas are subjected to frequent earthquakes and volcanic eruptions; a major earthquake in 1985 devastated Mexico City.

Much of the northern border with the United States is formed by the Rio Bravo del Norte (Rio Grande). From here the land arcs southeast toward the borders with Guatemala and Belize. The low-lying Yucatán Peninsula extends northward from the Belize border. The long and mountainous peninsula of Baja California runs parallel to the west coast of the Mexican mainland. Some 500 km (300 mi) out to sea is a group of volcanic islands called the Islas de Revillagigedo.

The broad coastal plain along the Gulf of Mexico is fringed by swamps, lagoons and sandbars, while the Pacific coastal plain is narrower and more interrupted. The area in between is occupied by the massive central plateau. This rises in the south to the Mesa Central, which is crossed by a series of mighty volcanoes, including Citlaltépetl and Popocatépetl. The plateau is bounded to the east, west and south by the mountain ranges of the Sierra Madre. In the south, beyond the Sierra Madre del Sur, the Isthmus of Tehuantepec links these highlands to the mountains on the Guatemalan border.

Climate

The climate varies both with latitude and altitude. In the far north low rainfall and extremes of temperature create desert landscapes. In the south most rain falls between May and August. Rainfall increases southeastward, with the heaviest falls along the Gulf coast and in the south.

The coastal plains and the foothills of the Sierra Madre are the *tierra caliente* ("hot land"), where temperatures are mostly uniformly high. Between August and October these areas are vulnerable to tropical hurricanes from the Gulf of Mexico and the Pacific. Above 1,000 m (3,000 ft) is the *tierra templada* ("temperate land"), which extends over much of the central plateau. The Mesa Central often reaches into the *tierra fria* ("cold land") above 2,000 m (6,000 ft). Above about 4,000 m (13,000 ft) is the *tierra helada* ("frozen land"), where temperatures are always below 10°C.

Plants and animals

Desert scrub predominates in the deserts of the north, with grasses, shrubs and succulents on higher ground. Farther south settlers have destroyed most of the ancient forests, but stands of conifers still remain on the higher slopes of the western mountains. Much of the central plateau is covered with grassland. Higher rainfall in the east allows tropical rainforest to flourish there, with temperate rainforest growing at higher altitudes. In southwestern areas , deciduous and semi-deciduous woodland grows along the Pacific coastline.

The rainforests of the south support a great diversity of animals, such as monkeys, parrots, jaguars, tapirs and anteaters. In the north, and in settled areas of the central plateau, domestic livestock have been introduced, taking the place of the native species. However, in the deserts and steppes native animals such as armadillos, deer, pumas and coyotes still thrive. The coastal seas abound with numerous varieties of fish.

SOCIETY

The cultural diversity and racial mix of modern Mexico owes much to its complex, colorful and often violent history.

History

Mexico had a succession of highly sophisticated early civilizations. The Olmecs reached their peak from 900–400 BC. They built impressive ceremonial centers elaborately decorated with mosaics and stone carvings. The Mayas rapidly expanded into the forests of southeastern Mexico from Guatemala, where they originated, in about 300 AD. They were a highly sophisticated, wealthy people; archaeological remains include not only majestic stone buildings and stepped pyramids, but also intricate jewelry, carved figures and hieroglyphic manuscripts that are not yet fully understood.

The Aztecs, originally a nomadic people from the arid northern highlands, migrated to Mexico's great Central Valley in about 1200. Here they became the dominant power and built their capital, Tenochtitlán, on the site of modern Mexico City, from which they established a vast empire. Their architecture, jewelry and textiles were outstanding; they used a complex calendar for time-keeping; and they practiced ritual human slaughter on a grand scale to propitiate their gods. The Aztec empire was largely destroyed by the 16th-century Spanish adventurer Hernán Cortés (1485–1547). Tenochtitlán fell in 1521, and in 1535 the conquered territory became New Spain.

Spanish rule was often harsh, especially in the 18th century, and independence was not achieved until 1821. Mexico then became a republic, adopting a federal constitution in 1824.

In 1845 Mexico tried and failed to resist the United States' annexation of Texas. A bitter war was to follow, and further territories were ceded by the Treaty of Guadalupe Hidalgo in 1848. In 1857 the introduction of a reformed constitution brought civil war. The reformers, led by Benito Juárez (1806–72), were victorious.

In 1910, following the overthrow of the dictatorial president Porfirio Díaz (1830–1915), there began a long power struggle by various revolutionary groups, who eventually formed a single National Revolutionary Party in 1929. In 1946 it

One for the road (*above*) Farm workers wearing traditional ponchos refresh themselves with a cup of pulque – a cloudy white beer made from the agave, a succulent plant.

Colonial riches (*left*) For centuries, gold and silver mines made the beautiful city of Guanajuato, with its fine baroque churches and colonial buildings, the wealthiest in Mexico. Today it is an important university town and cultural center.

NATIONAL DATA – MEXICO

Land area	1,923,040 sq km (742,490 sq mi)			
Climate		**Temperatures**		**Annual**
	Altitude m (ft)	January °C(°F)	July °C(°F)	precipitation mm (in)
Mexico City	2,306 (7,564)	14 (56)	17 (63	816 (32.1)

Major physical features highest point: Ciltlaltépetl 5,610 m (18,406 ft); longest river: Rio Grande (part) 3,060 km (1,900 mi)

Population (2006 est.) 107,449,525

Form of government federal multiparty republic with two legislative houses

Armed forces army 144,000; navy 37,000; air force 11,770

Largest cities Mexico City (capital - 8,658,576); Ecatepec* (1,884,585); Guadalajara (1,632,171); Ciudad Juarez (1,496,551); Tijuana (1,482,294); Puebla (1,441,685); Nezahualcóyotl (1,229,673); León (1,156,053); Monterrey (1,123,799); ZZopopan† (1,022,564) * part of Mexico City Metropolitan area † part of Guadalajara Metropolitan area

Official language Spanish

Ethnic composition Mestizo (Amerindian-Spanish) 60%; Amerindian or predominantly Amerindian 30%; White 9%; other 1%

Religious affiliations Roman Catholic 89%; Protestant 6%; other 5%

Currency 1 New Mexican peso (MXN) = 100 centavos

Gross domestic product (2006 est.) U.S. $1.134 trillion

Gross domestic product per capita (2006 est.) U.S. $10,600

Life expectancy at birth male 72.63 yr; female 78.33 yr

Major resources petroleum, silver, copper, gold, lead, zinc, natural gas, hydropower, timber; textiles; iron and steel; petrochemicals; beverages, fisheries, tourism, coffee, cotton, fruit and vegetables, aluminium, sorghum, sugarcane

was renamed the Party of the Institutionalized Revolution (PRI). This party carried out many important reforms, and remained the dominant political force during subsequent decades.

Government

According to the 1917 constitution, voting is mandatory for all adults over 18 years of age. The president is elected for one six-year term only. There are two legislative houses: a 500-member chamber of deputies and a 64-member senate. The deputies are elected for a three-year term – three-quarters of them directly and the rest according to the votes cast for individual parties. Senators are elected for a six-year term – two for each of the 31 states and two for the Federal District. Each state elects its own governor and legislature, but the appointment of the governor of the Federal District is the prerogative of the president.

People

About one-third of Mexico's population are Amerindians, while over half are mestizos of mixed European and native descent; there are several other minority groups. The official language is Spanish, but some 50 Amerindian languages are also spoken. Most of the population is Roman Catholic.

More than two-thirds of Mexicans live in or around the three major cities of Mexico City, Guadalajara and Monterrey. Mexico City, the capital, is probably the world's most populous city. Large-scale emigration across the northern border has caused difficulties with the United States, Mexico's chief trading partner.

ECONOMY

Mexico has little good agricultural land, and much food has to be imported. Its struggle for economic growth has been hampered by falling oil prices, the 1985 Mexico City earthquake and massive foreign debt. Inflation remains high, and economic controls have been only moderately successful. Emergency austerity measures were taken in 1995.

Agriculture

Most agricultural land is on the central plateau, though irrigation projects in the north have created new farmlands there. Since 1910 much of this land has been occupied by small subsistence farms. Export crops include coffee, sugar cane and cotton, while staples include maize, squash and kidney beans. Cattle ranching is largely concentrated in the north, and some meat is exported. Fishing is largely for export, the most important being tuna, anchovies, sardines and shrimps.

About one-fifth of Mexico is forested, mostly in the east and south. These forests are exploited for hard and fragrant woods, and for chicle, the base material for chewing gum. Many forested areas, however, are threatened by large-scale clearance for agriculture.

Industry

Mexico is the world's leading source of silver; other commercial minerals include zinc, copper and lead. Ore from Durango

Cultural revival (*above*) A member of an Aztec dance troupe helps to keep ancient traditions alive. Mexico's rich Amerindian culture inspires many modern artists, sculptors and architects.

Monte Albán (*left*), the center of ancient Zapotec culture, was built on a flattened hilltop rising 300 m (1000 ft) from the valleys below. This impressive site is ·a lasting tribute to the sophistication and skill of Mexico's indigenous civilizations.

in and around Mexico City. Products include industrial chemicals, machinery and transportation equipment.

Transportation and communications

Tourism is an important foreign-currency earner, and has stimulated the development of both road and air networks. The extensive railroad system is slow and unreliable, but the road network has almost doubled in extent since the 1940s and is continuing to expand. There are two state-owned airlines, Aeroméxico and Mexicana, with more than 75 airports serving major cities, resorts and international destinations.

Mexico has a free press, with several hundred daily newspapers.

Health and education

Subsidized health care is available to many, but poor facilities in rural areas are reflected in higher rates of death and disease, particularly malaria. Adequate housing is a major problem, particularly in the cities. There are minimum-wage laws, but rural poverty is widespread.

Education is free, and literacy levels are high as a result of an extensive campaign. However, rural areas lack facilities, and there are few secondary schools outside the cities.

in the west of the country is used for iron and steel production at Monterrey in the northeast, fueled by coking coal from the nearby Sabinas fields. Mexico's massive hydrocarbon resources now provide the majority of foreign-exchange earnings. The state-owned company Pemex, Mexico's largest employer, controls the exploration, production and marketing of natural gas and petrochemical products.

Petrochemicals are the chief energy source. Hydroelectricity is also significant, however, and geothermal energy is being increasingly exploited. There is a nuclear power plant at Laguna Verde. Manufacturing industry is concentrated

Guatemala

REPUBLIC OF GUATEMALA

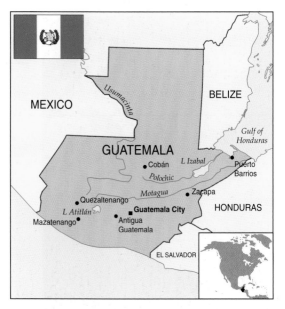

THE REPUBLIC OF GUATEMALA LIES ON THE southwestern border of Mexico, with coastlines on both the Atlantic and the Pacific. It also shares frontiers with Belize to the northeast, and Honduras and El Salvador to the southeast.

GEOGRAPHY

The heart of the country is crossed by two main mountain ranges. The Altos Cuchumatanes to the north are older and more eroded. The younger Sierra Madre range in the south includes 33 volcanoes, three of them active. The soils here are enriched by volcanic ash, making it the most fertile part of the country. In the north the highlands fall away to a large, flat, forested area called the Petén, and to the plains bordering the Gulf of Honduras.

The climate is mostly tropical with seasonal rains, though the mountains are both cooler and wetter. The lowland rainforests are rich in tropical hardwoods and rubber trees, with oaks and conifers higher up, and some mangroves along the coasts. Typical wildlife includes jaguars, peccaries and monkeys, while the rivers support crocodiles and manatees.

SOCIETY

For many centuries the area that is now Guatemala formed the heart of the ancient Maya civilization, which reached its peak from 300 to 900 AD. The empire then fell into decline, and the ruins of the Mayas' great stone cities can still be found in the remote northern lowlands. In 1523 the country was overrun by Spanish conquistadors under Pedro de Alvarado (c. 1485–1541). They set up large agricultural estates that were worked by Amerindian laborers. Between 1821 and 1847 the ruling landowners established Guatemala's independence. Their leader, Rafael Carrera (1841–65), became the first of a series of powerful dictator-presidents. A few were relatively liberal reformers whose aims were to modernize the country; Jorge Ubico (1878–1946), who ruled in the 1930s, was a protector of the Amerindians, but there were often periods of political repression.

The effects of World War II caused an economic decline, and subsequent decades were characterized by political extremism and constant divisive military coups. In 1986 democratic civilian rule was restored. However, several abortive coups and continued assassinations have shown that democracy is fragile. A new constitution was ratified in 1994 and in 1996 an accord was signed recognizing the rights of the indigenous population.

Over 50 percent of the population are of mixed Amerindian and Hispanic descent, and Roman Catholicism is the dominant religion in Guatemala.

ECONOMY

The economy is still largely agricultural. Coffee is the main crop and principal export. Bananas, cotton, sugar cane and maize are also important. Large numbers of cattle are raised on the Pacific coastal pastures. The forests provide timber and chicle, used in chewing gum. Guatemala has few mineral or energy reserves apart from small quantities of petroleum, copper and other metals. Industry is mostly confined to the processing of timber, sugar, tobacco and other agricultural products. Education is free, but resources are insufficient to provide adequate schooling or health care in rural areas.

NATIONAL DATA - GUATEMALA

Land area	108,430 sq km (41,865 sq mi)			

Climate		Temperatures		Annual
	Altitude m (ft)	January °C(°F)	July °C(°F)	precipitation mm (in)
Guatemala	1,480 (4,856)	18 (64)	20 (68)	1,186 (46.7)

Major physical features highest point: Tajumulco ken 4,220 m (13,845 ft)

Population (2006 est.) 12,293,545

Form of government multiparty republic with one legislative house

Armed forces army 27,000; navy 1,500; air force 700

Largest cities Guatemala City (capital - 1,024,101); Mixco* (540,348); Villa Nueva* (499,133); Quezaltenango (140,405)
*Port of Guatemala City Metropolitan area

Official language Spanish

Ethnic composition Mestizo (Ladino) and European 59.4%; K'iche 9.1%; Kaqchikel 8.4%; Mam 7.9%; Q'eqchi 6.3%; other Mayan 8.6%; indigenous non-Mayan 0.2%; other 0.1%

Religious affiliations Roman Catholic 75%; Protestant 25%

Currency 1 quetzal (Q) = 100 centavos

Gross domestic product (2006 est.) U.S. $60.57 billion

Gross domestic product per capita (2006 est.) U.S. $4,900

Life expectancy at birth male 67.65 yr; female 71.18 yr

Major resources petroleum, nickel, fisheries, chicle, hydropower, bananas, cardamom, cattle, coffee, cotton, lead, maize/corn, rice, sugar beet, timber, tobacco

Thriving Maya culture (*above*) Guatemala is famous for the colorful dress, folk art, village markets and religious festivals – which mingle traditional and Christian practices – of its large Maya population.

Lake Atitlán (*right*), fringed by three volcanoes, is one of Guatemala's most beautiful sights. On market days, people from miles around bring their goods in brightly colored bundles to sell in the villages along the lakeside.

Belize

City on water Many of the buildings in Belize are wooden and raised on stilts above the mangrove swamps that surround the town. Once the capital, Belize is still the major port and commercial center.

Belize, formerly British Honduras, is a small country on the Caribbean coast of Central America, bordered by Mexico and Guatemala.

GEOGRAPHY

Northern Belize is mostly a low-lying and often swampy plain drained by the Belize and Honda rivers; the latter forms the Mexican border. Southern Belize rises sharply from a narrow coastal strip to the Maya Mountains – a fragmented plateau of serrated ranges; the highest point is Victoria Peak. Along the coast a chain of small coral islands forms a long barrier reef, second only to Australia's.

The climate is subtropical with marked seasonal rains, usually between June and November. The coastal area is particularly vulnerable to hurricanes; in 1961 Belize City was destroyed, and the capital was moved inland to Belmopan. Dense tropical rainforests cover nearly half the country, while drier areas are characterized by open forest and savanna. The abundant wildlife includes pumas, jaguars, crocodiles and manatees.

SOCIETY

Belize was once part of the ancient Maya empire, whose remains are still to be found there. Spanish settlement did not extend into this area, and the first recorded European colonists were British logwood cutters in the 17th century. They gradually migrated into the interior, clashing with the indigenous Amerindian peoples and repelling Spanish incursions from Guatemala. Sugar plantations were later established, worked by imported African slaves. British Honduras was declared a crown colony in 1862, but newly independent Guatemala continued to claim the country. The colony became independent as Belize in 1981; Britain relinquished responsibility for external defence in 1993. The country remains within the Commonwealth, and has parliamentary government on British lines.

The official language is English, but the majority of people speak a Creole dialect that reflects the complex racial mix. The largest element are descendants of black Africans. Then there are Maya Indians, Black Caribs of mixed black and Amerindian descent, and a variety of Europeans and North Americans, including farmers of the Mennonite religious sect.

ECONOMY

Agriculture is the mainstay of the economy, employing more than one-quarter of the labor force, and is being actively developed. Sugar cane is the main crop, while citrus fruits and bananas are also important. Maize, rice, kidney beans and sweet potatoes are the main domestic staples. Livestock is less important, except for beef and dairy cattle, especially on Mennonite farms.

Forestry, once the principal activity, is today less important but still yields valuable rosewood, mahogany and chicle, used in chewing gum. A great deal of timber is produced not only for export but also as a chief source of fuel, and reforestation schemes are encouraged. Cooperative fisheries provide some items such as lobsters and shrimps for export.

Mineral and fuel reserves (with the exception of timber) are scanty, so both have to be imported in quantity. Manufacturing industry is small but growing, producing food products, furniture and some clothing for export. There are virtually no railroads, but there is a good road network in the north. Education, though government-funded, is mostly run by the Roman Catholic church; literacy levels are relatively high. Medical care is free, and extensive social-security schemes are in operation.

NATIONAL DATA – BELIZE

Land area	22,806 sq km (8,805 sq mi)			
Climate		Temperatures		Annual
	Altitude m (ft)	January °C(°F)	July °C(°F)	precipitation mm (in)
Belmopan	41 (135)	23 (73)	27 (81)	1,890 (74.4)

Major physical features	highest point: Victoria Peak 1,122 m (3,681 ft)

Population	(2006 est.) 287,730

Form of government multiparty constitutional monarchy with two legislative houses

Armed forces army 1,050

Largest cities Belize City (66,140); San Ignacio (18,334); Belmopan (capital - 15,940); Orange Walk (15,927); Dangriga (11,512)

Official language English

Ethnic composition Mestizo 48.7%; Creole 24.9%; Maya 10.6%; Garifuna 6.1%; other 9.7%

Religious affiliations Roman Catholic 49.6%; Protestant 27% (Pentecostal 7.4%; Anglican 5.3%; Seventh-Day Adventist 5.2%; Mennonite 4.1%; Methodist 3.5%; Jehovah's Witnesses 1.5%); other 14%; none 9.4%

Currency 1 Belize dollar (BZD) = 100 cents

Gross domestic product (2006 est.) U.S. $2.307 billion

Gross domestic product per capita (2006 est.) U.S. $8,400

Life expectancy at birth male 66.43 yr; female 70.26 yr

Major resources sugarcane,, tourism, timber, fisheries, hydropower, bananas, citrus fruits, coconuts, fish, maize/corn, rice

El Salvador

REPUBLIC OF EL SALVADOR

T HE SMALLEST AND MOST DENSELY POPU-lated country in Central America, El Salvador lies between Guatemala to the west and Honduras to the northeast.

The dangers of earthquakes Heavy debris from a collapsing building landed on a car during this quake in January 2001 in El Salvador. Like the rest of Central America, the country is prone to earthquakes.

GEOGRAPHY

In the south a narrow Pacific coastal plain rises to a mountain range including more than 20 volcanoes. Beyond a central plain, cut by the river valleys of the Lempa, are the Metapán and Chalatenango mountain chains near the Honduran border. The climate is mainly hot, with heavy rains between May and October. Temperatures are lower in the mountains, and rainfall there is generally much higher.

The coastal plains and foothills are covered in savanna grasslands or decid-uous forest. The trees become sparser in the valleys and the central plains. Higher up the mountains there are temperate grasslands, with scattered areas of pine and deciduous forest. Large areas of the land are cultivated, and mammals are therefore less prolific than elsewhere in Central America.

SOCIETY

Between 1524 and 1539 Spanish invaders overran the Pipil Indian kingdom of Cuzcatlán, creating the territories of San Salvador and Sonsonate. In 1821 the area became independent as part of the Mexican empire. On the empire's collapse in 1823, San Salvador and Sonsonate were united to form the state of El Salvador. However, even after independence, internal strife remained, and in 1931, when there was a brief move toward democracy, there followed a succession of military dictatorships.

From 1979 to 1992 the left-wing Fara-bundo Martí Front for National Libera-tion (FLMN) fought a guerrilla war against successive governments. The conflict led to the deaths of tens of thousands of people. A new constitution in 1983 brought in direct elections to a 60-member legislative assembly but failed to end the civil war. The FLMN boycotted the 1989 elections, which were won by the right-wing Arena party. In January 1992 the United Nations sponsored a successful peace conference in Mexico, and in the following January the last FLMN troops were demobilized.

Most of the population are mestizos of mixed Amerindian and European ances-try, and the majority are Roman Catholic. The church played a leading role in antigovernment movements in the early 1980s. The official language is Spanish.

ECONOMY

The economy is predominantly agricul-tural; the land has, in the main, been owned by a small elite of white people, but is now being slowly distributed by the government. Coffee is a vital export crop, along with cotton, maize and sugar cane.

There are few important mineral re-sources. Most power comes from a hydro-electric project on the Lempa river, east of the capital, San Salvador. Massive investment in the 1970s brought a host of new manufacturing industries, but civil war caused much damage to the infra-structure and the economy. In the early 1990s, fuel oil, petroleum and food pro-cessing were the major industries.

The El Salvador National Railroad oper-ates a narrow-gauge railroad system. There are three Pacific ports, and a road and rail link to Puerto Barrios in Guatemala.

The welfare system protects those in paid employment, but unemployment is high. Health and sanitation are often poor, and disease is common in many rural areas, where there are few trained medical personnel.

Primary education is free and compul-sory but the war reduced enrollment to nearly half. There are three state universi-ties and adult literacy had risen to 71.5 percent in 1995.

NATIONAL DATA – EL SALVADOR

Land area 20,720 sq km (8,000 sq mi)

Climate	Altitude m (ft)	Temperatures January °C(°F)	July °C(°F)	Annual precipitation mm (in)
San Salvador	682 (2,238)	23 (73)	25 (76)	1,734 (68.2)

Major physical features highest point: Izalco 2,386 m (7,828 ft); longest river: Lempa 320 km (200 mi)

Population (2006 est.) 6,822,378

Form of government multiparty republic with one legislative house

Armed forces army 13,850; navy 700; air force 950

Largest cities San Salvador (capital – 542,799); Soyapango* (340,110); Mejicanos* (164,407) * suburbs of San Salvador

Official language Spanish

Ethnic composition Mestizo 90%; White 9%; Amerindian 1%

Religious affiliations Roman Catholic 83%; other 17%

Currency 1 United States Dollar (USD) = 100 centavos

Gross domestic product (2006 est.) U.S. $33.2 billion

Gross domestic product per capita (2006 est.) U.S. $4,900

Life expectancy at birth male 67.88 yr; female 75.28 yr

Major resources hydropower, geothermal power, petroleum, coffee, cotton, maize/corn, sugarcane, beans, cattle, poultry, rice, silver, sorghum, timber

Honduras

REPUBLIC OF HONDURAS

MOUNTAINOUS HONDURAS LIES BETWEEN Guatemala and El Salvador to the west, and Nicaragua to the southeast.

GEOGRAPHY

From the Caribbean coastal plain in the north, several deep river valleys cut southwestward into the central uplands; the coastal plains become broader toward the east. South of the mountains, a narrow lowland strip borders the very short Pacific coastline.

The hot, humid climate is moderated by altitude inland. The mainly seasonal rainfall is highest in the north and east, where destructive summer hurricanes may also strike, and is lowest in sheltered inland valleys and basins.

In the east, coastal mangroves and palms give way to sandy plains, pine-savanna and dense rainforest near the Nicaraguan border. The highland areas support grassland and broadleaf forests, with further savanna forest and mangroves on the Pacific coast. Although many forest areas have been cleared for agriculture, there is still a profusion of wildlife: insects, birds and reptiles as well as peccaries, pumas, jaguars and tapirs.

SOCIETY

Honduras was once an important center of the Maya civilization. It was first settled by Europeans in 1522. After a savage war with the indigenous Amerindians, the Spanish made Honduras part of Guatemala. Independence as part of Mexico came in 1821, and full independence in 1838. Government became increasingly unstable in the early 20th century, and the United States sent in marines to protect their investments, particularly in the banana trade. The 1932 election produced a dictator, General Tiburcio Carlas Andino (1876–1969), who ruled until 1949 despite the frequent uprisings. Later governments have been generally short-lived, and the military has often kept effective control. Until 1989 Honduras sheltered the Contra rebels who, with American backing, were fighting the Nicaraguan government.

The Honduran constitution has been rewritten at least 16 times since 1821. That of 1982 provides for a 128-member legislative assembly and a president, both directly elected. In practice ultimate power rests with the army. The population is still mostly rural, consisting of mestizos of mixed Amerindian and European ancestry. The official language is Spanish, and the majority are Roman Catholic.

ECONOMY

Just over one-third of the work force are farmers. Bananas and coffee are the major export crops, but Hurricane Mitch caused extensive damage in Honduras in 1998, affecting economic recovery and future growth badly. Maize, beans and rice are the most important staples, but the majority of the population is still affected by malnutrition. Mineral resources are quite considerable, but have so far remained largely unexplored. Energy comes mainly from hydroelectricity; a small amount comes from imported fuels. Small-scale manufacturing produces a wide variety of goods such as furniture, textiles, footwear, chemicals and cement.

There are few railroads, and freight and passengers rely mostly on the extensive road network, reducing the importance of domestic air services. Welfare benefits are mostly limited to urban workers. Health and sanitation are poor, and tropical diseases are common, with few doctors on hand to treat them. Education is free and compulsory, but often inaccessible to rural communities; illiteracy is relatively high.

Legacy of the Maya (*above*) A carved head at the ruins of Copán. This great city, first inhabited in 2000 BC, was the heart of Honduras' Maya civilization.

"The last paradise" (*left*) So-named by enchanted tourists, the Islas de la Bahía, 50 km (30 mi) off the coast of Honduras, were once a base from which pirates launched their attacks on Spanish ships.

NATIONAL DATA – HONDURAS

Land area	111,890 sq km (43,201 sq mi)			
Climate	Altitude m (ft)	Temperatures January °C(°F)	July °C(°F)	Annual precipitation mm (in)
Tegucigalpa	1,004 (3,294	20 (68)	23 (73)	917 (36.1)

Major physical features highest point: Cerro Las Minas 2,827 m (9,275 ft); longest river: Coco (part) 685 km (425 mi)

Population (2006 est.) 7,326,496

Form of government multiparty republic with one legislative house

Armed forces army 8,300; navy 1,400; air force 2,300

Largest cities Tegucigalpa (joint capital with Comayaguela – 893,589); San Pedro Sula (516,608); Comayagua (61,576)

Official language Spanish

Ethnic composition Mestizo 90%; Amerindian 7%; Black 2%; White 1%

Religious affiliations Roman Catholic 97%; Protestant 3%

Currency 1 Honduran lempira (HNL) = 100 centavos

Gross domestic product (2006 est.) U.S. $22.13 billion

Gross domestic product per capita (2006 est.) U.S. $3,000

Life expectancy at birth male 67.75 yr; female 70.98 yr

Major resources timber, gold, silver, copper, lead, zinc, iron ore, antimony, coal, fisheries, hydropower, bananas, beans, cattle, coffee, fruits, maize/corn, rice, shellfish, sugarcane, timber, tin, tobacco

Nicaragua

REPUBLIC OF NICARAGUA

THE CENTRAL AMERICAN STATE OF NICA-ragua lies between Honduras to the northwest and Costa Rica to the south.

GEOGRAPHY

Its eastern plains are bordered with lagoons and swamps along the Caribbean or Mosquito Coast. The land rises westward toward a roughly triangular highland area bordering Honduras. Westward again are Lake Nicaragua and Lake Managua, and beyond them is a second, smaller range of about 40 volcanoes, flanked to the west by a narrow lowland belt. Some volcanoes are still active, and the area has severe earthquakes, as in 1992 when 16,000 people were made homeless.

Nicaragua's tropical climate is hotter and drier on the Pacific side, where the rainy season lasts from May to November. On the Caribbean side it lasts for at least nine months, and there is a greater risk of hurricanes. Temperatures are lower on the northern mountains. Most of the western half of the country is covered in savanna, and the eastern half in tropical rainforest.

SOCIETY

Nicaragua, like its neighbors, was conquered by Spain in the 16th century, and gained full independence in 1838. Later politics were dominated by Conservatives (the coffee and sugar plantation owners) and Liberals (artisans and small landowners). Great Britain and the United States both had strategic interests in Nicaragua, and the United States sent in military forces on two occasions to support Conservative regimes.

From 1934 until 1979 Nicaragua was ruled by a repressive military regime dominated by the Somoza family. However, in 1979 rebel forces seized power in a country devastated by civil war, and formed the left-wing Sandinista government. Until 1990 the United States supported the Contras – guerrilla opponents of the new Sandinista regime, trained in neighboring Honduras by United States advisers. Elections followed their withdrawal in 1990, in which the opposition coalition, the National Opposition Union (UNO), came to power. The president is elected by a majority, and the 92-member national assembly on a proportional basis. Constitutional deadlock ensued after the president blocked reforms passed by the national assembly in 1994.

Most of the population are mestizos of mixed Amerindian and European descent. Spanish is the official language, and most people are Roman Catholic.

ECONOMY

At the end of the 1990s, Nicaragua was showing signs of economic recovery from the effects of the civil war, but was devastated by Hurricane Mitch in 1998, estimated to cause US $1 billion damage. Despite international aid, growth will take some time to resume. The chief crops are cotton, coffee and bananas. Mineral resources are plentiful, but only gold and silver are intensively mined. Food processing, chemicals and textiles are the chief industries. Industrial development is limited.

Communications are poor except in the west, and even here only half the roads are passable when it rains. Railroads are also limited to the west, leaving eastern areas dependent on water and air transportation. Housing, sanitation and water supplies are inadequate, and tropical diseases are common. Health care is difficult to obtain despite reforms by the Sandinistas, and there are few trained doctors. There is limited social security for workers in Managua. Primary education is free, and enrollment is high. Few receive secondary education, but literacy levels are on the increase.

A fertile land Crops flourish on the rich volcanic soils of Nicaragua. Some volcanoes are active, but earthquakes are a worse threat; the capital Managua was almost totally destroyed in 1972.

NATIONAL DATA - NICARAGUA

Land area	120,254 sq km (46,430 sq mi)			
Climate	**Altitude** m (ft)	**Temperatures** January °C(°F)	July °C(°F)	**Annual precipitation** mm (in)
Managua	55 (180)	26 (79)	27 (81)	1,141 (44.9)
Major physical features	highest point: Cerro Mogotón 2,103 m (6,900 ft)			
Population	(2006 est.) 5,570,129			
Form of government	multiparty republic with one legislative house			
Armed forces	army 12,000; navy 800; air force 1,200			
Capital city	Managua (1,007,811)			
Official language	Spanish			
Ethnic composition	Mestizo 69%; White 17%; Black 9%; Amerindian 5%			
Religious affiliations	Roman Catholic 58.5%; Evangelical 21.5%; Moravian 1.5%; Jehovah's Witnesses 0.9%; other 1.6%; none 16%			
Currency	1 córdoba (C) = 100 centavos			
Gross domestic product	(2006 est.) U.S. $16.83 billion			
Gross domestic product per capita	(2006 est.) U.S. $3,000			
Life expectancy at birth	male 68.55 yr; female 72.81 yr			
Major resources	gold, silver, copper, tungsten, lead, zinc, timber, fisheries, bananas, cattle, coffee, cotton, maize/corn, rice, sugarcane			

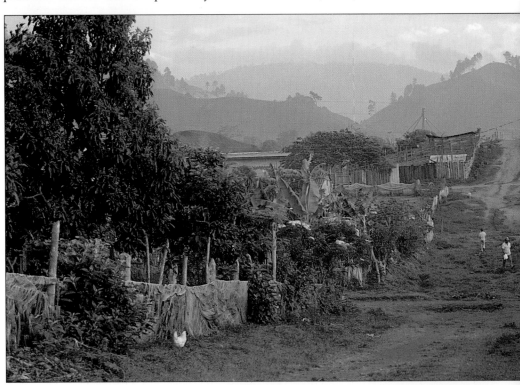

Costa Rica

REPUBLIC OF COSTA RICA

COSTA RICA LIES ON THE CENTRAL AMERICAN isthmus between Nicaragua to the north and Panama to the east.

GEOGRAPHY

From the Caribbean lowlands in the northeast the land rises gently to a mountainous backbone. The Cordillera de Guanacaste, including four major volcanoes, runs southeastward from the Nicaraguan border to meet the much higher Cordillera Central. Between this and the volcanic Cordillera de Talamanca in the southeast lies the Meseta or Valle Central, where the capital, San José, lies in an area also subject to both volcanic eruptions and earthquakes. The narrow Pacific coastal belt and the Caribbean coast are lined with mangrove swamps and white, sandy beaches.

The climate is generally hot and rainy, but there is a marked dry season on the Pacific coast, whereas the Caribbean lowlands have rain throughout the year. Both areas are heavily forested, with an enormous variety of birds, insects and reptiles, and a mixture of South and North American mammals. The higher Central and Talamanca ranges are warm and temperate, with dry seasons on the more sheltered southern slopes.

SOCIETY

Costa Rica was not fully controlled by Spain until 1570. Its later history was broadly similar to that of the rest of Central America until independence in 1838. Coffee exports put the country on a sound financial footing, and its isolation largely protected it from the more dramatic events in neighboring lands. In 1890 the free election of President José Joaquin Rodríguez set a pattern of democracy that survived revolutions in 1917 and 1948. The new constitution of 1949 disbanded the army to prevent any future military coups. The president and the 57-member legislative assembly are each directly elected for a four-year term. More recently Costa Rica has suffered the effects of eathquake damage, falling coffee prices, foreign debt and an influx of refugees from its troubled neighbors.

The population profile is unusual for Central America in that most people are of European descent, with only a minority of mestizos of mixed Amerindian, European and African extraction. Most are Roman Catholic, and Spanish is the official language.

ECONOMY

The main cash crops are coffee and bananas. Maize, beans and rice are the chief staples, but food still has to be imported. Timber resources have enormous potential, and there are significant bauxite deposits. Hydroelectricity is being increasingly developed, but much electrical power still comes from diesel plants. Manufactures for export include medicines, paper products, sheet metal and electrical machinery.

Apart from the Pan-American Highway, the limited road network centers on the Valle Central, which is also linked by railroad to ports on the Pacific and Caribbean coasts. There is an international airport at Juan Santa Maria, near the capital, and a few other airports serve domestic flights. There are five daily

A craftsman at work in Sarchí, Costa Rica's handicrafts center. This prosperous and peaceful country – it has no army – has largely managed to avoid the political conflict that has troubled so much else of Central America.

newspapers and a single television network; there is no government censorship.

Welfare benefits are provided for those in employment, and a preventive health program is in place. Rural areas, however, are less well served and malnutrition and disease are a problem. Education is free and compulsory, enrollment is high, and so are literacy levels.

NATIONAL DATA - COSTA RICA

Land area	50,660 sq km (19,600 sq mi)			

Climate	Altitude m (ft)	Temperatures January °C(°F)	July °C(°F)	Annual precipitation mm (in)
San José	1,146 (3,760)	19 (67)	21 (69)	1,866 (73.4)

Major physical features highest point: Mount Chirripó 3,820 m (12,533 ft)

Population (2006 est.) 4,075,261

Form of government multiparty republic with one legislative house

Armed forces paramilitary 8,400

Capital city San José (342,977)

Official language Spanish

Ethnic composition White (including Mestizo) 94%; Black 3%; Amerindian 1%; Chinese 1%; other 1%

Religious affiliations Roman Catholic 76.3%; Evangelical 13.7%; Jehovah's Witnesses 1.3%; other Protestant 0.7%; other 4.8%; none 3.2%

Currency 1 Costa Rican colón (CRC) = 100 céntimos

Gross domestic product (2006 est.) U.S. $48.77 billion

Gross domestic product per capita (2006 est.) U.S. $12,000

Life expectancy at birth male 74.43 yr; female 79.74 yr

Major resources hydropower, coffee, bananas, timber, bauxite, cattle, cocoa, gold, iron ore, maize/corn, oranges, rice, silver, sugarcane, sulfur

Panama

REPUBLIC OF PANAMA

PANAMA IS A STRATEGICALLY POSITIONED country forming an S-shaped neck of land that links together Central and South America.

GEOGRAPHY

At its narrowest point, in the center of the country, the Panama Canal links the Caribbean Sea and the Pacific Ocean. West of the canal the mountains of the Serrania de Tabasara are flanked by narrow coastal plains on either side. In the southeast of this area the land rises once again to the mountains of the Azuero Peninsula. East of the canal, two mountain arcs run parallel to the Pacific and Caribbean coastlines.

The climate is generally hot and wet

The Panama Canal (*above*) links the Pacific and the Atlantic Oceans. The trade generated by this great feat of engineering – a tourist attraction in itself – is the mainstay of the Panamanian economy.

throughout the year, but the mountains are cooler, and there is a marked dry season along the Pacific coast. Most of the country is forested, with tropical rainforest predominating. Plant and animal life are rich and abundant except in the mainly deforested Canal Zone.

SOCIETY

After 300 years of Spanish rule, Panama achieved freedom from Spain in 1821, though it remained a part of neighboring Colombia. The California Gold Rush of 1849 provided the incentive for building a canal across the Central American isthmus that would make the journey from the eastern United States to the booming western states easier and faster than overland travel. The project finally got under way in 1879, but foundered ten years later. In 1903 the United States helped Panama to secure independence from Colombia in return for perpetual control of the 16 km (10 mi) wide Canal Zone. The canal was finally opened to traffic in 1914.

In 1977, after many years of unrest, a process began whereby formal control of the Canal Zone was gradually to be returned to Panama. Ownership of the canal itself was transferred in 2000. The 1983 constitution provided for the election of a president, two vice-presidents and a 72-member legislative assembly, all of these for a five-year term. However, political unrest continued, and in 1989 General Manuel Noriega (b. 1940), who had assumed effective power, declared a "state of war" against the United States,

who had accused him of drug-running and election rigging. An American invasion in 1989 ousted Noriega, who in January 1990 was flown to Miami in the United States to stand trial on drug-trafficking charges. In 1992 he was found guilty and jailed.

Most Panamanians are mestizos or mulattos of mixed Amerindian, European and African descent; the official language is Spanish. The birth rate is the lowest in Central America, but economic growth is slower than population growth.

ECONOMY

The chief export crops are bananas, coffee, and sugar. Rice, beans and corn are the main staples. Livestock is also important, and Panama is largely self-sufficient in food. Fishing is encouraged, and shrimps are now a major export. Mineral resources are scanty except for copper ore, and power comes largely from hydroelectricity. The industrial base around Panama City manufactures consumer goods, building materials and also petroleum products, as a result of the construction of a trans-Panamanian oil pipeline.

Most of the country's revenue comes from the service industries, including finance, trade and tourism. The main earners are the Panama Canal and the Colón Free Zone, which is one of the largest trading centers in the world. There is an extensive road system, and a railroad runs parallel to the canal. There are good port facilities on both coasts.

Welfare services are well developed, and medical facilities are adequate in most parts of the country, though tuberculosis is common in urban slum areas. Education is free and compulsory, and higher education is available.

NATIONAL DATA – PANAMA

Land area	75,990 sq km (29,340 sq mi)			

Climate		Temperatures		Annual
	Altitude m (ft)	January °C(°F)	July °C(°F)	precipitation mm (in)
Panama City	36 (118)	26 (79)	27 (81)	2,907 (114.4)

Major physical features highest point: Volcán Barú 3,475 m (11,401 ft); largest lake: Gatún 430 sq km (166 sq mi)

Population (2006 est.) 3,191,319

Form of government multiparty republic with one legislative house

Armed forces paramilitary 11,800

Capital city Panama City (403,808)

Official language Spanish

Ethnic composition Mestizo 70%; Amerindian and mixed (West Indian) 14%; White 10%; Amerindian 6%

Religious affiliations Roman Catholic 85%; Protestant 15%

Currency 1 Panamanian balboa (PAB) = 100 centesimos

Gross domestic product (2006 est.) U.S. $25.29 billion

Gross domestic product per capita (2006 est.) U.S. $7,900

Life expectancy at birth male 72.68 yr; female 77.87 yr

Major resources copper, shrimps, hydropower, tourism, coffee, bananas, cattle, maize/corn, rice, sugarcane, timber

Cuba

REPUBLIC OF CUBA

CUBA IS AN ISLAND STATE IN THE CARIBBEAN lying directly south of Florida. It consists of a single long island, flanked by the much smaller Isla de la Juventud (Isle of Youth) to the south.

GEOGRAPHY

The main island of Cuba is characterized by wide stretches of lowland divided by three main mountain areas. The largest of these, in the east, comprises the Sierra del Cristal and the Sierra Maestra along the southeast coast, the highest point being Pico Turquino. The lower central chain crosses the island to the east of Cienfuegos, while the Sierra de Los Organos runs along the northwestern tip beyond the capital, Havana. The coastline is irregular, and is lined with numerous mangroves, beaches and coral reefs.

The climate is semitropical, with heavy seasonal rainfall and a tendency to hurricanes. Rivers also vary with the seasons, and only a few are navigable. Although there has been considerable deforestation, the island's forests are still extensive, including tropical near-jungle.

SOCIETY

Cuba was among the earliest Spanish settlements after Christopher Columbus (1451–1506) landed there in 1492. From the early 16th century it was divided into large estates, worked at first by enslaved Amerindians, then by African slaves until slavery was finally abolished in 1886. Independence from Spain was achieved in 1899 after two uprisings (1868–78, 1895–99), the latter led by Jose Marlí (1853–95) and in which US troops became involved – they finally left in 1902. In 1959 the decaying regime of Fulgencio Batista (1901–73) was ousted by the communist revolutionary movement led by Fidel Castro (b.1927). Castro remodeled the Cuban state as a personal dictatorship on Soviet lines, but significantly improved the living standards of the poor. The United States imposed a trade embargo, and supported an unsuccessful countercoup in the Bay of Pigs on Cuba's south coast in 1961. From then onward, Cuba became increasingly dependent on the former Soviet Union. First direct elections were held in 1994, but only approved Communist Party candidates stood. Further relaxation of hard-line policies were seen in January 1998 when the Pope visited Cuba .

ECONOMY

The suspension of Soviet financial aid and technical advice in the early 1990s severely affected the Cuban economy. Gross domestic product declined by 35 percent during the period 1989–93 and growth was only slowly resumed in 1994. However, severe hurricanes in 1993, 1994 and 1996 devastated the sugar cane harvest and the economy slumped badly. The main cash crops are sugar cane, tobacco, citrus fruits and coffee. There has been some relaxation of state control of agriculture and limited free trading is allowed. Food has to be imported to meet domestic needs. Fishing, chiefly for tuna, has been developed with government support.

Large nickel deposits supply Cuba's other main export. Other reserves include copper, chromite and iron ore. Some petroleum has been found, but most fuel has to be imported. The traditional industries are food and tobacco processing, but this sector has been greatly diversified to include such items as steel, fertilizers, machinery and electrical goods.

The transportation network is well developed, with an emphasis on public service. Social benefits and health care are available, and literacy is high.

Keeping the revolution alive (*above*) Fidel Castro – Central America's most charismatic and durable leader – holds the local crowd spellbound with his oratory.

Money to spend (*left*) Automobiles parked proudly outside their owners' homes attest to good wages; yet despite many advances in living standards, Cuba has failed to industrialize and diversify its economy.

NATIONAL DATA – CUBA

Land area	110,860 sq km (42,803 sq mi)			
Climate		**Temperatures**		**Annual**
	Altitude m (ft)	January °C(°F)	July °C(°F)	precipitation mm (in)
Havana	49 (161)	22 (72)	28 (82)	1,189 (46.8)

Major physical features highest point: Turquino 2,000 m (6,560 ft); longest river: Cauto 249 km (155 mi)

Population (2006 est.) 11,382,820

Form of government multiparty republic with one legislative house

Armed forces army 38,000; navy 3,000; air force 8,000

Largest cities Havana (capital – 2,164,362); Santiago de Cuba (570,503); Camagüey (355,027); Holguín (330,116); Guantánamo (280,024); Santa Clara (256,184);

Official language Spanish

Ethnic composition Mulatto 51%; White 37%; Black 11%; Chinese 1%

Religious affiliations nonreligious 55.2%; Roman Catholic 39.7%; Afro-Cuban 1.7%; Protestant 3.4%

Currency 1 Cuban peso (CUP) or Convertible peso (CUC) = 100 centavos

Gross domestic product (2006 est.) U.S. $44.54 billion

Gross domestic product per capita (2006 est.) U.S. $3,900

Life expectancy at birth male 75.11 yr; female 79.85 yr

Major resources cobalt, nickel, iron ore, chromium, copper, salt, timber, silica, petroleum, tourism, beans, cassava, coffee, livestock, maize, oranges, rice, sugarcane, sweet potatoes, tobacco, tropical fruits

Bahamas

COMMONWEALTH OF THE BAHAMAS

T HE BAHAMAS ARE A LONG ARCHIPELAGO OF some 700 islands and numerous cays (reefs) lying in the Caribbean Sea southeast of Florida, and north of Cuba.

GEOGRAPHY

The islands are very much alike; they are all exposed segments of a single coralline limestone shelf. They are low-lying, apart from a few hills formed by windblown sand, and their coastlines are fringed with mangroves, lagoons and coral reefs. The climate is mild and humid, but the porous limestone of the landscape means that rainfall is easily lost. Caribbean pine forest covers much of the larger islands, such as Andros, New Providence, Grand Bahama and Great Abaco. The teeming birds and colorful sea life along the reefs attract numerous tourists, who create considerable environmental problems.

SOCIETY

The native Arawak people were forcibly removed by Spanish slave raiders, who put them to work on Hispaniola. In the 17th century British settlers arrived from Bermuda, but the colony degenerated into a pirate haven. Numbers were later swelled by Loyalist refugees from the American War of Independence (1775–83). The islands thrived briefly on illicit trade with the United States during the Civil War (1861–67) and later during Prohibition in the 1920s, but they gained more lasting prosperity from tourism following World War II. Internal self-government was granted by Britain in 1967, and full independence in 1973. Government is along British lines, with a governor-general representing the sovereign as head of state.

ECONOMY

Since only a tiny percentage of the land is arable, there is little agriculture beyond fruit and vegetable growing. Salt is the main mineral produced, although refining of imported petroleum for export is a major industry; others include pharmaceuticals, rum distilling and fruit canning. International banking is important.

Tourism is the country's economic mainstay. Poor road provision is offset by good air and sea communications. Education, social benefits and health care are widely available.

Haiti

REPUBLIC OF HAITI

H AITI, ONE OF THE POOREST COUNTRIES IN the world, occupies the western third of the island of Hispaniola in the central Caribbean east of Cuba.

GEOGRAPHY

Two mountainous peninsulas enclose the central plain of the Artibonite river. The northern range, called the Massif du Nord, extends into the heart of the Dominican Republic, where it becomes the Cordillera Central. The climate is generally hot with seasonal rains, but temperature and rainfall patterns vary. Some areas are practically desert.

Echoes of Africa Over 90 percent of Haiti's people are descendants of African slaves. Haiti became the world's first black republic in 1804. Today, social services and health care are poor, and education, though free and compulsory, is poor outside the capital.

NATIONAL DATA – BAHAMAS				
Land area 10,070 sq km (3,888 sq mi)				

Climate	Altitude m (ft)	Temperatures January °C(°F)	July °C(°F)	Annual precipitation mm (in)
Nassau	4 (13)	21 (70)	28 (82)	1,389 (54.6)

Major physical features highest point: Cat Island 63 m (206 ft); largest island: Andros 5,957 sq km (2,300 sq mi)

Population (2006 est.) 303,770

Form of government multiparty constitutional monarchy with two legislative houses

Armed forces Royal Bahamian Defense Force 860

Capital city Nassau (235,105)

Official language English

Ethnic composition Black 85%; White 12%; Asian and Hispanic 3%

Religious affiliations Baptist 35.4%; Anglican 15.1%; Roman Catholic 13.5%; Pentecostal 8.1%; Church of God 4.8%; Methodist 4.2%; other Christian 15.2%; none or unspecified 2.9%; other 0.8%

Currency 1 Bahamian dollar (BSD) = 100 cents

Gross domestic product (2006 est.) $6.476 billion

Gross domestic product per capita (2006 est.) $21,300

Life expectancy at birth male 62.24 yr; female 69.03 yr

Major resources salt, calcium carbonate, timber, tourism, fruit and vegetables, limestone, okra, shellfish

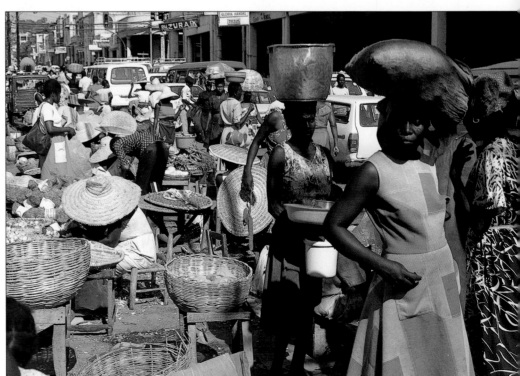

Jamaica

JAMAICA

T HE LUSH, TROPICAL ISLAND STATE OF Jamaica lies in the Caribbean south of Cuba and west of Haiti.

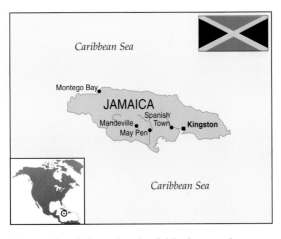

GEOGRAPHY

Apart from the irregular coastal plains and several fertile river basins, mostly in the south and west, the land rises sharply to a limestone plateau that is heavily eroded. This is especially true of the Cockpit Country toward the northwest, with its typical karst scenery of bare limestone pavements and underground cave systems. Several mountain ranges rise above the plateau, the highest being the Blue Mountains in the east.

The climate is mainly tropical, with even temperatures all the year round. The mostly seasonal rains fall heavily on the northeastern slopes. These are still clad in the tropical rainforest that once covered most of the island, but the southern lowlands in the rainshadow of the mountains, support only savanna scrub.

SOCIETY

Jamaica's indigenous Arawak peoples were wiped out by the original Spanish settlers, and African slaves were imported to work the land. The island was captured by Britain in 1655; it soon became an important base for buccaneers, who mounted attacks on Spanish ships. The Spanish, in the meantime, gave assistance to the Maroons – escaped slaves who had taken refuge in the Cockpit Country, and who harried the British for nearly two centuries. Jamaica became an immensely profitable colony through its slave-worked sugar cane plantations, but this prosperity ended with the abolition of slavery in 1833. In 1866 Jamaica became a crown colony, and banana-growing was introduced.

Local self-government was introduced after World War II, leading to full independence in 1962. Jamaica is a constitutional monarchy within the Commonwealth. The sovereign is represented by the governor-general, and a two-chamber parliament on the British model is headed by a prime minister and cabinet.

The population is mostly English-speaking and of African descent, but there are several ethnic minorities, including Indians and Chinese. Life for many of the inhabitants is harsh, and poverty-stricken slums sit side by side with the wealthy tourist areas. Rastafarianism, the religious-political movement that regards Ethiopia, in Africa, as the spiritual homeland of black people, originated in the island.

ECONOMY

Agriculture has declined but remains an important sector. Sugar cane and bananas are still the main cash crops, along with coffee, cocoa and fruit; peppers and ginger are also grown. Staples are grown mostly at subsistence levels and for local sale. Fisheries are significant. The main mineral resource is bauxite, and alumina processing is one of the main industries. Other varied export activities include printing, textile manufacture and food processing, including rum distilling and sugar production. Tourism provides the largest source of foreign income.

The transportation network is generally good, and health, welfare and housing are government priorities. Education is free and widely available at all levels.

SOCIETY

Hispaniola's indigenous Arawak peoples were wiped out by the first Spanish conquistadors, through enslavement and disease. In the 17th century the west of the island became the French colony of Saint Domingue. Hundreds of thousands of slaves were used to work the plantations. The majority of today's population are Creole-speaking descendants of these Africans. Their animist religions mixed with the Roman Catholic faith, giving rise to the religious cult known as Voodoo.

The French were driven out in 1804, but in the next century Haiti experienced numerous brutal dictatorships and revolutions until the United States assumed control from 1915 to 1934. More brief dictatorships followed, until in 1957 François "Papa Doc" Duvalier (1907–71) was elected president and established a police state. A military coup in 1986 deposed his son. In 1991 Jean Bertrand Aristide was elected president, but was overthrown by a military coup. He was restored to power in 1994 with US support. In 1995 a UN peacekeeping force took over and Renee Préval succeeded Aristide.

ECONOMY

Haiti's impoverished population depends largely on subsistence agriculture. Some coffee and sugar cane are grown for export. Small amounts of bauxite, copper and other metals are mined.

NATIONAL DATA - HAITI

Land area	27,560 sq km (10,641 sq mi)			
Climate		Temperatures		Annual precipitation
	Altitude m (ft)	January °C(°F)	July °C(°F)	mm (in)
Port-au-Prince	37 (121)	26 (79)	29 (84)	1,353 (53.2)

Population (2006 est.) 8,308,504

Form of government multiparty republic with two legislative houses

Armed forces no armed forces

Capital city Port-au-Prince (1,321,522)

Official language English

Ethnic composition Black 95%; Mulatto and White 5%

Religious affiliations Roman Catholic 80%; Protestant 16% (Baptist 10%; Pentecostal 4%; Adventist 1%; other 1%); none 1%; other 3% (approximately half the population practices Voodoo)

Currency 1 gourde (HTG) = 100 centimes

Gross domestic product (2006 est.) U.S. $14.56 billion

Gross domestic product per capita (2006 est.) U.S. $1,800

Life expectancy at birth male 51.89 yr; female 54.6 yr

Major resources bauxite, copper, calcium carbonate, gold, marble, hydropower, coffee, sugarcane, bananas, cassava, maize/corn, rice, sisal, sorghum, timber

---

NATIONAL DATA - JAMAICA

Land area	10,831 sq km (4,182 sq mi)			
Climate		Temperatures		Annual precipitation
	Altitude m (ft)	January °C(°F)	July °C(°F)	mm (in)
Kingston	34 (112)	26 (79)	29 (84)	813 (32)

Major physical features highest point: Blue Mountain Peak 2,256 m (7,402 ft)

Population (2006 est.) 2,758,124

Form of government constitutional monarchy with two legislative houses

Armed forces army 2,500; navy 190; air force 140

Capital city Kingston (586,630)

Official language English

Ethnic composition Black 90.9%; East Indian 1.3%; White 0.2%; Chinese 0.2%; Mixed 7.3%; other 0.1%

Religious affiliations Protestant 61.3% (Church of God 21.2%; Seventh-Day Adventist 9%; Baptist 8.8%; Pentecostal 7.6%; Anglican 5.5%; others 9.2%); Roman Catholic 4%; other including some spiritual cults 34.7%

Currency 1 Jamaican dollar (JMD) = 100 cents

Gross domestic product (2006 est.) U.S. $12.71 billion

Gross domestic product per capita (2006 est.) U.S. $4,600

Life expectancy at birth male 71.54 yr; female 75.03 yr

Major resources bauxite, gypsum, limestone, rum, coffee, bananas, cocoa, citrus fruits, coconuts, ginger, molasses, pimiento, sugarcane, tourism

Dominican Republic

DOMINICAN REPUBLIC

T HE DOMINICAN REPUBLIC OCCUPIES THE
eastern part of the Caribbean island of
Hispaniola. It is twice as large as its
western neighbor Haiti.

GEOGRAPHY

The landscape is mountainous but varied.
At its heart lies the mighty Cordillera
Central, which extends into Haiti as the
Massif du Nord. The highest peak, Pico
Duarte, is also the highest point in the
Caribbean islands. There are several
lesser ranges along the north coast, in the
east and in the southwest bordering Haiti.
These are interspersed by extensive low-
lands, including the long Cibao Valley
in the north, and much of the south coast

to the east of the capital, Santo Domingo.
  The climate is mainly tropical with
seasonal rains, but the mountainous terrain produces considerable variations in
temperature and rainfall. The north and
east have high rainfall, fertile soils and
lush evergreen forests. Low-lying areas in
the south are dry to the point of desert,
supporting only savanna scrub.

SOCIETY

Hispaniola was first colonized by Spain,
but in the 17th century the western third
was ceded to France. The rest remained
Spanish as Santo Domingo, but it later
came under French control too. The
French imported African slaves in great
numbers, and established a profitable
plantation economy. In 1821 Santo Domingo declared its independence as the
Dominican Republic. However, it was
immediately occupied by Haiti until Juan
Pablo Duarte (1813–76) secured independence for the Dominican Republic in
1844. In the 1860s the country became
Spanish again for a short time.
  The republic has largely been ruled by
dictators, with only brief periods of
democratic rule. From 1916 to 1924 it was
occupied by the United States, who did
much to modernize the country. In 1930
Rafael Trujillo Molina (1891–1961) seized
power, but his relentless suppression of
his opponents led to his assassination in
1961. In 1978, for the first time ever, the
government changed hands in a democratic election. Democracy has since been
maintained, but with much civil unrest.
  The majority of the rapidly growing
population are mestizos of mixed European and African descent, but there are
several other ethnic groups.

ECONOMY

The economy is still largely agricultural.
Exportable cash crops are grown on larger
plantations; these include cocoa, coffee,
tobacco and especially sugar cane. Staples
such as rice and corn are produced mostly
by low-output subsistence farming.
  Mineral reserves are varied, but only
rock salt, bauxite, nickel and some other
metals are exploited in quantity. Industries are mostly light, ranging from sugar
refining to textiles and fertilizers. Tourism
and telecommunications led to vigorous
economic growth in the late 1990s. A
social-security system offers reasonable
benefits, while health care is limited.
Education, though free, is not available in
isolated rural districts.

NATIONAL DATA – DOMINICAN REPUBLIC

Land area	48,380 sq km (18,680 sq mi)			
Climate		Temperatures		Annual
	Altitude m (ft)	January °C(°F)	July °C(°F)	precipitation mm (in)
Santo Domingo	17 (56)	24 (76)	27 (81)	1,447 (56.9)
Major physical features	highest point: Pico Duarte 3,175 m (10,417 ft)			
Population	(2006 est.) 9,183,984			
Form of government	multiparty republic with two legislative houses			
Armed forces	army 15,000; navy 4,000; air force 5,500			
Largest cities	Santo Domingo (capital – 2,306,138); Santiago (589,358); San Pedro de Macorís (235,467)			
Official language	Spanish			
Ethnic composition	Mixed 73%; White 16%; Black 11%			
Religious affiliations	Roman Catholic 95%; other 5%			
Currency	1 Dominican peso (DOP) = 100 centavos			
Gross domestic product	(2006 est.) U.S. $73.74 billion			
Gross domestic product per capita	(2006 est.) U.S. $8,000			
Life expectancy at birth	male 70.21 yr; female 73.33 yr			
Major resources	nickel, bauxite, gold, silver, tourism, ferronickel, cocoa, coffee, cotton, flowers, mangoes, maize/corn, oranges, platinum, rice, salt, sugar beet, tobacco, tomatoes			

Saint Kitts-Nevis

FEDERATION OF SAINT KITTS-NEVIS

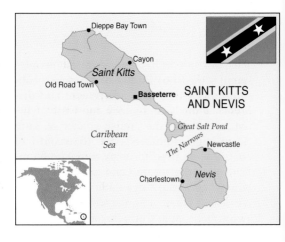

T HE ISLANDS OF SAINT KITTS AND NEVIS FORM
part of the western chain of the Leeward Islands in the eastern Caribbean.

GEOGRAPHY

The capital, Basseterre, is on Saint Kitts
(Saint Christopher) – a mountainous volcanic island rising to Mount Misery.
Nevis, to the south of a channel called The
Narrows, is similarly volcanic and surrounded by coral reefs. The warm, humid
climate is cooled by northeasterly trade
winds. The interior has lush vegetation,
including tropical rainforest on some
mountain slopes. The lower slopes have
mostly been cleared for cultivation.

The volcanic island (*below*) of Saint Kitts attracted
early sugar cane planters with its verdant vegetation.
Tourism now supplements the island's sugar economy.

Antigua and Barbuda

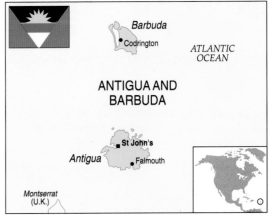

ATLANTIC OCEAN

ANTIGUA AND BARBUDA

Barbuda
Codrington

St John's
Antigua
Falmouth

Montserrat
(U.K.)

SOCIETY

The British colony of Saint Kitts was founded by Sir Thomas Warner in 1623, but ownership was disputed by the French until 1783. In 1882 Saint Kitts and the neighboring islands of Nevis and Anguilla were united in a federal state. Anguilla became unhappy with the Saint Kitts administration, and declared unilateral independence in 1967. Saint Kitts-Nevis became fully independent in 1983. The British sovereign, represented by a governor-general, is head of state. The prime minister heads a 15-member parliament; most are elected, but a few are appointed. Nevis enjoys a degree of autonomy. Most of the population are black or are mixed-race descendants of former slaves. The official language is English, and the main religious groups are Anglican and Methodist Christians.

ECONOMY

The main cash crops are sugar cane on Saint Kitts, cotton and coconuts on Nevis. Other major exports include footwear, fabrics and electronic goods. Tourism is a growing source of revenue, thanks to the fine beaches. There are good roads and ferry connections, a deep-water harbor at Basseterre and an airport on each island. Education is free and compulsory, and literacy levels are high.

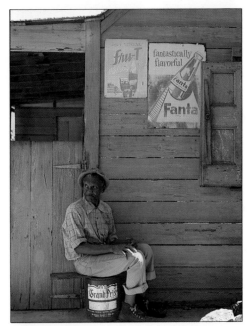

Whiling the time away (*above*) Now at leisure after a lifetime working in the canefields, an old man surveys the scene. The sugar cane industry, once the sole basis of Antigua's economy, has been closed down and is now supplanted by tourism.

THE ISLAND STATE OF ANTIGUA AND BARBUDA is part of the Leeward Islands group in the eastern Caribbean, and is of considerable strategic importance for the United States, which operates two military bases from the islands.

GEOGRAPHY

The low-lying island of Antigua has been mostly deforested, and includes an area of volcanic rock around Boggy Peak in the southwest. The coral island of Barbuda to the north is mostly flat and well wooded. The tiny islet of Redonda to the southwest is steep and totally uninhabited. The climate is warm and humid, but there is little fresh water, and the islands are subject to seasonal hurricanes.

SOCIETY

Barbuda was colonized by the British in 1628, and Antigua four years later. In 1685 the Codrington family leased Barbuda from the British, at a rent of one fat pig a year, but it reverted to the crown in the 19th century after the ending of slavery. In 1967 Antigua achieved internal self-government, but in 1979 demanded full independence. At the same time Barbuda sought secession from Antigua. The islands achieved their full independence together in 1981. The British sovereign, represented by a governor-general, is head of state. The parliament has two chambers and consists of an appointed senate and an elected lower house. Most of the population live on the island of Antigua. The official language is English, and the majority of the people are Anglican Christians.

ECONOMY

Tourism accounts for the majority of the islands' revenue; there are numerous white, sandy, palm-fringed beaches, a number of casinos, and some fine 18th-century architecture in the capital, St John's, which also has a splendid cathedral in the gothic style and a deep-water harbor. The roads are surfaced, and there are two airports. Good health-care and social-security schemes have been established. Basic education is free and compulsory, literacy levels are fairly high, and higher education is also available.

NATIONAL DATA – SAINT KITTS AND NEVIS

Land area	261 sq km (101 sq mi)			
Climate		Temperatures		Annual
	Altitude m (ft)	January °C(°F)	July °C(°F)	precipitation mm (in)
Basseterre	15 (49)	24 (75)	27 (81)	1,375 (54.2)
Major physical features highest point: Mount Misery 1,156 m (3,792 ft)				
Population (2006 est.) 39,129				
Form of government multiparty constitutional monarchy with one legislative house				
Armed forces no armed forces				
Capital city Basseterre (12,849)				
Official language English				
Ethnic composition Black 90.5%; Mixed 5.0%; Indian 3.0%; White 1.5%				
Religious affiliations Protestant 76.4%; Roman Catholic 10.7%; other 12.9%				
Currency 1 East Caribbean dollar (ECD) = 100 cents				
Gross domestic product (2002 est.) U.S. $339 million				
Gross domestic product per capita (2005 est.) U.S. $8,200				
Life expectancy at birth male 69.56 yr; female 75.42 yr				
Major resources tourism, coconuts, livestock, sugarcane, vegetables, cotton				

NATIONAL DATA – ANTIGUA AND BARBUDA

Land area	443 sq km (171 sq mi)			
Climate		Temperatures		Annual
	Altitude m (ft)	January °C(°F)	July °C(°F)	precipitation mm (in)
St. John's	10 (33)	25 (77)	28 (83)	1,052 (41.4)
Major physical features highest point: Boggy Peak 405 m (1,330 ft); largest island: Antigua 280 sq km (108 sq mi)				
Population (2006 est.) 69,108				
Form of government multiparty constitutional monarchy with two legislative houses				
Armed forces army 125; navy 45				
Capital city St. John's (25,498)				
Official language English				
Ethnic composition Black 94.4%; Mulatto 3.5%; White 1.3%; others 0.8%				
Religious affiliations Anglican 33%; other Protestants 31%; Roman Catholic 12%; other 23%; none 1%				
Currency 1 East Caribbean dollar (ECD) = 100 cents				
Gross domestic product (2002 est.) U.S. $750 million				
Gross domestic product per capita (2005 est.) U.S. $10,900				
Life expectancy at birth male 69.78 yr; female 74.66 yr				
Major resources tourism, cotton, rum, shellfish, sugarcane, vegetables, fruit, chickens				

Dominica

T HE ISLAND OF DOMINICA IS AN INDEPENDENT republic in the eastern Caribbean, situated in the Leeward Island chain between Guadeloupe and Martinique.

GEOGRAPHY

Dominica has a backbone of steep mountains, rising to Morne Diablotin. Vents and hot springs are a reminder of their volcanic origins. The climate is warm and humid, with a rainy season from June to October, when there is a risk of hurricanes. Rainfall is much higher in the mountains. The island is covered with lush tropical vegetation that supports abundant wildlife. The Morne Trois Pitons National Park is a wildlife preserve and an important tourist attraction.

NATIONAL DATA – DOMINICA

Land area	754 sq km (291 sq mi)			
Climate	Altitude m (ft)	Temperatures January °C(°F)	July °C(°F)	Annual precipitation mm (in)
Roseau	16 (60)	24 (75)	27 (81)	1,981 (77.9)

Major physical features highest point: Morne Diablotin 1,447 m (4,747 ft)
Population (2006 est.) 68,910
Form of government multiparty republic with one legislative house
Armed forces no armed forces
Capital city Roseau (16,582)
Official language English
Ethnic composition Black 91.2%; Mixed 6.6%; Amerindian 1.5%; White 0.5%; others 0.2%
Religious affiliations Roman Catholic 80%; Protestant 16%; other 4%
Currency 1 East Caribbean dollar (ECD) = 100 cents
Gross domestic product (2003 est.) U.S. $384 million
Gross domestic product per capita (2005 est.) U.S. $3,800
Life expectancy at birth male 71.95 yr; female 77.93 yr
Major resources timber, hydropower, bananas, cattle, cocoa, coconuts, citrus fruits, pigs, pumice, tourism, vanilla

Untamed beauty (*above*) Dominica's mountainous interior is covered with dense rainforest; the clouds that shroud its peaks send cascades of water into the valleys below. Some parts are still unexplored.

SOCIETY

For many years the British and the French fought for possession of Dominica, but the British eventually took over in 1783. The island was made a separate colony in 1771, but did not become self-governing until 1967; full independence followed in 1978. Soon afterward the island was devastated by a series of hurricanes. Despite two failed coup attempts, the 1980s were mainly a period of rebuilding. The president is head of state, and appoints a prime minister from the majority party. The president is appointed by parliament, which has a majority of elected members. The people are mostly the descendants of former African slaves, though several hundred Caribs (descendants of the original inhabitants) also live on the island. The official language is English, but a French patois is spoken, and there is a Roman Catholic majority with Protestants next.

ECONOMY

The chief cash crops are bananas, citrus fruits and coconuts, while cocoa, bay leaves and vanilla also produce revenue. Other major exports include galvanized sheets and coconut products. Tourism is an ever-increasing source of further revenue. Food, fuels and most manufactured goods have to be imported. The roads are good, and the capital, Roseau, has a port and two airfields nearby. Medical services are adequate, but disease continues to be a problem. Education is free, and literacy levels are high.

Saint Lucia

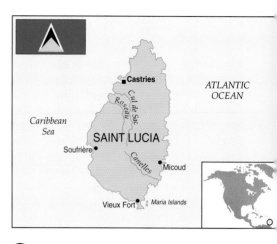

S AINT LUCIA IS A SMALL ISLAND STATE IN THE eastern Caribbean – part of the Lesser Antilles chain that stretches from Puerto Rico to Venezuela.

GEOGRAPHY

The island's rugged landscape was created by volcanic activity. The mountains at its heart rise to Mount Gimie, its highest point. Near the west coast is the volcano Qualibou, which was last active in 1766. It still vents gases, and is responsible for the boiling sulfurous springs that gave the neighboring town of Soufrière its name. The climate is hot and humid, and the volcanic soils are fertile. The lush rainforest that formerly covered the island has been cleared.

NATIONAL DATA – SAINT LUCIA

Land area	606 sq km (234 sq mi)			
Climate	Altitude m (ft)	Temperatures January °C(°F)	July °C(°F)	Annual precipitation mm (in)
Castries	3 (10	25 (77)	27 (81)	2,199 (86.5)

Major physical features highest point: Mount Gimie 950 m (3,117 ft)
Population (2006 est.) 168,458
Form of government multiparty constitutional monarchy with two legislative houses
Armed forces no armed forces
Capital city Castries (12,819)
Official language English
Ethnic composition Black 90%; Mixed 6%; East Indian 3%; White 1%
Religious affiliations Roman Catholic 75%; Protestant 13%; Hindu 1%; other 11%
Currency 1 East Caribbean dollar (ECD) = 100 cents
Gross domestic product (2002 est.) U.S. $866 million
Gross domestic product per capita (2005 est.) U.S. $4,800
Life expectancy at birth male 70.29 yr; female 77.65 yr
Major resources timber, minerals (pumice), geothermal potential, bananas, cocoa, coconuts, fruit, spices, vegetables, tourism

Saint Vincent and the Grenadines

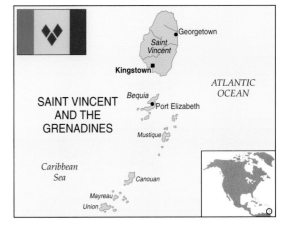

SOCIETY

The indigenous Arawak peoples were driven from Saint Lucia by the Caribs, who were themselves killed off by the diseases introduced by 17th-century French settlers. The island changed hands frequently between Britain and France, but was finally ceded to Britain in 1814. The population are mainly descended from African slaves. Saint Lucia became self-governing in 1967 and independent in 1979. The British sovereign remains head of state, but power lies with the prime minister and cabinet, responsible to an elected two-chamber parliament.

ECONOMY

Saint Lucia's economy depends chiefly on tourism and agriculture. Sugar cane has been replaced as the main cash crop by bananas, which provide some two-thirds of exports. Food crops such as cassava and sweet potatoes are produced by subsistence farmers, and much food has to be imported. Industry is developing, aided by the use of geothermal power. Primary education is free and compulsory.

The Pitons (*below*), twin peaks rising dramatically from the sea to a height of over 750 m (2,560 ft) on the southwest coast of Saint Lucia. Exposed by erosion, they are formed from the lava plugs of volcanoes.

THE ISLAND STATE OF SAINT VINCENT AND the Grenadines forms part of the Windward Islands in the Lesser Antilles and comprises Saint Vincent and all but a few of the Grenadines.

GEOGRAPHY

Saint Vincent, the main island, is a thickly forested cluster of volcanic mountains. Soufrière, its highest peak, is still very much active. It caused severe damage to the island in 1891 and 1902, and resumed activity in 1979.

The Grenadines are a string of low-lying coral islands, which extend south of Saint Vincent toward Grenada. The climate is hot with seasonal rainfall and occasional destructive hurricanes.

SOCIETY

When Christopher Columbus (1451–1506) visited the islands on Saint Vincent's Day in 1498, he found them inhabited by the fierce Caribs, who successfully resisted European settlement until the 18th century. After a brief period in French hands, Saint Vincent came under British rule from 1783. African slaves were brought in to work the fertile plantations. After the abolition of slavery, Indonesian and Portuguese laborers were also settled.

In 1958 Saint Vincent joined the Federation of the West Indies. In 1979 it became an independent constitutional monarchy, headed by a governor-general representing the British sovereign. The prime minister is the leader of the majority party.

ECONOMY

The economy of the islands rests largely on agriculture, mostly small-scale or subsistence farming on the lower slopes or terraces. Bananas are the chief cash crop and export, followed by arrowroot – used in the making of medicines, fine flour and computer paper. The 1979 volcanic eruption, followed by a hurricane in 1980, devastated agriculture and discouraged tourism, but both have since recovered. Industry is confined to food processing.

NATIONAL DATA – SAINT VINCENT AND THE GRENADINES

Land area	389 sq km (150 sq mi)			

Climate	Altitude m (ft)	Temperatures January °C(°F)	July °C(°F)	Annual precipitation mm (in)
Kingstown	1 (3)	26 (45)	28 (77)	1,524 (60)

Major physical features highest point: Soufrière 1,234 m (4,048 ft); largest island: Saint Vincent 344 sq km (133 sq mi)

Population (2006 est.) 117,848

Form of government multiparty constitutional monarchy with one legislative house

Armed forces no armed forces

Capital city Kingstown (18,323)

Official language English

Ethnic composition Black 66%; Mixed 19%; East Indian 6%; Carib Amerindian 2%; other 7%

Religious affiliations Anglican 47%; Methodist 28%; Roman Catholic 13%; Hindu, Seventh-Day Adventist; other Protestant 12%

Currency 1 East Caribbean dollar (ECD) = 100 cents

Gross domestic product (2002 est.) U.S. $342 million

Gross domestic product per capita (2005 est.) U.S. $3,600

Life expectancy at birth male 71.99 yr; female 75.77 yr

Major resources hydropower, arrowroot, starch, bananas, coconuts, spices, taro, timber, tourism

Barbados

BARBADOS

BARBADOS IS AN ISLAND STATE IN THE Caribbean, lying outside the Lesser Antilles some distance from Venezuela.

GEOGRAPHY

The island is founded on coral deposits around a rocky core. The rolling terrain rises to a hilly region in the north and center. The west and south coasts have extensive sandy beaches. The climate is hot with seasonal rains and occasional hurricanes. Trees are plentiful, chiefly mahogany, palm and other tropical varieties, but there are no large forests, and not much wildlife apart from birds.

SOCIETY

Arawak peoples may once have lived on Barbados, but modern settlement was begun by British sugar planters in the 17th century. Present-day Barbadians are

Colonial and modern (*above*) British colonial architecture rubs shoulders with modern office blocks and hotels in Bridgetown, the capital, main port and tourist center of Barbados.

descended from their slaves, liberated in the 1830s. Barbados is culturally the most British of the Caribbean states. From the 1930s onward, the island was given increasing self-government. It headed the short-lived West Indies Federation before becoming fully independent in 1966. The British sovereign, who is represented by a governor-general, remains the head of state, but power resides with the two-chamber parliament, which is constituted on British lines; it is headed by a prime minister and cabinet.

ECONOMY

Agriculture is dominated by sugar cane production, to which most of the flat land is devoted; despite a recent decline it remains by far the largest export. Sugar cane and its by-products, molasses and rum, are processed on the island. Diversification into other cash crops such as citrus fruits and cotton is officially encouraged in order to end the reliance on a single crop, as is fishing. Small petroleum and gas reserves are exploited. Light industries produce electrical components, clothing and many other items for export and for domestic use. Tourism, however, is now the single largest earner, visitors being attracted to the glittering white beaches and the coral reefs. Education is free and compulsory up to the age of 14, and available up to degree level. Social-security and health-care systems have also been introduced.

NATIONAL DATA – BARBADOS

Land area	431 sq km (166 sq mi)			
Climate		Temperatures		Annual
	Altitude m (ft)	January °C(°F)	July °C(°F)	precipitation mm (in)
Bridgetown	55 (180)	25 (77)	27 (81)	1,278 (50)

Major physical features highest point: Mount Hillaby 337 m (1,104 ft)

Population (2006 est.) 279,912

Form of government multiparty constitutional monarchy with two legislative houses

Armed forces army 500; navy 110

Capital city Bridgetown (98,947)

Official language English

Ethnic composition Black 90%; White 4%; Asian and mixed 6%

Religious affiliations Protestant 31%; other Anglican 29%; other Christian 24%; Roman Catholic 4%; other 12%

Currency 1 Barbadian dollar (BDS) = 100 cents

Gross domestic product (2006 est.) $5.108 billion

Gross domestic product per capita (2006 est.) $18,200

Life expectancy at birth male 70.79 yr; female 74.82 yr

Major resources bananas, fish, goats, limestone, natural gas, petroleum, onions, pigs, poultry, shellfish, sheep, sugarcane, vegetables, tourism, cotton

Grenada

GRENADA

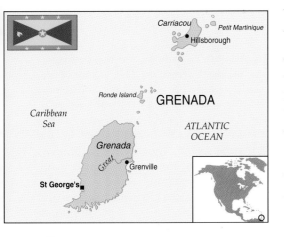

G RENADA IS AN ISLAND STATE IN THE LESSER Antilles, centered on the most southerly of the Windward Islands off the coast of Venezuela. The territory also includes the southernmost islands of the neighboring Grenadines group.

GEOGRAPHY

Grenada itself consists largely of a ridge of volcanic mountain peaks. Their slopes are thickly forested, thanks to the hot, humid climate and the fertile volcanic soils. The southern coasts have many bays and harbors, including that of the capital, Saint George's.

SOCIETY

Grenada's first inhabitants were the Arawak peoples, exterminated by the fierce Caribs whom Christopher Columbus (1451–1506) encountered in 1498. The island was settled by the French in 1650, but was seized by Britain in the 18th century. The people are mostly descended from former African slaves, and some still speak a French patois. Grenada became fully independent in 1974 with a parliamentary democracy based on the British system. In 1979 the New Jewel Movement, led by Maurice Bishop (1944–83), seized power in a left-wing coup. Bishop made some economic progress, as well as establishing links with Cuba, Libya and Algeria, but in 1983 he was deposed and murdered by rival Marxists. The United States, with some neighboring Caribbean states, launched a successful invasion. Constitutional government was restored, and the governor-general instituted free elections to the lower house of the two-chamber parliament. US troops finally withdrew in 1995.

ECONOMY

Grenada's economy depends almost entirely on agriculture and tourism. Agriculture is mostly small scale except for a few cooperatives. The main cash crops and exports are bananas, cocoa, and the spices such as nutmeg and mace that once earned Grenada the name Isle of Spice. There is little industry apart from various food-processing plants. Tourism, shattered by the 1979 coup, is gradually recovering. There is an effective health-care program and free education facilities.

St George's, Grenada's capital, is attractively set on wooded hillsides surrounding a horseshoe bay. Traditional exports of spices, bananas and cocoa still leave the island from St George's port.

NATIONAL DATA – GRENADA

Land area	344 sq km (133 sq mi)			
Climate	Altitude m (ft)	Temperatures January °C(°F)	July °C(°F)	Annual precipitation mm (in)
St. George's	1 (3)	25 (77)	27 (81)	1,560 (61.4)
Major physical features	highest point: Mount St. Catherine 840 m (2,757 ft)			
Population	(2006 est.) 89,703			
Form of government	multiparty constitutional monarchy with two legislative houses			
Armed forces	no armed forces			
Capital city	St. George's (4,298)			
Official language	English			
Ethnic composition	Black 82%; Mixed Black and European 13%; European and East Indian 5%			
Religious affiliations	Roman Catholic 53%; Anglican 13.8%; other Protestant 33.2%			
Currency	1 East Caribbean dollar (ECD) = 100 cents			
Gross domestic product	(2002 est.) U.S. $440 million			
Gross domestic product per capita	(2005 est.) U.S. $3,900			
Life expectancy at birth	male 63.06 yr; female 66.68 yr			
Major resources	timber, tropical fruits, cocoa, vegetables, mace, nutmeg, tourism			

Trinidad and Tobago

REPUBLIC OF TRINIDAD AND TOBAGO

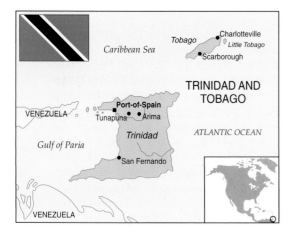

TRINIDAD AND TOBAGO IS MADE UP OF TWO islands lying between the Caribbean Sea and the Atlantic Ocean just off the coast of Venezuela.

GEOGRAPHY

Venezuela's coastal mountains effectively continue into Trinidad's Northern Range and Tobago's Main Ridge. The Northern Range rises steeply to the Cerro del Aripo, and its rivers create some spectacular waterfalls. Otherwise Trinidad is mostly rolling or flat, with many swampy areas along the coast. South of Tobago's Main Ridge is a coral plain that extends out to sea, creating some magnificent reefs. The climate is hot and humid, and much of Trinidad is still covered with thick rainforest. Both islands are rich in bird life, especially flamingoes and the scarlet ibis.

NATIONAL DATA - TRINIDAD AND TOBAGO

Land area	5,128 sq km (1,980 sq mi)			
Climate		Temperatures		Annual
	Altitude m (ft)	January °C(°F)	July °C(°F)	precipitation mm (in)
Port-of-Spain	20 (66)	26 (77)	27 (81)	1,408 (55.4)

Major physical features highest point: Mount Aripo 940 m (3,084 ft)

Population (2006 est.) 1,065,842

Form of government multiparty republic with two legislative houses

Armed forces army 2,000 navy 700

Capital city Port-of-Spain (49,865)

Official language English

Ethnic composition Indian (South Asian) 40%; African 37.5%; Mixed 20.5%; other 1.2%; unspecified 0.8%

Religious affiliations Roman Catholic 26%; Hindu 22.5%; Anglican 7.8%; Baptist 7.2%; Pentecostal 6.8%; other Christian 5.8%; Muslim 5.8%; Seventh-Day Adventist 4%; other 10.8%; unspecified 1.4%; none 1.9%

Currency 1 Trinidad and Tobago dollar (TTD) = 100 cents

Gross domestic product (2006 est.) U.S. $20.99 billion

Gross domestic product per capita (2006 est.) U.S. $19,700

Life expectancy at birth male 65.71 yr; female 67.86 yr

Major resources asphalt, bananas, citrus fruits, cocoa, coffee, oil and natural gas, rice, sugarcane, tourism

SOCIETY

The original inhabitants were the Arawak peoples, exterminated on Tobago by the fierce Caribs. After colonization by Spain and France, both peoples were wiped out by disease and enslavement on tobacco and sugar plantations; they were replaced by African slaves. In the early 19th century both islands were ceded to Britain. With the ending of slavery laborers were brought in from Asia to work the plantations. Full independence was granted in 1962. In 1976 the islands became a Commonwealth republic, which survived an abortive coup led by Asian Trinidadians in 1980. The islands' ethnic mix of peoples has created a colorful local culture including carnival and calypso music. There is a large Hindu minority.

ECONOMY

The islands are among the most prosperous in the Caribbean. The economy depends heavily on the refining and export of the extensive petroleum reserves, but a fall in petroleum revenues has led to increased diversification in the industrial sector. Agriculture concentrates on cash crops for export – principally sugar cane, cocoa and coffee – with less emphasis on staple crops such as rice and vegetables. Tourism is an important source of revenue, and there are good education and health care facilities.

Tropical mix Trinidad is full of variety. Industrial in parts, it also has fertile agricultural land. In the towns, mosques and temples, English churches, the sounds of calypso and steel bands, all combine to form a vibrant cosmopolitan society.

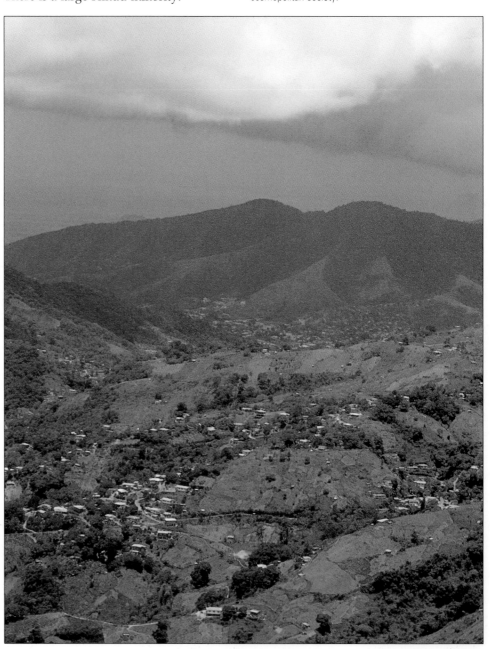

Dependencies in the region

TURKS AND CAICOS ISLANDS
UNITED KINGDOM

The Turks and Caicos Islands are a British colony consisting of a string of about 30 small islands at the end of the Bahamas group, which they resemble in landscape and climate. Rainfall, however, is scantier except during hurricanes. The terrain is dry, and the vegetation is scrubby except in the coastal saltmarshes.

The inhabitants are descended from African slaves imported by the British in the 17th century to work saltpans. The islands have on several occasions been annexed both to the Bahamas and to Jamaica. In 1976 they were given self-government under the British crown, but their declared goal of full independence has so far been delayed indefinitely.

Salt production has become uneconomic, and agriculture is hindered by lack of water. Thus the island's economy rests largely on fishing, chiefly crayfish and conch, and on the development of tourism. Many offshore financial companies are also based here, taking advantage of the islands' status as a tax haven. Education and health care are government priorities, and are generally good.

BERMUDA
UNITED KINGDOM

The British colony of Bermuda occupies a string of coral islands in the North Atlantic. About 20 are inhabited, the largest being Great Bermuda. The climate is warm, mild and humid, with moderate rainfall. The vegetation is luxuriant, with whispering pines and many flowering plants. Lizards are the only reptiles, and there are no native mammals.

In 1609 the crew of a shipwrecked English vessel landed in the islands, and in 1612 Bermuda became a colony of the British Virginia Company. It passed to the crown in 1684, but remained self-governing. In 1941 Bermuda leased air and naval stations to the United States. The 1968 constitution allowed greater self-government. The 40-member House of Assembly is directly elected, but the 11-member senate is chosen by the governor, who is appointed by the British crown. The majority of the people are descended from former African slaves, others from British or Portuguese settlers. The official language is English, but some Portuguese is spoken. Christianity is the dominant religion; the majority of people belong to one of the denominations of the Protestant churches.

There is little scope for much agriculture or for large-scale industry, and food and fuel must be imported. The colony relies almost entirely on the income from tourism, and from military and financial interests. The standard of living is high. There are no railroads, but there is a good road network, which links the islands by bridges and ferries. The main port at Hamilton is well served by international shipping lines, and the American airport base at Kindley Field serves both military and international traffic. Health care is readily available, and there is social security for all citizens. Education is free and compulsory, and literacy levels are high. Overseas scholarships are available for higher education.

CAYMAN ISLANDS
UNITED KINGDOM

The Cayman Islands are a British colony in the Caribbean to the northwest of Jamaica. Grand Cayman, Cayman Brac and Little Cayman are low-lying coral islands on a rocky base. The coasts are lined with magnificent beaches and reefs, and the tropical climate is pleasantly cooled by the sea.

Though visited by Christopher Columbus (1451–1506), the islands remained no more than a pirate lair until ceded to Britain with Jamaica in 1670. Later settlers from Jamaica lived mostly from turtle fishing. The Caymans became a separate colony after Jamaica's independence in 1962. They enjoy some local autonomy under a governor, but have yet to become fully independent.

A thriving British colony, Bermuda's picturesque towns are a magnet for tourists; lenient tax laws bolster the economy by attracting foreign investors.

There is little agriculture or industry, although some turtle farming is being developed. Tourism is one major earner, and offshore finance another, encouraged by liberal tax and confidentiality laws. The Caymans now house hundreds of financial companies. These help to create high living standards, and good health and education services.

PUERTO RICO
UNITED STATES

Puerto Rico, an island state in the West Indies, is a self-governing commonwealth in association with the United States.

Most of the main island is upland, rising from the narrow southern coastal plains to the Cordillera Central, a mountain range that runs along the island like a backbone; Cerro del Punta is the highest peak. To the north it flattens out into hill country, through which the island's main rivers flow down to the much wider northern coastal plain. The two adjoining islands to the east are similarly rugged.

The climate is mainly warm and humid, but rainfall is unevenly distributed. The north receives most of the island's rainfall, and is dotted with marshlands, small lakes and mangrove swamps. By contrast the south, in the rainshadow of the mountains, is too dry for farming without irrigation, and in the southwest there are several saltpans. The island was once covered in tropical hardwood rainforest, but so much of this has been cleared that it now survives only in the Caribbean National Forest enclave. Wildlife consists mostly of small reptiles and birds.

When Christopher Columbus (1451–1506) reached Puerto Rico, the islands were already inhabited by the Arawaks and Caribs. Spanish settlers wiped out both these peoples, by disease and en-slavement in gold mines and plantations. Puerto Rico, unlike nearby Hispaniola, remained Spanish throughout the 19th century. African slaves were introduced to work the plantations until slavery was abolished in the 19th century.

The island was granted self-govern-ment in 1897, only to be occupied and ceded to the United States in the Spanish–American war (1898); its people were given United States citizenship in 1919. Various independence movements developed, but disagreed on whether to become wholly independent or seek statehood within the United States. In spite of strong American influences, the culture has remained primarily Hispanic.

After World War II the office of gover-nor was made elective, and in 1952 a commonwealth was established. How-ever, a fierce debate still continues re-garding possible independence or state-hood. The island returns a nonvoting member to the United States House of Representatives, but elects its own gover-nor, who appoints an executive cabinet.

El Morro (*above*) One of the two 16th-century forts on the clifftops of San Juan, Puerto Rico, this is an impressive reminder of the days of Spanish control.

Holiday bliss (*right*) The clear blue waters and sandy beaches of the United States Virgin Islands have brought a thriving tourist trade. The fine harbor at St Thomas, once frequented by pirates, is now the haunt of luxury cruisers.

Puerto Rico's close links with the United States have made it one of the most prosperous Caribbean countries. Since World War II investment policies have made industry more important than agriculture. The main cash crops are sugar cane, coffee and tobacco, along with dairy products and livestock, but large amounts of food have to be im-ported. There are few mineral resources, although copper and nickel deposits have been found. Industry is based on a wide range of products, from petrochemicals, plastics and pharmaceuticals to textiles, machinery and electronics. Considerable amounts are exported, chiefly to the United States, but there is also a high volume of imports. Tourism, however, has become a major earner.

Social-security and welfare benefits operate to United States standards, and health care is improving rapidly toward that goal. Education is a priority, freely available at all levels, up to and including university.

VIRGIN ISLANDS OF THE UNITED STATES
UNITED STATES

The United States Virgin Islands are an unincorporated territory of the United States, comprising Saint Croix, Saint John and Saint Thomas and some 50 smaller islets of the Virgin Islands group to the east of Puerto Rico. Their rugged land-scapes are formed by the peaks of an undersea mountain range.

Christopher Columbus (1451–1506) named the islands after Saint Ursula's martyred virgins. They were settled by the Spanish and French before being acquired by Denmark in the 17th and 18th centuries. African slaves were introduced to work the sugar cane plantations. In 1917 the islands were bought by the United States. In 1981 a constitution, adopted by referendum, introduced self-government to the islands by an elected governor and senate.

were driven out by the Caribs, who were in turn wiped out by the Spanish. The islands were used as havens for piracy until they were annexed by Britain in the 17th century. They now have effective self-government through an elected council headed by an appointed governor.

Tourism is expanding to dominate the economy; the sheltered waters around the islands attract a great number of visitors who stay on yachts. Agriculture produces livestock for local use, and cash and food crops such as bananas and sugar cane. Fishing is also important, but manufacturing industry is very small. Despite water shortages, health care is excellent, as is primary education.

ANGUILLA
UNITED KINGDOM

Anguilla is the northernmost of the Leeward Islands in the eastern Caribbean. It consists of a long, thin coral island with a bare landscape of scrubland and occasional fruit plantations. The climate is warm and humid, and the island is subject to hurricanes.

Anguilla has been a British colony since 1650. For several years it was part of the federal state of St Kitts–Nevis–Anguilla, but since 1980 it has been a dependent territory of the United Kingdom. The governor is appointed by the crown, but internal affairs are managed by an elected assembly. Most Anguillans are descended from former African slaves. English is the chief language, and the main religions are Anglican and Methodist Christianity.

The chief exports are fish, salt and livestock. There are significant saltpeter deposits, and tourism is being developed, along with transportation and communications. Housing, health care and educational facilities are good.

MONTSERRAT
UNITED KINGDOM

The British colony of Montserrat is one of the Leeward Islands in the eastern Caribbean. Volcanic peaks, several of them active, dominate a mountainous landscape of pastureland and lush vegetation. The climate is warm and humid, with seasonal hurricanes.

Montserrat was colonized in the 17th century by Irish settlers from nearby Saint Kitts. It belonged to the Federation of the West Indies between 1958 and 1962, and hopes to gain full independence from the United Kingdom. The governor, who is appointed by the crown, is head of an

executive council and a partly elected legislative council. Most people are descendants of former African slaves. English is the chief language, and Christianity, with its various denominations, is the predominant religion.

The chief exports are vegetables and cotton, but tourism is a growing source of revenue. There are surfaced roads, a seaport at the capital, Plymouth, and an air link with neighboring Antigua. Sanitation is good and health care is improving. The island was devastated in June 1997 when the Soufriere Hills volcano erupted.

ARUBA
THE NETHERLANDS

The island of Aruba lies at the southwestern end of the Lesser Antilles off the Venezuelan coast. The terrain is mostly flat, but is generously scattered with volcanic boulders. The climate is warm and humid, but rainfall is generally poor, and vegetation is sparse apart from cacti and succulents.

Aruba became part of the Netherlands Antilles in 1634, but seceded in 1986, becoming an internally self-governing part of the Netherlands. The monarch is represented by a governor, but the 21 members of the States of Aruba are directly elected. Arubans are mainly of African, European and Asian descent, and most are Roman Catholic. Dutch is the official language, but Papiamento, a Spanish creole, is widely spoken.

Tourism is almost the sole source of revenue. Aruba has surfaced roads, deepwater harbors and an international airport. Most drinking water is distilled from seawater. Education is not compulsory, but literacy levels are high.

THE NETHERLANDS ANTILLES
THE NETHERLANDS

The Netherlands Antilles are made up of two island groups within the Lesser Antilles. The southern group includes Curaçao and Bonaire – and formerly also Aruba – off the Venezuelan coast. The northern group is part of the Leeward Islands, and comprises Sint Eustatius, Saba and the southern part of Sint Maarten, the rest of which is French. The southern islands are generally flat, but those in the north are mountainous and volcanic. The climate is warm and humid; Curaçao and Bonaire have poor rainfall, while the northern islands are subject to seasonal hurricanes. The vegetation is sparse apart from cacti and succulents.

Agriculture has shifted from sugar cane to food crops and livestock, mostly for domestic use. Manufacturing ranges from rum distilling to petroleum refining. Tourism has been developed since 1945, and now dominates the economy, with game fishing being one of the major attractions. Education and health care are widely available, reflecting the islands' relative prosperity.

BRITISH VIRGIN ISLANDS
UNITED KINGDOM

The British Virgin Islands are a British colony comprising Tortola, Anegada, Virgin Gorda, Jost Van Dyke and more than 30 islets of the Virgin Islands group in the Caribbean. Most are the peaks of a submerged mountain chain, and share a subtropical climate. The landscape is mainly hilly, with sandy beaches and coral reefs around the coasts. Anegada, however, is very flat.

The original inhabitants, the Arawaks,

South America

COUNTRIES IN THE REGION

COLOMBIA · VENEZUELA · GUYANA · SURINAME

BOLIVIA · ECUADOR · PERU · BRAZIL · PARAGUAY

CHILE · ARGENTINA · URUGUAY

DEPENDENCIES IN THE REGION

FALKLAND ISLANDS · FRENCH GUIANA

- ■ capital city
- ● major town

height of land (meters)

5000
3000
2000
1000
500
200
0

- ▲ mountain peak

South American specialities Vicuna in Lanca National Park, Chile. Vicuna are the smallest of the South American members of the camel family, which includes llamas, guanacos and alpacas. South America contains an enormous variety of animals and plants that are found nowhere else in the world.

103

Colombia

REPUBLIC OF COLOMBIA

COLOMBIA IN THE NORTHWEST OF SOUTH America takes its name from the explorer Christopher Columbus (1451–1506). This beautiful country has a rich colonial heritage, but is plagued by internal strife.

GEOGRAPHY

Colombia's short frontier with Panama to the northwest divides the western (Pacific) coast from the northern (Caribbean) shoreline. Snowcapped Pico Cristóbal Colón forms part of an isolated massif overlooking the Caribbean to the north and a swampy basin to the west.

The two main rivers that flow into the basin from the south, through deep trenches separating the three main ranges of the mighty Colombian Andes, are the Magdalena and the Cauca. The ranges, which run roughly northeastward from the southeastern border with Ecuador, are the Cordillera Occidental, overlooking the narrow Pacific coastal plain, the volcanic Cordillera Central east of the Cauca river, and farther east still, beyond the broader Magdalena valley, the Cordillera Oriental. This divides at the Venezuelan border. Recent earthquakes and a devastating eruption in 1985 show that the area is geologically active.

More than half the country consists of a sparsely inhabited plain that drops gently southeast from the snowcapped Andean ranges toward the Peruvian and Brazilian borders. The northern lowlands – the *llanos* – are drained by the Orinoco river, which runs along the western border with Venezuela. The southern lowlands form part of the Amazon river basin.

Colombia's climate is very varied. The Caribbean lowlands are relatively dry with two short rainy seasons, though they are vulnerable to hurricanes. Rainfall is highest on the Pacific coast, in the central valleys and in the Amazonian lowlands, where tropical rainforest is characteristic. This gives way to savanna on the *llanos* and in the northern valleys. The vegetation changes with altitude, giving way on higher mountain slopes to cloud forests and tundra.

Human settlement and agriculture have stripped away much of the original forest cover, but the moist mountain forests contain many orchids and epiphytes, as well as balatas, from which nonelastic rubber is obtained, and ivory nuts, used for making ornaments. Animal life is extremely varied. Typical examples include tapirs, sloths, anteaters, jaguars and howler monkeys. Hummingbirds tap the nectar of innumerable flowers, while the rivers support caimans and capybaras, large, semiaquatic rodents.

SOCIETY

In the early 16th century Spanish invaders overcame the sophisticated farming and trading culture of the indigenous Chibcha peoples, craftsmen highly skilled at working gold and embellishing their artifacts with jewels. The colonists founded the city of Bogotá in the Cordillera Oriental, and Colombia became the heart of the Viceroyalty of New Granada. Its independence was gained after the Spanish defeat at Boyacá in 1819 and until 1830 it was part of the Republic of Gran Colombia, established by Simón Bolívar (1783–1830), which included modern Ecuador, Venezuela and Panama.

The modern history of Colombia has been one of continual unrest, beginning with a revolt in 1840 and the emergence of bitterly opposed liberal and conservative

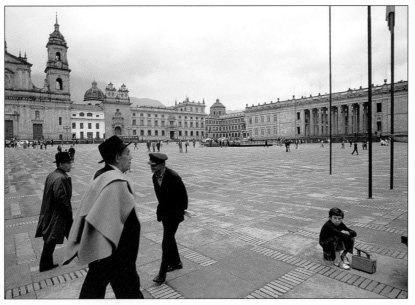

The beauty of the high Andes (*left*) The river Cauca has carved a deep valley through the mountains on its journey north to enter the Caribbean Sea.

A pre-Columbian statue (*below*) from high in the Andes. Its fangs, and the tiny figure clasped in its hands, are indications that it is a sacred figure.

Despite their impressive civic architecture – a legacy from colonial days – Colombia's cities are beset by unemployment and housing shortages. The crime and violence that arise from the illegal drug trade cast a long shadow over the lives of ordinary people.

ural gas reserves and produces 60 percent of the world's emeralds. In the late 1990s, the government began to privatize coal, oil and electricity production. Production and investment by foreign companies are seriously affected by guerrilla activity. Manufacturing declined in the late 1990s after import subsidies were largely removed. Coffee and food processing are important, followed by chemicals and motor vehicles. Colombia has a well-developed telecommunications sector and this was expected to expand strongly into the 2000s.

Overland transportation is hampered by the difficult terrain, and most roads are unsurfaced, but buses and trucks carry more passengers and freight than the limited railroad system. The thriving airline network is heavily used, with six international airports.

Newspapers are the dominant news media. They are generally free of government censorship, and tend to reflect party loyalties. Radio and television broadcasting are subject to stricter controls. Welfare provision includes a range of unemployment, maternity and disability benefits, and retirement pensions. Health facilities are limited, but there is an emphasis on health education and improving sanitation. Among the poor, however, malnutrition and inadequate housing threaten general health. Primary education is free and compulsory for five years, and literacy is relatively high.

Colombia's colonial past is reflected in today's population, the majority of whom are of mixed ancestry. Most of these are mestizos of European and Amerindian descent, but mulattoes (of mixed African and European origin) form the majority along the coast. There is a large white minority, and small numbers of blacks and Amerindians, the latter divided into as many as 400 different groups. Though Spanish is the official language, many Amerindian languages are also spoken. The vast majority of the population are Roman Catholic.

ECONOMY

The Colombian economy is founded largely on private enterprise, and though mainly agricultural it is becoming increasingly industrialized.

Only a small portion of the land is cultivated, but the crops produced are vital to the economy. The wide range of climate and terrain gives rise to a great range of food crops, from bananas and cassava in the lowlands to maize and potatoes in the mountains. Coffee accounts for one-third of all official export revenue, but cannabis and coca, from which cocaine is derived, provide the drug cartels with twice this amount.

Nearly half the land area is forested, and the lumber industry is growing fast. Much of the remaining available land is occupied by pasture, mostly for cattle. River fishing, once plentiful, has suffered from silting and pollution, and the potential for offshore fisheries has not yet been fully developed. Colombia has the largest coal reserves in Latin America, huge nat-

political factions: a liberal uprising in 1899 triggered the murderous War of the Thousand Days, which lasted until 1903. In 1948 La Violencia broke out – a long struggle between the factions that led in 1958 to the creation of the National Front. In this uneasy coalition, presidents from each faction were elected in rotation. Multiparty elections since 1974 have brought some prosperity but little peace. Guerrilla groups and paramilitaries have grown in strength through the 1990s and have been responsible for large-scale violence and thousands of civilian deaths. Much of their income comes from drug trafficking and extortion.

Legislative power rests with a congress consisting of a senate and a house of representatives, each of which is directly elected every four years. The head of state is the president, elected for a single four-year term; he appoints the cabinet.

NATIONAL DATA – COLOMBIA

Land area	1,038,700 sq km (401,044 sq mi)			

Climate		Temperatures		Annual
	Altitude m (ft)	January °C(°F)	July °C(°F)	precipitation mm (in)
Bogotá	2,645 (8,678)	11 (55)	11 (55)	799 (31.4)

Major physical features	highest point: Pico Cristóbal Colón 5,800 m (19,029 ft)

Population	(2006 est.) 43,593,035

Form of government	multiparty republic with two legislative houses

Armed forces	army 178,000; navy 22,000; air force 7,000

Largest cities	Bogotá (capital - 7,363, 492); Cali (2,498,074); Medellín (2,042,093); Barranquilla (1,429,031)

Official language	Spanish

Ethnic composition	Mestizo 58%; White 20%; Mulatto 14%; Black 4%; Mixed Black-Amerindian 3%; Amerindian 1%

Religious affiliations	Roman Catholic 90%; other 10%

Currency	1 Colombian peso (COP) = 100 centavos

Gross domestic product	(2006) U.S. $366.7 billion

Gross domestic product per capita	(2006) U.S. $8,400

Life expectancy at birth	male 68.15 yr; female 75.96 yr

Major resources	petroleum, natural gas, coal, iron ore, nickel, gold, copper, , hydropower, bananas, cassava, cattle, coffee, cut flowers, emeralds, maize/corn, platinum, potatoes, rice, silver, sugarcane, sorghum, soybeans, timber

Venezuela

REPUBLIC OF VENEZUELA

V ENEZUELA IS A FEDERAL REPUBLIC ON THE northern (Caribbean) coast of South America, bordering Colombia to the west, Brazil to the south and Guyana to the east. Once a poor country, it has been radically transformed by the harnessing of its vast petroleum reserves – though not all sections of the population have shared in this newfound wealth.

GEOGRAPHY

The landscape of northwestern Venezuela is shaped by two northern branches of the Andes. The western branch, the Sierra de Perijá, defines part of the Colombian frontier. The eastern branch, the Cordillera de Mérida, enters Venezuela from the south and swings northeast toward the island-fringed Caribbean coast; it includes Venezuela's highest peak, Pico Bolívar. Between the two Andean ranges lie the swampy Maracaibo Lowlands, surrounding the huge, shallow Lake Maracaibo – an irregular extension of the Gulf of Venezuela.

Further mountain ranges run eastward along the coast from the Cordillera de Mérida. The basins and valleys between them shelter several important cities, including the capital, Caracas. South of the mountains are the extensive lowland plains, or *llanos*, of the Orinoco basin, which cross the country from Colombia to the Orinoco Delta area near the Guyanan border. Farther south is the vast, irregular granite plateau of the Guiana Highlands, where the Orinoco river rises.

The climate is generally tropical with a marked rainy season from April to October. However, the mountain areas are cooler, and rainfall varies enormously. The driest areas are in the lee of mountains, especially on the coast; the Orinoco Delta receives the highest rainfall. The *llanos* are flooded during the rainy season and arid for much of the rest of the year.

Forests still cover much of the country, ranging from semitropical evergreen in the northern mountains to tropical rainforest around the Orinoco Delta and in the far south. Wild grassland covers most of the rest of the land, including the extensive *llanos* areas, the Guiana Highlands and the high alpine meadows of the Andean mountain ranges.

The range of altitudes has created an enormous diversity in both the plant and the animal life. Human settlement has confined many animals to more sparsely populated areas, but they remain varied nonetheless, and include bears, caimans, ocelots, opossums and peccaries. Among the more remarkable bird species is the strange cave-dwelling oilbird. Manatees and dolphins are found in coastal waters. The threat to the many rare species has long been recognized, and hunting is forbidden to all except Amerindians, who depend on it for their way of life.

SOCIETY

When European explorers first arrived in the 15th century, they found a number of different Amerindian peoples in the area, including a sophisticated farming culture in the Andean highlands. In 1523 the Spaniards founded the first European settlement, situated at Cumaná, and a colonial economy was established using local slave labor.

Until the beginning of the 19th century, Venezuela was effectively in the hands of Spanish colonial administrators and priests. It was in Venezuela that Francisco Miranda (1750–1816) and his successor Simón Bolívar (1783–1830) began South America's struggle for independence. After the final defeat of the Spanish at Carabobo in 1821, Venezuela, Colombia and Ecuador were briefly united in the Republic of Gran Colombia. Venezuela left the union in 1829, and for more than a century was ruled by military dictators.

The establishment of a major petroleum industry in the 1920s initiated an era of greater prosperity and growth. Between 1945 and 1948, Rómulo Betancourt (1908–81) paved the way for a short-lived democratic radical government. In 1959, after a further period of dictatorship, he established a second, more moderate administration, which laid the foundations for lasting democracy. In 1990 the government agreed to United Nations mediation regarding Venezuela's claim to parts of neighboring Guyana. An austerity program was announced in 1994.

The Guiana Highlands, Venezuela (*above*) This vast area of rolling savanna and sandstone outcrops – occupying nearly half the country – is largely uninhabited and unexplored. Eighty-five percent of Venezuelans live in urban areas.

The petroleum industry (*below*) is the mainstay of the Venezuelan economy, producing about three-quarters of all government revenue. The country's largest oilfields are located in Lake Maracaibo, a shallow inland sea on the northwest coast.

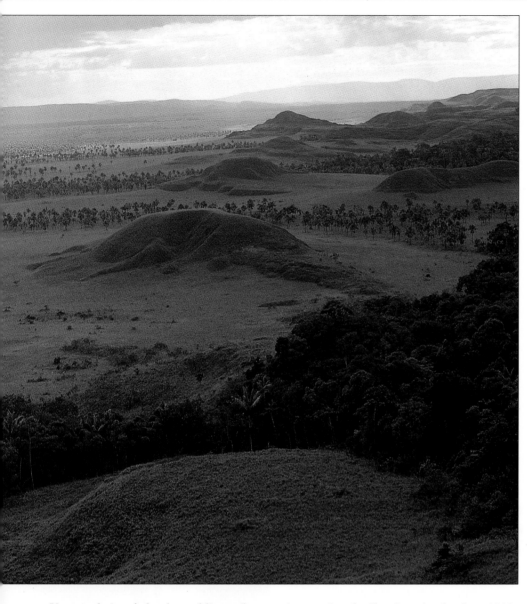

the *llanos* provide good grazing.

Fishing potential is considerable but largely unexploited. However, anchovies, sardines and shellfish are caught off the coast by local fishing fleets. Most forestry land is in government ownership.

The road network is well developed in the north, with three major highways. Railroads make only a small contribution, chiefly to the iron and steel industries. Shipping carries most overseas trade, and there are dredged channels into Lake Maracaibo and the Orinoco river. There are three national airlines and some 30 airports operating scheduled flights, including internal links to areas that would otherwise be inaccessible. The press, radio and television are privately owned and operated, and are not subject to government censorship.

The welfare system is well established, with social security for the old, disabled and unemployed, and increasing public health provision. The quality of health care is good, but the rising population and a general move to urban centers have created shanty towns with inadequate housing and sanitation.

Primary education is free and compulsory, but secondary education is less well developed; however, adult literacy is very high. Higher education is free to students who qualify to enter a university, teacher-training college or polytechnic institute.

Venezuela is a federal republic made up of 23 states, a federal district (around Caracas), two federal territories and 34 dependencies. Legislative power rests with the two-chamber parliament, whose members belong to several political parties. The president, the 49 senators and the 201 deputies are each directly elected every five years.

The majority of the population are mestizos of mixed European and Amerindian origin, and there are substantial white and black minorities. Few pure Amerindians remain, but they have preserved as many as 25 native languages. Although Spanish is the official language, English and Italian are also spoken. The Venezuelan constitution guarantees freedom of religion to its citizens, but the great majority of the population belong to the Roman Catholic church.

ECONOMY

Venezuela's economy is heavily dependent on its petroleum and natural resources, which account for the majority of its export income. Sharply fluctuating international oil prices in the late 1990s had severe economic effects, forcing the government to impose austerity measures and hampering efforts to modernize the economy through diversification and free market policies. Petroleum, natural gas and hydroelectricity enable Venezuela to be self-sufficient in meeting industrial and domestic power needs. The petrochemical industry is based mainly around Morón, to the west of Caracas, and near the Gulf of Venezuela and Lake Maracaibo. The city of Maracaibo on its western shore produces food, pharmaceuticals, machinery and electrical equipment. Ciudad Guayana in the east is a major center for the processing of local iron ores. Significant amounts of diamonds and gold are also extracted.

Only a small proportion of the land is used for agriculture. Despite various agrarian reforms, modernization programs and extensive irrigation in the *llanos*, some food still has to be imported. The principal food crops include bananas, sorghum and maize, while cash crops include coffee, sugar cane and tobacco. Cattle are the most important livestock;

NATIONAL DATA – VENEZUELA

Land area	882,050 sq km (340,561 sq mi)			
Climate		Temperatures		Annual
	Altitude m (ft)	January °C(°F)	July °C(°F)	precipitation mm (in)
Caracas	1,042 (3,419)	26 (79)	24 (75)	914 (35.9)

Major physical features highest point: Pico Bolívar 5,007 m (16,427 ft); longest river: Orinoco 2,061 km (1,281 mi)

Population (2006 est.) 25,730,435

Form of government federal multiparty republic with two legislative houses

Armed forces army 34,000; navy 18,300; air force 7,000

Largest cities Caracas (capital – 1,801,562); Maracaibo (2,054,039); Valencia (1,457,912); Barquisimeto (833,338); Ciudad Guayana (792,508); Maracay (399,446); Petare (386,712); Barcelona (459,148); Maturin (445,451); Turmero (373,871)

Official language Spanish

Ethnic composition Mestizo 69%; White 20%; Black 9%; Amerindian 2%

Religious affiliations Roman Catholic 96%; Protestant 2%; other 2%

Currency 1 bolivar (VEB) = 100 centimos

Gross domestic product (2006) U.S. $176.4 billion

Gross domestic product per capita (2006) U.S. $6,900

Life expectancy at birth male 71.49 yr; female 77.81 yr

Major resources petroleum, natural gas, iron ore, gold, bauxite, manganese, other minerals, hydropower, diamonds, bananas, coal, cocoa, coffee, livestock, maize/corn, rice, sorghum, sugarcane, tobacco, vegetables

Guyana

COOPERATIVE REPUBLIC OF GUYANA

Guyana, formerly British Guiana, is a small state on the north coast of South America, lying between Venezuela to the northwest, Brazil to the west and south, and Suriname to the east across the Courantyne river. The country's name is derived from an Amerindian word meaning "land of waters".

GEOGRAPHY

Apart from the narrow coastal plain, which is mostly reclaimed land protected by dikes and crisscrossed by canals, most of the country consists of an irregular, thickly forested plateau. Straddling the western border are the ancient Pakaraima Mountains, including Mount Roraima –

inspiration for the famous novel *The Lost World* by the British writer Sir Arthur Conan Doyle (1859–1930). The Rupununi area in the southwest of the country supports savanna grassland interrupted by low mountain ranges.

The climate is hot and often humid. Heavy seasonal rains bring flooding to the poorly drained soils. The forests are mostly tropical hardwoods. Among the varied wildlife, manatees and capybaras are found in and beside the rivers, sloths and jaguars in the forests, and giant anteaters and armadillos on the savanna. The rich bird life includes toucans and brilliantly plumaged hummingbirds.

SOCIETY

The area was originally inhabited by Arawaks and Caribs. The first European settlers were the Dutch, who brought in thousands of African slaves to work the sugar plantations. The British occupied the area around the Demerara river during the Napoleonic Wars (1796–1815), and secured most of present-day Guyana in 1814. With the abolition of slavery, immigrant laborers were imported from Asia. A border dispute with Venezuela was temporarily settled in 1899, and in 1923 British Guiana became a crown colony. In 1953 it was granted home rule along British parliamentary lines, and in 1966 was granted full independence as Guyana. In 1970 it became a republic under Prime Minister Forbes Burnham (1923–85), who was made the country's executive president in 1980.

Guyanans are remarkable for their racial diversity. The great majority are descendants of Asian Indians or Africans. Amerindians are a minority, but are the dominant group in the villages of the sparsely populated interior. There are also small European and Chinese communities. Religion and language reflect the cultural mix: English is the main language, but creole and Hindi are also spoken and there are many Hindus and Muslims as well as Christians.

ECONOMY

Guyana's mixed economy experienced steady growth in the late 1990s following expansion in agriculture and mining. Agriculture is important, but is mostly confined to coastal areas. Plantation sugar cane is the principal crop, followed by rice and coconuts. There is some livestock ranching, chiefly of cattle, in the Rupununi valley in the southwest.

Guyana's industry is dominated by its world-class deposits of bauxite, which provide the chief export item. Other activities include sugar processing, gold mining and river dredging for diamonds. The main imports are fuel and manufactured goods. There is great hydroelectric potential, but its development has been hampered by poor road and railroad links in all but coastal areas. Light aircraft is the usual mode of transportation inland.

Health conditions are generally good, and endemic diseases are under control. Social welfare schemes are comprehensive, but inadequate housing remains a problem. Good education is generally available, and literacy levels are high.

The Kaieteur Falls, formed where the Potaro river plunges over an escarpment in the highlands of west-central Guyana. Some 230 m (750 ft) high and 110 m (350 ft) wide, the falls lie at the center of the Kaieteur National Park.

NATIONAL DATA – GUYANA

Land area	196,850 sq km (76,004 sq mi)			
Climate		**Temperatures**		**Annual**
	Altitude m (ft)	January °C(°F)	July °C(°F)	precipitation mm (in)
Georgetown	2 (7)	26 (79)	27 (80)	2,160 (85)

Major physical features highest point: Roraima 2,810m (9,219 ft); longest river: Essequibo 1,014 km (630 mi)

Population (2006 est.) 767,245

Form of government multiparty republic with one legislative house

Armed forces army 900; navy 100; air force 100

Capital city Georgetown (238,747)

Official language English

Ethnic composition East Indian 50%; Black 36%; Amerindian 7%; White, Chinese, and mixed 7%

Religious affiliations Christian 50%; Hindu 35%; Muslim 10%; other 5%

Currency 1 Guyanese dollar (GYD) = 100 cents

Gross domestic product (2006) U.S. $3.62 billion

Gross domestic product per capita (2006) U.S. $4,700

Life expectancy at birth male 63.21 yr; female 68.65 yr

Major resources bauxite, gold, diamonds, hardwood timber, shrimp, fish, coconuts, oranges, rice, semiprecious gems, sugarcane

Suriname

REPUBLIC OF SURINAME

SURINAME, FORMERLY THE COLONY OF DUTCH
Guiana, is a small state on the northeastern coast of South America, lying
between Guyana to the west, French
Guiana to the east and Brazil to the south.

GEOGRAPHY

Suriname's landscape falls into three main
areas. The narrow coastal plain is low and
swampy, consisting largely of artificially
drained or reclaimed land. The coastal
soils are rich and fertile, and most of the
population live here. Farther inland the
land rises to a low savanna plateau. In the
south the plateau becomes higher and
more rugged, with low mountains and
thick forests. Several rivers rise near the
Brazilian border and flow northward, the
Corantijn forming the border with
Guyana, while the Marowijne and Litani
straddle that with French Guiana.

The climate is hot and very humid, tem-
pered only by light coastal breezes, and
there is heavy seasonal rainfall. The
forests are chiefly of tropical hardwoods
inland, with some mangroves along the
coast. Wildlife is plentiful, and includes
tapirs, sloths and ocelots, together with
many unusual bats, snakes and birds.

SOCIETY

The Surinen peoples, after whom the
country is named, had mostly been ousted
by other indigenous groups before the
arrival of European colonists. In 1667 the
Dutch secured the colony from Britain in
exchange for New Amsterdam (later New
York). From 1682 the Dutch West India
company brought in African slaves to
work its tea and coffee plantations. Many
escaped to the interior, where their
descendants are known as bush blacks.
When slavery was abolished in 1863,
laborers from India, China and Java were
brought over to work the land. In 1954

Suriname's largest lake – Professor van Blommestein
Lake – was created by the damming of the river
Suriname at Afobaka. The many small islands in the
lake are covered in the tropical hardwood forest typical
of northern Suriname.

Suriname was granted internal self-
government; independence followed in
1975. Racial conflicts led to a military
coup in 1980 under Lieutenant-Colonel
Dési Bouterse. Foreign aid was withdrawn, the economy collapsed, and the
Jungle Commandos, a bush blacks organization led by Ronnie Brunswijk (b.1961),
began a guerrilla campaign. Bouterse contested elections in 1988 and lost. The new
civilian government concluded a tentative
peace with the Jungle Commandos, but
Bouterse staged another coup in 1990. A
peace accord was signed in 1992 and elections were held in 1996. The country's
many political parties reflect its strong
racial divisions. Bush blacks form the
majority in the interior, with a small
Amerindian minority. The two main
groups, however, are Asian Indians and
mestizos (of mixed descent). There are
also significant Javanese and Chinese
minorities. The official language is Dutch,
but many others are spoken, including a
lingua franca called Sranan. The main religions are Christianity, Hinduism and
Islam.

ECONOMY

Suriname's economic potential is high.
Agriculture is mainly concentrated on the
coastal plain. The main crop is rice, some
of which is exported; other crops include

sugar cane and bananas. Livestock raising
is small-scale, but forestry and fishing fulfill both domestic and export needs. Most
exports, however, are provided by the
country's rich bauxite deposits, which are
smelted to aluminum using local hydroelectric power. The road network is mainly limited to coastal areas, and transportation elsewhere is mostly by water or air.
Welfare and health-care standards are
generally high, and free education is
available to university level.

NATIONAL DATA – SURINAME

Land area	161,470 sq km (62,344 sq mi)			
Climate	Altitude m (ft)	Temperatures January °C(°F)	July °C(°F)	Annual precipitation mm (in)
Paramaribo	4 (13)	26 (79)	27 (80)	2,312 (91)

Major physical features	highest point: Juliana Top 1,230 m (4,035 ft)
Population	(2006 est.) 439,117
Form of government	multiparty republic with one legislative house
Armed forces	army 1,400; navy 240; air force 200
Capital city	Paramaribo (226,124)
Official language	Dutch
Ethnic composition	Hindustani 37%; Creole 31%; Javanese 15%; "Maroons" 10%; Amerindian 2%; Chinese 2%; white 1%; other 2%
Religious affiliations	Hindu 27.4%; Protestant 25.2% (mainly Moravian); Roman Catholic 22.8%; Muslim 19.6%; traditional beliefs 5%
Currency	1 Surinam dollar (SRD) = 100 cents
Gross domestic product	(2006) U.S. $3.098 billion
Gross domestic product per capita	(2006) U.S. $7,100
Life expectancy at birth	male 66.66 yr; female 71.47 yr
Major resources	timber, hydropower, fish, kaolin, shrimp, bauxite, gold, nickel, copper, platinum, iron ore, bananas, cattle, citrus fruits, cocoa, coconuts, coffee, rice, sugarcane

Bolivia

REPUBLIC OF BOLIVIA

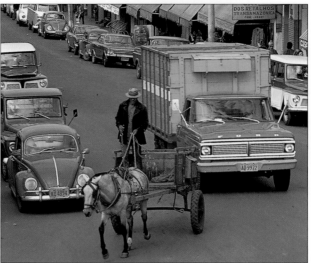

A **devil-mask** (*above*) shows the mix of Roman Catholic and pre-Columbian traditions in Bolivia.

A **horse and cart** (*left*) weaves a way through La Paz – at 3,627 m (11,900 ft) the world's highest capital.

BOLIVIA IS A LANDLOCKED STATE IN CENTRAL South America. It is a country with a rich culture and plentiful natural resources, but geographical isolation has hampered its development.

GEOGRAPHY

The population of Bolivia is mostly concentrated in the west on the Altiplano – a high plateau sandwiched between the two lofty ranges of the Andes. The Cordillera Occidental on the Chilean frontier includes several active volcanoes, while the Cordillera Oriental to the east is thickly forested. At the northern end of the Altiplano is Lake Titicaca. The southern Altiplano is an arid tableland scattered with salt flats. The Cordillera Oriental descends sharply to the northeast through the cliffs and cloud forests of the Yungas to the vast lowlands of the Oriente, which occupy the north and east.

Bolivia is a tropical country, but local climate varies enormously. Seasonal rainfall is high in the rainforests of the northern Oriente, moderate near Lake Titicaca and very sparse in the south.

SOCIETY

In the 7th century AD Tiahuanaco to the south of Lake Titicaca became the center of the first great Andean empire – that of the Aymará people. The Aymará culture survived despite the later arrival of the Quechua-speaking Incas in the 13th century and of the Spanish in the 16th. In 1545 the world's largest silver deposits were found at Potosí in the southwest.

In 1809 uprisings in Chuquisaca (Sucre) and La Paz signaled the start of South America's struggle for independence. In 1824 the rebel leader Antonio José de Sucre (1795–1830), a lieutenant of the revolutionary commander Simón Bolívar (1783–1830), routed the Spanish at Ayacucho in Peru, bringing independence the following year. The new country was named Bolivia after Bolívar, and Sucre became the first president.

Between 1836–39 Bolivia formed a confederation with Peru, but this was brought to an end after Chile declared war on it. Further disputes with Chile over nitrate extraction led in 1879 to the War of the Pacific; Bolivia and Peru were again defeated, and Bolivia lost the province of Atacama – its only outlet to the sea. By 1900 tin had replaced silver as the chief export, bringing substantial international investment. A disastrous war with Paraguay in the 1930s brought a further loss of territory in the southwest.

A series of bloody coups and countercoups began in 1943, but in 1952 the military were ousted by civilian forces promising a new social order. For the first time Amerindians were given the vote. A succession of military regimes followed until civilian rule was restored in 1982. The constitution allows for a two-chamber multiparty assembly. The president is directly elected for a five-year term, as are the 27 senators and 65 of the 130 deputies, but in 1993 no majority candidate emerged and the president was elected by Congress.

Bolivia's ethnic and cultural mix is complex. The largest group are mestizos of mixed Spanish and Amerindian ancestry, but they are outnumbered by the Amerindian Quechua and Aymará together.

ECONOMY

Almost half the labor force work on the land; many are subsistence farmers. Coffee and sugar are the chief export crops. Coca, chiefly from the Yungas, is easy to grow, accounting for a large proportion of the world's cocaine supply. Though illegal, this trade has given a massive boost to the ailing economy. Bolivia has considerable mineral resources – notably tin, tungsten, zinc and antimony – but natural gas is the main legal export. Petroleum production meets local needs. Most electricity comes from hydroelectric plants.

Roads are largely confined to the Altiplano, and few are surfaced. Railroad links with Peru and Chile carry Andean minerals to Pacific ports. La Paz and Santa Cruz have international airports.

Health care is adequate in the cities but under-resourced elsewhere. Malaria and Chagas' disease are common in the Oriente. Primary education is free and compulsory between the ages of 6 and 14.

NATIONAL DATA - BOLIVIA				
Land area	1,084,390 sq km (418,685 sq mi)			
Climate		Temperatures		Annual
	Altitude m (ft)	January °C(°F)	July °C(°F)	precipitation mm (in)
La Paz	4,103 (13,461)	12 (72)	9 (82)	571 (22.4)
Major physical features highest point: Sajama 6,520 m (21,391 ft); largest lake: Lake Titicaca (part) 3,810 sq km (3,200 sq mi)				
Population (2006 est.) 8,989,046				
Form of government multiparty republic with two legislative houses				
Armed forces army 25,000; navy 5,000; air force 3,000				
Largest cities Santa Cruz (1,468,658); Cochabamba (966,049); El Alto (911,525); La Paz (administrative capital - 820, 888); Sucre (judicial capital - 240,604); Oruro (211,217)				
Official languages Spanish, Quechua, Aymará				
Ethnic composition Quechua 30%; Mestizo (mixed White and Amerindian) 30%; Aymará 25%; White 15%				
Religious affiliations Roman Catholic 95%; Protestant (Evangelical Methodist) 5%				
Currency 1 boliviano (BOB) = 100 centavos				
Gross domestic product (2006) U.S. $27.21 billion				
Gross domestic product per capita (2006) U.S. $3,000				
Life expectancy at birth male 63.21 yr; female 68.61 yr				
Major resources tin, natural gas, petroleum, zinc, tungsten, antimony, silver, iron, lead, gold, timber, hydropower, alpacas, barley, bismuth, cassava, cocoa, coffee, copper, cotton, fruits, llamas, maize/corn, potatoes, quinol, rice, sheep, sugar beet, sugarcane, sulfur, tobacco				

Ecuador

REPUBLIC OF ECUADOR

ECUADOR AS ITS NAME SUGGESTS, LIES astride the Equator on the west coast of South America between Colombia to the north and Peru to the south and east. It also includes the Galapagos Islands some 1,000 km (600 mi) out in the Pacific.

GEOGRAPHY

Mainland Ecuador is a country of high mountains that is prone to earthquakes. The coastal lowlands consist largely of fertile alluvial plains interspersed with volcanic hills. Inland the massive wall of the Andes crosses the country from north to south. It is made up of two main parallel chains divided by high valleys, one of which is the site of the capital, Quito. Many of the highest peaks are volcanic, notably Cotopaxi, the world's tallest active volcano. The eastern slopes of the Andes drop into the impenetrable rainforests of the Amazon basin. In this wild, little-known area the exact borders with Peru are disputed.

The climate is generally hot and humid, especially in the east, but the mountainous areas are much colder. The coastal lowlands are cooled by the Humboldt Current, which also reduces rainfall along parts of the coast. The forest here is sparser than on the mountain slopes and in the east, where wildlife is abundant.

The Galapagos are a complex archipelago formed by the cones of young volcanoes. Their shores consist of arid lava rocks, and only the peaks are forested. Kept cool and dry by the Humboldt Current around their shores, the islands have a unique wildlife, including marine iguanas, giant tortoises, flightless cormorants and 13 species of Darwin's finch, whose variations inspired the evolutionary theories of Charles Darwin (1809–82). The islands are protected for research, but have a limited tourist trade.

SOCIETY

In the 15th century a short-lived Amerindian kingdom on the mainland was conquered by the Incas, the ruling class from Peru. Civil war later weakened the Inca empire, which was quickly overrun by the Spanish conquistadors in the 1530s. Spanish settlers divided the country into large estates worked by landless Amerindian laborers, or peons. The situation remained unchanged up until the 19th century, when independence movements grew up among the emergent middle classes. Ecuador won its independence in 1830 after some two decades of war.

A divide remained between the conservative landowners of Quito and the more liberal commercial classes of Guayaquil, the chief port. The resulting instability brought about a series of ruthless dictatorships that lasted until World War II, when Ecuador lost some territory to Peru. The postwar period was dominated by the charismatic president, José María Velasco Ibarra (1893–1979), whose inconsistent policies sometimes hampered the country's economic development. Military rule in the 1970s was followed by a series of more liberal civilian presidents. An inconclusive four-week border war was fought with Peru in 1995.

The 1976 constitution provides for an executive president and vice president for a four-year term. There is a single legislative National Congress which is also elected for four years.

Spanish is the main language, although a large portion of the population are Quechua-speaking Amerindians. Most of the rest are mestizos (of mixed descent), with small African and European minorities. The majority are at least nominally Roman Catholic, though much Amerindian culture survives.

ECONOMY

Economically, Ecuador remains largely agricultural. Bananas are the principal cash crop, along with sugar cane, coffee and cacao. Rice and the indigenous potato are among the main staples. Despite land reforms, much food is still grown on massive hacienda estates. Livestock is raised both along the coast and in the mountain valleys. The great hardwood forests remain relatively unexploited, while the fishing industry is one of South America's largest. The chief mineral resources are petroleum and natural gas, which have overtaken bananas and shrimps as the main export.

Transportation is hindered by the difficult terrain. Road and rail networks are limited, and the mule is often the most effective means of land transportation. Social welfare schemes are in place, but there is much rural poverty, and health care and living conditions are often poor. Education is available to university level.

Women in Quito (*above*) carry their burdens in bundles on their backs. The Ecuadorean capital – which retains the air of a Spanish colonial town – lies almost on the Equator, situated in a high valley about 3,000 m (9,000 ft) above sea level.

NATIONAL DATA – ECUADOR

Land area	276,840 sq km (106,889 sq mi)			
Climate		Temperatures		Annual
	Altitude m (ft)	January °C(°F)	July °C(°F)	precipitation mm (in)
Quito	2,879 (9,446)	15 (59)	15 (59)	1,116 (43.9)

Major physical features highest point: Chimborazo 6,267 m (20,561 ft); longest river: Napo (part) 1,100 km (700 mi)

Population (2006 est.) 13,547,510

Form of government multiparty republic with one legislative house

Armed forces army 37,500; navy 5,500; air force 4,000

Largest cities Guayaquil (1,952,029); Quito (capital – 1,399,814); Cuenca (276,974); Machala (198,123); Santo Domingo de los Colorados (200,421)

Official language Spanish

Ethnic composition Mestizo (mixed Amerindian and White) 65%; Amerindian 25%; Spanish and others 7%; Black 3%

Religious affiliations Roman Catholic 95%; other 5%

Currency 1 U.S. dollar (USD) = 100 cents

Gross domestic product (2006) U.S. $60.48 billion

Gross domestic product per capita (2006) U.S. $4,500

Life expectancy at birth male 73.55 yr; female 79.43 yr

Major resources petroleum, fish, timber, hydropower, bananas, cassava, cacao, coffee, natural gas, gold, limestone, livestock, maize/corn, oranges, potatoes, rice, sugarcane

Peru

REPUBLIC OF PERU

P ERU, WHICH LIES ASTRIDE THE ANDES ON THE western coast of South America, derives its name from a Quechua word implying abundance, recalling the former opulence of the Inca and Spanish empires. However, its modern development has been hindered by the hostile, though often spectacular terrain.

GEOGRAPHY

Peru is divided into three distinct geographical areas. The narrow coastal plain, or Costa, runs along the whole length of the Pacific coast from Ecuador in the northwest to the Chilean border in the south. Much is arid desert, but the Andean rivers have laid down patches of fertile alluvial soil. Most of the population lives in this part of the country.

The lofty Andes, or Sierra, form a continuous backbone that enters from Ecuador and broadens in the southeast on the border with Chile and Bolivia. The mountains are most dramatic in central Peru, where peaks such as Huascarán, the highest, tower above deep mountain trenches. In the south the two main ranges run either side of an extensive high plateau called the Altiplano, at the center of which is the great Lake Titicaca straddling the Bolivian border. The whole country is liable to earthquakes, and several volcanoes in the south are dormant rather than extinct.

The Andes fall away northeast to the Montaña – the deep, forest-clad valleys that lead out into the dense rainforests of the Amazon basin. This wild, undulating country bordering Colombia and Brazil is almost uninhabited.

The climate of the Amazon basin is hot and humid, typical of its latitude, but the

mountains are much colder with seasonal rain and snow. The coastal areas are kept dry by cold offshore currents; cloud and fog are common, but little rainfall normally occurs. However, an occasional switch from cold to warm sea currents creates a climatic phenomenon known as El Niño, causing severe flooding on land and a reduction in offshore marine plankton, leading to loss of coastal fish and birds.

Plant and animal life vary tremendously. Coastal vegetation ranges from cacti to hardy spiny shrubs; the highlands support grasses and puna scrub; and the east has typical hardwoods such as mahogany and wild rubber. Wildlife includes tapirs and sloths in the forests, and chinchillas, vicunas and condors in the mountains.

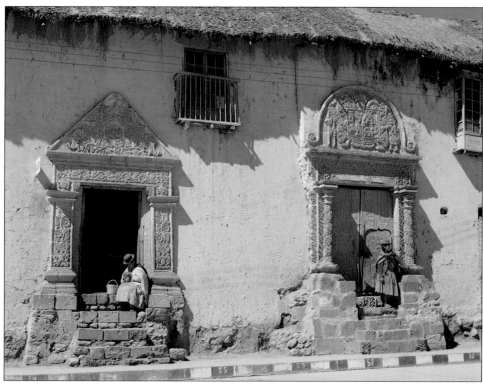

Mestizo-style architecture (*above*) at Juli on the western shore of Lake Titicaca, Peru. The elaborate carving is Amerindian in style, but the subjects – the coat of arms above the door on the right and the floral motifs – are colonial.

A Peruvian river valley (*left*) shows the legacies of the Inca empire. Terraced fields make subsistence farming possible on the steep hillsides, while the road running parallel to the river was once part of a sophisticated communications network.

In 1963 Fernando Belaúnde Terry (b.1912) was elected president, and launched a program of reforms. In 1968 he was ousted by an army junta, which in turn was toppled by another junta in 1975, but in 1980 Belaúnde was returned to power in new elections. His attempts at reform were frustrated by economic difficulties, a devastating El Niño and ever-increasing

valley-dwelling Nasca, the Chimú and the Huari. In the 15th century, however, the whole country was taken over by the Quechua-speaking Incas. They built massive stone cities, the most famous of which are at Cuzco – their capital and the center of their religion, which was based on sun-worship – and mountain-top Machu Picchu. Weakened by civil war, however, their wealthy empire was overthrown by the Spanish adventurer Francisco Pizarro (c. 1475–1541) in 1531. This marked the beginning of the Amerindian peoples' long history of exploitation.

Peru, with its immense natural wealth, remained Spain's treasurehouse for some three centuries, until its liberation by the Argentine general José de San Martín (1778–1850) in 1821. A brief federation with Bolivia collapsed in 1839. After the liberal presidency of General Ramón Castilla (1797–1867) military juntas alternated with civilian politicians, who failed in their attempts to stabilize the economy. Such problems were exacerbated by the War of the Pacific (1879–84), in which valuable territory was lost to Chile.

The cold offshore seas support rich fish populations and the many birds that feed on them such as pelicans and boobies.

SOCIETY

Various cultures once flourished among Peru's original Amerindian inhabitants, including most notably the Chavín, the

NATIONAL DATA – PERU

Land area	1,280,000 sq km (494,210 sq mi)			
Climate		**Temperatures**		**Annual**
	Altitude m (ft)	January °C(°F)	July °C(°F)	precipitation mm (in)
Lima	120 (394)	22 (77)	17 (61)	13 (0.5)

Major physical features	highest point: Huascarán 6,768 m (22,205 ft)

Population	(2006 est.) 28,302,603

Form of government	multiparty republic

Armed forces	army 40,000; navy 25,000; air force 15,000

Largest cities Lima (capital - 7,979,965); Arequipa (878,129); Trujillo (788,911); Chiclayo (607,121); Iquitos (468,687)

Official languages	Spanish, Quechua

Ethnic composition Amerindian 45%; Mestizo (mixed Amerindian and White) 37%; White 15%; Black, Japanese, Chinese, and other 3%

Religious affiliations Roman Catholic 81%; Seventh Day Adventist 1.4%; other Christian 0.7%; other 0.6%; unspecified or none 16.3%

Currency	1 nuevo sol (PEN) = 100 centimos

Gross domestic product	(2006) U.S. $181.8 billion

Gross domestic product per capita	(2006) U.S. $6,400

Life expectancy at birth	male 68.05 yr; female 71.71 yr

Major resources copper, silver, gold, petroleum, timber, fish, iron ore, coal, phosphate, potash, hydropower, natural gas, cattle, cotton, fruit, lead, livestock, potatoes, rice, sugarcane, tungsten, uranium, vegetables, wheat, zinc

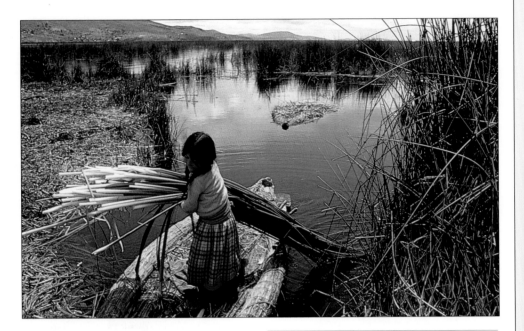

Reed-gathering (*above*) on the shores of Lake Titicaca. The Uru – an ancient Amerindian people – live in floating villages made from huge rafts of totara, a reedlike papyrus. Their crescent-shaped boats are also fashioned from the reeds.

guerrilla activity. The two main guerrilla groups involved were the Tupac Amarú Revolutionary Movement (MRTA) and the neo-Maoist group known as Sendero Luminoso (Shining Path). The MRTA came to world attention in 1997 after besieging the Japanese embassy for 126 days.

President Alan García, elected in 1985, fared no better with the economy, the guerrillas or drug trafficking and lost the 1990 election to Alberto Fujimori. In April 1992 Fujimori appropriated dictatorial powers to enforce economic reforms and defeat the guerrillas, but their activity continues despite the capture of Shining Path leader Abimael Guzmán, with Shining Path joining forces with drug organizations in 1995 to provide funding. Fujimori had been reelected in November 1992, ending his dictatorship, and again in 1995, but was dismissed by Congress in November 2000 as being morally unfit to govern after he was implicated in a serious corruption scandal. He fled to Japan, where he remained as Peru prepared criminal charges. New elections in April 2001 led to a run-off between former president Garcia and Alejandro Toledo.

Peru's instability is partly due to ethnic and social divisions between the Spanish-speaking mestizos of the Costa and the Quechua-speaking Amerindians of the Sierra. Other minority groups include some whites, Aymará and other Amerindian peoples. Most of the population is Roman Catholic, but the influence of Amerindian traditions remains strong.

ECONOMY

Peru's economy is heavily dependent on the export of its rich mineral and other resources. Agriculture is limited by a lack of arable land, most of which is found in coastal areas, where irrigation is needed to grow crops such as rice and sugar cane. Highland farming is mainly at subsistence level. Crops here include wheat, potatoes and coca, which supplies an illicit cocaine trade. Sheep, llamas and alpacas are the main livestock animals, bred chiefly for their wool. There is some forestry in parts of the east, but the area is inaccessibe and ecologically fragile. The vast fishing industry periodically suffers drastic reductions in catch size due to the effects of El Niño.

Industry is powered mainly by hydroelectricity. Copper ore is the major mineral, but petroleum is becoming almost as important. These are the two main exports, followed by uranium and other metals such as iron and silver. Manufacturing industries are concentrated around the capital, Lima, and nearby Callao, the main port; many manufactured goods still have to be imported.

Transportation is limited by the rugged terrain and the climate, and there are few railroads. Air travel is hampered by the difficulty of constructing airstrips in swampy forest or steep mountain terrain.

Welfare provision and health care is available in the cities. However, housing and health conditions remain inadequate from impoverished mountain areas to the outskirts of cities such as Lima. Where education can be obtained it is free and compulsory, but it is often unavailable to poor or isolated rural communities.

The Incas

The popular image of the Inca empire, the last of the great South American indigenous civilizations, is of a society as durable and mysterious as the mighty Inca fortifications that survive today. Yet the Inca empire was relatively short-lived. It was established only a few years before the birth of Christopher Columbus (1451–1506), and lasted little more than a century. The real mystery is how the Incas achieved so much in that short time.

They began as a band of Andean mountain dwellers who left their little and colonization began; Cuzco became the capital of the growing Inca territory. His son Yupanqui (1438–71), who assumed the name Pachacuti, continued the expansion. To govern his large territories he created an entire administrative class of honorary Incas and undertook what may have been the Incas' most amazing achievement – an elaborate road system, thousands of kilometers long, that included tunnels and suspension bridges. Along it at regular intervals ran a system of relay stations that could send government communications as far as 250 km (150 mi) a day. The Incas achieved this without the use of either the wheel or the horse; messengers ran between the stations on foot. They had no writing system either, so messengers had to relay their messages from memory, with only knotted-string patterns called *quipu* to aid them.

Pachacuti and his successor Topa (1471–93) conquered an enormous area of western South America from present-day Ecuador in the north to Chile and Argentina in the south. Everywhere they passed, they built not only roads but towns, canals, roadside forts, resthouses and irrigation schemes in order to increase food production. Topa's successor, Huayna Cápac (1493–1525), continued the process, building a newer and more splendid capital on the site of the present-day Ecuadorean capital of Quito.

What made all this possible was sheer manpower; Huayna ruled some 12 million people, and every province owed the Inca not only tribute but levies of soldiers and workers. Service was a religious as well as a social duty, and may well have been given gladly. The results were impressive. The mountain-top city of Machu Picchu was surrounded by wide terraced fields, all the soil for which must have been carried up the mountainside on men's backs.

Every man was bound to his father's occupation, but no one starved; relief was generous, and storehouses provided support during famines. Although the result appeared to be a stable if authoritarian society, Huayna's sudden death precipitated a war between his two sons. No sooner had a winner emerged than he was seized and murdered by the Spanish conquistadors. Without a ruler at the top, the entire Inca edifice crumbled almost at once.

Machu Picchu (*above*), perhaps the best-known of Peru's Inca cities, was probably built no earlier than the 15th century, and is perched high above the Urubamba valley on the eastern flanks of the Andes.

Inca artifacts (*left*) An Inca stirrup-cup in the form of a jaguar savaging its victim, and a rare ceremonial gold figurine, one of the few pieces of gold to have survived the marauding Spanish conquistadors.

village to found a city at Cuzco in what is now south-central Peru. They grew into a dominant city-state that raided and feuded with its neighbors. Their chieftains – the Incas, for whom the empire was later named – became the prime focus of a divine cult.

Under the eighth Inca, Viracocha (reigned 1410–38) a process of conquest

Brazil

FEDERAL REPUBLIC OF BRAZIL

BRAZIL IS THE LARGEST COUNTRY IN SOUTH America, the fifth largest in the world and the sixth most populous To the northeast and southeast it faces the Atlantic along a 7,400-km (4,600-mi) coastline, while to the west it shares borders with most other South American countries. A land of wealth and opportunity, it has attracted settlers from all over the world. However, its development as a modern industrial state has brought with it problems of inadequate finance, overpopulation and environmental damage.

GEOGRAPHY

Brazil divides into two main areas. The vast Amazon basin to the north contains the largest river system in the world, while the highland areas of the south and east are the most populous.

The land
Brazil's northern frontier runs through an area of mountains and plateaus known as the Guiana Highlands, which extend into Venezuela, Guyana, Suriname and French Guiana. They include Brazil's highest mountain, Pico da Neblina, discovered as recently as 1962.

Immediately to the south is the great Amazon basin, which occupies most of the north of the country. Its landscape is generally low-lying, apart from the hillier fringes, and some flatter areas are flooded every year. The main river is some 6,570 km (4,080 mi) long – second in the world only to the Nile – and has over 1,000 known tributaries; most of the larger ones rise farther west in neighboring Colombia, Peru and Bolivia. To the east, the Amazon enters the Atlantic via a broad delta, and the lowlands extend eastward along the northern coast.

The Brazilian Highlands cover most of the rest of the country, rising gently from the Amazon basin along a line roughly from Bolivia to Cape São Roque, Brazil's northeasternmost point. The highlands consist mostly of broad tablelands cut by deep river valleys. Along the eastern seaboard they rise more sharply to form a steep escarpment overlooking a narrow coastal plain. Among the highest peaks is Pico das Agulhas Negras near Rio de Janeiro and São Paulo, the two largest cities. Most of the rivers drain southwest into neighboring Paraguay, Argentina and Uruguay. The Paraná and Uruguay rivers run along sections of the border, creating spectacular waterfalls as they leave the plateau.

The only other extensive lowland area is in the west bordering Bolivia and northern Paraguay, where the swampy plains of the Pantanal form the northernmost extension of the Gran Chaco.

Climate
The climate is generally warm and humid with seasonal rains, but varies somewhat across the country. Temperatures are higher around the Amazon basin, where they remain largely stable throughout the year. Conditions are similar along the east coast, while the Brazilian Highlands and the south are both cooler, with occasional winter frosts.

In the northeast the São Francisco basin has a unique microclimate of its own. This notorious drought pocket, known as the drought polygon, has virtually no rain in the dry season, and far less than the surrounding areas at other times. Elsewhere, rainfall is heaviest in parts of the Amazon basin and along the flanks of the eastern mountains.

Plants
Much of Brazil is forested; the tropical evergreen rainforests of the Amazon basin have a greater variety of plant species than any other habitat on earth. There are similar forests along the southeastern coast, and semideciduous forests on the heights of the escarpment. Inland, the highlands show a mixture of savanna and deciduous woodland, and in the far south are some of the original grass prairies. The dry northeastern areas are characterized by *caatinga* – a type of thorny woodland.

There has been international concern over the mass destruction of the Amazon rainforests for cattle ranching and mining. Some effort has been made to slow down the process of uncontrolled ex-

The many waters of the Amazon (*above*) The lushness of Amazonian vegetation reflects the vast amounts of fertile silt carried down through the world's largest river system, which receives up to one-quarter of the earth's run-off water every year.

The famous statue of Christ the Redeemer (*below*) with arms outstretched in blessing above the city of Rio de Janeiro. It stands on Corcovado peak, 787 m (2,310 ft) high, and looks across to the unmistakable outline of the Pão de Açúcar, or Sugar Loaf mountain.

Dutch briefly captured Bahia in 1624, and settled at Pernambuco (modern Recife) in 1630. They were driven out by Brazilian forces in 1654, 14 years after the Portuguese had once again wrested control of their homeland from Spain.

During the 17th and 18th centuries, numerous expeditions set out from São Paulo to capture Amerindians for slaves and to exploit the resources of the vast interior. The first rebellion against Portuguese rule came as early as 1789, but independence was to take another three decades. In 1808 the French invaded Portugal, and the prince regent, Dom John (1767–1826), took refuge in Brazil. He took a strong interest in his new home, and in 1815 he proclaimed the United Kingdom of Portugal, Brazil and the Algarve. A year later he became King John VI. In 1821 he returned to Portugal, leaving behind his son, Dom Pedro (1798–1834), as regent. The Portuguese court sought to return Brazil to its former colonial status, and Dom Pedro was faced with an angry population. In 1822 he declared independence, and became Brazil's first emperor.

His reign was troubled: war with Argentina (1825–28) led to the loss of Uruguay, and in 1831 he abdicated in

ploitation, but more effective conservation measures are essential if this fragile ecosystem is to be preserved.

Animals

Animal life in the Amazon Basin is as colorful and variegated as the plant life. Numerous species of monkeys, snakes and tree frogs share the trees with hundreds of different bird species. The rivers teem with caimans, capybaras, river dolphins and as many as 1,500 known species of fish.

Outside the Amazon basin, much of Brazil's once-abundant wildlife has all but vanished. There are still wildfowl in the marshlands of the Pantanal, and rheas (flightless birds) roam the tablelands, as do bustard-like seriemas. Snakes, lizards and armadillos abound, while Giant anteaters feed on termites in their pillar-like nests. Ocelots and jaguars, the predators of the Brazilian savanna, have been hunted to virtual extinction.

SOCIETY

Brazil has produced a remarkably integrated society since the early days of Portuguese colonial settlement. The country was named after its first export, *brasil*, a red dye made from wood. As the interior was opened up, other valuable items were exploited, including gold, diamonds and coffee. After World War II, Brazil's predominantly rural society was dramatically changed into an urban, industrial society beset with the problems common to many developing countries.

History

Pedro Álvares Cabral (1467/8–1520), a Portuguese naval commander, was probably the first European to set eyes on Brazil. He recorded his discovery in 1500, but it was not until 1532 that Martim Afonso de Sousa (c.1500–64) founded the Portuguese colony of São Vicente – the first of 15 hereditary captaincies. The colonists created large estates, worked by Amerindians and later by African slaves.

In 1549 King John III of Portugal (1502–57) appointed a governor-general, who controlled the captaincies from his base at Bahia (later Salvador). Meanwhile, other colonial countries began to take an interest in Brazil. The French established a settlement at Rio de Janeiro in 1555, and it took the Portuguese five years to evict them. By 1580 the Spanish had gained control of Portugal, and colonial administration in Brazil was being neglected. The

Land area	8,456,510 sq km (3,265,077 sq mi)			
Climate		**Temperatures**		**Annual**
	Altitude m (ft)	January °C(°F)	July °C(°F)	precipitation mm (in)
Manaus	83 (272)	27 (81)	27 (81)	2,386 (93.9)
Brasilia	1,006 (3,301)	22 (72)	19 (66)	1,632 (64.2)
Rio de Janeiro	61 (200)	27 (81)	22 (72)	1,173 (46.1)

Major physical features highest point: Pico de Neblina 3,014 m (9,889 ft); longest river: Amazon (part) 6,436 km (4,000 mi)

Population (2006 est.) 188,078,227

Form of government federal multiparty republic with two legislative houses

Armed forces army 189,000; navy 32,850; air force 65,309

Largest cities Sao Paulo (10,095,244); Rio de Janeiro (6,086,917); Salvador (2,810,334); Belo Horizonte (2,424,579); Fortaleza (2,390,058); Brasilia (capital – 2,314,172); Curitiba (1,776,074); Manaus (1,690,331); Recife (1,500,293); Belem (1,465,670)

Official language Portuguese

Ethnic composition White 53.7%; Mulatto (mixed White and Black) 38.5%; Black 6.2%; other (includes Japanese, Arab, Amerindian) 0.9%; unspecified 0.7%

Religious affiliations Roman Catholic 73.6%; Protestant 15.4%; Spiritualist 1.3%; Bantu/voodoo 0.3%; other 1.8%; unspecified 0.2%; none 7.4%

Currency 1 real (BRL) = 100 centavos

Gross domestic product (2006) U.S. $1.616 trillion

Gross domestic product per capita (2006) U.S. $8,600

Life expectancy at birth male 68.02 yr; female 76.12 yr

Major resources bauxite, gold, iron ore, manganese, nickel, phosphates, platinum, tin, uranium, petroleum, hydropower, timber, asbestos, bananas, beryllium, cassava, cereals, chromium, coffee, coal, cocoa, cotton, crude oil, diamonds, fish, graphite, livestock, natural gas, oranges, quartz crystal, rice, rubber, salt, silver, soybeans, sugarcane, titanium, tobacco, tourism, tungsten, zinc

Dependencies in the region

FALKLAND ISLANDS
UNITED KINGDOM

The Falkland Islands, or Islas Malvinas, are a self-governing British colony in the South Atlantic consisting of an archipelago off the southeastern coast of South America. The colony also administers a series of island dependencies. These include the volcanic South Sandwich Islands and South Georgia – the largest of the dependencies – a whaling settlement some 1,300 km (800 mi) to the southeast.

The islands of East and West Falkland comprise most of the land area. Both are heavily indented, hilly and often swept by cold westerly winds. The landscape is mostly peat-covered moorland, with low grasses, scrubby vegetation and no native trees. Animal life includes elephant seals, sea lions and penguins.

These previously uninhabited islands were first sighted by British navigators in 1592. It was not until 1765, however, that they were settled by a British garrison. This was removed in 1774, but Britain maintained its claim to the islands. Spain kept a settlement there until 1811, and in 1820 the newly independent Argentina claimed the islands on the basis of previous Spanish claims – marking the start of a long-running conflict between the two nations over the islands. An Argentinian attempt to settle the islands in 1832 soon collapsed in mutiny, and in 1833 a British garrison was reestablished. An almost entirely British population became established with a limited economy based, as it is now, on fishing and sheep-rearing. In 1892 colonial government was introduced, with a governor appointed by the British monarch.

Argentina, however, continued to uphold its claim to the islands. Intervention by the United Nations in 1964 led to talks, in which Britain sought to settle the sovereignty issue on the basis of the islanders' wishes to remain British. Nevertheless, in 1982 Argentina invaded the islands. British forces were brought in, and after six weeks of fierce fighting they expelled the occupying forces. Hostilities were formally ended in 1989, but Argentina still claims the colony.

The local economy and way of life have been greatly affected by the war and its aftermath. An exclusion zone was enforced around the islands to prevent overfishing, and fishing licenses have now become the Falkland Islands' largest source of income. A 320 km (200 mi) oil exploration zone was imposed in 1993.

Port Stanley (*above*) on East Falkland, the largest of the Falkland Islands. In June 1982 fighting between British and Argentinian forces raged around this little town – the colony's principal settlement and main harbor – during the last days of a short war that was precipitated, in part, by the prospect of rich offshore oil fields.

Parasols in the jungle (*right*) These children belong to one of several Asian groups that have settled in French Guiana, notably the Hmong people from Laos, formerly a French protectorate in Southeast Asia. The Hmong, refugees from war-torn Indochina, were settled here in the late 1970s.

A tropical idyll (*far right*) French Guiana was once the home of the infamous penal colony of Devil's Island, to which France until 1945 exiled its worst offenders for long sentences of hard labor. Today this small tropical island is becoming a tourist resort.

FRENCH GUIANA
FRANCE

French Guiana is a French overseas *département* lying just north of the Equator on the northeast coast of South America. It is bordered by Suriname to the west, and Brazil to the south and east.

There are two main geographical areas: the low-lying coastal plains and the forested plateau of the interior, which rises to form mountains along the southern border. The climate is hot and humid, with heavy seasonal rains. Most people live along the coast, which is fertile but often swampy, with savanna in the west. Much of the country is covered by impenetrable hardwood forests that support a wide variety of wildlife.

French merchants first founded Cayenne, now the capital, in 1643, but their

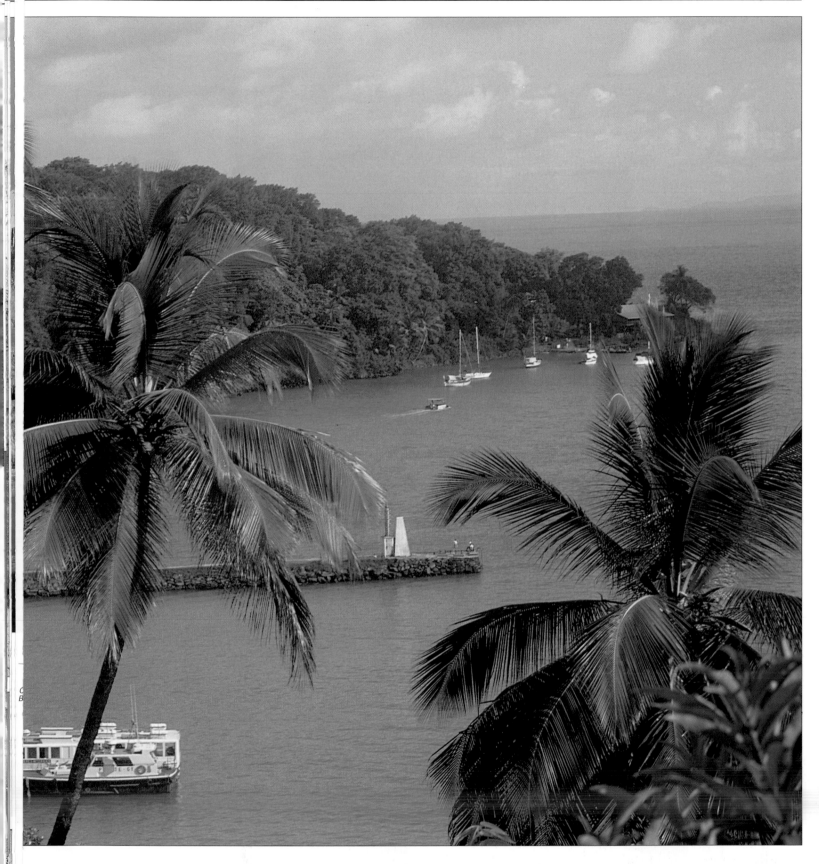

initial attempts at colonization were frustrated by other European colonists. However, by the late 17th century the French were fully established, importing African slaves to work their plantations. After the French Revolution, political prisoners were sent here to die in appalling conditions. In the 1850s the abolition of slavery caused economic collapse in spite of the introduction of indentured laborers from India. French missionaries, however, set up educational institutions for freed slaves. Penal settlements were again established, including the notorious Devil's Island; they were not abolished until after World War II. French Guianans have had French citizenship since 1848 and parliamentary representation since 1870; moves toward independence have had little success. Most people are of mixed descent, speaking French and Creole; most are Roman Catholic.

French Guiana remains economically dependent on France. Agriculture is concentrated in coastal areas, producing both staples and cash crops such as bananas and sugar cane. Fish are the other main export apart from minerals, particularly gold and bauxite. The European Space Agency has a rocket-launching base at Kourou. The roads are few, and most internal transportation is by air. Welfare and healthcare are to French standards.

Mauritania

ISLAMIC REPUBLIC OF MAURITANIA

MAURITANIA EXTENDS INLAND FROM THE Atlantic coast of West Africa and northward into the Sahara. It borders Western Sahara and Algeria to the north, with Mali and Senegal to the south.

GEOGRAPHY

Behind the low-lying coastal plains are slightly higher tableland plateaus broken only by occasional isolated peaks and westward scarps. Farther inland, occupying most of the country, are the sand dunes of the Sahara, and in the extreme south the semiarid Sahel scrubland. The climate is hot and dry apart from late summer rains in the south.

Most of the fertile land lies along the coast and in the catchment area of the Senegal river, which forms the southwestern border with Senegal. This is grassy savanna country, with baobab and palmyra trees. In the Sahel vegetation is sparse, and in the desert it is confined to oases and semidry riverbeds, or wadis. Despite conservation measures, hunting and habitat destruction have greatly reduced the numbers of wildlife; this includes antelopes, Senegalese lions, warthogs, African elephants, ostriches, gazelles and crocodiles.

SOCIETY

In medieval times Mauritania was part of the Almoravid empire. It was then overrun by Arabs, or Moors, from the north, who subjugated the various black Sudanic peoples that still live in the south. The Moors (*Maures* in French) gave their name to the country, and the population today is three-fourths Moorish. The Moors are divided into two very distinct groups: the *bidan* or "white Moors", of Arab descent; and the mixed-race *harratin*, or "black Moors", descended from slaves (slavery was not outlawed until 1970, and persists today). Most Moors are nomadic pastoralists, and much of the population is Muslim.

In the 19th century the area was gradually taken over by France, becoming part of French West Africa. In 1960 it won full independence under President Moktar Ould Daddah (b.1924). In 1978 he failed to suppress Polisario guerrillas in Mauritania's Western Sahara territory (now mainly occupied by Morocco), and was deposed by a military coup. Sid'Ahmed Ould Taya, a former prime minister, took over as president. A new constitution to give greater political freedom and allow for multiparty elections was approved by referendum in 1991. However, the opposition parties boycotted elections in 1992 and 1997 on grounds of vote rigging, and on each occasion Taya was reelected president with a substantial majority.

ECONOMY

Agriculture accounts for 25 percent of GDP, and consists mostly of livestock farming, including sheep and goats, with cattle in the south. Arable crops include millet, sorghum, yams, cotton, maize and dates. Some areas are irrigated, but the Sahel droughts have cut production by as much as two-thirds. Fishing accounts for nearly 60 percent of total exports, the waters off Mauritania contain some of the world's richest fishing grounds although these are threatened by overexploitation.

The major industries are fish processing and mining. Iron ore makes up 40 percent of exports and there are also reserves of copper, gold, phosphate and gypsum. Roads are poor, making transportation difficult. There are international airports at Nouâdhibou and Némy. The only railroad runs from Zouérate to the port of Point-Central and is mostly used for the transport of iron ore.

A Mauritanian fisherman carries his nets ashore for repair. Fishing is the backbone of the Mauritanian economy. Foreign investment has developed the industry to the point where it now accounts for almost half the country's exports.

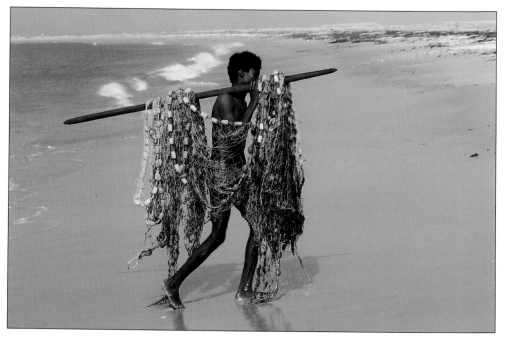

NATIONAL DATA – MAURITANIA

Land area	1,030,400 sq km (397, 840 sq mi)			
Climate		Temperatures		Annual
	Altitude m (ft)	January °C(°F)	July °C(°F)	precipitation mm (in)
Nouakchott	1 (3)	22 (72)	28 (83)	159 (6.2)

Major physical features	highest point: Kediat Idjil 915 m (3,002 ft)
Population	(2006 est.) 3,177,388
Form of government	multiparty republic with two legislative houses
Armed forces	army 15,000; navy 620; air force 250
Capital city	Nouakchott (881,000)
Official language	Arabic
Ethnic composition	Mixed Maur/black 40%; Moor 30%; Black 30%
Official religion	Islam
Religious affiliations	100% Sunni Muslim
Currency	1 ouguiya (MRO) = 5 Khoums
Gross domestic product	(2006 est.) U.S. $8.397 billion
Gross domestic product per capita	(2006 est.) U.S. $2,600
Life expectancy at birth	male 50.88 yr; female 55.42 yr
Major resources	iron ore, gypsum, copper, phosphate, diamonds, gold, oil, fish, dates, livestock, millet, rice, salt, sorghum, wheat

Mali

REPUBLIC OF MALI

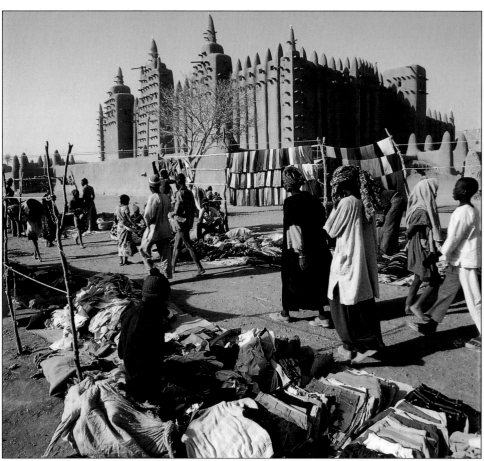

A **colorful market** in the shadow of a mosque at Djenne in central Mali. Mali was a great trading nation in the 13th and 14th centuries, renowned for its gold, but it now has to contend with famine, disease, droughts and the encroaching Sahara.

THE LANDLOCKED REPUBLIC OF MALI LIES IN the Sahel area of northwest Africa on the fringes of the Sahara.

GEOGRAPHY

Mali's landscape is monotonously flat, consisting mostly of plains and sandstone plateaus or hills. The great northern plains are part of the Sahara, with a hot, almost rainless climate characterized by cold nights. In recent years rainfall has declined in the semiarid Sahel area in the heart of Mali, bringing devastating droughts. Only the south, crossed by the great Niger river and its tributaries, has enough rainfall to support much plant or animal life.

All the arable land is in the south, covering less than one-fiftieth of the total area. Otherwise the south is mostly savanna, with mahogany, kapok and baobab trees, replaced farther north by palm species and eventually by scrub. Animal life includes characteristic African species such as lions, antelopes, hyenas, jackals and many smaller creatures. A national park has been established to protect them.

SOCIETY

Although inhabited since the Stone Age, Mali derives its name from the Muslim empire of the Malinke or Mandingo people that flourished here in the 13th and 14th centuries. The area was a rich center of trade and Islamic scholarship until conquered by the Moors of Morocco. It was colonized by the French in the 19th century as part of French West Africa, and in 1946 became the overseas territory of French Sudan. In 1958 it became autonomous as the Sudanese Republic; with Senegal it formed the Federation of Mali in 1959 and became fully independent in 1960.

Military coups in 1968 and 1991 were followed by the installation of a multiparty constitution in 1992. Two elections have been held but have been boycotted by opposition parties. Former president Moussa Traore and his wife were sentenced to death in the late 1990s for corruption of state funds.

Mali is home to a variety of ethnic groups, ranging from the Tuareg and Moors in the north to the Bambara, Fulani and Malinke in the south. Each people has its own language, but French and Arabic are widely spoken. Most of the population are Muslim.

ECONOMY

Most people are employed in agriculture, on which the economy largely rests. The country was self-sufficient in food production until the Sahel droughts (1968 onward), which affected cattle and other livestock particularly badly. Staple crops include grains, cotton and groundnuts (peanuts).

Industry is small-scale, and energy is generated by thermal and hydroelectric projects built with foreign aid. Mineral resources include gold, uranium and phosphates, iron and bauxite; however, inadequate transportation services, including poor roads, contributes to their underexploitation. Foreign aid and borrowing have supported the economy, and debt remains a serious handicap despite substantial debt relief. Mali's economy was showing signs of growth following an IMF structural readjustment program.

NATIONAL DATA – MALI

Land area	1,220,000 sq km (471,045 sq mi)			
Climate		Temperatures	Annual	
	Altitude m (ft)	January °C(°F)	July °C(°F)	precipitation mm (in)
Bamako	331 (1,086)	25 (77)	27 (81)	991 (39)

Major physical features highest point: Hombori Tondo 1,155 m (3,789 ft); longest river: Niger (part) 4,200 km (2,600 mi)

Population (2006 est.) 11,716,829

Form of government multiparty republic with one legislative house

Armed forces army 7,350; paramilitary 4,800; inactive militia 3,000

Largest cities Bamako (capital – 1, 690,471); Sikasso (154,823); Mopti (112,248); Koutiala (106,857); Kayes (103,797)

Official language French

Ethnic composition Mande 50% (Bambara, Malinke, Soninke); Peul 17%; Voltaic 12%; Songhai 6%; Tuareg and Moor 10%; other 5%

Religious affiliations Muslim 90%; indigenous beliefs 9%; Christian 1%

Currency 1 Communauté Financière Africaine franc (XOF) = 100 centimes

Gross domestic product (2006 est.) U.S. $14.59 billion

Gross domestic product per capita (2006 est.) U.S. $1,200

Life expectancy at birth male 47.05 yr; female 51.01 yr

Major resources gold, phosphates, kaolin, salt, limestone, uranium, gypsum, granite, hydropower, bauxite, iron ore, manganese, tin, copper

Niger

REPUBLIC OF NIGER

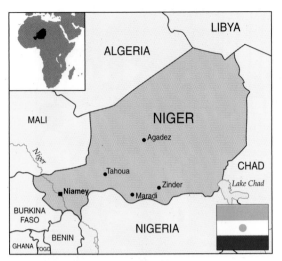

THE REPUBLIC OF NIGER LIES AT THE HEART of northwestern Africa. This land-locked country borders on no fewer than seven other nations. With them it forms the Sahel area, notorious for its repeated droughts since the late 1960s.

GEOGRAPHY

Two-thirds of Niger forms part of the vast Sahara, though the hot, mostly dry climate varies sharply according to the season and the area. The entire north of the country is an arid highland that forms part of a mountain chain stretching from Algeria to Chad. To the east lies desert country, with sand dunes and vast areas of stony ground. Only in the southwest, where the Niger river crosses the country, is there rich arable soil, which becomes flooded in the brief rainy season. In the southeast of the country around Lake Chad lies the Manga wasteland, which mostly supports savanna grassland.

There is a great diversity of plant life in the country, including mahogany, kapok and baobab trees, and date palms near the desert margins. Animal life is rich, but many species are threatened, and conservation is minimal.

SOCIETY

Niger's original settlers were the nomadic Fulani and Tuareg peoples from the north and the Hausa from the east. These remain the chief ethnic groups today, and more than half the population is Hausa. In 1804 the strictly Muslim Fulani subjugated the Hausa – also nominally Muslim – in a holy war (jihad) and established the Sokoto empire.

In the late 19th century France gradually assumed control of the country. In 1922 Niger became part of French West Africa, and in 1946 it became a French overseas territory. After independence in 1960 the first president, Hamani Diori (b. 1916), maintained economic links with France. In 1974 Colonel Seyni Kountché (1931–87) staged a coup and established a military dictatorship. He died in 1987 after a series of coup attempts by his colleagues, and was succeeded by the more moderate and popular General Ali Seybou (b. 1940).

The first multiparty elections (1993) were won by a left-wing coalition, Alliance of the Forces for Change (AFC), with Mahamane Ousmane as president. He was toppled in 1996 in a military coup led by Ibrahim Barre Maïnassara, who was assassinated in 1999. Daouda Mallam Wanké now heads a military dictatorship. Most donor countries suspended aid payments causing Niger economic difficulties.

ECONOMY

Agriculture is Niger's principal employer, and accounts for 39 percent of GDP. The country is virtually self-sufficient in food but suffers from periodic drought. The main crops include grains, with rice in the flooded areas. Groundnuts (peanuts), cotton and cattle are exported.

Most manufacturing is in agricultural products, textiles and chemicals. Energy resources are limited; electricity is generated, but fuels are imported. The main natural resource is uranium ore. It represents over one-third of gross national product, but the economy was badly destabilized by a decline in worldwide uranium prices during the 1980s, and Niger is now burdened with high levels of foreign debt. Surfaced roads have now been constructed across the desert to the mines, and there is an extensive domestic air service.

Social services consist largely of vaccination and disease-elimination campaigns. An education system has been established with French aid, but illiteracy remains high. There is one university, at Niamey, the capital, in the southwest.

Niger has few roads, is subject to repeated droughts, and consists largely of desert. With so few surfaced roads, it is scarcely surprising that the traditional camel train remains a vital means of carrying goods.

NATIONAL DATA – NIGER

Land area	1,266,700 sq km (489,076 sq mi)			
Climate		**Temperatures**		**Annual**
	Altitude m (ft)	**January** °C(°F)	**July** °C(°F)	**precipitation** mm (in)
Niamey	222 (728)	26 (80)	29 (84)	541 (21.2)

Major physical features highest point: Mont Gréboun 1,944 m (6,378 ft); longest river: Niger (part) 4,200 km (2,600 mi)

Population (2006 est.) 12,525,094

Form of government multiparty republic with one legislative house

Armed forces army 5,200; air force 100

Largest cities Niamey (capital – 707,951); Zinder (170,575); Maradi (148,017); Agadez (78,289); Arlit (69,485)

Official language French

Ethnic composition Hausa 56%; Djerma 22%; Fula 8.5%; Tuareg 8%; Beri Beri (Kanouri) 4.3%; Arab, Toubou, and Gourmantche 1.2%

Religious affiliations Muslim 80%; traditional beliefs and Christian 20%

Currency 1 Communauté Financière Africaine franc (XOF) = 100 centimes

Gross domestic product (2006 est.) U.S. $12.23 billion

Gross domestic product per capita (2006 est.) U.S. $1,000

Life expectancy at birth male 43.8 yr; female 43.73 yr

Major resources uranium, coal, iron ore, tin, phosphates, gold, molybdenum, gypsum, salt, petroleum, cotton, groundnuts, livestock, millet, rice, sorghum

Chad

REPUBLIC OF CHAD

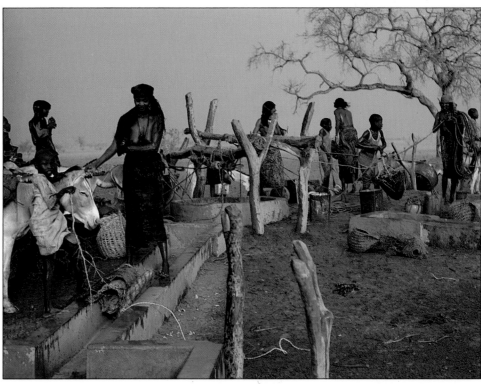

Villagers take water from a well in Chad. Here, at the very heart of the Sahel drought area, water can never be taken for granted, while war and famine are a constant threat.

The REPUBLIC OF CHAD STRADDLES THE drought-stricken area of the Sahel. Since the mid-1960s it has been gripped by civil wars, aggravated by conflict with its powerful neighbor Libya.

GEOGRAPHY

Chad is landlocked, and the only large body of water is Lake Chad in the southwest. Shallow and marshy, it once covered a larger area, but is shrinking as the desert advances. The only significant rivers feeding it, the Chari and the Logone, flow from the southeast to meet at the lake. Through it run the country's frontiers with Niger, Nigeria and Cameroon. From the area of the Chad Basin the land gradually rises to the mountains along the opposite frontiers: the volcanic Tibesti mountains to the north; and the Ennedi plateau and Ouaddai mountains that lie to the east.

The climate is generally tropical and hot, but rainfall varies widely. Summer rains support fertile savanna grasslands and broadleaf trees in the south (as much as one-fifth of the land is forested), but rainfall dwindles farther north. The entire northern third of the country consists of arid scrubland fading into barren true desert, with only a few palm oases. The land to the south supports an abundance of wildlife, including African elephants, cheetahs, giraffes, lions and warthogs, in a relatively unspoilt environment.

SOCIETY

Chad is a melting-pot of peoples and cultures, and more than 100 different languages are spoken. Arab settlers from the north founded the Islamic kingdom of Kanem-Bornu, which reached its peak in the 16th century, commanding the main trans-Saharan caravan route to Libya and the Mediterranean. The Muslim slave traders preyed heavily on the people of the south, engendering present-day conflicts. In the 19th century the whole area was conquered by a Sudanese warlord, who in turn was ousted by the French. Chad became part of French Equatorial Africa in 1908, and in 1960 gained independence under President François Tombalbaye (1918–75).

Within several years Libyan-backed guerrilla organizations had raised a revolt in the north. Tombalbaye called in French troops, but in 1975 he was assassinated. Soon afterward Libya annexed the Aouzou Strip along the northern border of Chad. Fighting continued between various guerrilla factions. The most powerful guerrilla leader, Hissen Habré (b. 1942), seized the presidency with French backing in 1984, and in 1986 drove out Libyan incursions. Habré won the 1990 elections, but within six months he was overthrown by Idriss Déby. The approval of a new constitution and the holding of multiparty elections in 1997 and 1998 brought some peace to Chad although further fighting broke out in the north in 1998.

ECONOMY

Agriculture employs about four-fifths of the work force and provides about half of gross domestic product. The country is self-sufficient in food when there are no droughts. The chief food crops are millet, sorghum and rice, with cotton and groundnuts (peanuts) for export. Livestock is the next most important export.

Petroleum is the major resource. Exploitation of oil in the Doba Basin has foundered on environmental objections to a new pipeline requiring World Bank funding. Industry is confined to agricultural processing and minor manufacturing. Energy generation and transportation are extremely limited. Literacy is low as less than half the children are able to attend school. Health services are poor and confined mainly to the south.

NATIONAL DATA – CHAD				
Land area 1,259,200 sq km (486,180 sq mi)				

Climate	Altitude m (ft)	Temperatures January °C(°F)	July °C(°F)	Annual precipitation mm (in)
N'Djamena	295 (968)	21 (70)	26 (79)	510 (20)

Major physical features highest point: Emi Koussi 3,415 m (11,204 ft); longest river: Chari 949 km (590 mi)	
Population (2006 est.) 9,944,201	
Form of government multiparty republic with one legislative house	
Armed forces army 25,000; air force 350	
Capital city N'Djamena (753,791)	
Official languages Arabic, French	
Ethnic composition Sara/Bagirmi/Kreish 30.5%; Sudanic Arab 26.1%; Teda 7.3%; Mbum 6.5%; Masalit/Maba/Mimi 6.3%; Mubi 4.2%; Kanuri 2.3%; Hausa 2.3%; Masa 2.3%; Kotoko 2.1%; others 10.1%	
Religious affiliations Muslim 51%; Christian 35%; animist 7%; other 7%	
Currency 1 Communauté Financière Africaine franc (XAF) = 100 centimes	
Gross domestic product (2006 est.) U.S. $15.26 billion	
Gross domestic product per capita (2006 est.) U.S. $1,500	
Life expectancy at birth male 45.88 yr; female 49.21 yr	
Major resources petroleum, uranium, natron, kaolin, fish (Lake Chad), gold, limestone, sand and gravel, salt, cassava, cattle, cotton, dates, gum Arabic, millet, groundnuts, rice, sorghum, sweet potatoes, yams	

Sudan

REPUBLIC OF SUDAN

SUDAN IS AFRICA'S LARGEST STATE, EXTENDing from just south of the Tropic of Cancer to just north of the Equator. Its landscape ranges from harsh desert in the north to dense rainforests in the south. For thousands of years it formed a trade bridge between Southern Africa and the Mediterranean, yet today it is one of the least-developed countries in the world.

GEOGRAPHY

In southern Sudan the land rises all along the borders with Kenya, Uganda, Congo (D.R.O.) and the Central African Republic, reaching its highest point at mount Kinyeti on the Ugandan border. The White Nile enters Sudan at a place called Nimule, also on the Ugandan border, and flows northward into a clay plain that occupies most of southern and central Sudan. The plain is dotted with isolated hills (inselbergs), along with the Nuba Mountains, and in the east it rises to the foothills of the Ethiopian Highlands. Here the Blue Nile flows northwestward to meet the White Nile at the capital, Khartoum. Between the two rivers is a huge marshy area known as the Sudd.

On the western side of Sudan, along the Chad border, the volcanic massif of the Jebel Marra rises from the sand dunes of the Qawz. To the north the dunes gradually give way to the rock desert of the Sahara, which continues to the northern borders with Libya and Egypt. This barren landscape is interrupted only by a thin line of cultivation along the Nile. In the northeast the Red Sea mountains rise to about 2,000 m (6,500 ft) above a narrow coastal plain.

The heat in Sudan is moderated only along the Red Sea coast. Rainfall, vegetation and wildlife decrease gradually toward the north. The far south has an equatorial climate with very high rainfall and a long rainy season. Tropical forests flourish here, and animals include chimpanzees, hippopotamuses, elephants, zebras, giraffes, leopards, lions and many species of antelope. The central southern area supports savanna and semiarid shrubland; papyrus swamps grow in the Sudd marshes. The north has a tropical continental climate with almost no rainfall, creating desert or semidesert.

SOCIETY

The ancient Egyptians knew the Sudan as Nubia, but from the 11th to the 4th century BC it was part of the kingdom of Kush. Coptic Christian missionaries in the 6th century AD found three kingdoms here: Nobatia, Makurra and Alwah. In the 7th century Makurra, which by then included Nobatia, stood firm against Islam and Arab invaders, but in the 14th century Mameluke Turks devastated the country and it gradually came under Arab control. Alwah in the south, the last area to stand out against the Arabs, fell in about 1500, but ethnic and religious conflicts remain here.

In the 16th century the Funj people carved out a kingdom between the Ethiopian Highlands and the Blue Nile, but in 1821 it was shattered when Muhammad Ali (1769–1849), viceroy of Egypt under the Ottoman Turks, invaded the Sudan. In 1881 Muhammad Ahmed (1848–85), known as the Mahdi or "divine guide", led a triumphant Islamic revolution to drive out the British and the Egyptians; this ended in 1885 with the fall of Khartoum, and the death of the British General Charles George Gordon (1833–85), governor-general of Sudan. An Anglo-Egyptian army under the command of General Horatio Herbert Kitchener (1850–1916) invaded the Sudan in 1897, and the area remained in British hands until 1951, when the Egyptian government seized control. In 1953 the Sudanese were granted self-government. Three years later they declared independence, but there was continued civil war because the underprivileged south resented the Islamic bias of the new regime in the north.

Parliamentary democracy was shortlived: in 1959 a bloodless army coup put General Ibrahim Abbud (b. 1900) in power. A half-hearted return to democracy in 1964 ended in 1969 with a coup led by Colonel Gaafar Mohamed el-Nimeiry (b. 1930), and in 1972 the civil war was temporarily halted. It erupted again in 1983, leaving the Sudanese Popular Liberation Army largely in control of the south. Two years later Nimeiry was overthrown by his defense minister, General Abd ar-Rahman Siwar ad-Dahab (b. 1934), who once again attempted to restore democracy. But in 1989 Lieutenant-General Omar Hassan Ahmad al-Bashir

NATIONAL DATA – SUDAN				
Land area	2,376,000 sq km (917,379 sq mi)			

| Climate | Altitude m (ft) | Temperatures | | Annual precipitation mm (in) |
		January °C(°F)	July °C(°F)	
Khartoum	380 (1,247)	23 (74)	32 (90)	162 (6.3)

Major physical features highest point: Kinyeti 3,187 m (10,456 ft); longest river: Nile (part) 6,690 km (4,160 mi)

Population (2006 est.) 41,236,378

Form of government multiparty republic with one legislative house

Armed forces army 100,000; navy 1,800; air force 3,000

Largest cities Omdurman (3,127,802); Khartoum (capital - 2,207,794); Khartoum Bahri (1,725,570)

Official language Arabic

Ethnic composition Black 52%; Arab 39%; Beja 6%; other 3%

Religious affiliations Sunni Muslim 70% (in north); traditional beliefs 25%; Christian 5%

Currency 1 Sudanese dinar (SDD) = 10 Sudanese pounds, which are each further subdivided into 100 piastres

Gross domestic product (2006 est.) U.S. $96.01 billion

Gross domestic product per capita (2006 est.) U.S. $2,300

Life expectancy at birth male 57.69 yr; female 60.21 yr

Major resources petroleum, iron ore, copper, chromium ore, zinc, tungsten, mica, silver, gold, hydropower, beans, cotton, natural gas, dates, groundnuts, gum Arabic, livestock, sesame seeds, sorghum, wheat

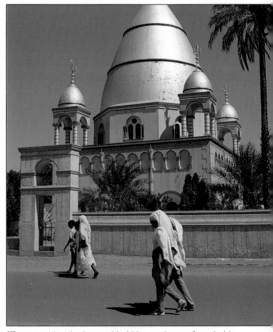

Khartoum has had a troubled history: it was founded in 1823 by an Egyptian ruler, destroyed by the Mahdi, and rebuilt in 1898 by Lord Kitchener. Fine mosques such as this one stand near grand colonial buildings, and there is also a large industrial quarter.

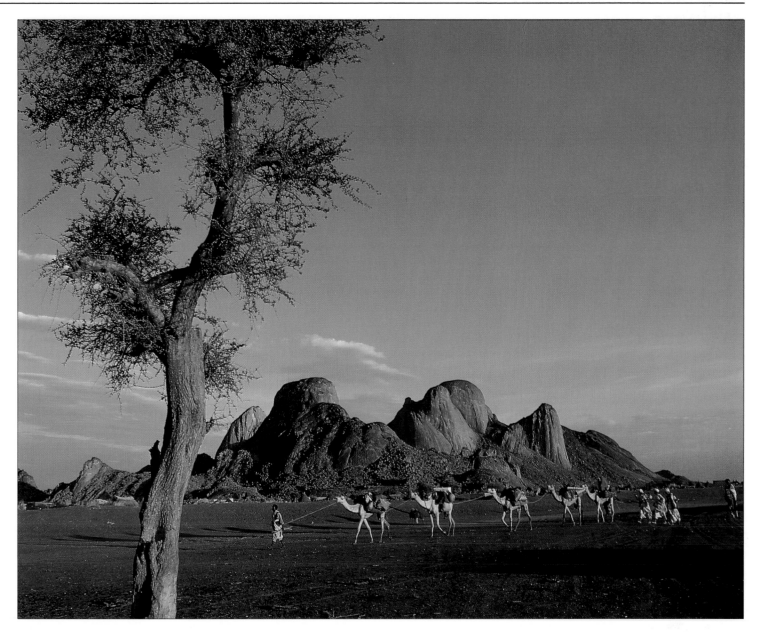

(b. 1944) seized power and dissolved all political institutions on behalf of the National Islamic Front. Multiparty elections were held in 1996 but the civil war continued. Up to two million people have died in the fighting or through war-induced famine according to estimates made in 1999.

The main internal divisions are between the Arab and the Nubian ethnic groups in the north (mostly Muslim), and the Nilotic and the Sudanic peoples in the south (mostly non-Muslim). During the 1970s and 1980s the southern population has been swelled by vast numbers of refugees from Ethiopia, Eritrea, Uganda and Chad. The official language is Arabic, spoken by about half the population, and the dominant religious group are the Sunni Muslims.

ECONOMY

Most of the work force are involved in agriculture. The area under cultivation varies with the changing rainfall patterns, and much depends on elaborate irrigation schemes. Cotton is a cash crop, usually at the expense of food for the domestic market (agriculture accounts for nearly one-third of the gross national product). Other crops include sorghum (the staple), millet, wheat, barley and groundnuts.

Forests cover about one-fifth of the land area, and most productive woodland is state-owned. A further one-fifth is given over to seasonal pasture for goats, cattle and other livestock such as camels and sheep (both exported). Fishing along the Nile and off the Red Sea coast mostly supplies local demand.

Gold, chromite, salt, manganese and mica are extracted, and oil production started in earnest in 1998. An oil pipeline and refinery were under construction in 1999. Over 50 percent of the Sudan's power production comes from hydroelectricity while the rest of the power comes from fossil fuels. Manufacturing, chiefly of textiles, shoes, beverages and soap, provides just under one-tenth of the gross national product.

A camel train approaches Kassala, 400 km (250 mi) east of the Sudanese capital, Khartoum, and near the border with Ethiopia. Kassala today is a market town noted for its fruit gardens, but its oldest building, constructed in 1840, was an Egyptian fort.

Saudi Arabia is Sudan's chief trading partner, followed by European Community countries. Foreign debt is crippling, and the balance of trade is very unfavorable. Most of the roads are unsurfaced, and impassable after rain. The state-run railroad, with some 5,000 km (3,000 mi) of track, employs one-fifth of the country's work force. Steamer services operate on the White Nile, but Port Sudan in the northeast is the only important outlet to the sea. There are 10 airports operating scheduled flights. Radio, television and both national newspapers are state-controlled.

Social–welfare services are available on a very limited basis, but there are few health care facilities. Infectious diseases (including malaria), malnutrition, poor housing and inadequate sanitation remain serious problems.

Ethiopia

FEDERAL DEMOCRATIC REPUBLIC OF ETHIOPIA

ETHIOPIA IS THE MODERN DESCENDANT OF a powerful and ancient empire, the fabled land of Prester John in medieval European legend. Today it is one of the world's poorest countries, scarred and torn by drought, famine and civil war.

GEOGRAPHY

Ethiopia is a massive plateau cut in half from north to south by the East African Rift Valley. To the west, along the Sudanese border, the land rises as high as 3,700 m (12,000 ft). At the heart of the western highlands lies the capital, Addis Ababa, some 2,500 m (8,000 ft) above sea level. From Lake Tana in the northwest, the Blue Nile gorge cuts a winding course, first to the southeast and then westward. In the north the Simen Mountains rise to Ras Dashan, Ethiopia's highest peak, then drop away to the narrow Red Sea coastal plain.

The Rift Valley averages some 50 km (30 mi) in width between its eastern and western flanks. At the southern end, near the Kenyan border, lakes, lava flows and volcanic cones are characteristic. To the north the valley widens into the broad plains of the Awash river. Northward again, in the Danakil Plains, there are hot springs, active volcanoes, salt lakes and the Kobar Sink, a large desert area below sea level. The border with Djibouti lies to the southeast of this area.

The eastern highlands climb to more than 4,300 m (14,000 ft) along the eastern side of the Rift Valley. Southeastward they descend to the Bale plateau and again to the Harerge plain in the Ogaden, which stretches east and south to the disputed border with Somalia.

The high plateaus enjoy a temperate climate that becomes increasingly hot in lower-lying areas. Rainfall is highest in the south and west. In recent years prolonged drought in the north has brought disease, destitution and famine.

There is an enormous variety of vegetation, from tropical rainforest and savanna grasslands in low-lying areas, to alpine plants in the high mountains. Animals range from ibex and nyala (an antelope) in the mountains to hippopotamuses and crocodiles in lakes and rivers.

SOCIETY

The origins of modern Ethiopia lie in the kingdom of Aksum (2nd century AD), which became Christian in the 4th century. It was cut off from Europe by the Muslim conquests of the 7th century. Not until the middle of the 15th century, when Ethiopian monks seeking to unite their church with Rome were believed to have come from the legendary priest-king, Prester John, was European awareness of Ethiopia resumed.

In 1520 the Portuguese made contact, and gave some aid to Ethiopia's continuing struggle against the Turks and insurrection in its own Muslim provinces in the southeast. However, in 1633 all Jesuit missionaries were expelled. A cultural renaissance followed, but the archaic social system remained unchanged.

The old empire was finally reunified under Menelik II (1844–1913) near the end of the 19th century, though by now the Italians had laid claim to the northern area of Eritrea. Modernization began

An Ethiopian woman makes a traditional headdress by stringing together colored beads.

under Emperor Haile Selassie (1892–1975) in 1930, but five years later the Italians invaded. They were driven out in 1941, and in 1950 Eritrea became a federated state of Ethiopia. Its full incorporation in 1962 sparked a rebellious conflict.

In 1974 a military coup deposed the emperor, triggering a chain of events that ended in civil war. By 1984 the conflict had become so serious that it hampered the international aid agencies bringing much-needed famine relief to the area. In April 1993 Eritrea won its independence from Ethiopia and was recognized as a new state by the United Nations in May. Fighting again erupted along the border between Ethiopia and Eritrea in early 1999. The Ethiopian people are a complex intermixture of ethnic groups, the most powerful are the Amhara, who make up some 30 percent of the population and live mostly in the north and the heartlands. The Oromo constitute about 40 percent of the population, and live mostly in eastern and southern areas. The official language is Amharic, though some 70 other languages are spoken. About half the population are thought to be Muslim. Most people live in rural areas, and many are nomadic or seminomadic.

ECONOMY

Agriculture accounts for some four-fifths of the work force, but output has been severely reduced by war and drought. The highest areas (above 3,000 m; 10,000 ft) support alpine pastures. Below them is a temperate zone where most of the population live. Here maize, wheat, barley and other cereals are grown, and horses, cattle

NATIONAL DATA – ETHIOPIA

Land area	1,119,683 sq km (432,312 sq mi)			
Climate		**Temperatures**		Annual
	Altitude m (ft)	January °C(°F)	July °C(°F)	precipitation mm (in)
Addis Ababa	2,360 (7,741)	20 (67)	20 (67)	1,020 (40.1)

Major physical features	highest point: Ras Dashan 4,620 m (15,158 ft)

Population	(2006 est.) 74,777,981

Form of government	one-party federal republic with one legislative house

Armed forces	army 180,000; air force 2,500

Largest cities Addis Ababa (capital - 2,973,000); Dire Dawa (281,800); Nazret (228,600); Bahir Dar (167,300)

Official language	Amharic

Ethnic composition Oromo 40%; Amhara and Tigre 32%; Sidamo 9%; Shankella 6%; Somali 6%; Afar 4%; Gurage 2%; other 1%

Religious affiliations Muslim 45%–50%; Ethiopian Orthodox 35%–40%; animist 12%; other 3%–8%

Currency	1 Ethiopian birr (ETB) = 100 cents

Gross domestic product	(2006 est.) U.S. $71.63 billion

Gross domestic product per capita	(2006 est.) U.S. $1,000

Life expectancy at birth	male 47.86 yr; female 50.24 yr

Major resources small reserves of gold, platinum, copper, natural gas, hydropower, barley, beans and peas, cattle, coffee, cotton, maize/corn, millet, potash, salt, sorghum, sugarcane, timber, wheat and sheep are raised. Below 1,800 m (6,000 ft) the climate is tropical, and high-rainfall areas are suitable for coffee (the main export crop) and sugar cane.

Ethiopia has reserves of gold, platinum and copper, and valuable limestone deposits. Most electricity comes from hydroelectric plants; there is scope for many more, and for geothermal stations along the Rift Valley. Textiles and food processing dominate the small, largely nationalized manufacturing industry.

Famine, drought and civil war drained the economy and disrupted export production in the 1980s, and the border war with Eritrea, which ended in December 2000 under UN auspices, further depressed the economy. Ethiopia maintains the largest standing army in Africa, and its overseas debts are enormous. Major trading partners are Germany, Saudi Arabia, the United States, Italy and Japan.

Surfaced roads link Addis Ababa with provincial capitals. The railroad from Addis Ababa to the port of Djibouti on the Gulf of Aden provides the country's main link to the sea. Waterways have not been developed to any significant extent. There are three international airports. In 1999 there were 25 television stations and one Internet service provider, but only 320,000 televisions.

Few of the population benefit from the four-tiered educational system or from modern health care. Contagious and tropical diseases are rife and average life expectancy is among the world's lowest.

The Tis Isat Falls in full flow during the rainy season at the outflow of the Blue Nile from Lake Tana in northwestern Ethiopia.

Eritrea

STATE OF ERITREA

ON 28 MAY 1993 ERITREA GAINED UNITED Nations recognition as an independent country after a 30-year war with Ethiopia that left the new nation famine-stricken and impoverished. A former Italian colony, and under British rule 1942–52, Eritrea became part of Ethiopia in 1962 against the wishes of its people, whose struggle for independence was won in 1991 when other discontented groups in Ethiopia joined with the Eritrean People's Liberation Front to bring down the Mengistu government.

Eritrea is currently controlled by a transitional government. A constitution, ratified in 1997, has not yet been put into effect, and parliamentary and presidential elections, promised for that year, have been postponed indefinitely following the border war with Ethiopia in 1998.

Eritrea is a small, desperately poor country. Eighty percent of the population are subsistence farmers, and there is a tiny, outmoded industrial sector. Worker remittances from overseas substantially augment the GDP. In the long term Eritrea has plans to develop tourism along the Red Sea as well as an offshore oil industry, and offshore fishing.

NATIONAL DATA - ERITREA	
Land area	121,320 sq km (46,842 sq mi)
Climate	hot, dry desert strip
Major physical features	highest point: Monte Soira 2,989 m (9,807 ft)
Population	(2006 est.) 4,786,994
Form of government	transitional
Armed forces	army 200,000; navy 1,400; air force 350
Capital city	Asmara (500,600)
Official languages	Tigrinya and Arabic
Ethnic composition	ethnic Tigrinya 50%; Tigre and Kunama 40%; others 10%
Religious affiliations	Muslim; Coptic Christian; Roman Catholic; Protestant
Currency	1 nafka (ERN) = 100 cents
Gross domestic product	(2005 est.) U.S. $4.471 billion
Gross domestic product per capita	(2005 est.) U.S. $1,000
Life expectancy at birth	male 57.44 yr; female 60.66 yr
Major resources	gold, potash, zinc, copper, salt, unexploited oil and natural gas, fish, barley, beans and peas, cattle, coffee, cotton, maize/corn, millet, platinum

Djibouti

REPUBLIC OF DJIBOUTI

D JIBOUTI – THE SMALLEST COUNTRY IN THE region – is a strategically sited desert republic on the east coast of Africa between the Red Sea and the Indian Ocean. It faces the Gulf of Aden and borders Somalia and Ethiopia.

GEOGRAPHY

The north of Djibouti is moderately mountainous, while the heartland and the south are dominated by volcanic plateaus; faults parallel to the Red Sea have produced deep troughs occupied by saline lakes and the Hanlé Plains. The Gulf of Tadjoura cuts deeply into the eastern coastal plain of Djibouti.

Djibouti is among the hottest countries in the world. Most of its scanty precipitation falls in the northern mountains between late summer and the end of March. The predominant vegetation is desert scrub, with mangroves along the coast and a few wooded areas in the mountains. The animal life is well adapted to the extreme conditions, and includes lynx, jackals and gazelles.

SOCIETY

Djibouti's two main ethnic groups both speak a Hamitic language and share a common religion – Islam. The Afars are descended from 3rd-century settlers, whereas the Issas were later arrivals who came from Somalia.

In 1862 Obock (on the northern side of the Gulf of Tadjoura) was sold to the French by the Sultanate of Tadjoura. Thirty years later the town of Djibouti across the gulf became the capital of French Somaliland (formerly the French Territory of the Afars and Issas). France protected its investment heavily, but a 1977 plebiscite showed 85 percent in favor of independence, and Hassan Gouled Aptidon (b. 1916) became the first president of the new republic. French is the official language (together with Arabic), and French military aid preserves Djibouti's independence and neutrality.

After a referendum, a new constitution was approved in 1992 which permits up to four political parties, each with an ethnic balance of membership. In the subsequent 1992 election the ruling party (Rassemblement Populaire pour le Progrès) won all 65 seats in the Chamber of Deputies and Aptidon was re-elected president. However, other parties boycotted the election.

ECONOMY

Agriculture meets only one-quarter of Djibouti's needs; its development is hampered by the scarcity of water for irrigation. Livestock are raised, but less than one-tenth of the land area is under pasture. Fishing is of increasing importance.

Djibouti's resources are limited and the economy depends on services and external trade, particularly trans-shipment of Ethiopian and Somalian exports and imports. Recession, civil war and high population growth rate have only added to the problems Djibouti faces and the country is heavily dependent on foreign aid. The capital's docks are well developed, and linked by rail with Addis Ababa in Ethiopia. Foreign assistance from countries such as France, Italy and Saudi Arabia is readily forthcoming because of its strategic position. Few of the roads are surfaced. An international airport links the capital with France, Ethiopia, Madagascar, Egypt and other east African countries. There are radio broadcasts in four languages, and one television station.

There are a number of dispensaries and medical posts in rural areas, and two hospitals. Malnutrition, dehydration and poisoning account for almost half the recorded deaths of children up to the age of 10.

Fishermen from Djibouti (*left*) catch tuna in the Bab al Mandab – the strait that links the Red Sea with the Gulf of Aden and the Indian Ocean. A program to develop fishing was started in 1980.

NATIONAL DATA – DJIBOUTI				
Land area 22,980 sq km (8,873 sq mi)				
Climate		Temperatures		Annual
	Altitude m (ft)	January °C(°F)	July °C(°F)	precipitation mm (in)
Djibouti	8 (26)	25 (74)	36 (97)	164 (6.4)
Major physical features highest point: Musa Ali Terara 2,063 m (6,768 ft); lowest point: Lake Assal -150 m (-492 ft)				
Population (2006 est.) 486,530				
Form of government multiparty republic with one legislative house				
Armed forces army 8,000; navy 200; air force 250				
Capital city Djibouti (317,000)				
Official languages Arabic, French				
Ethnic composition Somali 60%; Afar 35%; French, Arab, Ethiopian, and Italian 5%				
Religious affiliations Muslim 94%; Christian 6%				
Currency Djiboutian franc (DJF) = 100 centimes				
Gross domestic product (2002 est.) U.S. $619 million				
Gross domestic product per capita (2005 est.) U.S. $1,000				
Life expectancy at birth male 41.86 yr; female 44.52 yr				
Major resources geothermal power, gold, clay, granite, limestone, marble, salt, diatomite, gypsum, pumice, petroleum, cattle, dates, fish, fruit and vegetables, goats, sheep				

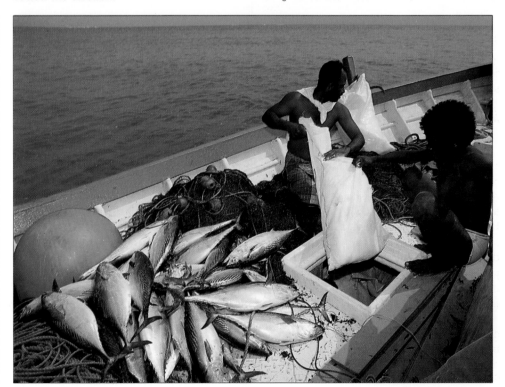

Somalia

SOMALI DEMOCRATIC REPUBLIC

SOMALIA LIES ALONG THE HORN OF AFRICA, a large promontory on Africa's northeastern coast. In recent years it has been torn apart by civil war and drought.

GEOGRAPHY

In the north lie the Ogo Highlands, where rugged cliffs and mountains of ancient rock extend from neighboring Ethiopia along the Gulf of Aden. The narrow coastal plain is a hot, barren scrubland called the Guban.

South of the mountains lies the Haud Plateau, beyond which the land slopes gradually down toward the Indian Ocean. The two main rivers, the Juba and the Shebelle, flow across the country from Ethiopia, providing a vital water resource in an almost constantly hot and inhospitable arid climate.

Only in the south and on northern slopes is there enough rainfall to support small forests and grassland. Otherwise Somalia is covered by semidesert or dry savanna vegetation. Elephants, zebras, giraffes, antelopes and leopards are among the wildlife typical of the region.

SOCIETY

Somalia was probably the "Land of Punt" referred to in the Bible. Its modern people are descended from Arab or Persian traders along the seacoast and the nomadic Somali tribesmen they came to dominate. More than two-thirds of the population are still nomadic; there are three main tribal groups, but Somali is the dominant language.

In the 19th century the area was colonized by Britain in the north and Italy in the east, though Muslim resistance remained. In 1946 Italian and British Somaliland were united to form a United Nations trust territory, which became independent in 1960.

In 1969 the president was assassinated in a Soviet-backed military coup and replaced by the repressive rule of Major-General Siyad Barre (1921–95). In 1977 he invaded the Ogaden area of Ethiopia, provoking a split with the Soviets and a bloody civil war that was fueled by rivalries between different groups of indigenous peoples. Coup attempts, massacres and growing civil unrest followed, leading in 1991 to the overthrow of Barre's government. Continuing unrest prevented international aid reaching refugee camps in famine areas, and United States troops were sent in December 1992 to assist. They withdrew after a truce in 1994, but clan-based fighting continued, and all UN troops withdrew in 1995.

ECONOMY

Arable farming has traditionally been confined to a small area lying between the two main rivers, which covers less than one-fiftieth of the overall area. Crops include maize, sorghum and groundnuts (peanuts), with bananas and sugar cane along the river banks. About half the land supports the nomadic herding of cattle, sheep and goats and exports consist mainly of livestock, skins, hides and bananas.

Industry accounted for only one-tenth of GDP before civil war and was based mainly on agricultural products and petroleum refining. In the late 1990s, few industries were operating. There are unexploited deposits of chromium, coal, copper, gold and other minerals.

Transportation is limited, with no railroads and limited roads, few of which are surfaced. Port and airport facilities are chiefly at Mogadishu, the capital. State radio stations broadcast in four languages;

Refugees wait for food at a Somalian camp. In the 1990s continued civil war, floods and drought made an estimated 1.5 million people dependent on foreign aid but movement of food aid only became effective in 1992 when American troops were used to police distribution.

the television station was destroyed in 1991. State education exists to university level, but is not compulsory and is open to only a few. Literacy levels are low due to the war. Health care, which lacks both staff and basic supplies, is not widely available.

NATIONAL DATA – SOMALIA

Land area 627,337 sq km (242,216 sq mi)

Climate	Altitude m (ft)	Temperatures January °C(°F)	July °C(°F)	Annual precipitation mm (in)
Mogadishu	17 (56)	27 (81)	26 (79)	427 (17)

Major physical features highest point: Surud Ad 2,406 m (7,894 ft); longest river: Shebelle (part) 2,010 km (1,250 mi)

Population (2006 est.) 8,863,338

Form of government transitional

Armed forces no national armed forces

Largest cities Mogadishu (capital – 2,855,805); Hargeysa (522,508); Marka (350,859); Berbera (263,095)

Official language Somali

Ethnic composition Somali 85%; Bantu and other non-Somali 15% (including Arabs 30,000)

Religious affiliations 100% Sunni Muslim

Currency 1 Somali shilling (SOS) = 100 cents

Gross domestic product (2006 est.) U.S. $5.023 billion

Gross domestic product per capita (2006 est.) U.S. $600

Life expectancy at birth male 46.71 yr; female 50.28 yr

Major resources uranium and largely unexploited reserves of iron ore, tin, gypsum, bauxite, copper, salt, natural gas, possible oil reserves, textiles, bananas, camels, cotton, goats, groundnuts, maize/corn, millet, sheep, sugarcane

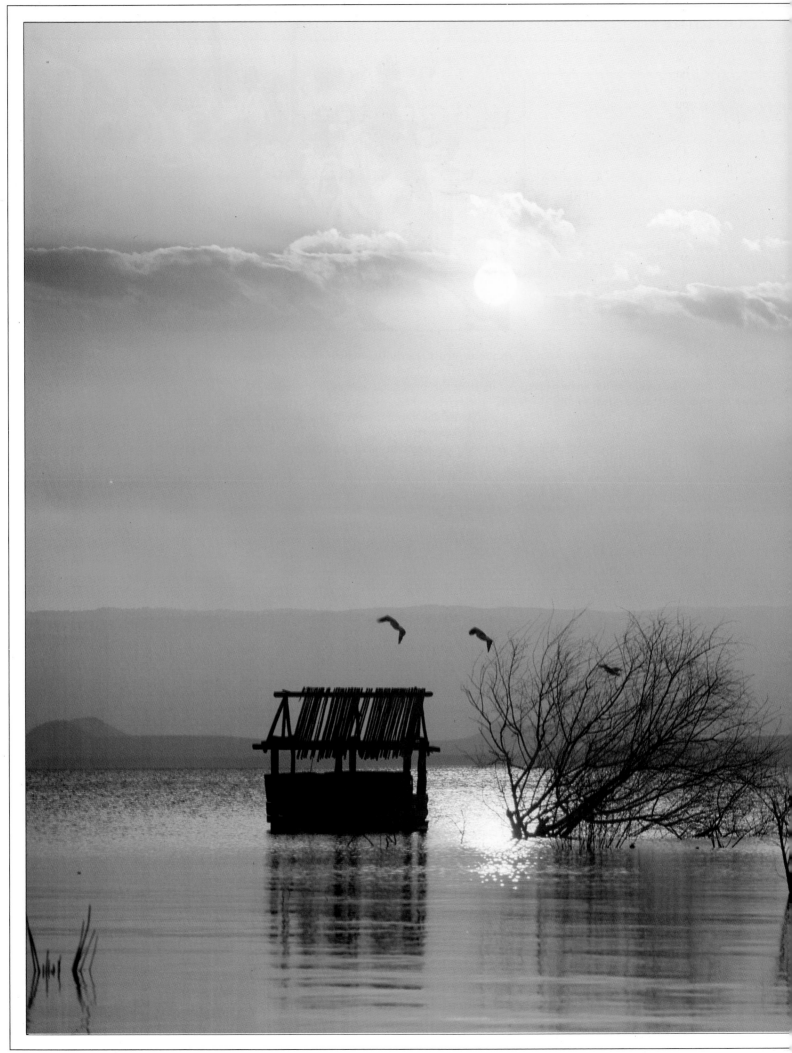

Central Africa

COUNTRIES IN THE REGION

SENEGAL · GAMBIA · GUINEA · GUINEA-BISSAU

SIERRA LEONE · LIBERIA · CÔTE D'IVOIRE · BURKINA FASO

GHANA · TOGO · BENIN · NIGERIA · CAMEROON

CENTRAL AFRICAN REPUBLIC · EQUATORIAL GUINEA

GABON · CONGO, Republic of

CONGO, Democratic Republic of (formerly Zaire)

BURUNDI · RWANDA · UGANDA · KENYA · TANZANIA

SEYCHELLES · SÃO TOMÉ AND PRÍNCIPE

CAPE VERDE

Dawn at Lake Naivasha, Kenya The lake is one of many lying in the eastern arm of the Great Rift Valley, the vast geological fault that runs 6,450 km (4,000 mi) down Africa, as it cuts through Kenya's highlands. Its fresh waters are a haven for wildlife, and the lake is much visited by fishermen and birdwatchers.

- ■ capital city
- ● major town

height of land (meters)

- 5000
- 3000
- 2000
- 1000
- 500
- 200
- 0

▲ mountain peak

ATLANTIC OCEAN

INDIAN OCEAN

Senegal

REPUBLIC OF SENEGAL

SENEGAL LIES ON THE WEST COAST OF WEST Africa, bordered by Mauritania, Mali and Guinea, and surrounding Gambia.

GEOGRAPHY

Senegal's landscape consists mostly of low-lying plains. The only higher ground is in the east, bordering the plateau areas of Mali and Guinea. The plains are crossed by several rivers, including the Gambia, the Casamance and the Senegal, which forms the northern border with Mauritania. The climate is hot with seasonal rains, and the landscape ranges from arid semidesert in the north to lush tropical savanna and rainforest in the south, where there are mangrove swamps along the coast. There is a wide variety of wildlife, including lions, antelopes, monkeys and warthogs, with crocodiles and hippopotamuses in the rivers.

SOCIETY

Senegal is the home of many peoples, notably the Wolof, the Fulani, the Serer and the Tukulor. It is predominantly Muslim. In the 17th and 18th centuries, both the French and the English had a stake in the country. It was fully colonized by France in the 19th century, after the abolition of a prosperous slave trade. In 1958 it gained internal autonomy, and in 1960 it became independent. The first president was the poet Léopold Sédar Senghor (b. 1906), who finally resigned in 1980. In 1981 his successor Abdou Diouf (b. 1935) sent troops into Gambia to help put down a revolt. Diouf was re-elected in 1993, followed by a legislative election. Senegal was ruled by a coalition in 1995.

ECONOMY

Senegal's economy is heavily dependent on agriculture. Groundnuts (peanuts) are the chief crop, followed by millet, beans, rice, sugar cane, sorghum and cotton. Fishing is also important. Large phosphate reserves are mined, and important petroleum and iron ore deposits have been found. Manufacturing, chiefly of

A colorful display of bracelets in beadwork, wood and snakeskin brightens up a busy market stall in Dakar, the capital of Senegal and one of the largest industrial centers in west Africa. Skilled work such as this reflects the country's enduring tradition of folk art.

agricultural products, especially groundnut (peanut) oil, is under development. A booming service sector based on information technology was providing 60 percent of GDP in the late 1990s. The transportation network is relatively good, especially around the modern capital of Dakar on the Cape Vert peninsula. Outside Dakar, however, health care, welfare and even education are relatively backward. However, the country has a strong tradition of art and literature.

The beach at Ngor, a fishing village on the coast of Senegal. Sea fishing continues to play an important role in the Senegalese economy, but small local boats are giving way to deep-sea trawlers, which now provide the greater part of the catch.

NATIONAL DATA – SENEGAL

Land area	192,000 sq km (74,132 sq mi)			
Climate		Temperatures	Annual	
	Altitude m (ft)	January °C(°F)	July °C(°F)	precipitation mm (in)
Dakar	23 (75)	22 (72)	26 (79)	514 (20.2)
Major physical features highest point: Futa Jalon (edge) 500 m (1,640 ft); longest river: Senegal (part) 1,633 km (1,015 mi)				
Population (2006 est.) 11,987,121				
Form of government Multiparty republic with one legislative house				
Armed forces army 11,900; navy 950; air force 770				
Capital city Dakar (2,564,900)				
Official language French				
Ethnic composition Wolof 43.3%; Pular 23.8%; Serer 14.7%; Jola 3.7%; Mandinka 3%; Soninke 1.1%; European and Lebanese 1%; other 9.4%				
Religious affiliations Muslim 94%, Christian 5% (mainly Roman Catholic), traditional beliefs 1%				
Currency 1 Communauté Financière Africaine franc (XOF) = 100 centimes				
Gross domestic product (2006 est.) U.S. $22.01 billion				
Gross domestic product per capita (2006 est.) U.S. $1,800				
Life expectancy at birth male 57.7 yr; female 60.85 yr				
Major resources fisheries, phosphates, iron ore, food processing, fertilizer production, petroleum refining, construction materials, textiles, groundnuts, maize/corn, millet, rice, sorghum, timber				

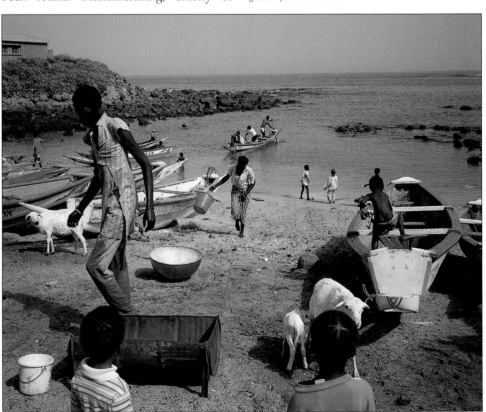

Gambia

REPUBLIC OF GAMBIA

At the mouth of the Gambia river, tall fan palms stand atop a cliff that is being rapidly undercut and gullied by the combined action of waves and water runoff. This stretch of coastline was first mapped by Portuguese explorers in the mid 15th century.

T HE TINY WEST AFRICAN STATE OF GAMBIA forms a narrow enclave along the banks and estuary of the Gambia river, surrounded on three sides by Senegal.

GEOGRAPHY

The river dominates the country, whose landscape essentially consists of its wide valley, flanked by low hills and sandy plateaus. The climate is tropical, with heavy seasonal rains. The floodplain supports seasonally flooded grasslands known as *banto faros*, while the river itself is lined with swamps. The hills are largely covered with savanna and open forest, now mostly cleared for agriculture. The river attracts much wildlife, and is particularly rich in birds. Other animals include leopards, eland, antelopes and various species of monkey, together with crocodiles and hippopotamuses in the river.

SOCIETY

The Gambia is inhabited mainly by the Malinke, Fulani, Wolof and Yola peoples, and is distinct from Senegal because it came under British influence from the 17th century onward.

The Gambia river area became a formal colony in 1888. Independence was declared in 1965, and in 1970 the Gambia became a multiparty republic within the British Commonwealth.

An attempted coup in 1981 led to the formation of the federation of Senegambia with Senegal. This ended in 1989 following which the two countries signed a treaty of friendship. A coup in 1994 installed a military council and banned political activity, however, in 1996 a new constitution and presidential elections returned Gambia to civilian rule.

About 80 percent of the population live outside the towns; most are Muslims. A seasonal influx of farm workers from Senegal causes population fluctuations.

ECONOMY

Around 75 percent of the population depend on agriculture. Groundnuts (peanuts) are the major crop far outweighing in importance others which include palm kernels, cotton, millet and rice. Commercial fishing is being developed.

Apart from kaolin and tin, there are no significant mineral resources. Industry is mostly limited to food processing and the chief export market is the European Union. Transportation is mainly by river. Tourism is an important foreign exchange earner.

Health care is limited, but there are programs in place to combat leprosy, malaria and other endemic tropical diseases. Primary education is free but not compulsory, and illiteracy remains extremely high.

NATIONAL DATA – GAMBIA				
Land area 10,000 sq km (3,861 sq mi)				

Climate		Temperatures		Annual
	Altitude m (ft)	January °C(°F)	July °C(°F)	precipitation mm (in)
Banjul	2 (7)	24 (72)	27 (79)	977 (38.4)

Major physical features longest river: Gambia (part) 1,100 km (700 mi)

Population (2006 est.) 1,641,564

Form of government multiparty republic with one legislative house

Armed forces army 800

Capital city Banjul (34,828)

Official language English

Ethnic composition African 99% (Mandinka 42%; Fula 18%; Wolof 16%; Jola 10%; Serahuli 9%; other 4%); non-African 1%

Religious affiliations Muslim 90%; Christian 9%; traditional beliefs 1%

Currency 1 dalasi (GMD) = 100 butut

Gross domestic product (2006 est.) U.S. $3.25 billion

Gross domestic product per capita (2006 est.) U.S. $2,000

Life expectancy at birth male 52.3 yr; female 56.03 yr

Major resources fisheries, titanium (rutile and ilmenite), tin, zircon, silica sand, clay, petroleum, tourism, cassava, cattle, cotton, groundnuts, millet, palm kernels, rice, timber

Guinea

REPUBLIC OF GUINEA

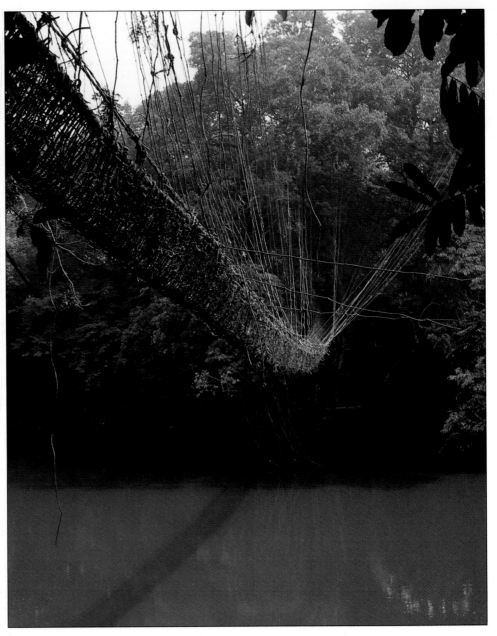

GUINEA LIES ON AFRICA'S WEST COAST, sandwiched between Guinea-Bissau and Sierra Leone, and inland bordering Senegal, Mali, Ivory Coast and Liberia.

GEOGRAPHY

Most of inland Guinea consists of uplands, from the savanna plains around the Niger river in the northeast to the Guinea Highlands in the southeast and the central ranges of the Futa Jalon plateau. The great Niger, Senegal and Gambia rivers all rise here, but flow through many other countries before reaching the Atlantic. From the Futa Jalon the land drops sharply to a relatively short coastline, lined with estuaries and mangrove swamps. The climate is hot, with heavy but variable seasonal rains. The Guinea Highlands are covered with dense tropical rainforest, including hardwood trees such as teak, mahogany and ebony. Much of the rest is savanna grassland. Typical wildlife includes lions, leopards and chimpanzees, with hippopotamuses and crocodiles in the rivers.

A bridge of lianas spans a river near Nzérékoré, in the Guinea Highlands – an ingenious way of exploiting natural resources to solve transportation problems. The country's tropical climate and rich vegetation support a remarkable variety of wildlife.

SOCIETY

Guinea, once a kingdom of the Islamic Fulani people, was colonized by France in the 19th century. In 1958 French Guinea became independent under President Ahmed Sékou Touré (1922–84), becoming the People's Revolutionary Republic of Guinea. Serious economic difficulties and increased repression led to a military coup in 1984. The new president, Brigadier-General Lansana Conté (b. 1934), instituted economic reform and political liberalization. Despite a number of attempted coups, Conté has remained in power and, in 1993, reintroduced multiparty politics.

He was re-elected in 1998 in a disputed election. There are 24 ethnic groups and most people are Muslim. There are eight official languages apart from French.

ECONOMY

Agriculture employs most of the workforce. Rice, bananas and pineapples, along with oil palms and corn, are grown in the wet coastal southwest, some on reclaimed land. Rice and other grains are grown beside the Niger river. Cattle are raised on the plateau grasslands. The economy, however, rests largely on the mining of immense reserves of bauxite, of which Guinea is a major producer. Otherwise manufacturing is poorly developed, even with foreign aid. Health care is elementary, and although education is compulsory, literacy levels are low.

NATIONAL DATA – GUINEA

Land area	245,857 sq km (94,926 sq mi)			
Climate		**Temperatures**		**Annual precipitation**
	Altitude m (ft)	January °C(°F)	July °C(°F)	mm (in)
Conakry	46 (151)	26 (78)	25 (77)	3,805 (69)

Major physical features highest point: Mount Nimba 1,850 m (6,069 ft); longest river: Niger (part) 4,200 km (2,600 mi)

Population (2006 est.) 9,690,222

Form of government multiparty republic with one legislative assembly

Armed forces army 8,500; navy 400; air force 800

Largest cities Conakry (capital – 2,064,236); Nzerekore (138,862)

Official language French

Ethnic composition Peuhl 40%; Malinke 30%; Soussou 20%; other ethnic groups 10%

Religious affiliations Muslim 85%; Christian 8%; traditional beliefs 7%

Currency 1 Guinean franc (GNF) = 100 centimes

Gross domestic product (2006 est.) U.S. $19.4 billion

Gross domestic product per capita (2006 est.) U.S. $2,000

Life expectancy at birth male 48.34 yr; female 50.7 yr

Major resources bauxite, iron ore, diamonds, gold, uranium, hydropower, fish, salt, tropical fruits, timber

Guinea-Bissau

REPUBLIC OF GUINEA-BISSAU

GUINEA-BISSAU, FORMERLY PORTUGUESE Guinea, lies on the west African coast between Senegal and Guinea. The country also includes the neighboring islands of the Bijagós Archipelago.

GEOGRAPHY

The only area of upland savanna in the country lies on the southeastern borders. Here, deep river valleys cut through the foothills of Guinea's Futa Jalon plateau. The Geba, Corubal and Cacheu rivers meander across the forested central plains to broad estuaries on a coast lined with mangrove swamps. Seasonal rainfall, high everywhere, is torrential here, while temperatures are tropical over most of the country. Riverine wildlife is particularly rich, and includes crocodiles, flamingoes and pelicans.

SOCIETY

From the 15th century the country was exploited by Portugal as a source of slaves. It was colonized as the slave trade declined in the 19th century, and became a Portuguese overseas territory. In 1961 the Partido Africano da Independência da Guiné e Cabo Verde (PAIGC) began a guerrilla campaign, which led to independence in 1974. In 1980, after the failure of a planned merger with the Cape Verde Islands, a military coup installed a Revolutionary Council which eventually gave way to a democratically elected government in 1994. An army uprising in 1998 was followed by a civil war and another coup in 1999. Further elections later that year appointed Koumba Yalla to the presidency.

ECONOMY

Civil war in 1998 damaged Guinea-Bissau's already ailing economy. Most of the population is employed in agriculture which is hampered by periodic droughts and locust swarms. Rice is the major crop and staple food. Cashew and groundnuts form the major cash crop followed by coconuts, palm kernels and timber. Frozen fish and shrimps are also exported.

Food processing and textiles are the only significant industries. Sizeable reserves of bauxite and phosphate remain unexploited as do offshore oil reserves.

Transportation is mainly by waterway. Health care is limited but around 60 percent of children receive primary education and adult literacy is over 50 percent. There is one television station and three daily newspapers.

A village near Farim in Guinea-Bissau belonging to the Malinke people – one of the forty ethnic groups in the country. Livestock shelters stand alongside the villagers' huts.

NATIONAL DATA – GUINEA-BISSAU

Land area	28,000 sq km (10,811 sq mi)			
Climate		Temperatures		Annual
	Altitude m (ft)	January °C(°F)	July °C(°F)	precipitation mm (in)
Bissau	21 (69)	26 (78)	26 (78)	1,756 (69)

Population	(2006 est.) 1,442,029

Form of government	multiparty republic with one legislative house

Armed forces	army 6,800; navy 350; air force 100

Capital city	Bissau (420,412)

Official language	Portuguese

Ethnic composition	African 99% (includes Balanta 30%; Fula 20%; Manjaca 14%; Mandinga 13%; Papel 7%); European and mulatto 1%

Religious affiliations	traditional beliefs 50%; Muslim 45%, Christian 5%

Currency	1 Communauté Financière Africaine franc (XOF) = 100 centimes

Gross domestic product	(2006 est.) U.S. $1.244 billion

Gross domestic product per capita	(2006 est.) U.S. $900

Life expectancy at birth	male 45.05 yr; female 48.75 yr

Major resources	fish, timber, bauxite, clay, granite, limestone, unexploited deposits of petroleum and phosphates, cashew nuts, groundnuts, palm kernels, rice, cereals, coconuts, cotton

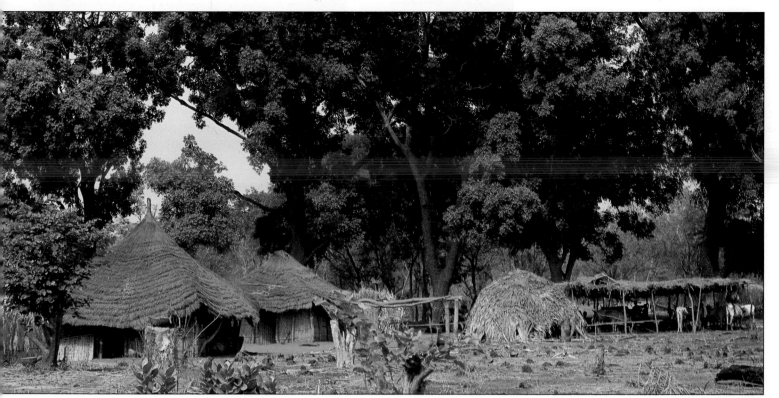

Sierra Leone

REPUBLIC OF SIERRA LEONE

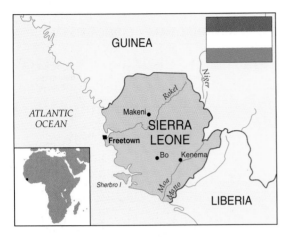

SIERRA LEONE LIES ON THE AFRICAN WEST coast bordering Guinea to the north and Liberia to the southeast along the Mano and Moro rivers.

GEOGRAPHY

A narrow coastal belt of swamps and mangroves is interrupted by the thickly wooded mountains of the Sierra Leone peninsula, site of the capital, Freetown. Forest trees include the palm, exploited for its oil and kernels. The southern plains of the interior are wooded, while those farther north are covered with rolling savanna grasslands. These are known as the Bolilands and are subject to seasonal flooding. To the northeast, the Loma Mountains rise to Bintimani Peak, and the Tingi Hills rise to Sankanbiriwa. Many rivers flow through Sierra Leone from across the Guinean border. Game animals are rare, but monkeys and chimpanzees are common in the forests.

Sierra Leone has a hot, tropical climate with a rainy season from May to October. Rainfall decreases farther north, and temperature ranges are greater in the northeastern highlands. The dry harmattan wind blows from the Sahara in the northeast between November and April.

SOCIETY

Early Portuguese explorers found a cluster of small kingdoms around the Sierra Leone peninsula (the name means "Lion Mountains"). In 1787 homeless freed slaves were settled on the site of Freetown by the British; their descendants are called Creoles. A British protectorate was declared in 1896. In 1961 the country achieved independence under Sir Milton Margai (1895–1964). In 1971 Sierra Leone became a republic, with Siaka Stevens (b. 1905) as its first president. In 1978 a new constitution made the All People's Congress the sole legal party. In 1991 a referendum endorsed multiparty politics but coups in 1992, 1996 and 1997 led to intervention by a Nigerian-led peacekeeping force in 1998. The civil war, which had been raging since 1991, continued with the loss of many lives and the displacement of several million people. Despite the backing of Nigerian troops the government lost control of the diamond fields in 1999. A peace agreement was signed with the rebels in July 1999 and UN peacekeeping troops were deployed but the situation was far from resolved.

The largest of the 18 ethnic groups are the Mende and the Temne. The official language is English, but most people communicate in Krio, a Creole pidgin language. About half the population hold traditional African beliefs, and about two-fifths are Muslims.

ECONOMY

The majority of the people work on the land. Palm kernels, cocoa and coffee are the main cash crops, and rice is the main staple. Mining is Sierra Leone's key industry; iron ore, diamonds, chromite, rutile, bauxite and gold are exported, mainly to Europe and the United States. Manufacturing centers on Freetown, and is mostly of consumer goods.

The single railroad was finally dismantled in 1975, but the road system is being expanded. The many waterways transport cargo and the main port is at Freetown. The international airport is at Lungi. Radio, television and the press are government-controlled.

Government health-care programs have reduced infant mortality and raised life expectancy, but sleeping sickness and leprosy remain a problem. Primary education is free, but low attendance means that literacy levels are slow to rise.

NATIONAL DATA – SIERRA LEONE

Land area	71,620 sq km (27,653 sq mi)			

Climate		Temperatures		Annual
	Altitude m (ft)	January °C(°F)	July °C(°F)	precipitation mm (in)
Freetown	11 (36)	27 (80)	26 (78)	2,946 (116)

Major physical features	highest point: Bintimani Peak 1,948 m (6,391 ft)

Population	(2006 est.) 6,005,250

Form of government	multiparty republic with one legislative house

Armed forces	joint forces 12–13,000

Capital city	Freetown (835,108)

Official language	English

Ethnic composition	Native African tribes 90% (Temne 30%, Mende 30%, other 30%), Creole (Krio) 10%; small numbers of Europeans, Lebanese, Pakistanis, and Indians

Religious affiliations	Muslim 60%; traditional beliefs 30%; Christian 10%

Currency	1 leone (SLL) = 100 cents

Gross domestic product	(2006 est.) U.S. $5.38 billion

Gross domestic product per capita	(2006 est.) U.S. $900

Life expectancy at birth	male 38.05 yr; female 42.46 yr

Major resources	diamonds, titanium ore, bauxite, iron ore, gold, chromite, cassava, cocoa, coffee, fish, groundnuts, palm nuts, rice, timber

A masked person from the Kanadevil sect (*above*) wears an elaborate headdress. The paleness of the mask suggests it may represent a "white man devil".

Palm-fringed beaches (*right*) are not just a draw for tourists. The palm fruits are highly valued for their kernels and oil, and the trunks make good dug-out canoes.

Liberia

REPUBLIC OF LIBERIA

Goods are loaded on board a freighter in Monrovia, Liberia's principal port. Before the civil war in 1990, Liberia had trading links with many countries, but its economy is likely to need time to recover.

L IBERIA, AFRICA'S OLDEST INDEPENDENT republic, borders Sierra Leone to the northwest (along the Mano and Moro rivers), Guinea to the north and Côte d'Ivoire to the east (along the Cavally river). The capital, Monrovia, lies on the coast.

GEOGRAPHY

Swamps and mangroves line Liberia's sandy coast. Inland, rolling hills covered in tropical rainforest rise to a high plateau. The mountains in the north reach the southwestern flanks of Mount Nimba, where the borders with Guinea and Côte d'Ivoire meet. Semideciduous rainforest and elephant grass are the characteristic vegetation in the north. Chimpanzees, pygmy hippopotamuses and crocodiles are found, but game animals are becoming increasingly rare.

Liberia has a hot, humid climate. The rainy season from May to October is interrupted by a short dry season from mid-July to August. Rainfall diminishes inland, and the greatest amount falls on the Cape Mount promontory near the Sierra Leone border. In December the hot, dry harmattan wind blows down from the Sahara.

SOCIETY

In 1822 a group of liberated American slaves landed in Liberia to found a colony that had been negotiated with the indigenous peoples by the American Colonization Society. In 1847 Liberia declared itself an independent republic. The 1920s saw heavy investment by the Firestone Tire and Rubber Company of the United States. Liberian rubber was later vital to the Allied victory in World War II.

The economy suffered when rubber prices fell in the 1960s and 1970s. In 1980 President William Tolbert (1913–80) was shot during a coup led by Master Sergeant Samuel Kanyon Doe (1951–90). In 1990 a civil war broke out, several million people were made homeless and Doe was assassinated by rebel forces. Despite the intervention of a peacekeeping force from the Economic Community of West African States (ECOWAS), the civil war continued until 1996. In 1997 Charles Taylor, leader of the National Patriotic Party, was elected president but fighting between government forces and rebels continued.

Liberia's peoples belong to three different language groups – Mande, Kwa and West Atlantic – but the official language remains English. Many of the population retain traditional African beliefs.

ECONOMY

Liberia's economy has been devastated by civil war, many businessmen fled the country and there are massive international debts. Registration of foreign-owned shipping provides substantial foreign-exchange income. Rubber is the main cash crop. Rice is the main staple, but some food must still be imported. Livestock includes cattle, goats and pigs. Timber is a significant export. The catch from coastal fisheries is supplemented from inland fish farms. Iron ore is Liberia's principal export. Other mineral resources include diamonds and gold. Energy is supplied by hydroelectric and oil-fired stations. Most manufacturing industries serve the local market.

During the civil war, many roads and railroads were destroyed. The two major airports are close to Monrovia. Epidemic diseases are rife and health care and education are poor. Literacy levels are low and infant mortality high.

NATIONAL DATA – LIBERIA

Land area	96,320 sq km (37,189 sq mi)			
Climate		Temperatures		Annual
	Altitude m (ft)	January °C(°F)	July °C(°F)	precipitation mm (in)
Monrovia	23 (75)	27 (81)	25 (79)	5,140 (202.3)

Major physical features highest point: Mount Wutivi 1,380 m (4,528 ft); longest river: Cavally (part) 515 km (320 mi)

Population (2006 est.) 3,042,004

Form of government multiparty republic with two legislative houses

Armed forces Joint forces 15,000

Capital city Monrovia (967,841)

Official language English

Ethnic composition Indigenous African tribes 95% (including Kpelle, Bassa, Gio, Kru, Grebo, Mano, Krahn, Gola, Gbandi, Loma, Kissi, Vai, Dei, Bella, Mandingo, and Mende); Americo-Liberians (descendants of repatriated slaves) 5%

Religious affiliations traditional beliefs 40%; Christian 40%; Muslim 20%

Currency 1 Liberian dollar (LRD) = 100 cents

Gross domestic product (2006 est.) U.S. $2.911 billion

Gross domestic product per capita (2006 est.) U.S. $1,000

Life expectancy at birth male 37.99 yr; female 41.35 yr

Major resources iron ore, timber, diamonds, gold, hydropower, fish, rubber, cocoa, coffee, bananas, cassava, maize/corn, palm kernels, rice

Côte d'Ivoire

REPUBLIC OF CÔTE D'IVOIRE

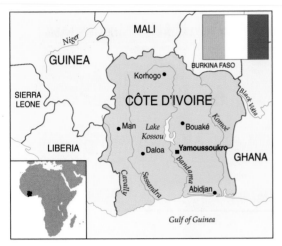

CÔTE D'IVOIRE (IVORY COAST) FACES THE Gulf of Guinea on the west coast of Africa, with Mali and Burkina Faso to the north, and Ghana to the east. It also borders Liberia and Guinea to the west; the three frontiers meet at Mount Nimba.

GEOGRAPHY

The land rises gently northward from the lagoons and sandbars of the coast. Tropical rainforest, much of it cleared for plantations, occupies the southern third of the country, yielding valuable timber. The thinly populated northern plateau is covered in savanna grasslands, while the highlands in the northwest rise to Mount Nimba. The country is dissected by four major north–south rivers: the Cavally, the Sassandra, the Bandama and the Komoé. Typical wildlife includes elephants and pygmy hippopotamuses. There are three major national parks.

The climate is hot and tropical. Along the coast, and to a lesser extent in the forest belt, there are two distinct rainy seasons: May–July and October–November. On the plateau the temperatures are less hot, and rainfall is lower, with a single rainy season. The dry, northeasterly harmattan wind blows here between the months of December and February. The western mountains have no dry season.

SOCIETY

The Portuguese first explored the country in the 15th century, and in the 17th century the French established trading posts here. In the 19th century they concluded protectorate agreements, and in 1893 Louis-Gustave Binger (1856–1936) became the first governor of a new Ivory Coast colony. During the first half of World War II the colony remained under the control of Vichy France (the part of France not occupied by German troops) until 1942. In 1946 it became a territory in the French Union. Upper Volta (Burkina Faso) was separated in 1947. Autonomy came in 1958, and independence in 1960, when Félix Houphouët-Boigny (1905–93) became the first president.

Multiparty politics were reestablished in 1990 and Houphouët-Boigny remained president until his death in 1993. Henri Konan Bedie took over the presidency until deposed in a military coup in 1999.

Fluctuating market prices for coffee and cocoa caused financial problems in the 1990s. There was also an influx of refugees from Liberia. French is the official language but there are 60 ethnic groups each with their own dialect. Most people hold traditional African beliefs often in combination with Muslim and Christian beliefs.

ECONOMY

The majority of the labor force work on the land. Cocoa and coffee are the main cash crops, while yams, manioc, plantains, corn and rice are the chief staples. Timber and fish are important resources; both are exported. Mineral resources include iron ore, copper, nickel, uranium, bauxite, cobalt, diamonds and offshore oil and gas. Power from hydroelectric and oil-fired stations supplies light industries, including textiles, food processing and vehicle assembly. Tourism is developing.

Communications center on the railroad to Burkina Faso. There is a good road network around the former capital of Abidjan, where the Vridi Canal gives access to the sea. The international airport at Port Bouet also connects with small airfields inland. Health care is poor: malaria, dysentery, yaws and tuberculosis are fairly common. Primary education is compulsory, but not all children are enrolled.

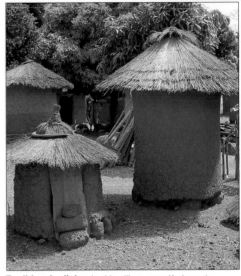

Traditional religion In this village near Korhogo in northern Côte d'Ivoire, fetish huts are set aside by the local people for use by ancestral spirits. The villagers leave food offerings inside the huts, and supplicate the spirit for good health and harvest.

NATIONAL DATA – CÔTE D'IVOIRE

Land area	318,000 sq km (122, 780 sq mi)			
Climate		Temperatures		Annual
	Altitude m (ft)	January °C(°F)	July °C(°F)	precipitation mm (in)
Abidjan	7 (23)	27 (81)	26 (79)	1,959 (77.1)

Major physical features highest point: Mount Nimba 1,850 m (6,069 ft)

Population (2006 est.) 17,654,843

Form of government multiparty republic with one legislative house

Armed forces army 6,500; navy 900; air force 700

Largest cities Abidjan (3,918,000); Bouaké (597,507); Yamoussoukro (capital - 205,723)

Official language French

Ethnic composition Akan 42.1%; Voltaiques or Gur 17.6%; Northern Mandes 16.5%; Krous 11%; Southern Mandes 10%; other 2.8%

Religious affiliations Muslim 35–40%; traditional beliefs 25–40%; Christian 20–30%

Currency 1 Communauté Financière Africaine franc (XOF) = 100 centimes

Gross domestic product (2006 est.) U.S. $28.47 billion

Gross domestic product per capita (2006 est.) U.S. $1,600

Life expectancy at birth male 46.24 yr; female 51.48 yr

Major resources petroleum, natural gas, diamonds, manganese, iron ore, cobalt, bauxite, copper, gold, nickel, tantalum, silica sand, clay, cocoa beans, coffee, maize, palm oil, rice, yams, hydropower

Burkina Faso

REPUBLIC OF BURKINA FASO

BURKINA FASO, UNTIL 1984 UPPER VOLTA, LIES ON the southern fringe of the Sahara. It is surrounded by Côte d'Ivoire, Ghana, Togo, Benin, Niger and Mali.

GEOGRAPHY

Most of Burkina Faso is a wide plateau, interrupted in the west by an escarpment called the Falaise de Banfora. The three main rivers, the Red, White and Black Volta, have cut deep valleys as they flow southward into Ghana. The climate is hot, with unreliable seasonal rains. The north of the country is typical of the arid Sahel, with scrubland and semidesert. The south is more tropical, ranging from savanna grassland to forest. The country has suffered severely from recurrent drought. Wildlife is abundant, but areas near rivers are plagued by tsetse and simulium flies, carriers of sleeping sickness and river blindness.

SOCIETY

Burkina Faso, once the kingdom of the Mossi people, was colonized by France in the late 19th century. Independence was gained in 1960, but the country was plagued by instability, characterized by frequent and bloody coups in the 1970s and 1980s. Captain Blaise Compaoré has remained president since the coup of 1987. There is an 111-member elected assembly and a 178-member chamber of appointed representatives. There are several ethnic groups, but the official language is French. Most people either follow Islam or hold traditional African beliefs.

ECONOMY

Most people work on the land. Cash crops include groundnuts (peanuts), cotton (the main export) and sesame, while the main food crops are sorghum, millet and sugar cane. Cattle and goats are raised mostly in the north and east. Some gold quartz is mined, along with manganese and limestone. Industry is undeveloped, and is based mostly on agriculture. Power is generated from imported fuels. The transportation networks are good, but health facilities are very poor, with much malnutrition and disease. Education, though compulsory, is available only to a minority, and both press and television are government-controlled.

NATIONAL DATA – BURKINA FASO

Land area	273,800 sq km (105,715 sq mi)			
Climate	Altitude m (ft)	Temperatures January °C(°F)	July °C(°F)	Annual precipitation mm (in)
Abidjan	304 (997)	25 (77)	28 (79)	895 (35.2)

Major physical features	highest point: Téna Kouron 747 m (2,451 ft)

Population	(2006 est.) 13,902,972

Form of government	multiparty republic with two legislative houses

Armed forces	army 6,400; air force 200

Largest cities	Ouagadougou (capital – 1,152,402); Koudougou (91,468); Ouahigouya (63,598)

Official language	French

Ethnic composition	Mossi 47.9%; Mande 8.8%; Fulani 8.3%; Lobi 6.9%; Bobo 6.8%; Senufo 5.3%; Grosi 5.1%; Gurma 4.8%; Tuareg 3.3%; others 2.4%

Religious affiliations	Muslim 50%; traditional beliefs 40%; Christian (mainly Roman Catholic) 10%

Currency	1 Communauté Financière Africaine franc (XOF) = 100 centimes

Gross domestic product	(2006 est.) U.S. $17.87 billion

Gross domestic product per capita	(2006 est.) U.S. $1,300

Life expectancy at birth	male 47.33 yr; female 50.42 yr

Major resources	manganese, limestone, marble; small deposits of gold, phosphates, pumice, salt, antimony, bauxite, copper, cotton, groundnuts, lead, livestock, maize/corn, millet, nickel, rice, sesame, shea, sorghum, sugarcane

Colorfully adorned (*above*) this Pau woman lives in Burkina Faso, one of the poorest countries in the world. Her own health seems good, but the prospects for her children are poor in a country plagued by malnutrition and disease.

Rock formations at Fabedougou (*left*) near Banfora in the southwest of Burkina Faso. These spectacular formations are the result of rapid erosion. Floodwaters released during flash storms have eaten away at the soft sedimentary rocks, creating this beautifully sculpted landscape.

Ghana

REPUBLIC OF GHANA

G HANA, FORMERLY THE GOLD COAST, FACES south across the Gulf of Guinea. More than half the country is occupied by the Volta basin.

GEOGRAPHY

In the north the Red and White Volta rivers meet just south of the Burkina Faso border. In the northwest the Black Volta marks the western border with Burkina Faso and part of the border with Côte d'Ivoire, before turning westward. Much of the basin has been flooded, even above the confluence of the Black and White Volta, by the construction of the Akosombo Dam, creating Lake Volta. East of the basin, the Akwapim-Togo Mountains run north–south along the border with Togo, then turn southwest toward the coastal capital, Accra. South of the Akosombo Dam, the Volta crosses a fertile coastal plain before flowing into a series of marshes and lagoons.

Most of Ghana is low-lying, so that the hot, tropical climate remains unmodified by altitude. Temperatures are highest from February to April, and are lowest near the coast, where sea mists have a cooling effect. The south has two rainy seasons (May–June and September–October), whereas in the north there is only one, peaking in August and September.

The coastal vegetation of mangrove swamps rapidly gives way to savanna grassland, and then to tropical forest (much of it cleared for agriculture), covering the southern third of the country. The savanna farther north has fewer trees and shorter grasses. Animal life is still plentiful, and includes lions, leopards, hyenas, antelopes and elephants. The rivers are the home of crocodiles, manatees and hippopotamuses. The largest of the many game reserves is the Mole, near Damongo in the northwest.

SOCIETY

Ghana's interior was settled by many different ethnic groups, but by the 17th century the Akan-speaking Ashanti were dominant. The Portuguese had discovered the Gulf of Guinea in 1471. They built forts along the coast, which was named the Gold Coast on account of the gold to be found there. The 17th century brought other colonists, attracted there by the Ashanti-controlled slave trade, but in the 19th century the British took control. By 1901 the Gold Coast was a British colony, the Ashanti had been subdued and the northern territories had become a

British protectorate. After World War I part of neighboring German Togoland also came under British control.

In 1957 the newly renamed state of Ghana won independence – the first state in sub-Saharan Africa to do so – and in 1960 it was declared a republic, with Prime Minister Kwame Nkrumah (1909–72) as its life president. Nkrumah's regime was overthrown by a military coup in 1966. The civilian government elected in 1969 was ousted by a second coup in 1972. Another brief period of civilian rule (1979–81) ended with a coup led by Flight Lieutenant Jerry Rawlings (b. 1947). Rawlings' Provisional National from Lake Volta) are encouraged, and provide an important supplement to the Ghanaian diet.

Gold is still an important export, as are diamonds, manganese and bauxite. Oil and natural gas provide another significant export. Hydroelectric power from Akosombo supplies domestic needs; the surplus is exported. Manufacturing is limited mainly to supplying local demand, but tourism is increasing.

Road and rail transportation are best in the south, especially in cocoa-growing areas near the coast. About one-fourth of roads are paved, and aid-financed improvements are in hand. The railroad system is largely used for freight. The main port, Tema, and the international airport, Kotoka, are both near Accra. The state airline is Ghana Airways.

Health care has been hindered by rapid population growth, bad sanitation and poor nutrition. Major diseases such as malaria, pneumonia and gastroenteritis have been partly controlled, but infant mortality is high and life expectancy is relatively short. River blindness (onchocerciasis), a disease causing blindness, has had a great impact on local people. Housing is a growing problem, especially in towns. Basic education is free and can be followed by secondary vocational training or preparation for one of the three universities. However, school attendance is still relatively low, which affects literacy levels.

Defense Council created a stable economy although the main exports, gold and cocoa, are highly susceptible to market fluctuations. Democracy returned in 1992. In 2000, Rawlings stepped down and John Kufuor became president. Simultaneous parliamentary elections resulted in victory for the National People's Party ousting the ruling National Democratic Congress.

Ghana has 75 ethnic groups; however, only 10 are of significant size, the largest being the Akan and the Mossi. English is the official language. The majority of people are either Muslims or Christians; a substantial amount though hold traditional African beliefs.

The harbor town of Dixcove on Ghana's Gold Coast. In the 15th century the Portuguese discovered alluvial gold in this area, washed down by the Volta river. However, for hundreds of years the most lucrative trade was in slaves.

ECONOMY

Most people are employed in agriculture, which is the mainstay of the economy. Cocoa and timber are the chief export crops. In the north yams and cereals are grown and cattle are kept, while the forests produce shea and kola nuts. Despite government efforts to increase food production, Ghana is still importing food. Marine and freshwater fishing (notably

---

NATIONAL DATA – GHANA

Land area 230,940 sq km (89,166 sq mi)

Climate	Altitude m (ft)	Temperatures January °C(°F)	July °C(°F)	Annual precipitation mm (in)
Accra	65 (213)	27 (81)	25 (77)	725 (28.5)

Major physical features highest point: Mount Afadjoto 885 m (2,903 ft); largest lake: Lake Volta 8,482 sq km (3,275 sq mi)

Population (2006 est.) 22,409,572

Form of government republic with multiparty parliament

Armed forces army 5,000; navy 1,000; air force 1,000

Largest cities Accra (capital - 2,096,653); Kumasi (1,604,909); Tamale (390,730); Takoradi (260,651); Ashiaman (228,509); Tema (161,106)

Official language English

Ethnic composition Black African 98.5% (major tribes - Akan 44%; Moshi-Dagomba 16%; Ewe 13%; Ga 8%; Gurma 3%; Yoruba 1%); other 1.5%

Religious affiliations Christian 63%; Muslim 16%; traditional beliefs 21%

Currency 1 cedi (GHC) = 100 pesewas

Gross domestic product (2006 est.) U.S. $59.15 billion

Gross domestic product per capita (2006 est.) U.S. $2,600

Life expectancy at birth male 58.07 yr; female 59.69 yr

Major resources gold, timber, industrial diamonds, bauxite, manganese, fish, rubber, hydropower, petroleum, natural gas, silver, salt, limestone, bananas, cassava, cocoa, maize/corn, sorghum, taro, yams, tourism

---

Togo

REPUBLIC OF TOGO

T HE WEST AFRICAN REPUBLIC OF TOGO IS A long, narrow country sandwiched between Ghana to the west and Benin to the east, with Burkina Faso to the north.

GEOGRAPHY

From coastal lagoons the Ouatchi Plateau extends toward a higher tableland. The Togo Mountains, which are covered with tropical forest in the southwest, extend northeast across the Benin border. Savanna dominates the lower-lying areas, where wildlife includes antelopes and elephants. The humid, tropical climate has two rainy seasons (April–June and September–October) in the south and one (June–September) in the north, where the dry and dusty harmattan wind blows for the rest of the year.

NATIONAL DATA – TOGO

Land area	54,385 sq km (20,998 sq mi)			
Climate		Temperatures		Annual
	Altitude m (ft)	January °C(°F)	July °C(°F)	precipitation mm (in)
Lomé	20 (66)	27 (81)	25 (77)	1,577 (62)

Major physical features highest point: Pic Baumann 986 m (3,235 ft); longest river: Mono 400 km (250 mi)

Population (2006 est.) 5,548,702

Form of government Republic under transition to multiparty democratic rule

Armed forces army 8,100; navy 200; air force 250

Capital city Lomé (756,122)

Official language French

Ethnic composition Native African (37 tribes; largest are Ewe, Mina, and Kabre) 99%; European and Syrian-Lebanese less than 1%

Religious affiliations traditional beliefs 51%; Christian 29%; Muslim 20%

Currency 1 Communauté Financière Africaine franc (XOF) = 100 centimes

Gross domestic product (2006 est.) U.S. $9.248 billion

Gross domestic product per capita (2006 est.) U.S. $1,700

Life expectancy at birth male 55.41 yr; female 59.49 yr

Major resources phosphates, limestone, marble, livestock, food processing, cement, textiles, beverages, handicrafts, coffee, copra, cocoa, cotton

SOCIETY

Togo became a German protectorate in 1885. After World War I it was administered by France and Britain. In 1956 the British section was united with the Gold Coast (Ghana), while French Togoland became autonomous, achieving independence in 1960. The first president, Sylvanus Olympio (1902–63), was assassinated in 1963, and the second was overthrown in 1967 by Colonel Gnassingbe Eyadéma (b. 1935). Multiparty rule was instituted in 1992 but the military still dominates the government. Although there are some 30 ethnic groups in the country, French is the official language. There are a large number of Christians, and over half the population retain traditional African beliefs.

ECONOMY

Phosphates account for half Togo's exports, but agriculture is important too. Cash crops include cocoa, coffee and cotton; staples include corn, cassava, rice and yams. Cattle, sheep and pigs are raised in the north. There is an effective road and rail network for carrying food and other resources between the coast and the interior. There are sea ports at Lomé and Kpémé, and an international airport at Lomé. Health conditions and housing are poor. Diseases such as malaria, yellow fever and tuberculosis are still common. Primary education is free, but secondary-school attendance is low, and this is reflected in literacy levels.

Men dancing in traditional costume (*above*) at a village fete in Togo, where the majority of the people hold traditional religious beliefs. The performance of such dances serves to preserve and strengthen the cultural identity of the ethnic group.

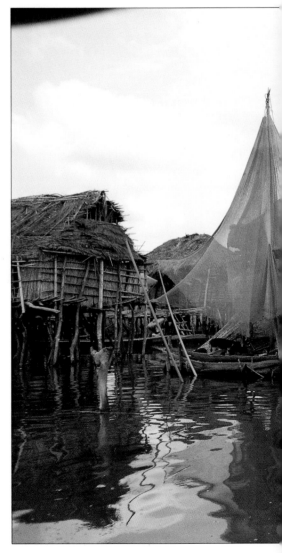

Benin

REPUBLIC OF BENIN

B ENIN (FORMERLY DAHOMEY) BORDERS THE Gulf of Guinea, and is bounded by Togo to the west, Burkina Faso and Niger to the north, and Nigeria to the east. Though fully independent since 1960, Benin has only recently seen a change toward full multiparty democracy.

GEOGRAPHY

Inland from the coastline are coastal lagoons, behind which the land rises to a fertile, heavily cultivated plateau, and then to the Atakora Mountains – a continuation of the Togo Mountains. The northeastern plains slope down to the valley of the Niger river. Northern areas are covered in mixed forest and savanna, which support a diversity of wildlife including leopards, hyenas, antelopes, elephants and monkeys. The climate is humid and tropical. In the south there are two rainy seasons (March–July and September–November), but in the north there is only one (May–September). The dry harmattan wind blows from the northeast between December and March.

SOCIETY

The French colony of Dahomey won full independence in 1960, but ethnic and religious divisions between north and south contributed to political instability.

NATIONAL DATA – BENIN

Land area 110,620 sq km (42,710 sq mi)				
Climate		**Temperatures**		**Annual**
	Altitude m (ft)	January °C(°F)	July °C(°F)	precipitation mm (in)
Cotonou	5 (16)	27 (81)	26 (79)	1,308 (51.4)

Major physical features   highest point: Atakora Massif 641 m (2,103 ft)

Population   (2006 est.) 7,862,944

Form of government   multiparty republic with one legislative house

Armed forces   army 4,300; navy 100; air force 150

Largest cities   Cotonou (711,600); Abomey-Calavi (443,670); Porto Novo (capital – 242,187); Djougou (217,360)

Official language   French

Ethnic composition   African 99% (42 ethnic groups, largest are Fon, Adja, Yoruba, Bariba); Europeans 5,500

Religious affiliations   traditional beliefs 50%; Christian 30%; Muslim 20%

Currency   1 Communauté Financière Africaine franc (XOF) = 100 centimes

Gross domestic product   (2006 est.) U.S. $8.931 billion

Gross domestic product per capita   (2006 est.) U.S. $1,100

Life expectancy at birth   male 51.9 yr; female 54.22 yr

Major resources   small offshore oil deposits, limestone (cement), marble, timber, beans, cassava, coffee, cotton, groundnuts, maize/corn, palm products, sorghum, yams

In 1972, after the last of a series of coups, Lieutenant-Colonel Mathieu Kerekou (b. 1933) was named as president. He changed the country's name to Benin in 1975. A new constitution in 1990 paved the way for multiparty elections in March 1991 and again in 1995. French is the official language, but most people speak African languages and retain traditional beliefs.

ECONOMY

About half the population are dependent on agriculture; many are subsistence farmers. Cash crops include cotton, cocoa, coffee and palm kernels, while corn, cassava, yams and beans are the main subsistence crops. Benin has reserves of limestone, gravel, gold and offshore oil, and exports energy from its geothermal stations.

There are two principal roads into Nigeria and Niger, and three railroad lines. The main port and the international airport are at the *de facto* capital, Cotonou, on the coast. Health conditions and housing are poor; major diseases include leprosy, meningitis and malaria. Primary education is compulsory but many children have difficulty in attending and literacy levels are low.

In the marshy coastlands of Benin the best place to build a house is on stilts well above the high-water mark. Fishing nets have been hung out to dry in this quiet southern village.

Nigeria

FEDERAL REPUBLIC OF NIGERIA

NIGERIA, BORDERED BY CAMEROON, CHAD, Niger and Benin, is the largest country on the coast of west Africa, and the most populous in the whole continent.

GEOGRAPHY

At the heart of Nigeria lies the high Jos Plateau, from which the land falls away northward toward the Niger border. There are lower-lying areas to the northwest and to the northeast bordering Lake Chad, where the land becomes swampy in the rainy season. The plateaus become lower farther south, where they are split by shallow river valleys. The broadest of these are formed by the great Niger river and its main tributary, the Benue, which flow from west and east respectively to meet in the center. The Niger continues southward across the broad coastal plains bordering the Gulf of Guinea, where it spreads out into a vast delta. The plateaus to the west continue into Benin. The mountain chains along the eastern border are a continuation of the volcanic Adamawa Highlands. They include the lofty Shebshi Mountains, and the Mandara Mountains farther north toward Lake Chad.

The climate is tropical with seasonal rains, but varies enormously from south to north. The coast is constantly humid, with an extremely high rainfall. Farther north the wet season decreases in length and intensity, and the far north has the relatively arid climate of the Sahel, with an eight-month dry season and more marked variations in temperature.

In the arid margins of Lake Chad, only hardy scrub vegetation such as acacia and doum palms flourish. Farther south there is savanna grassland, with wildlife such as lions, giraffes and antelopes. Farther south still this becomes savanna woodland, with its distinctive baobab trees. Toward the coast there is a broad rainforest belt, supporting wildlife such as elephants, chimpanzees and the Red river hog. Although a valuable source of timber such as mahogany and iroko, the rainforest has been cleared in many places for cultivation, often of the naturally occurring oil palm. Beyond the cultivated areas, especially along the channels of the Niger delta, lie broad areas of mangrove swamp, giving way to the sandy beaches and lagoons of the coastline.

SOCIETY

Nigeria is home to more than 200 peoples, producing a rich and varied culture. The Hausa and Fulani are dominant in the north, with the Yoruba, Edo and Ibo peoples forming the largest groups in the south. Just under half the population are Muslim. The rest are mostly Christian or follow traditional African beliefs.

There were many ancient city-states in Nigeria, including the Yoruba city of Ife, the mainly Hausa Kanem-Bornu kingdom, and the powerful Benin kingdom in the southwest, famous for its bronze sculptures. Europeans and Americans bought slaves from these kingdoms, but made no attempt to conquer them. In the 19th century the largely Islamic Fulani launched a jihad (holy war), and conquered most of the country.

It was the suppression of the slave trade

that led Britain to take control of the country, formally annexing Lagos in the southwest in 1861. Further areas were later added, and Nigeria became a crown colony and protectorate in 1914. In 1960 it was granted independence as a federal state made up of four regions. In 1961 the northern half of British Cameroon voted to join the federation.

In 1966 Ibo army officers seized the government in a coup. This led to the massacre of thousands of Ibo in the Northern Region. Following a Hausa countercoup, Lieutenant-Colonel Yakubu Gowon (b. 1934) was made president and worked for reconciliation. Nevertheless, in 1967 the Ibo-dominated Eastern Region seceded under Lieutenant-Colonel C. Odumegwu Ojukwu (b. 1933) as the state

---

NATIONAL DATA – NIGERIA

Land area	910,768 sq km (351,649 sq mi)			

Climate		Temperatures		Annual
	Altitude m (ft)	January °C(°F)	July °C(°F)	precipitation mm (in)
Lagos	3 (10)	27 (81)	2 (77)	1,538 (60.5)

Major physical features highest point: Dimiong 2,042 m (6,700 ft); longest river: Niger (part) 4,200 km (2,600 mi)

Population (2006 est.) 131,859,731

Form of government federal republic with military government

Armed forces army 62,000; navy 7,000; air force 9,500

Largest cities Lagos (9,229,944); Kano (3,848,885); Ibadan (3,847,472); Kaduna (1,652,844); Abuja (capital since 1991 - 1,405,201)

Official language English

Ethnic composition Hausa 21.3%; Yoruba 21.3%; Ibo 18.0%; Fulani 11.2%; Ibibio 5.6%; Kanuri 4.2%; Edo 3.4%; Tiv 2.2%; Ijaw 1.8%; others 11%

Religious affiliations Muslim 50%; Christian 40%; traditional beliefs or none 10%

Currency 1 naira (NGN) = 100 kobo

Gross domestic product (2006 est.) U.S. $188.5 billion

Gross domestic product per capita (2006 est.) U.S. $1,400

Life expectancy at birth male 46.52 yr; female 47.66 yr

Major resources natural gas, petroleum, tin, iron ore, lead, zinc, coal, fisheries, palm kernels, cassava, cocoa, cotton, groundnuts, livestock, maize/corn, millet, rice, rubber, sorghum, timber, yams

A Tuareg herdsman (*above*) The Tuareg are traditionally nomadic, ranging as far south as northern Nigeria. They pay no heed to national boundaries, and have a distinct culture of their own.

The long journey to the coast (*left*) In the absence of good roads, the long rivers of Africa, which often flow through several countries, provide the principal means of transporting goods from the heart of the continent to the ports. Here logs have been roped together to be floated down the river Niger to a sawmill on the coast, where they will be prepared for export.

of Biafra. After a bloody war, in which thousands of Ibo died, Biafra collapsed in 1970. Gowon refrained from making reprisals, but was deposed in 1975. Further military coups thwarted subsequent attempts to restore a stable, civilian government.

Multiparty and presidential elections in 1992 and 1993 were followed by the reimposition of military rule. General Sani Abacha dissolved all political parties and established a provisional ruling council headed by himself. Elected president, Chief Moshood Abiola was imprisoned and died a day before his expected release in 1998. Abacha died in June 1998 and was replaced by General Abdulsam Abubakar who released political detainees and paved the way for a return to civilian rule. In 1999 presidential elections resulted in the coming to power of Olusegan Obasanjo. In 1999, the stability of the country looked threatened by the widespread adoption of Islamic sharia law which was seen as underlining the ethnic and religious split between the northern Muslim states and their southern Christian counterparts. Amidst ethnic and religious clashes and violence there were demands for the secession of the south.

ECONOMY

Agriculture still employs the majority of the labor force, but has declined in relation to other sectors. Activities vary between different areas and among the various ethnic groups. The Hausa, for example, tend toward crop growing, whereas the Fulani prefer to raise cattle. Cocoa and rubber, the main export crops, are grown mostly in the forested south, and palm kernels mainly in the east. Groundnuts (peanuts) are another important cash crop. The main staples are cassava, yams and rice in the south, and millet and sorghum in the north, which is also the main livestock area. Production has not kept pace with population growth, and much food must be imported. Fisheries have some potential, but forestry has suffered from overclearance.

Nigeria is rich in minerals including coal, iron ore, lead and zinc. Oil provides 20 percent of GDP and 95 percent of foreign exchange earnings. Manufacturing contributes around 9 percent of GDP and includes paper products, cement, cigarettes and a variety of consumer goods. Nigeria is one of the most indebted of the African countries, in 1997 it owed US$34 billion to the international banks.

The railroads are not as vital as the extensive road network but there are inland waterways and the principal ports are Lagos, Port Harcourt, Warri and Calabar. Lagos, Port Harcourt and Kano have international airports.

Primary education is available to all but female participation is low. There is a growing higher education sector featuring 13 universities and other institutions.

Cameroon

REPUBLIC OF CAMEROON

CAMEROON LIES ON THE GULF OF GUINEA between west and central Africa. It shares borders with Equatorial Guinea, Gabon, Congo (R.O.), the Central African Republic, Chad and Nigeria.

GEOGRAPHY

Inland Cameroon consists mostly of low plateau country, rising northward to the Adamawa Highlands before sloping gradually down toward Lake Chad in the far north. The west is more mountainous volcanic country, dominated near the coast by the active volcano Mount Cameroon, the highest point in West Africa. The southern plateaus are covered with rainforest, and they fall away to a wide band of plains along the coast.

The climate is tropical, with heavy seasonal rainfall and high temperatures that are moderated by altitude. In the southern rainforests many kinds of tree flourish, including ebony, mahogany and sapele. They also support an immense variety of other plant and animal life, including orchids, ferns, chimpanzees, many species of monkey, and a rich bird life. The northern savanna country supports elephants, lions and leopards. A national park has been established here.

SOCIETY

Cameroon was probably originally inhabited by the Pygmies. Later various groups of Bantu and Sudanic peoples moved in. A source of slaves for Dutch and Portuguese traders, the country was colonized by Germany in the 19th century in opposition to Britain. In World War I it was invaded by France and Britain, and was later divided between them under a League of Nations mandate.

In 1960 French Cameroon became independent. In 1961 the southern half of British Cameroon joined it in a federation, while the north joined Nigeria. In 1972 a referendum replaced the federation with a unified government under Ahmadou Ahidjo (b. 1924), the president since independence. In 1982 he resigned and was replaced by Paul Biya (b. 1933), who defeated a military coup in 1984. In 1990 a multiparty constitution was adopted; Biya was re-elected in 1997.

ECONOMY

The economy has benefited from liberal management and the exploitation of oil reserves. It has continued to grow since independence, although it still depends on foreign aid. Agriculture has been less affected by drought than in countries to the north. A wide range of crops such as coffee, bananas and rubber are grown for export, alongside staple crops such as corn, millet, sorghum, groundnuts (peanuts) and cassava. Timber production is now growing.

Industry is also developing fast. Factories range from a large aluminum plant at Edéa, driven by power from the Sanaga river hydroelectric scheme, to a paper mill and tire factory based on local supplies of timber and rubber. Road transportation is often hampered by the rainy season, but rail and air links are well developed, especially to Douala, the main port. Social security is limited, and health care is poor except in the cities. Education is free, and is fast expanding. The media are under government control.

A cattle-breeders' village in Cameroon. The wild and barren landscape around this village seems to offer little in the way of good grazing, but Cameroon has been less affected by periodic drought than many of its neighbors to the north.

NATIONAL DATA –CAMEROON

Land area	469,440 sq km (181,252 sq mi)			
Climate		Temperatures		Annual
	Altitude m (ft)	January °C(°F)	July °C(°F)	precipitation mm (in)
Yaoundé	760 (2,493)	24 (75)	23 (73)	657 (25.9)

Major physical features	highest point: Mount Cameroon 4,070 m (13,353 ft)
Population	(2006 est.) 17,340,702
Form of government	multiparty republic with one legislative house
Armed forces	army 12.500; navy 1,300; air force 300
Largest cities	Douala (1,404,831); Yaoundé (capital – 1,390,516)
Official languages	French, English
Ethnic composition	Cameroon Highlanders 31%; Equatorial Bantu 19%; Kirdi 11%; Fulani 10%; Northwestern Bantu 8%; Eastern Nigritic 7%; other African 13%; non-African 1%
Religious affiliations	traditional beliefs 40%; Christian 40%; Muslim 20%
Currency	1 Communauté Financière Africaine franc (XAF) = 100 centimes
Gross domestic product	(2006 est.) U.S. $42.2 billion
Gross domestic product per capita	(2006 est.) U.S. $2,400
Life expectancy at birth	male 50.98 yr; female 51.34 yr
Major resources	petroleum, bauxite, iron ore, hydropower, cocoa, cassava, coffee, cotton, groundnuts, gold, livestock, maize, millet, natural gas, palm oil, plantains, rubber, sorghum, sweet potatoes, timber, tin, yams

Cent. African Rep.

CENTRAL AFRICAN REPUBLIC

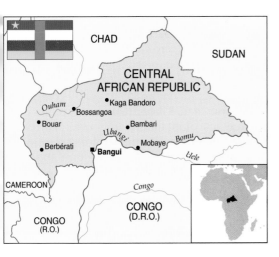

A<small>S ITS NAME SUGGESTS, THE CENTRAL</small> African Republic is a landlocked country close to the continent's center; it is situated between Cameroon, Congo (R.O.), Congo (D.R.O.), Sudan and Chad.

GEOGRAPHY

Most of the country consists of an undulating plateau, flanked by the higher Bongo Massif in the northeast and the Monts Karre to the west. The plateau forms a major divide between the watershed of Lake Chad, some distance away to the northwest, and that of the Ubangi river, the Republic's southern border. The climate ranges from near-desert in the far north to lush equatorial in the south, with seasonal rainfall (March–October) so heavy that it practically cuts

off the southeast. Here there is dense rainforest – one of the last habitats of the lowland gorilla. However, most of the country is covered with savanna grassland and open forest. Wildlife includes elephants, hyenas, antelopes and buffaloes. Both crocodiles and hippopotamuses are found in the rivers.

SOCIETY

By the 19th century the original Pygmy population had been practically wiped out by slave traders, mostly from Sudan. Bantu and Azande peoples settled in their place. The slave trade was suppressed by the French, who took formal control of the country in 1911, calling it the French Congo or Ubangi-Shari. Exploitation by ruthless commercial concessionary companies caused a major scandal. In 1946 the country became a French overseas territory, and in 1958 it gained full independence under its present name. The national language is Sango, a lingua franca, but French is still used for official communications.

In 1965 the second president, David Dacko (b. 1930), was deposed in a military coup led by Colonel Jean-Bédel Bokassa (1921–96), who ruled as a dictator. In 1975 he renamed the country the Central African Empire, crowning himself emperor. In 1979 he was deposed with French help, and was later tried on charges that included murder and cannibalism. General André Kolingba took power in 1981 but it was not until 1993 that free elections returned a civilian government headed by president Ange-Félix Patassé. In 1996 a government of national unity was formed following an army mutiny and French intervention.

ECONOMY

The Republic's economy is now being rebuilt following social unrest and years of dictatorship. Agriculture is mostly at subsistence level, and less than one-thirtieth of the land is cultivated. Groundnuts (peanuts), cotton, coffee and oil palms are grown for export, while food crops include millet, rice and cassava. Conditions rule out livestock raising.

Diamonds are the Republic's main natural resource, accounting for a substantial proportion of exports. Uranium mining has begun, but is hampered by seasonal flooding. The manufacturing sector is mainly small-scale, but includes among other things an automobile-assembly plant. Most energy comes from

Sunrise over the Ubangi river on the southern border of the Central African Republic. Not far away is the capital, Bangui, which was established by the French in 1889. On the other side of the river are the tropical rainforests of the Democratic Republic of Congo.

the hydroelectric plant at Boali in the southwest of the country.

The Ubangi river provides transportation, with Bangui, the capital, as the main port. A road network remains from colonial days, but is often flooded in the rainy season. A limited social-welfare system exists, but health care is inadequate to cope with sleeping sickness, tuberculosis, leprosy and other endemic diseases. Education is free, and literacy levels are improving. The media are government-controlled.

NATIONAL DATA – CENTRAL AFRICAN REPUBLIC

Land area	622,984 sq km (240,535 sq mi)			
Climate	Altitude m (ft)	**Temperatures** January °C(°F)	July °C(°F)	Annual precipitation mm (in)
Bangui	381 (1,250)	26 (79)	25 (77)	1,545 (60.8)

Major physical features	highest point: Kayagangiri 1,420 m (4,660 ft)
Population	(2006 est.) 4,303,356
Form of government	multiparty republic with one legislative house
Armed forces	army 1,400; air force 150
Capital city	Bangui (552,904)
Official languages	French, Sango
Ethnic composition	Baya 33%; Banda 27%; Mandjia 13%; Sara 10%; Mboum 7%; M'Baka 4%; Yakoma 4%; other 2%
Religious affiliations	traditional beliefs 35%; Protestant 25%; Roman Catholic 25%; Muslim 15%
Currency	1 Communauté Financière Africaine franc (XAF) = 100 centimes
Gross domestic product	(2006 est.) U.S. $4.913 billion
Gross domestic product per capita	(2006 est.) U.S. $1,100
Life expectancy at birth	male 43.46 yr; female 43.62 yr
Major resources	diamonds, uranium, timber, gold, oil, hydropower, bananas, cassava, coffee, cotton, groundnuts, livestock, maize/corn, millet, plantains, sweet potatoes

Equatorial Guinea

REPUBLIC OF EQUATORIAL GUINEA

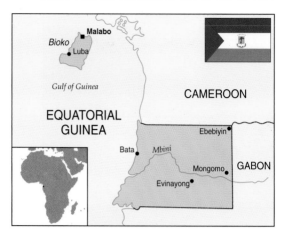

T HE MAINLAND PART OF EQUATORIAL GUINEA faces the Gulf of Guinea. Formerly called Río Muni, it is bounded by Cameroon to the north and Gabon to the south and west. The country also includes five islands, of which volcanic Bioko (the largest) and Pagalu (Annobón) are widely separated across the gulf.

GEOGRAPHY

The mainland rises from a narrow coastal plain to inland plateaus and hills that are extensions of Gabon's Monts de Cristal. Half this area is cloaked in forests, and the animal life is rich and varied. The equatorial climate is hot and humid, with two rainy seasons (February–June and September–December).

NATIONAL DATA – EQUATORIAL GUINEA

Land area	28,051 sq km (10,831 sq mi)			
Climate		Temperatures		Annual
	Altitude m (ft)	January °C(°F)	July °C(°F)	precipitation mm (in)
Malabo	50 (164)	25 (77)	25 (77)	1,799 (70.8)

Major physical features highest point: Pico de Santa Isabel 3,007 m (9,685 ft); largest island: Bioko 2,017 sq km (779 sq mi)

Population (2006 est.) 540,109

Form of government multiparty republic but ruled more as a dictatorship

Armed forces army 1,100; navy 120; air force 100

Capital city Malabo (166,535)

Official languages Spanish, French

Ethnic composition Fang 72.0%; Bubi 14.7%; Duala 2.7%; Ibibio 1.3%; Maka 1.3%; others 8.0%

Religious affiliations Roman Catholic 88.8%; traditional beliefs 4.6%; Muslim 0.5%; others 0.2%; none 5.9%

Currency 1 Communauté Financière Africaine franc (XAF) = 100 centimes

Gross domestic product (2005 est.) U.S. $25.69 billion

Gross domestic product per capita (2006 est.) U.S. $6,572

Life expectancy at birth male 48 yr; female 51.13 yr

Major resources petroleum, natural gas, timber, gold, bauxite, diamonds, tantalum, sand, gravel, clay, bananas, cassava, cocoa, coconuts, coffee, sweet potatoes

SOCIETY

Bioko (formerly Fernando Póo) was discovered by the Portuguese around 1472, and ceded first to Spain and then to Britain. The Spanish reoccupied Fernando Póo in 1858, and began to colonize the mainland. In 1963 Spanish Guinea became Equatorial Guinea, which gained independence in 1968.

A period of military rule lasted from a 1979 coup until the first multiparty elections were held in 1993 as a result of which Brigadier General Obiang Nuguema Mbasogo (b. 1943), who had seized power in 1979, was re-elected president. Obiang was re-elected again in 1996 amidst accusations of fraud.

The Fang form the largest ethnic group. The official language is Spanish and most people are nominally Roman Catholic.

ECONOMY

Oil production started in 1992 and in 2000, production was over 100,000 bbls a day and investment by US-based oil companies was high. There are also unexploited reserves of titanium, iron ore, manganese, uranium and alluvial gold. Agriculture still supports 75 percent of the population in subsistence farming while food processing is really the only industry. Corruption is rife and has led to the suspension of international aid packages. The country is underdeveloped: roads are poor but there are ports at Bata, Luba, and Malabo which also has an international airport. Health care is poor but the literacy rate is the second highest in Africa.

A palm-fringed beach in Equatorial Guinea. The black sand here has been formed by the marine erosion of iron-rich volcanic rock.

Gabon

GABONESE REPUBLIC

G ABON SITS ASTRIDE THE EQUATOR ON Africa's west coast, between Equatorial Guinea, Cameroon and Congo (R.O.).

GEOGRAPHY

The country is centered around the great Ogooué river, which rises in western Congo. The Ogooué and its tributaries flow westward, cutting deep valleys through Gabon's uneven plateaus, then spreading out to form a wide delta across the coastal plain. South of the Ogooué the land rises to the Massif du Chaillu. The climate is hot and humid, with heavy rains between October and May. Dense rainforest covers most of the country. Wildlife flourishes here.

SOCIETY

The Pygmies were Gabon's original inhabitants, but Bantu peoples, especially the Fang, later moved into the area. Gabon was discovered by the Portuguese, and for centuries was a source of slaves. It was colonized by France in the late 19th century, and was harshly exploited by concessionary companies. It became independent in 1960. In 1967 Omar Bongo (b. 1935) became the republic's second president. His liberal economic policies heralded a period of stability and some prosperity. In 1990, after over 20 years of one-party rule, there were elections to a new multiparty National Assembly. Multiparty presidential elections were won by Omar Bongo in 1993. Parliamentary elections were held in late 1996.

ECONOMY

Agriculture is still mostly subsistence farming. Coffee, cocoa and oil palms have been encouraged as cash crops, while cassava remains the chief food crop. Forestry has become more important, especially of the gaboon or Gabon mahogany tree. As coastal forests have become depleted, so the TransGabon Railroad has been built to facilitate the exploitation of the interior.

Mining has greatly contributed to the country's prosperity. Coastal and offshore petroleum deposits provide most exports; timber, manganese and uranium provide much of the rest, leaving a substantial surplus. Manufacturing is mostly oil refining and timber processing, and power is provided by the Kinguélé hydroelectric station in the northwest. Social welfare and medical care are fairly good. Education is free and compulsory, and is available to most children.

Lambaréné, in western Gabon, is a small town on an island in the Ogooué river. It is most famous for its hospital, founded by the German theologian, musician and doctor Albert Schweitzer (1875–1965).

NATIONAL DATA - GABON

Land area	257,667 sq km (99,486 sq mi)		
Climate		Temperatures	Annual

Climate	Altitude m (ft)	January °C(°F)	July °C(°F)	Annual precipitation mm (in)
Libreville	9 (30)	27 (81)	24 (75)	2,509 (98.7)

Major physical features highest point: Mont Milondo 1,020 m (3,346 ft)

Population (2006 est.) 1,424,906

Form of government multiparty republic with one legislative house

Armed forces army 3,200; navy 500; air force 1,000

Capital city Libreville (604,265)

Official language French

Ethnic composition Fang 35.5%; Mpongwe 15.1%; Mbete 14.2%; Punu 1.5%; others 23.7%

Religious affiliations Roman Catholic 65.2%; Protestant 18.8%; African Christian 12.1%; traditional beliefs 2.9%; Muslim 0.8%; others 0.2%

Currency 1 Communauté Financière Africaine franc (XAF) = 100 centimes

Gross domestic product (2006 est.) U.S. $10.21 billion

Gross domestic product per capita (2006 est.) U.S. $7,200

Life expectancy at birth male 53.21 yr; female 55.81 yr

Major resources petroleum, natural gas, diamonds, niobium, manganese, uranium, gold, timber, iron ore, hydropower, bananas, cassava, cocoa, coffee, livestock, maize/corn, palm oil, plantains, rice, sugarcane

Congo

REPUBLIC OF CONGO

CONGO EXTENDS INTO THE HEART OF AFRICA from the Atlantic coast and shares borders with Congo (D.R.O.), the Central African Republic, Cameroon and Gabon.

GEOGRAPHY

The narrow coastal plain runs between neighboring Gabon and the Angolan enclave of Cabinda. Immediately inland is the Massif du Mayombé, followed by the Niari valley. Farther east the land rises again to a series of plateaus that form the western rim of the great Congo (Zaire) river basin. This covers the northeast of the country – an area crisscrossed by rivers and swamps. The Congo and its great tributary the Ubangi mark Congo's eastern border.

The climate is tropical throughout, but the season of heavy rains varies according to latitude. There is savanna grassland to the east, but most of the country is covered with dense rainforest. Forest animals flourish here, ranging from buffaloes to the rare okapi and gorilla. The savanna areas support other animals such as giraffes and rhinoceroses.

SOCIETY

The Congo area was first settled by the Pygmies. These were later followed by the Kongo peoples, whose kingdoms flourished on the slave trade with Europe. In the late 19th century the area was explored by a French–Italian called Pierre Savorgnan de Brazza (1852–1905). It was colonized by the French, and became French Congo in 1891. Until the 1930s the country was harshly exploited by concessionary companies.

Independence came in 1960, but unrest in 1963 forced the first president to resign. In 1968 his successor was ousted and the new president attempted to remodel the state on Marxist lines. From 1969–91 Congo was renamed the People's Republic of the Congo. In 1979 Colonel Denis Sassou-Nguesso was appointed president. Multiparty elections were held in 1992 and Pascal Lissouba was elected president. The results were strongly disputed and politically-inspired violence erupted into civil war in 1997 and again in 1999. Sassou-Nguesso returned to power and promised to restore full democracy when the conflict ended.

ECONOMY

Agriculture consists mostly of low-yield subsistence farming. Cassava is the main food crop, while sugar cane and coffee are grown for export. Forestry is carried out in the southern rainforests. Most exports are of crude petroleum, which is mined both inland and offshore. Other substantial mineral resources include zinc and lead. Manufacturing includes food processing, textiles, cement, metal goods and chemicals. The civil war had a deleterious effect on industry and infrastructure as well as on the economy as a whole.

Congo has a good rail network in the south. The roads, however, are poor, and are subject to seasonal flooding. Major rivers provide the best transportation in the north. Health care is generally inadequate, especially in rural areas, where tropical diseases are rife. Education is free and compulsory from the ages of 6 to 16.

NATIONAL DATA – CONGO (R.O.)

Land area	341,500 sq km (131,854 sq mi)			
Climate		**Temperatures**		**Annual**
	Altitude m (ft)	January °C(°F)	July °C(°F)	precipitation mm (in)
Brazzaville	314 (1,030)	26 (79)	23 (73)	1,493 (58.7)

Major physical features highest point: Monts de la Lékéti 1,040 m (3,410 ft); longest river: Congo (part) 4,630 km (2,880 mi)

Population (2006 est.) 3,702,314

Form of government multiparty republic with one legislative house

Armed forces army 8,000; navy 800; air force 1,200

Largest cities Brazzaville (capital – 1,370,612); Pointe Noire (708,772)

Official language French

Ethnic composition Kongo 48%; Sangha 20%; M'Bochi 12%; Teke 17%; other 3%

Religious affiliations Christian 50%; animist 48%; Muslim 2%

Currency 1 Communauté Financière Africaine franc (XAF) = 100 centimes

Gross domestic product (2006 est.) U.S. $4.958 billion

Gross domestic product per capita (2006 est.) U.S. $1,300

Life expectancy at birth male 51.65 yr; female 53.98 yr

Major resources petroleum, timber, potash, lead, zinc, uranium, copper, phosphates, gold, magnesium, natural gas, hydropower, bananas, cassava, cocoa, coffee, groundnuts, palm oil, rice, sugarcane, sweet potatoes

Congo (formerly Zaire)

DEMOCRATIC REPUBLIC OF CONGO

CONGO (D.R.O.) STRADDLES THE EQUATOR IN the very heart of the African continent, and is bordered on all sides by as many as nine different countries. It is landlocked apart from a 50-km (30-mi) strip of coastline, and is the third-largest country in Africa after Algeria and Sudan.

ENVIRONMENT

The great Congo river rises in the west, and flows north and west in a great horse-shoe, meeting the Ubangi on the Congo (R.O.) border. The vast Congo Basin occupies central and northwestern Congo, surrounded by higher plateau country. This becomes mountainous along the eastern border, where the Great Rift Valley is marked by a number of lakes such as Lake Tanganyika, and by many ranges of volcanic mountains. To the south rise the peaks of the Shaba plateau. To the west, the Congo river runs between further plateaus to enter the Atlantic.

As well as vast swamps, the Congo Basin contains one of the world's greatest expanses of tropical rainforest, encouraged by torrential rainfall and consistently high temperatures. The forest is one of the last homes of rare and endangered

Traditional farming in Congo (D.R.O.). The country's massive agricultural potential has yet to be fully exploited, and most farmers operate at subsistence level. Livestock raising is an underdeveloped resource.

animals such as the okapi and the mountain gorilla. In the east and south, the forests give way to savanna woodlands and grasslands supported by heavy seasonal rainfall. Wildlife here includes giraffes, lions, antelopes, and the endangered black and white rhinoceroses. Various national parks have been created to conserve wildlife.

SOCIETY

Congo (D.R.O.) is the home of over 200 (mostly Bantu) ethnic groups, many of whom had formed kingdoms before the arrival of Europeans. In the 1870s King Leopold II of Belgium (1835–1909), keen to join the European scramble for African possessions, commissioned the British-born American explorer Henry Morton Stanley (1841–1904) to penetrate the little-known Congo Basin area. Leopold established the so-called Congo Free State at his own private expense, and ran it as his own domain. Widespread atrocities practiced by concessionary companies aroused international protests, and in 1908 Leopold was forced to grant a proper colonial status to the Belgian Congo.

After many decades of centralized rule from Belgium, growing nationalist parties demanded independence. It was granted in 1960, and Patrice Lumumba (1925–61) was the first prime minister. But an army mutiny broke out, led by Colonel Joseph-Désiré Mobutu (b. 1930). The southeastern province of Katanga (later Shaba) seceded under the leadership of Moise Tshombe (1919–69). The government called in United Nations peacekeeping forces, but Lumumba was murdered and the United Nations Secretary-General, Dag Hammarskjöld (1905–61), was killed in a plane crash while visiting the country.

When United Nations forces left in 1964, further revolts were suppressed by Tshombe, now president of the whole country. In 1965 Mobutu seized the presidency, promoting intensely nationalist policies. The country's name was changed to Zaire, and its citizens had to take African names; the president himself became Mobutu Sese Seko.

Mobutu's administration brought stability to Zaire and multiparty elections were promised for 1992. However, economic problems led to rioting and mutiny of the army in 1991. Political and economic problems continued exacerbated by an influx of half a million refugees from Rwanda. In 1997 Mobutu was ousted by rebel forces headed by Laurent Kabila who declared himself president and

NATIONAL DATA – CONGO (D.R.O.)

Land area	2,267,600 sq km (875,525 sq mi)			
Climate		Temperatures		Annual precipitation mm (in)
	Altitude m (ft)	January °C(°F)	July °C(°F)	
Kinshasa	311 (1,020)	26 (79)	23 (73)	1,358 (53.4)

Major physical features highest point: Mount Stanley 5,109 m (16,763 ft); largest lake: :Lake Tanganyika (part) 32,900 sq km (12,700 sq mi)

Population (2006 est.) 62,660,551

Form of government transitional multiparty republic

Armed forces army 60,000; navy 1,800; air force 3,000

Capital city Kinshasa (8,418,819)

Official language French

Ethnic composition Luba 18.0%; Kongo 16.1%; Mongo 13.5%; Rwanda 10.3%; Azande 6.1%; Bangi/Ngale 5.8%; Rundi 3.8%; Teke 2.7%; Boa 2.3%; Chokwe 1.8%; Lugbara 1.6%; Banda 1.4%; others 16.6%

Religious affiliations Roman Catholic 50%; Protestant 20%; Kimbanguist 10%; Muslim 10%; other syncretic sects and traditional beliefs 10%

Currency 1 Congo franc (CDF) = 100 centimes

Gross domestic product (2006 est.) U.S. $44.6 billion

Gross domestic product per capita (2006 est.) U.S. $700

Life expectancy at birth male 50.01 yr; female 52.94 yr

Major resources cobalt, copper, niobium, tantalum, petroleum, industrial and gem diamonds, gold, silver, zinc, manganese, tin, uranium, coal, hydropower, timber, bananas, cassava, cocoa, coffee, cotton, groundnuts, maize/corn, millet, natural gas, plantains, palm oil and kernels, rice, rubber, sugarcane, tea

renamed Zaire the Democratic Republic of Congo. Violence continued in 1998 and 1999, underlain by ethnic rivalry; in 2001 Kabila was assassinated by his own bodyguard and his son took over the country.

ECONOMY

The civil war had a serious impact on the economy: in 1998 GDP was estimated at 65 percent lower than in 1960, and inflation reached 240 percent. Congo's (D.R.O.) agriculture is potentially rich, but apart from some large plantations it remains mostly at subsistence level. Cash crops include coffee, cocoa, rubber and cotton. The main staples are cassava, corn, rice and millet. The vast forests are rapidly disappearing.

Congo (D.R.O.) enjoys great mineral wealth. A major copper belt extends across the southern border into Congo's southeastern province of Shaba, with secondary deposits of cobalt, both among the world's largest reserves. Uranium, gold and silver ores are extracted and there are petroleum reserves off the coast. Ore smelting and petroleum refining are the main industries. Congo (D.R.O.) also produces over half the world output of low-grade industrial diamonds.

Transportation depends heavily on the river and railroad networks. Social welfare and health facilities are limited. Primary education is good and literacy levels are relatively high.

Burundi

REPUBLIC OF BURUNDI

BURUNDI IS A SMALL, LANDLOCKED AND densely populated country in Central Africa on the northeastern shore of Lake Tanganyika. It is bordered by Congo (D.R.O.), Tanzania and Rwanda, with which it was formerly federated.

GEOGRAPHY

Along Burundi's western border, the Ruzizi river flows along part of the Great Rift Valley into Lake Tanganyika. Above here the land rises steeply to a mountainous ridge that extends into Rwanda, forming the ultimate divide between the great Congo and Nile river systems. Beyond the mountains the land falls away into an irregular plateau, cut through by the valley of the Ruvubu river. The tropical climate is moderated by the high altitude, and the seasonal rainfall pattern leads to both floods and droughts. Forests in the mountains give way to wooded country and then grassland savanna.

SOCIETY

The Twa, the original Pygmy inhabitants of the area, and the Hutu peoples that later settled there, were subjugated by the less numerous Tutsi, who ruled as a feudal aristocracy through their control of cattle, the primary source of wealth. Burundi, a former German colony mandated to Belgium, was granted independence in 1962 as a Tutsi kingdom, but became a one-party republic after two separate coups in 1966. A series of coups followed involving both Tutsi and Hutu.

Political and ethnic unrest between Hutu and Tutsi has continued unceasingly up to the present day, displacing hundreds of thousands of people and killing an estimated quarter of a million more between 1972 and 1999. In the 1990s a number of attempted coups and political assassinations led to a further coup in 1996 which installed Major Pierre Buyoya as president. Kurundi and French are the two official languages of the country.

ECONOMY

The economy rests largely on subsistence agriculture and has suffered from the effects of continued ethnic violence. Coffee is by far the most important export but cotton is exported as well. Staple crops include cassava, yams and grains such as corn and millet. Livestock includes cattle – reared in quantity for status and wealth – together with pigs, goats and sheep. Cattle hides are also a significant export.

Mining is small scale and manufacturing is limited. There are no railroads, and few good roads across the mountainous terrain. Social welfare and health care provide a very limited service; malaria and tuberculosis are widespread. Education reaches only a minority of the population.

A Tutsi man (*below*) sits within a house compound in the Burundi village of Gisozo. The Tutsi, who are thought to be of Nilotic origin, are only a minority in Burundi, but have dominated the country for many years. Most Burundians belong to the Hutu group – an agricultural people of Bantu origin.

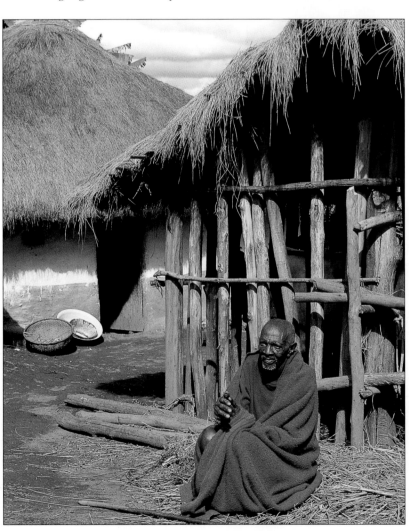

NATIONAL DATA - BURUNDI

Land area	25,650 sq km (9,904 sq mi)			
Climate		Temperatures		Annual
	Altitude m (ft)	January °C(°F)	July °C(°F)	precipitation mm (in)
Bujumbura	815 (2,674)	24 (75)	23 (73)	848 (33.3)

Major physical features largest lake: Lake Tanganyika (part) 32,900 sq km (12,700 sq mi)

Population (2006 est.) 8,090,068

Form of government republic

Armed forces army 45.000

Capital city Bujumbura (342,449)

Official languages Kirundi, French

Ethnic composition Hutu (Bantu) 85%; Tutsi (Hamitic) 14%; Twa (Pygmy) 1%; Europeans 3,000, South Asians 2,000

Religious affiliations Christian 67% (Roman Catholic 62%; Protestant 5%); traditional beliefs 23%; Muslim 10%

Currency 1 Burundi franc (BIF) = 100 centimes

Gross domestic product (2006 est.) U.S. $5.744 billion

Gross domestic product per capita (2006 est.) U.S. $700

Life expectancy at birth male 50.07 yr; female 51.58 yr

Major resources nickel, uranium, rare earth oxides, peat, cobalt, copper, platinum, vanadium, hydropower, niobium, tantalum, gold, tin, tungsten, kaolin, limestone, bananas, beans, cassava, coffee, coconuts, columbium, maize/corn, peat, phosphates, sweet potatoes, tea, timber

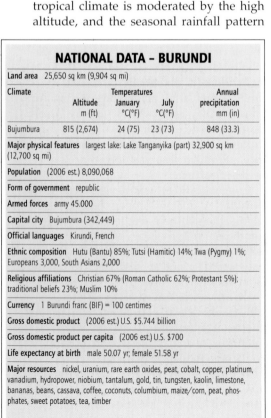

Rwanda

REPUBLIC OF RWANDA

THE MOUNTAINOUS, LANDLOCKED COUNTRY OF Rwanda in Central Africa shares its borders with Congo (D.R.O.), Uganda and Tanzania, and to the south with Burundi, with which it was formerly federated.

GEOGRAPHY

In the west of Rwanda, Lake Kivu on the Congo (D.R.O.) border is part of the Great Rift Valley. East of here the land rises steeply to the mountains of the Congo-Nile divide. The Virunga Mountains to the north include two active volcanoes. The high plateaus to the east fall away toward the Tanzanian border. The tropical climate is moderated by altitude, with heavy seasonal rains. The mountain forests are the last home of the mountain gorilla. The plateaus are covered with woodland and savanna.

A smallholding in Rwanda Subsistence agriculture like this forms the backbone of the Rwandan economy. Bananas and other fruits are being grown, and food crops are likely to include cassava, yams and grains.

SOCIETY

The original inhabitants were Twa Pygmies. The Hutu settled later in the area, and both were conquered by invading Tutsi. The country was claimed by Germany, and was mandated to Belgium after World War I. In 1959 the Hutu rebelled, ousting the Tutsi king and many of his people. In 1962 Rwanda became an independent republic. Interethnic clashes between Hutu and Tutsi began almost immediately, leading eventually to a military coup in 1973. A period of relative stability ended with the breaking out of civil war in 1990. The war culminated, in April 1994, in a genocide in which around one million Tutsis and moderate Hutus were killed. International intervention halted the genocide but approximately two million Hutu refugees – many fearing Tutsi retribution – fled the country. Civil strife continued into the 2000s.

ECONOMY

The economy is based on subsistence agriculture. Cash crops include coffee (the main export), tea and sugar cane. Cassava, corn, plantains and vegetables are the main food crops. The Tutsi raise cattle, which confer social status, whereas the Hutu raise sheep and goats. Some tin ore is mined. Social welfare is very limited.

NATIONAL DATA – RWANDA

Land area 24,948 sq km (9,632 sq mi)

Climate	Altitude m (ft)	Temperatures January °C(°F)	July °C(°F)	Annual precipitation mm (in)
Kigali	1,545 (5,069)	19 (67)	21 (70)	1,028 (40.4)

Major physical features highest point: Mount Karisimbi 4,507 m (14,787 ft); largest lake: Lake Kivu (part) 2,699 sq km (1,042 sq mi)

Population (2006 est.) 8,648,248

Form of government presidential multiparty republic. New Constitution adopted in 2003.

Armed forces army 40,000; air force 1,000

Capital city Kigali (857,719)

Official languages Kinyarwanda, French, English

Ethnic composition Hutu 84%; Tutsi 15%; Twa (Pygmoid) 1%

Religious affiliations Roman Catholic 56.5%; Protestant 26%; Adventist 11.1%; Muslim 4.6%; traditional beliefs 0.1%;, none 1.7%

Currency 1 Rwandan franc (RWF) = 100 centimes

Gross domestic product (2006) U.S. $13.54 billion

Gross domestic product per capita (2006) U.S. $1,600

Life expectancy at birth male 46.26 yr; female 48.38 yr

Major resources gold, cassiterite (tin ore), wolframite (tungsten ore), methane, hydropower, coffee, tea, sugarcane, bananas, beans and peas, cassava, cattle, groundnuts, goats, potatoes, pyrethrum, pigs, sheep, sorghum, sweet potatoes

Uganda

REPUBLIC OF UGANDA

T HE EAST AFRICAN REPUBLIC OF UGANDA HAS been torn apart by war in recent decades, but since 1986 it has achieved a slow return to stability.

GEOGRAPHY

The country is landlocked, bordering Rwanda and Tanzania to the south and Kenya to the east, with Sudan to the north and Congo (D.R.O.) to the west. Most of it occupies a northward-sloping plateau north and west of Lake Victoria. The capital, Kampala, lies near Lake Victoria's northern shore. In the far southwest, the border with Rwanda runs through the volcanic Virunga mountains. The western border with Congo runs parallel to the western branch of the Great Rift Valley,

features of which include Lake Edward, the Ruwenzori range (rising to Mount Stanley) and Lake Albert. The river Nile flows north from Lake Victoria through Lake Kyoga, then north and west into Lake Albert, and finally north into Sudan.

Uganda's equatorial climate is moderated by its altitude, and by large areas of lake and swamp that occupy one-sixth of the total area. The north has a single wet season from April to November, whereas the south has two rainy seasons (April–May and October–November). Rainfall patterns are reflected in the vegetation, with modified tropical forest around Lake Victoria and wooded savanna parkland in the center and the north. Uganda's national parks and game reserves protect numerous animals and habitats.

SOCIETY

Oral tradition describes several African kingdoms in the area, including the Nilotic state of Bunyoro-Kitara. Buganda, a former satellite of Bunyoro, rose to dominance in the 18th century. In 1894 it became a British protectorate, and in 1896 it was extended to include Bunyoro and neighboring kingdoms. Independence under a federal constitution was granted in 1963, with Mutesa II (1924–69), *kabaka* (king) of Buganda, as the first president. In 1966 Mutesa was ousted by prime minister Milton Obote (b. 1924), who declared a republic. Obote himself was deposed by Idi Amin (b. 1925) in 1971.

Under Amin's dictatorship Asians and other skilled professionals were expelled, the majority Christian population was persecuted, and at least 300,000 people were murdered or disappeared. The economy began to collapse. In 1978 Amin invaded Tanzania, only to be overthrown by a combined force of Tanzanians and Ugandan exiles. In 1980 Milton Obote became president of a shattered country. However, after coups in 1985 and 1986 Yoweri Museveni (b. 1944) took the presidency, facing civil war, troubled relations with Kenya and the burden of some 100,000 refugees.

In 1994 a new constitution created a directly-elected 278-member single assembly and multiparty democracy is planned early in the 2000s. The monarchy was reinstated in 1993 with Ronald Muwenda Mutebi II (b. 1955). Television and radio broadcasting is state-controlled, but some press freedom has been reintroduced.

Uganda has some 40 different ethnic groups, the largest being the Bantu-speaking Ganda (about one-fifth of the

population). Other large groups include the Teso, Nkole, Soga and Gisu. English is the official language, but Swahili is widely spoken.

ECONOMY

During the 1990s the economy increasingly stabilized, inflation reduced and there was sizeable investment in roads and railroads and in projects which will increase agricultural production and manufacturing. Around 80 percent of the workforce is involved with agriculture which accounted for almost half of GDP in 1996. Coffee, cotton, tea and tobacco are the principal exports. Spain, France and Germany are the main export markets. Uganda is rich in minerals including copper, gold and cobalt. Power is chiefly supplied by the Owen Falls Dam on the river Nile.

The road network is undergoing an internationally-funded 10-year improvement plan and there is a project to establish a direct rail link between Kampala and Johannesburg, South Africa. There is an international airport at Entebbe.

An emphasis on preventive medicine has had some success, and there are some rural clinics as well as centralized medical facilities. However, malaria, hookworm, AIDS and gastric disorders are common and housing is in short supply. There is a contributory social security system. Secondary education is limited, and is closely linked to economic needs.

Boys at the pump (*above*) in the Ugandan village of Bugiri, not far from Lake Victoria. Only one-eighth of the population has access to clean drinking water. One-tenth of Ugandan children die in infancy, and AIDS has become a serious problem in the region.

NATIONAL DATA - UGANDA

Land area	199,710 sq km (77,108 sq mi)			
Climate		**Temperatures**		**Annual**
	Altitude m (ft)	**January** °C(°F)	**July** °C(°F)	**precipitation** mm (in)
Kampala	1,150 (3,773)	23 (67)	21 (70)	1,150 (46.2)

Major physical features highest point: Mount Stanley 5,109 m (16,763 ft); largest lake: Lake Victoria (part) 62,940 sq km (24,300 sq mi)

Population (2006 est.) 28,195,754

Form of government republic with one legislative house part-elected by popular vote

Armed forces joint forces 45,000

Largest cities Kampala (1,455,027)

Official languages English, Swahili

Ethnic composition Baganda 17%; Ankole 8%; Basoga 8%; Iteso 8%; Bakiga 7%; Langi 6%; Rwanda 6%; Bagisu 5%; Acholi 4%; Lugbara 4%; Batoro 3%; Bunyoro 3%; Alur 2%; Bagwere 2%; Bakonjo 2%; Jopodhola 2%; Karamojong 2%; Rundi 2%; non-African (European, Asian, Arab) 1%; other 8%

Religious affiliations Roman Catholic 33%; Protestant 33%; Muslim 16%; traditional beliefs 18%

Currency 1 Ugandan shilling (UGX) = 100 cents

Gross domestic product (2006 est.) U.S. $51.89 billion

Gross domestic product per capita (2006 est.) U.S. $1,800

Life expectancy at birth male 51.68 yr; female 53.69 yr

Major resources copper, cobalt, gold, hydropower, limestone, salt, coffee, cotton, tea, tobacco, bananas, fish, livestock, maize/corn, millet, phosphates, sorghum, sugarcane, timber, yams, tourism

Kenya

REPUBLIC OF KENYA

Masai herders in the arid landscape of Kenya's Rift Valley. Livestock raising is mainly confined to the drier parts of the country. Most agricultural land is in the wetter areas, where it is communally owned and used to grow cash crops or staples.

KENYA STRADDLES THE EQUATOR ON THE eastern coast of Africa. Like many African countries, it is a melting pot of different peoples and cultures.

GEOGRAPHY

The coastal plain is narrow in the south, but broadens northeastward toward the Somali border. Inland, the Nyika Plain rises northwestward from the coast. The volcanic western highlands are split in two by the north–south line of the Great Rift Valley. On its western side, the Mau and Elgeyo Escarpments stretch from the Tanzanian border in the south to the Cherangany Hills near Mount Elgon on the Ugandan border. East of the rift are the Laikipia Escarpment and the Aberdare Range, at the southern end of which lies the capital, Nairobi. East of the Aberdare Range is the country's highest mountain, Mount Kenya. The two major rivers, the Tana and the Galana, drain southeastward from the highlands across the Nyika Plain. West of the highlands are grasslands bordering Lake Victoria.

The climate is equatorial and temperatures are similar throughout the year. Large areas have been developed as game reserves, notably the Nairobi and Tsavo National Parks. However, ivory poachers are a continuing threat, especially along the border with Somalia.

SOCIETY

Arab traders settled along the Kenyan coast from about the 9th century onward. Masai invaders moved south in the 18th century, but in about 1830 their progress was checked by resistance from the Kikuyu. In 1887 the Sultan of Zanzibar leased part of the coast to the British, and by 1901 a railroad had been built from Mombasa to Lake Victoria. White settlers found the highlands congenial, and in 1920 Kenya became a British colony.

After World War II rising African nationalism in the 1950s led to a campaign of violence by the Mau Mau, a Kikuyu anticolonial organization. Following several years of unrest, Kenya finally gained independence in 1963, and in 1964 a republic was declared, with the Kikuyu leader Jomo Kenyatta (1889–1978) as its first president. In 1992 the first multiparty elections were held after a series of violent demonstrations which legalized opposition parties. During the last few decades, however, growth has been marred by inflation, droughts, and a rapidly rising population leading to unemployment.

Most Kenyans are Africans of Bantu, Nilotic and Kushitic origin. Out of the 70 or more distinct ethnic groups, the largest and most influential are the Kikuyu and the Luo. Apart from Nairobi and the port of Mombasa, the few large settlements are in the highlands and along the railroad.

ECONOMY

Most Kenyans are still agriculturalists living and working on communally owned land. Agriculture is the chief export earner, but farming is largely restricted to high-rainfall areas of the highlands, central Rift Valley and around Lake Victoria. Coffee and tea are the main cash crops. Staples include corn, potatoes, pulses, millet and bananas. Horticulture is a major growth area. Tourism, an invaluable source of foreign exchange, has been damaged by political violence.

Wood from very limited forests is the only natural source of fuel. Major industrial activities include cement production and petroleum processing based on refineries at Mombasa. Hydroelectricity from the Tana river was supplemented by a geothermal station in 1981.

Road connections are good in the more populated regions, and a highway links Nairobi with Addis Ababa (Ethiopia). Mombasa is the principal port. Nairobi's airport is important in East Africa.

A variety of health-care services is provided by central and local government, voluntary agencies and private doctors. Diseases include malaria, sexually transmitted diseases (particularly AIDS), dysentery and trachoma. Self-help welfare activities are encouraged. Education is not compulsory, but the facilities are improving. Seven years of primary education can lead, after an examination, to four years of secondary education.

NATIONAL DATA – KENYA

Land area	569,250 sq km (219,789 sq mi)			
Climate		**Temperatures**		**Annual**
	Altitude m (ft)	January °C(°F)	July °C(°F)	precipitation mm (in)
Nairobi	1,616 (5,302)	18 (67)	15 (70)	1,024 (40.3)

Major physical features highest point: Mount Kenya 5,199 m (17,058 ft); longest river: Tana 708 km (440 mi)

Population (2006 est.) 34,707,817

Form of government multiparty republic with one legislative house

Armed forces army 20,000; navy 1,620; air force 2,500

Largest cities Nairobi (capital – 2,982,226); Mombasa (847,626); Nakuru (273,039); Eldoret (237,398)

Official languages Swahili, English

Ethnic composition Kikuyu 22%; Luhya 14%; Luo 13%; Kalenjin 12%; Kamba 11%; Kisii 6%; Meru 6%; other African 15%; non-African 1%

Religious affiliations Protestant 45%; Roman Catholic 33%; traditional beliefs 10%; Muslim 10%; other 2%

Currency 1 Kenyan shilling (KES) = 100 cents

Gross domestic product (2006 est.) U.S. $40.77 billion

Gross domestic product per capita (2006 est.) U.S. $1,200

Life expectancy at birth male 49.78 yr; female 48.07 yr

Major resources limestone, soda ash, salt, gemstones, fluorspar, zinc, diatomite, gypsum, wildlife, hydropower, tourism, bananas, beans, cattle, cassava, coffee, cotton, maize/corn, millet, potatoes, pyrethrum, sisal, sugarcane, sweet potatoes, tea

Tanzania

UNITED REPUBLIC OF TANZANIA

THE EAST AFRICAN REPUBLIC OF TANZANIA was created in 1964 by the union of Tanganyika and Zanzibar, which consists of the islands of Zanzibar and Pemba.

GEOGRAPHY

Most of Tanzania consists of a plateau traversed by a sprawling network of mountains and depressions. A narrow coastal plain runs south from the Kenyan border toward Mozambique.

In the southwest Lake Malawi marks the Malawian border. From here the Zambian frontier runs northwestward to the southern end of Lake Tanganyika. The northwestern corner of the country

borders Burundi, Rwanda and Uganda. The Ugandan border cuts across Lake Victoria, meeting the Kenyan border near its eastern shore. In the many scattered mountain ranges are several volcanoes flanking the Great Rift Valley. These include the Ngorongoro Crater and also mount Kilimanjaro, which is Africa's highest peak.

The climate is equatorial, with a fairly narrow range of temperatures; rainfall depending largely on altitude and aspect. There are two rainy seasons in the north, whereas the south has a single, much longer season.

Much of the country is covered in grassland with scattered woodland but tsetse-fly infestation makes a great deal of this land unsuitable for cattle or human settlement. Alpine desert is found at higher altitudes with semidesert on some north-facing slopes. Coastal vegetation ranges from mangroves to reedy swampland. The abundant wildlife includes elephants, rhinoceroses and lions, and there are many fine national parks, notably the Serengeti in the north.

SOCIETY

Excavation of the Olduvai Gorge in the northeast has revealed some of the oldest known human remains, dating back 2 million years. Coastal trade with Arabia and India began in about the 1st century AD. An Arab civilization flourished on the coast from the 8th century until the Portuguese took over in the 16th century. Arabs later regained control, and their trade in slaves and ivory spread inland. In 1886 Germany and Britain both established claims in the area. The Germans controlled Tanganyika, and the British made Zanzibar a protectorate.

During World War I Britain occupied Tanganyika, but in 1947 it was placed under United Nations trusteeship. The 1950s saw the emergence of nationalism, with the foundation of the Tanganyika African National Union (TANU), leading to independence in 1961. In 1962 a republic was declared, with Julius Nyerere (1922–99) as president. Zanzibar gained independence in December 1963. A month later a violent revolution replaced the mainly Arab government with a republic controlled by the Afro-Shirazi Party (ASP). A similar revolution in Tanganyika was quelled with British help. In 1964 Tanganyika and Zanzibar united.

TANU and the ASP united in 1977 to form the Revolutionary Party of Tanzania (Chama Cha Mapinduzi or CCM), which

Tanzania's abundant wildlife is protected in its many national parks, where lions such as this one generally live in family groups. The females are usually the hunters, though the long-maned males are perfectly able to kill for themselves.

became Tanzania's only political party. In 1992 the CCM promised to abolish one-party rule; first multiparty elections were held late in 1995. The president is the head of state, and his cabinet includes a vice-president and prime minister. Zanzibar's internal affairs are administered by its own elected council and president.

Tanzania is one of the least urbanized countries in Africa. There are some 120 ethnic groups, the largest being the Nyamwezi and the Sukuma. Swahili is Zanzibar's principal language. Christianity and Islam each claim about one-third of the population, but Islam dominates in Zanzibar.

ECONOMY

Agricultural products (notably coffee and cotton) are the major export earners, but low selling prices create a massive trade gap, and overseas aid is vital. Cattle are the main livestock, and staple crops include grains, fruit and vegetables. Production is susceptible to drought and fuel shortages. Many mineral resources are unexploited, but salt and gems are exported. Energy comes from hydroelectric and geothermal sources, and petroleum products are exported to Zambia. Other industries include food processing, textiles and tourism.

The infrastructure is slowly improving, with a new highway being built into Zambia. Dar es Salaam is the main port, and steamers operate on the Tanganyika and Victoria Lakes. The main airports are at Dar es Salaam and Kilimanjaro. Malaria and sleeping sickness are being eradicated, but malnutrition is a problem. Primary education is free and compulsory, and secondary education is oriented toward practical training.

NATIONAL DATA – TANZANIA

Land area	886,037 sq km (342,101 sq mi)			
Climate		**Temperatures**		**Annual**
	Altitude m (ft)	January °C(°F)	July °C(°F)	precipitation mm (in)
Salaam	58 (190)	27 (81)	23 (74)	1,056 (41.6)

Major physical features highest point: Kilimanjaro 5,895 m (19,341 ft); largest lake: Victoria (part) 62,940 sq km (24,300 sq mi)

Population (2006 est.) 37,445,392

Form of government multiparty republic with one legislative house

Armed forces army 23,000; navy 1,000; air force 3,000

Largest cities Dar es Salaam (2,915,878); Mwanza (480,279); Dodoma (capital - 196,006)

Official languages Swahili, English

Ethnic composition African 99% (of which 95% are Bantu consisting of more than 130 tribes); other 1%

Religious affiliations mainland - Christian 30%, Muslim 35%, traditional beliefs 35%; Zanzibar more than 99% Muslim

Currency 1 Tanzanian shilling (TZS) = 100 cents

Gross domestic product (2006 est.) U.S. $29.25 billion

Gross domestic product per capita (2006 est.) U.S. $800

Life expectancy at birth male 44.93 yr; female 46.37 yr

Major resources hydropower, tin, phosphates, iron ore, coal, diamonds, gemstones, gold, natural gas, nickel, beans, cassava, cloves, coconuts, coffee, cotton, maize/corn, sisal, tobacco

Seychelles

REPUBLIC OF SEYCHELLES

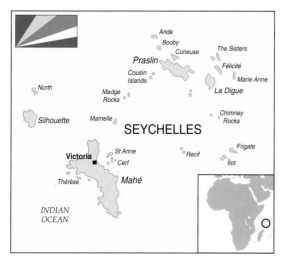

THE SEYCHELLES FORM A WIDELY SCATTERED archipelago of 100 or so islands, plus innumerable islets and cays, in the Indian Ocean northeast of Madagascar.

GEOGRAPHY

The islands fall into two main physical types. The islands of the central group, including the main island of Mahé, are founded on granite, and consist of a mountainous heart surrounded by a flat coastal strip. The outer islands are made up of coral accretions at various stages of formation, from reefs to atolls. These are generally smaller and almost entirely flat, only a few meters above sea level. Most of these islands have virtually no water, and few are inhabited.

The main islands enjoy high humidity and rainfall, feeding the remaining lush tropical rainforest of hardwoods and palms such as coco-de-mer that once totally covered them. Wildlife includes a rare giant land turtle and colorful reef animals such as the Green sea turtle. This island paradise has suffered from human incursions, but is now valued as a major tourist attraction; some coral islands are bird sanctuaries.

SOCIETY

Occupying a strategic position on the Europe–India sea route, the Seychelles were annexed by France in 1756 to support its colony on Mauritius. It passed to Britain after the Napoleonic Wars. Most of the people are of mixed colonial descent and speak a Creole dialect. The Seychelles became independent in 1976 with James R. Mancham (b. 1939) as president and Albert René (b. 1935) as prime minister. In 1979, while Mancham was abroad, René seized the presidency in a coup. In 1993 he was re-elected in the first multiparty elections.

ECONOMY

The mainstay of the economy is tourism, served by an international airport on Mahé and many hotels. The islands' agriculture no longer supports the expanding population. Copra, tobacco, cinnamon, vanilla and coconuts are grown for export, and much food is imported.

Other important exports include fish (especially canned tuna), guano and re-exported petroleum products. Industry is largely confined to food processing. Health care is excellent, and education is available to most children.

NATIONAL DATA – SEYCHELLES

Land area	455 sq km (176 sq mi)			
Climate		**Temperatures**		**Annual precipitation**
	Altitude m (ft)	January °C(°F)	July °C(°F)	mm (in)
Victoria	3 (10)	27 (80)	26 (56)	2,172 (85.5)

Major physical features largest island: Mahé 153 sq km (59 sq mi)

Population (2006 est.) 81,541

Form of government multiparty republic with one legislative house

Armed forces army 200

Capital city Victoria (22,336)

Official languages English, French, Creole

Ethnic composition Seychellois Creole (Asian/African/European) 89.1%; Indian 4.7%; Malagasy 3.1%; Chinese 1.6%; English 1.5%

Religious affiliations Roman Catholic 82.3%, Anglican 6.4%, Seventh Day Adventist 1.1%, other Christian 3.4%, Hindu 2.1%, Muslim 1.1%, other non-Christian 1.5%, unspecified 1.5%, none 0.6%

Currency 1 Seychelles rupee (SCR) = 100 cents

Gross domestic product (2002 est.) U.S. $626 million

Gross domestic product per capita (2002 est.) U.S. $7,800

Life expectancy at birth male 66.69 yr; female 77.63 yr

Major resources fish, coconuts, copra, cinnamon, tobacco, vanilla, tourism, fruit

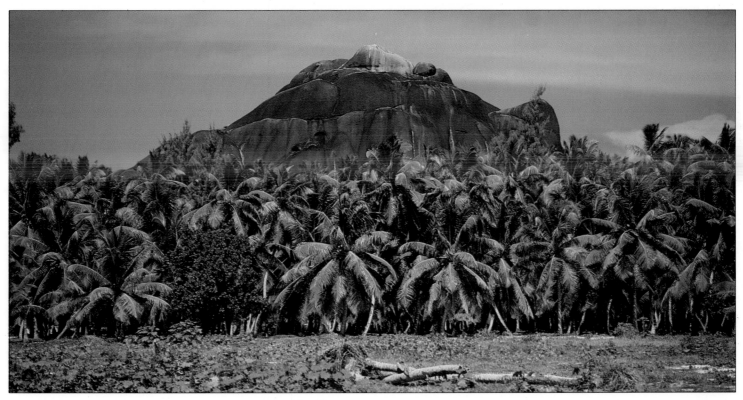

La Digue, the third-most populous island of the Seychelles archipelago after Mahé and Praslin, has an area of only 15 sq km (6 sq mi). High temperatures and rainfall have eroded this granite outcrop to create an isolated dome or inselberg.

São Tomé & Príncipe

DEMOCRATIC REPUBLIC OF SÃO TOMÉ AND PRÍNCIPE

T HE VOLCANIC ISLAND GROUP OF SÃO TOMÉ and Príncipe forms an independent republic in the Gulf of Guinea off the west coast of Africa.

GEOGRAPHY

São Tomé is the largest and most populous island, rising at the center to the Pico de Tomé. Rainforest clothes the upper slopes of the mountains. The smaller rectangular island of Príncipe lies some 150 km (100 mi) to the northeast; it is equally mountainous. Other islets in the group include the Pedras Tinhosas and the Ilha das Rôlas, which lies directly astride the Equator. The hot, humid equatorial climate is moderated by the surrounding ocean and the cold Benguela Current. The mountain slopes enjoy somewhat cooler temperatures.

SOCIETY

São Tomé was discovered by European explorers in the 1470s. It was used as a settlement for convicts and Jewish refugees banished to the island by the Portuguese, who later brought slaves from the mainland to work on their sugar plantations. The islands became independent from Portugal in 1975. The Movement for the Liberation of São Tomé and Príncipe (MLSTP) became the sole legal party in the 51-member National People's Assembly. A plan for full democracy was launched in 1988, and in 1990 the MLSTP was defeated in multiparty elections, but returned to power in 1994, and again in 1995 after a military coup failed. The inhabitants are largely Portuguese-speaking Africans, including immigrant workers from Angola and Mozambique.

ECONOMY

Agriculture, fishing and tourism form the backbone of the economy, and cocoa is the main export crop. About one-third of the land is cultivated, mostly for export, leaving little opportunity for staples; dependence on imported food has led to widespread malnutrition. São Tomé has one railroad and several roads. It is linked by air to Príncipe and Luanda (Angola), and by boat to the other islands of the group. Education and health care are free.

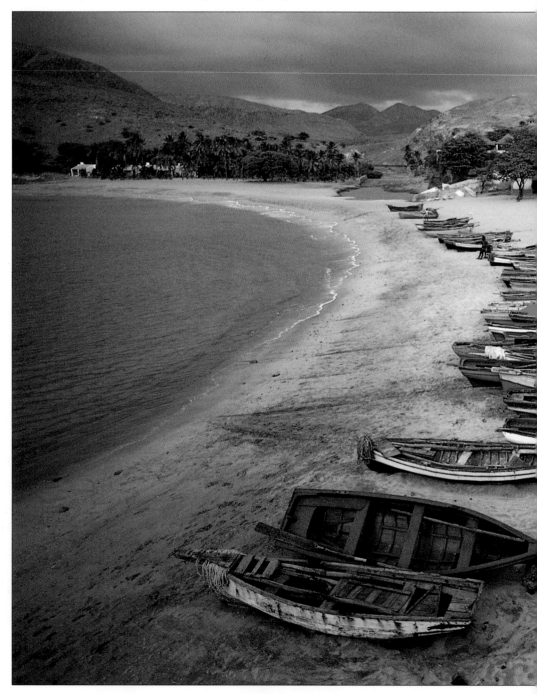

Cape Verde

REPUBLIC OF CAPE VERDE

CAPE VERDE CONSISTS OF TWO SMALL archipelagos lying some 600 km (400 mi) off the coast of west Africa.

GEOGRAPHY

All the islands are volcanic, ringed by steep rocky cliffs and reefs. Most have been eroded into jagged shapes by wind-blown sand. The island of Fogo has an active volcano. The climate is stable and moderately warm, but extremely arid. Agriculture is only possible where sea

A fishing village on one of the many volcanic islands of Cape Verde – one of the smallest nations in the world. The white painted church and buildings in the background are evidence of the country's long association with Portugal.

mists linger, and periodic droughts have devastated the islands.

SOCIETY

The islands were settled by Portugal as early as the 15th century. They became a slave-trading center, and in the 19th century they were an important staging post for transatlantic sea traffic. The majority of the people are mulattos, descended from West African slaves and speaking a Creole dialect called Crioulo. In 1951 the islands became an overseas province of Portugal. In 1975 they became independent under President Aristides Pereira (b. 1924). Moves to unite the country with Guinea-Bissau did not come to fruition. In 1990 the ruling party was defeated in the first multiparty elections.

ECONOMY

The economy rests mainly on agriculture. The main crops include bananas, coconuts and sugar cane; fishing is being developed. Industries are based mainly on local food produce. Other resources include salt and a volcanic rock used in cement. Transportation is mainly by sea and air, and tourism is increasing. There are limited health and welfare programs. Education is freely available, but jobs are scarce and many islanders are forced to emigrate in order to find work. Most of the population is under the age of 20.

NATIONAL DATA – CAPE VERDE

Land area	4,033 sq km (159 sq mi)			
Climate		Temperatures		Annual
	Altitude m (ft)	January °C(°F)	July °C(°F)	precipitation mm (in)
Praia	27 (89)	23 (73)	26 (79)	2,172 (85.5)

Major physical features highest point: Pico do Cano (Fogo) 2,829 m (9,281 ft); largest island: São Tiago 991 sq km (383 sq mi)

Population (2006 est.) 420,979

Form of government multiparty republic with one legislative house

Armed forces army 1.000; coast guard 100; air force 100

Capital city Praia (121,393)

Official language Portuguese

Ethnic composition Creole (mulatto) 71%; African 28%; European 1%

Religious affiliations Roman Catholic 97.8%; others 2.2%

Currency 1 Cape Verdean escudo (CVE) = 100 centavos

Gross domestic product (2006 est.) U.S. $3.129 billion

Gross domestic product per capita (2006 est.) U.S. $6,000

Life expectancy at birth male 67.41 yr; female 74.15 yr

Major resources salt, basalt rock, limestone, kaolin, fisheries, clay, gypsum, bananas, coffee, coconuts

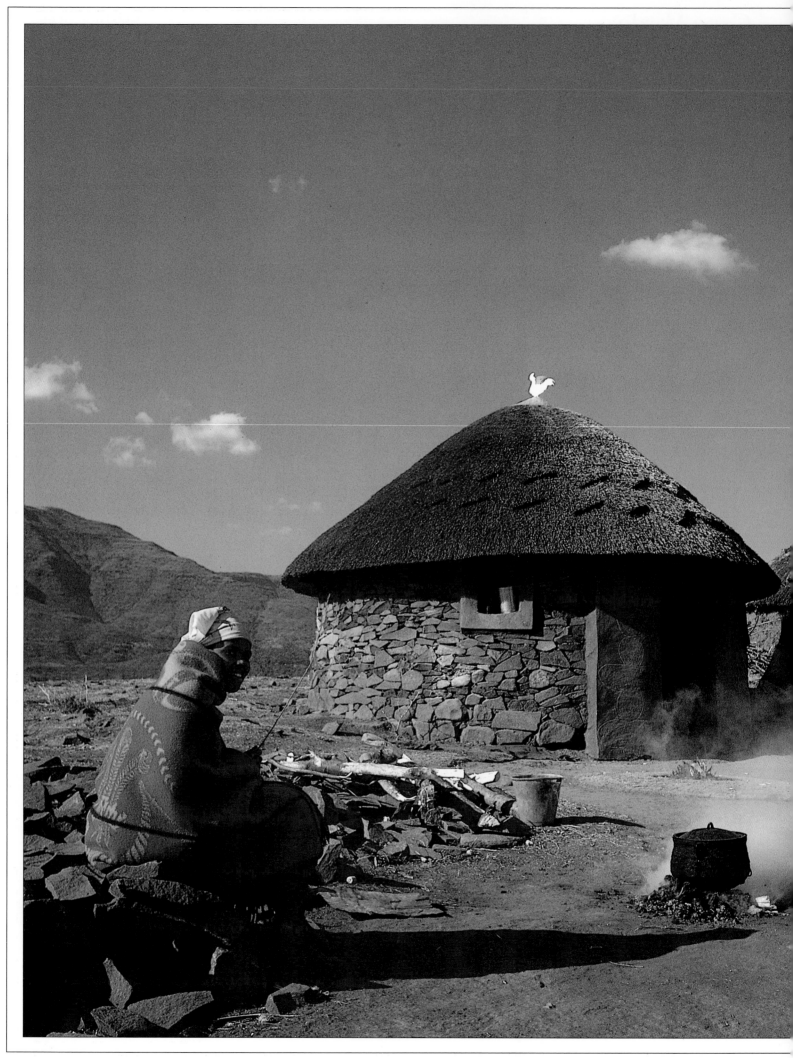

Southern Africa

COUNTRIES IN THE REGION

ANGOLA · ZAMBIA · MALAWI · MOZAMBIQUE
NAMIBIA · BOTSWANA · ZIMBABWE · SOUTH AFRICA
LESOTHO · SWAZILAND · MADAGASCAR
MAURITIUS · COMOROS

DEPENDENCIES IN THE REGION

ST HELENA · MAYOTTE · REUNION

Mountain kingdom The people of Lesotho inhabit a landlocked country of snowcapped peaks and windswept uplands cut by deep gorges. Thick, colorful blankets are worn to withstand cold winter temperatures and frequent summer storms.

capital city
major town

height of land (meters)
3000
2000
1000
500
200
0 (sea level)

mountain peak

Luanda
Cuanza · Malanje
ANGOLA · Luena
Lobito · Huambo
Bié Plateau
Cuando
Cubango
Cunene
Ovamboland
Namib Desert
Etosha Pan · Tsumeb
NAMIBIA
Walvis Bay · Windhoek
Tropic of Capricorn
Keetmanshoop
Kalahari Desert
ATLANTIC OCEAN
Orange
Great Karroo
Cape Town
Cape of Good Hope
Little Karroo
SOUTH AFRICA
Stormberg
Drakensberg
East London
Port Elizabeth

Lake Mweru
Kasama
Lake Bangweulu
Chililabomwe
Ndola
ZAMBIA
Mongu
Kafue · Lusaka
Zambezi
Lake Kariba
Livingstone
Victoria Falls
Okavango Delta
Makgadikgadi Pan
BOTSWANA · Serowe
Gaborone
Lobatse
Pretoria
Johannesburg
Vaal
Kimberley
Bloemfontein · Maseru
LESOTHO
Orange
Thabana-Ntlenyana 3482
Durban

Lake Tanganyika
Muchinga Mts
Rift Valley
Mzuzu
Lake Malawi
MALAWI
Lilongwe
Lake Cabora Bassa
Tete
Blantyre
Mt Mulanje 3000
Harare
Mutare
ZIMBABWE
Bulawayo
Matopo Hills
Save
Limpopo
Mbabane
Lobamba
Maputo
SWAZILAND
Lurio
Nampula · Nacala
MOZAMBIQUE
Beira
INDIAN OCEAN

Moroni
COMOROS
Mayotte (France)

Madagascar inset

Mozambique Channel
Antsiranana
Maromokotro 2876
Mahajanga
Antananarivo
Toamasina
Morondava
Ikopa
Mangoky
MADAGASCAR
Fianarantsoa
Toliara
Tolanaro
INDIAN OCEAN
Tropic of Capricorn

185

Angola

REPUBLIC OF ANGOLA

ANGOLA IS A LARGE STATE ON AFRICA'S southwest coast, flanked by Namibia to the south, Zambia to the east and Congo (D.R.O.) to the north. The coastal enclave of Cabinda is divided from the rest of Angola by a strip of Congo (D.R.O.), and borders Congo (R.O.) to the north. Until 1974 Angola was an overseas colony of Portugal, and since independence in 1975 it has suffered civil war.

GEOGRAPHY

From the narrow coastal plains, the land rises to an extensive tableland that covers most of the country. The uplands are fairly level in the south, but become mountainous in southwestern and central Angola, where the highest point is Mount Moco. In the north the plateau becomes more fragmented; the valleys and coastal plains are broader toward the mouth of the Congo river and in the Cabinda enclave beyond. A network of rivers fans out from the central highlands; some flow into the Atlantic, while others feed the rivers of adjoining countries.

The climate is generally hot except in the central highlands and along the coast, where temperatures are moderated by the cold Benguela Current that flows north from the colder southern oceans. Rainfall varies widely across the country and according to the season. The north receives heavy rains over a seven-month period, whereas the west and south are subject to long droughts.

The vegetation varies according to climate, from dense tropical rainforest in Cabinda and the north to savanna woodland and then more arid grassland, dotted with euphorbia, acacia and baobab trees. Along the southwest coast a strip of pure desert extends north from Namibia.

Angola has a rich variety of African wildlife: cheetahs, leopards, elephants and rhinoceroses. Many of these are threatened by the destruction of their habitats for farming land, by the disruptions of civil war and by the illicit trade in ivory and rhinoceros horns. National parks exist, but conservation measures have so far been ineffective.

SOCIETY

The earliest inhabitants of Angola, hunter–gatherers related to the Khoisan-speaking San (Bushmen) of present-day Namibia, were displaced by the influx of Bantu peoples from the north, who introduced cereal agriculture and ironworking to the area. By the time the Portuguese arrived in the 16th century, the Bantu had established a number of kingdoms, some of which were highly organized. The largest, that of the Kongo, included the northwest of modern Angola, while the Ndongo kingdom occupied the south.

The Portuguese called the country Angola after the name Ngola, which the Ndongo gave to their kings. At first the Bantu kingdoms enjoyed friendly relations with the Portuguese, exploiting the slave trade in order to expand their own power. Nevertheless, the Kongo kingdom was overrun by a neighboring group in 1568. Meanwhile, the Ndongo thrived until the Portuguese decided on a policy of subjugation. Despite this they were not finally conquered until a century later.

Angola became the slave center for the main European powers, leaving the area largely depopulated and with no chance of thriving even as a colony. Slavery was abolished in the early 19th century, only to be replaced by indentured labor. Portugal established its right to colonize the whole area, and initiated projects such as

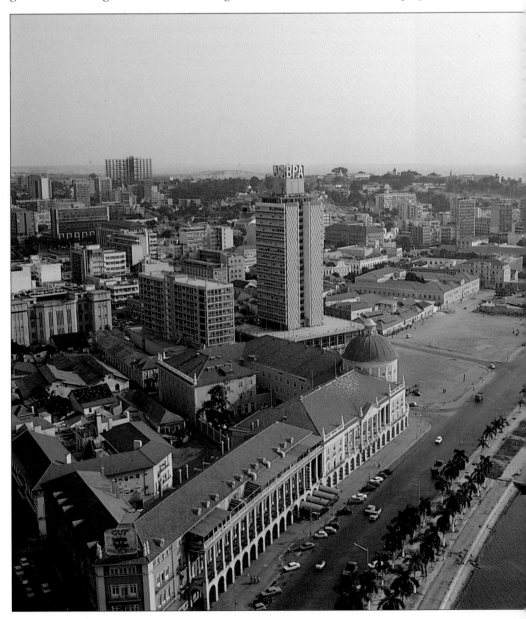

railroads and diamond mines, but the borders of Angola were not formally drawn until as late as 1926. No rights were granted to indigenous Africans, who continued to remain impoverished, often as forced laborers.

Growing nationalist feeling in the 1960s gave rise to three different guerrilla campaigns: the primarily Marxist-oriented Angolan People's Liberation Movement (Movimento Popular de Libertação de Angola; MPLA), the Kongo-led Angolan National Liberation Front (Frente Nacional de Libertação de Angola; FNLA), and in the south the National Union for Total Angolan Independence (União Nacional para a Independência Total de Angola; UNITA). All three were fiercely resisted by the Portuguese. In 1974, however, the new regime in Lisbon abruptly withdrew the Portuguese troops, paving the way for full independence the following year.

Luanda, capital of Angola (*left*), was founded as a trading post by the Portuguese in 1576. It remains a busy port, and today modern towers of concrete and glass rise up alongside the spacious buildings of the colonial past.

Village life (*below*) Women carry out the traditional tasks of food preparation in a small village in the countryside south of Luanda. Agriculture suffered during the years of fighting, which disrupted the lives of Angola's peasant farmers.

War immediately broke out. UNITA and FNLA, with South African backing, confronted the Soviet-backed MPLA, supported by Cuban troops. The MPLA took over the government, creating a one-party republic. However, the war continued for many years, with large areas under UNITA control. The withdrawal of Cuban troops (and of Soviet backing) in 1989 opened the way for peace negotiations, which were concluded in 1991. But hopes for peace were dashed when fighting resumed in November 1992 after UNITA lost an election. Further peace agreements in 1994 and 1998 provided temporary respite from the fighting which resumed again in 1999.

Portuguese is the official language, although spoken only by a minority; the status of the native Bantu languages is being improved. The largest of the various ethnic groups are the Ovimbundu in the central highlands, the Mbundu along the coast around the capital, Luanda, and the Kongo in the north. Most people are Roman Catholic.

ECONOMY

Despite many years of incessant civil war, Angola's economy has begun to improve in some respects, thanks partly to abundant natural resources. A wide variety of crops are grown, especially in the more fertile north. The main cash crops are coffee, sugar cane, cotton and oil palm. Maize is the main staple, together with cassava, yams and millet. The hardwoods of the Cabinda forests are also an important source of revenue, and the coastal fisheries bring in mackerel and sardines.

Up until the 1960s diamonds were the only mineral resource to be exploited. Since then substantial deposits of iron and other metal ores have been opened up. Extensive petroleum and natural-gas fields, mostly off the coast of Cabinda, now provide the main source of foreign income, and are the only energy source apart from hydroelectricity.

The main manufacturing industries are petroleum and steel processing. Other concerns are on a smaller scale, and are mostly for domestic markets; products include food, textiles, building materials and plastics. Petroleum products account for the vast majority of exports, together with diamonds and food products; the main imports include machinery and chemicals. The road system remains poor, and railroads are used mainly for transporting freight and raw materials.

Health and welfare services are in theory available to all, but there is little provision in rural areas, where disease and malnutrition are rife. Primary education is free, but has been hampered by the civil war. However, literacy levels are rising thanks to the greater use of African languages.

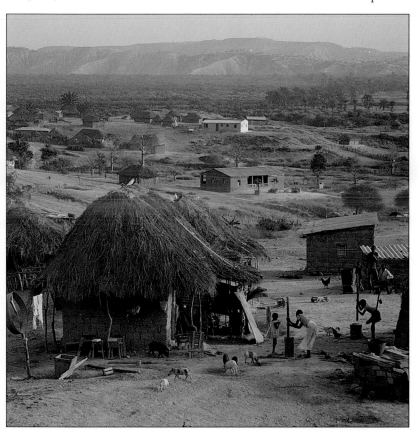

NATIONAL DATA – ANGOLA

Land area	1,246,700 sq km (481,354 sq mi)			
Climate		**Temperatures**		**Annual**
	Altitude m (ft)	January °C(°F)	July °C(°F)	precipitation mm (in)
Luanda	44 (144)	26 (67)	21 (70)	323 (12.7)

Major physical features highest point: Mount Moco 2,559 m (8,397 ft)

Population (2006 est.) 12,127,071

Form of government presidential multiparty republic

Armed forces army 100,000; navy 1,000; air force 6,000

Capital city Luanda (2,977,212)

Official language Portuguese

Ethnic composition Ovimbundu 37%; Kimbundu 25%; Bakongo 13%; Mestico (mixed European and native African) 2%; European 1%; other 22%

Religious affiliations traditional beliefs 47%; Roman Catholic 38%; Protestant 15%

Currency 1 kwanza (AOA) = 100 centimos

Gross domestic product (2006 est.) U.S. $51.95 billion

Gross domestic product per capita (2006 est.) U.S. $4,300

Life expectancy at birth male 37.47 yr; female 39.83 yr

Major resources petroleum, natural gas, diamonds, iron ore, phosphates, copper, feldspar, gold, bauxite, uranium, asphalt, bananas, cassava, citrus fruits, coffee, cotton, fish, maize/corn, palm oil, salt, sisal, sugar beet, sweet potatoes, timber

Zambia

REPUBLIC OF ZAMBIA

ZAMBIA IS A LANDLOCKED COUNTRY IN south-central Africa. Congo (DRO) to the north almost divides it in two. It also shares borders with many other states, and its history has been closely linked with Zimbabwe to the south.

GEOGRAPHY

Zambia occupies a broad, rolling plateau, broken only by scattered mountains and deep valleys. The Zambezi river rises in Zambia, flows through part of Angola and back into Zambia forming the country's southern frontier with Namibia and Zimbabwe. It has been dammed at Kariba to form the great Lake Kariba. The Kafue river rises farther east, spreading out into wide floodplains to the south of the capital, Lusaka, before joining the Zambezi river via a gorge to the north of Kariba. The Luangwa river rises in the eastern mountains on the border with Malawi, then runs southwest along a broad rift valley to join the Zambezi river at the Mozambique border. North of here, across the Muchinga Escarpment, is a swampy basin around Lake Bangweulu. This is drained by the Luapula river, which runs north along the Congo border to enter Lake Mweru in another deep rift valley. The southern end of Lake Tanganyika occupies the far northeast of Zambia.

The climate is subequatorial, but is moderated by the country's altitude and location. The warm wet season (from November to April) is characterized by high humidity and heavy rain, often with thunderstorms. The dry season is cooler, with night frosts in mountainous areas, but becomes very hot soon before the next rains begin.

Most of the country is covered by "bush" – a mixture of woodland and savanna grassland. Larger forests grow in the southwest, consisting chiefly of valuable Rhodesian teak. The floodplains

are covered with richer grassland, and the lake and swamp areas are thick with papyrus reeds. Wildlife is varied, including characteristic African grassland species such as giraffes, elephants, leopards and lions. However, it has been decimated by hunting, and destruction of habitat for livestock grazing. Game reserves have been established to preserve the wildlife, including the Kafue and Luangwa National Parks, but there are wide variations in the protection and conservation measures that are used.

SOCIETY

Humankind's early ancestors lived in the Zambia area millions of years ago, and it has been inhabited by various peoples ever since. Most of the present ethnic groups, however, arrived from the north, probably in the 17th century, not long before the first Europeans. Two small groups, the Ngoni and the Kololo, arrived as late as 1835, fleeing from the Zulus in the far south.

In the late 19th century Arab slave

In Lusaka a striking monument commemorates the 1970 Conference of Non-Aligned Countries, opposed to colonialism and the dominance of the major world powers. It stands in sharp contrast to the pillared colonial facade of the High Court building.

traders began to raid the extreme east of what is now Zambia. The country first came prominently to European notice with the explorations of the Scottish missionary David Livingstone (1813–73) in the Upper Zambezi area, where he discovered and named the Victoria Falls in 1855. Later the British South Africa Company, led by the British-born South African politician Cecil Rhodes (1853–1902), began to expand into the area from the south, concluding treaties with many local peoples. In 1911 the country was renamed Northern Rhodesia, with its capital at Livingstone. European settlement began to increase with the establishment of copper mines, and in 1924 the country was formally made a British protectorate.

After World War II a nationalist movement grew up, headed by the United National Independence Party (UNIP) and

David Livingstone's statue stands gazing toward the Victoria Falls on the Zambezi river, bordering Zambia and Zimbabwe. A Scottish missionary and explorer, Livingstone was the first European to see these falls, and it was he who gave them their present name.

retained its legislative function. After unrest in 1990 multiparty politics returned, and Frederick Chiluba, a former trade unionist, was elected president in democratic elections in 1991 and 1996.

English is the official language, but the majority speak a variety of Bantu languages, the most important being Bemba in the far northeast, Nyanja in the east and Tonga in the south. Although most people are nominally Christian, many also hold to their traditional (animistic) African beliefs.

ECONOMY

Agriculture employs a majority of the work force, but most production is absorbed by the domestic market. Most export crops are grown on large European-run farms. They include tobacco, sugar cane, cotton and groundnuts (peanuts). Most farmers are subsistence producers. Maize is the main food crop, along with millet, cassava, sorghum and various vegetables. Cattle are the main livestock, but are limited by the local incidence of tsetse fly. The teak forests in the southwest are extensively exploited, and there are newer plantations in the Copperbelt along the border with Congo (DRO). The many lakes support a flourishing fishing industry.

Zambia has effectively a single-commodity economy, based on copper. The extensive ore deposits in the Copperbelt have made Zambia the world's fourth-largest producer. This has funded much of the country's development, but has also made it vulnerable to world price fluctuations. Other metal ores and precious stones are also mined. Some coal is extracted, but hydroelectricity is the main energy source. Manufacturing benefited somewhat from the break in Rhodesian trade, which encouraged local enterprise. Among the major industries are copper fabrication and the refining of imported petroleum, obtained via a Tanzanian pipeline. Other products include textiles, tires, fertilizers and explosives.

Rhodesia's UDI cut the main transportation links, so a new northeastern railroad was built through Tanzania. The road system is still mostly unpaved, but is gradually being improved. Aero Zambia is the national carrier and there is an international airport at Lusaka. Health provision has improved despite staff shortages, and headway has been made in combating endemic diseases. Primary education is available to most children, and further education is being developed.

its rival, the African National Congress (ANC). In 1953 the country formed a federation with Southern Rhodesia and Nyasaland (now Zimbabwe and Malawi), but in 1963 an African majority gained control of the government and dissolved the federation. In 1964 full independence was granted, and the UNIP leader, Kenneth Kaunda (b. 1924), became Zambia's first president.

Zambia relied heavily on its southern neighbor Rhodesia (now Zimbabwe) for trade and communications. Consequently Rhodesia's unilateral declaration of independence (UDI) in 1965 severely disrupted the Zambian economy until 1980, when links were restored with the new Zimbabwean state. From the 1960s until 1989 Zambia provided a refuge for guerrilla forces operating in Rhodesia, Angola, Mozambique and finally South Africa. Meanwhile inter-party violence, mostly reflecting differences between the ethnic groups, led to the creation in 1972 of a one-party state. A new constitution in 1973 gave more power to the president, although the single-chamber parliament

NATIONAL DATA – ZAMBIA

Land area	740,724 sq km (285,995 sq mi)			
Climate		**Temperatures**		**Annual**
	Altitude m (ft)	January °C(°F)	July °C(°F)	precipitation mm (in)
Lusaka	1,297 (4,195)	22 (71	16 (61)	843 (33.1)

Major physical features	highest point: Nyika Plateau (part) 2,164 m (7,100 ft)

Population	(2006 est.) 11,502,010

Form of government	multiparty republic with one legislative house

Armed forces	army 13,500; navy 1,600; air force 1,400

Capital city	Lusaka (1,346,522)

Official language	English

Ethnic composition	Bemba 36.2%; Maravi 17.6%; Tonga 15.1%; Northwestern tribes 10.1%; Barotze 8.2%; Mambwe 4.6%; Tumbuka 4.6%; others 3.6%

Religious affiliations	Christian 50%–75%, Muslim and Hindu 24%–49%, indigenous beliefs 1%

Currency	1 Zambian kwacha (ZMK) = 100 ngwee

Gross domestic product	(2006 est.) U.S. $11.51 billion

Gross domestic product per capita	(2006 est.) U.S. $1,000

Life expectancy at birth	male 39.76 yr; female 40.31 yr

Major resources	copper, cobalt, zinc, lead, coal, emeralds, gold, silver, uranium, hydropower, fisheries, cassava, cattle, groundnuts, maize/corn, tobacco, sugarcane

Malawi

REPUBLIC OF MALAWI

M ALAWI LIES AT THE SOUTHERN END OF THE East African Rift Valley, bordering Zambia to the west, Tanzania to the northeast, and Mozambique to the east and to the southwest.

GEOGRAPHY

Northeastern Malawi is occupied by Lake Malawi, which covers nearly one-fifth of the total area. Its southern outflow, the Shire river, runs through a deep, often swampy valley flanked by mountains to the east. Central and western Malawi consists mostly of plateaus, with high mountains in the north.

The tropical climate is modified by the country's landlocked position and by variations in relief. Temperatures are highest just before the summer rainy season in the Rift Valley, while rainfall is greatest in the northern highlands.

Savanna grasslands in the valleys give way to open woodland higher up, and then to grassland and evergreen forest in the mountains. Much of the original woodland has been cleared for cultivation. Larger animals are mainly confined to the reserves, although hippopotamuses are found in Lake Malawi.

SOCIETY

Bantu speakers first settled in the area many centuries ago. Muslim slave traders arrived in the early 19th century, closely followed by Christian missionaries. In 1891 Britain created the Nyasaland Districts Protectorate, which became Nyasaland in 1907. In 1953 the country joined a federation with Northern and Southern Rhodesia, where it remained until 1963.

Malawi achieved full independence in 1964, becoming a single-party republic two years later. In 1971 its first president,

Hastings Banda (1905–1997), was proclaimed president for life but this was repealed in 1993 following a referendum which introduced a multiparty democracy. In the first multiparty election held in 1994 the United Democratic Front led by Bakili Muluzi won an overwhelming victory over Banda's Malawi Congress Party. Malawi has had differences with its African neighbors but during Mozambique's civil war, it took in refugees.

Almost all the population are Bantu-speaking; English, though an official language, is spoken by a minority. About half the population are Christian, but traditional beliefs are also widely held, and there is a strong Muslim minority.

ECONOMY

Maize is the chief staple, while cash crops, notably tobacco, sugar cane and

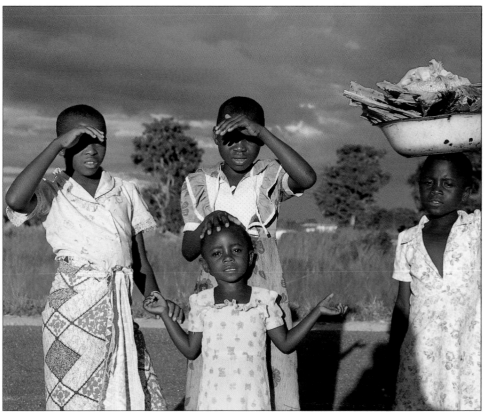

Happy faces (*above*) Young girls at Nkhotakota on the shore of Lake Malawi. The spectacular woodland scenery, glimmering lakes and the dramatic waterfalls in the south make Malawi a tourist paradise.

The Nyika Plateau (*left*) in northern Malawi. Flowers enliven the bleak highland landscape, which is one of the wildest in the country. Much of this area on the Zambian border has been declared a nature reserve.

The road system is limited, and most roads are unsurfaced. Railroad links with Mozambique give access to the sea at Beira and Nacala, but in the 1980s these were effectively cut off by the civil war in Mozambique. Malawi now spends over 40 percent of gross national product on transportation by truck through Zambia, Botswana and Zimbabwe to South African ports. The Shire river is navigable, and there are passenger and cargo services on Lake Malawi. The national airline, Air Malawi, operates an international service from Lilongwe, the capital city since 1975, in the center of the country. Press and broadcasting are controlled by the government.

Some free medical treatment is provided, but preventive health care is limited. Major diseases include malaria, bilharzia, gastrointestinal diseases and leprosy. Housing is in short supply, but welfare help is available for children, the handicapped and the poor. Limited resources mean that less than half the school-age population have access to education, and literacy is correspondingly low.

tea, account for the majority of export revenue. New forest plantations have created a surplus for export. Fishing, mostly from Lake Malawi, is also an important industry. Mineral resources are scarce, but there is abundant water and hydroelectric potential. Most power comes from this source, reducing import bills for fuel. Many consumer products are now manufactured, but basic products and machinery are imported.

NATIONAL DATA - MALAWI

Land area 94,080 sq km (36,324 sq mi)

Climate		Temperatures		Annual
	Altitude m (ft)	January °C(°F)	July °C(°F)	precipitation mm (in)
Lilongwe	1,095 (3,593)	22 (71)	15 (59)	775 (30.5)

Major physical features highest point: Mount Mulanje 3,000 m (9,843 ft); longest river: Shire (part) 400 km (250 mi); largest lake: Lake Malawi 29,600 sq km (11,400 sq mi)

Population (2006 est.) 13,013,926

Form of government multiparty republic with one legislative house

Armed forces army 5,300

Largest cities Blantyre (618,153); Lilongwe (capital - 721,988); Mzuzu (143,320); Zomba (86.217)

Official language English

Ethnic composition Maravi 58.3%; Lomwe 18.4%; Yao 13.2%; Ngoni 6.7%; others 3.4%

Religious affiliations Christian 79.9%; Muslim 12.8%; other 3%; none 4.3%

Currency 1 Malawian kwacha (MWK) = 100 tambala

Gross domestic product (2006 est.) U.S. $8.038 billion

Gross domestic product per capita (2006 est.) U.S. $600

Life expectancy at birth male 41.93 yr; female 41.45 yr

Major resources limestone (marble), hydropower, unexploited uranium, coal and bauxite, beans, cassava, cotton, groundnuts, maize/corn, millet, sugarcane, sorghum, tobacco, tea

Mozambique

REPUBLIC OF MOZAMBIQUE

MOZAMBIQUE IS A LARGE, IRREGULARLY shaped country on Africa's southeast coast. It is bordered by South Africa and Swaziland to the southwest, Zimbabwe and Zambia to the west, and Tanzania to the north; southern Malawi is sandwiched between western and northern Mozambique. The east coast faces the island of Madagascar across the Mozambique Channel.

GEOGRAPHY

The wide coastal plain occupies most of the south, but becomes narrower farther north. Away from the sea the land rises in a series of steps to the tablelands of the High Veld. The granite plateaus of the north, beyond the Zambezi river, are higher than the sandstone plateaus of the south. In the far northwest the land drops sharply to the shores of Lake Malawi.

The coastal plain is crossed by numerous rivers. These include the great Zambezi, the Lugenda, the Save and the Limpopo. The capital, Maputo (formerly Lourenço Marques), lies in the far south on Delagoa Bay, which is the finest natural harbor on Africa's east coast. North of here the coastline features sandbars and mangrove swamps, but beyond the Zambezi river it is lined by sandy beaches interspersed with rocky cliffs and small islands.

The climate is subtropical, with a very marked division between rainy and dry seasons. However, it is surprisingly dry in view of the prevailing onshore winds and the warm offshore Mozambique Current. Many areas suffer from drought, especially south of the Zambezi river, and high temperatures increase water loss. Plateau areas enjoy cooler temperatures and higher rainfall, especially in the north.

Savanna woodland and grassland are the dominant vegetation types, with mangroves and palms (especially coconut palms) along the coast. Typical trees include ironwood, ebony and khaya, with mopani and baobab on the dry savanna. Bamboo and spear grass are common in river and swamp areas. Wildlife is still abundant despite excessive hunting. Mammals include White rhinoceroses, zebras, warthogs and various species of monkey. Bird life is similarly varied, with flamingoes, cranes, ibis and many other species.

SOCIETY

When the Portuguese navigator Vasco da Gama (1460–1524) reached Mozambique in 1498, he found Arab and Swahili traders already active there. The Portuguese founded several settlements, both along the coast and up the course of the Zambezi river, where they sought gold from the nearby empire of the Mwene Matapa. However, they administered the colony from Goa in India until 1752. Slave trading then became a bulwark of its economy, and continued, albeit illegally, until the 1850s. At first the inland areas were controlled by powerful landholders, but these were eventually replaced by private companies. From the late 19th century the government increasingly took over the administration.

In 1930 limited autonomy was replaced by centralized government, and in 1951 Mozambique was declared a Portuguese overseas province. A nationalist movement led by the Mozambique Liberation Front (Frente de Libertação de Moçambique, or Frelimo) erupted into violence in 1964. The war was long and costly, and contributed to the fall of the government in Portugal in 1973. Independence in 1975 brought Frelimo at last to power.

The resulting one-party state was headed by a president, who in turn was chosen by the People's Assembly. Election to this body was by delegation from regional and local levels of government.

The government's authority was very quickly challenged by rebel movements, resulting in a prolonged civil war. The main armed opposition was from the Mozambique National Resistance (MNR), also known as Renamo (Resistência Nacional Moçambicana), which was backed by South Africa. The late 1980s saw an improvement in relations between the Frelimo government and South Africa, with promises of cooperation and joint development in return for greater democracy. However, the Nkomati Accords – a non-aggression pact signed between South Africa and Mozambique – were violated soon after the signing ceremony. In 1992 the United Nations agreed to provide troops to supervise demobilization of warring armies and oversee future elections. First troops arrived in 1993. Presidential and multiparty legislative elections were finally held in 1994.

Most Mozambiquans belong to one of ten major ethnic groups, the largest of which is the Makua-lomwe, accounting for almost half the population. There are tiny minorities of Europeans and Asians. The official language is Portuguese, but few can actually speak it. Several distinct Bantu languages are spoken in different parts of the country, and many people speak Swahili. Almost half the population hold traditional African beliefs. The two other main religious groups are Muslims, mostly in the north, and Christians, mostly in the south.

ECONOMY

Mozambique is one of the poorest countries in southern Africa. It suffered disastrous floods in 1998 and again in 2000. The Mozambiquan economy is centrally planned, and chiefly depends on agriculture, fishing and light industry. Cash crops include cashew nuts, cotton, sugar cane, copra and citrus fruits. Shrimps and lobsters are also exported. Maize is the

NATIONAL DATA - MOZAMBIQUE

Land area   784,090 sq km (302,739 sq mi)

Climate		Temperatures		Annual
	Altitude m (ft)	January °C(°F)	July °C(°F)	precipitation mm (in)
Maputo	60 (197)	27 (80)	19 (56)	814 (32)

Major physical features   highest point: Monte Binga 2,436 m (7,992 ft)

Population   (2006 est.) 19,686,505

Form of government   multiparty republic with one legislative house

Armed forces   army 10,000; navy 200; air force 1,000

Largest cities   Maputo (capital - 1,249,309); Matola (570,242); Beira (561,151); Nampula (407,338)

Official language   Portuguese

Ethnic composition   Traditional tribal groups 99.66% (Makhuwa, Tsonga, Lomwe, Sena, and others); Europeans 0.06%; Euro-Africans 0.2%; Indians 0.08%

Religious affiliations   Catholic 23.8%; Muslim 17.8%; Zionist Christian 17.5%; other 17.8%; none 23.1%

Currency   1 metical (MZM) = 100 centavos

Gross domestic product   (2006 est.) U.S. $29.32 billion

Gross domestic product per capita   (2006 est.) U.S. $1,500

Life expectancy at birth   male 39.53 yr; female 40.13 yr

Major resources   coal, titanium, natural gas, hydropower, tantalum, graphite, clothing, bananas, cashew nuts, cassava, cereals, coconuts, sisal, tea

Colorful Mozambiquans (*above*) The introduction of modern dyes has transformed the dress of African women by making colored fabrics – until the 20th century a luxury only for the rich – easily available.

Rural tranquility (*left*) in the Niassa province of northern Mozambique. The children running down this country track are more fortunate than those thousands, mainly in the south, who have been injured or made homeless in the civil war.

chief staple, but production cannot fully meet demand, and food remains a major import. Tsetse flies, which transmit sleeping sickness to cattle, prevent livestock rearing in many areas.

Mozambique's great mineral potential has so far remained largely untapped, and production has mostly been limited to coal, limestone and salt. However, natural gas and iron ore are now being exploited, together with small amounts of tantalite (a rare mineral). Manufacturing industry is mainly based on the processing of local raw materials. Machinery, chemicals and spare parts have to be imported. Electricity is generated both from thermal stations and from hydro-electric plants, most notably the Cabora Bassa Dam on the Zambezi river.

Mozambique also supplies thousands of migrant workers to South Africa in return for payment in gold at the official rate. It then sells the gold at the much higher free-market rates in order to reduce the overall economic deficit.

The transportation system was originally designed for colonial exploitation rather than to meet local needs. The three major ports of Maputo in the south, Beira in the center and Nacala in the north are the focal points for both road and rail. Much of the country is still without roads, and only a small part of the existing network is surfaced. Zambia and Zimbabwe are working with Mozambique to improve and expand the railroad network, which provides a link for these two countries to the coast, reducing their reliance on South Africa. There are international airports at Maputo, Beira and Nampula in the north, as well as many smaller airports and rural airstrips.

Health care has been nationalized and medicine is free, although private health-care was introduced in 1992. Government efforts have so far failed to combat major diseases such as malaria, tuberculosis, hepatitis and pneumonia. Education too has been nationalized, and is compulsory up to the age of 14. An adult literacy scheme has been established but enrollment is low, and this is reflected in the lower literacy levels.

Namibia

REPUBLIC OF NAMIBIA

THE NEWLY INDEPENDENT COUNTRY OF Namibia on the southwest coast of Africa borders South Africa to the south, Botswana to the east and Angola to the north. From its northeastern corner, the narrow Caprivi Strip extends its territory eastward as far as Zambia. The main port, Walvis Bay, was South African territory until 1994.

GEOGRAPHY

The sand dunes of the Namib Desert extend from the coast toward the foot of the Great Escarpment that rims the great central plateau. The mountains of the escarpment are highest in the central

NATIONAL DATA – NAMIBIA

Land area	825,418 sq km (318,696 sq mi)			

Climate		Temperatures		Annual
	Altitude m (ft)	January °C(°F)	July °C(°F)	precipitation mm (in)
Windhoek	1,738 (5,669)	12 (54)	13 (56)	362 (14.2)

Major physical features	highest point: Mount Brand 2,574 m (8,445 ft)
Population	(2006 est.) 2,044,147
Form of government	multiparty republic with two legislative houses
Armed forces	army 9,000; navy 200
Capital city	Windhoek (286,762)
Official language	English
Ethnic composition	Ovambo 49.8%; Kavango 9.3%; Herero 7.5%; Damara 7.5%; European 6.4%; Nama (Hottentots) 4.8%; others 14.7%
Religious affiliations	Christian 80%-90% (Lutheran 50% at least); traditional beliefs 10%-20%
Currency	1 Namibian dollar (NAD) = 100 cents and 1 South African rand (ZAR) = 100 cents
Gross domestic product	(2006 est.) U.S. $15.04 billion
Gross domestic product per capita	(2006 est.) U.S. $7,400
Life expectancy at birth	male 44.46 yr; female 42.29 yr
Major resources	diamonds, copper, uranium, gold, silver, lead, tin, tungsten, lithium, cadmium, zinc, salt, hydropower, fish, possible deposits of oil, coal, and iron ore, cattle, maize/corn, millet, sorghum

portion around the capital, Windhoek. The plateau itself occupies the eastern part of the country, and is covered by the dry scrub of the Kalahari Desert.

The climate is generally hot and very dry. Seasonal rainfall is largely confined to the north. The Namib Desert supports little vegetation apart from succulents and the extraordinary welwitschia plant, which absorbs moisture from the fog that rolls in from the sea. The welwitschia, often described as a living fossil, can survive to a great age; some specimens are thought to be up to 2,000 years old.

Inland areas are less barren, however, with many drought-resistant grasses and shrubs, and savanna woodland in the extreme north. The rich wildlife is typical of southern Africa, and includes baboons, kudu antelopes, elephants, giraffes, zebras, lions and flamingoes. The Etosha National Park is among the world's largest game parks.

SOCIETY

South West Africa, as Namibia was formerly known, became a German protectorate in 1884. During World War I it was occupied by South Africa, to which it was mandated in 1920 by the League of Nations as a trust territory. In 1946 it came under the United Nations trusteeship system, despite South Africa's stated intention to annex the country. In 1966 the United Nations called for total South African withdrawal, and the South West Africa People's Organization (SWAPO) began a lengthy guerrilla war against the continuing South African occupation. Free elections in 1989 brought SWAPO to power, and independence followed soon afterward in 1990.

The multiparty government is headed by a president, who is accountable to a two-chamber parliament. Both the president and the lower house are directly elected every five years. Namibia is still largely economically dependent on South Africa, with which it remains in customs union; but it is gradually becoming more independent.

The two oldest racial groups are the desert-dwelling San (Bushmen) and Nama (Hottentots). Bantu peoples include the Kavango and the Ovambo – the largest ethnic group, who occupy the more populated northern areas. Europeans are mainly confined to the towns; Afrikaans and English are the two official languages, but German is spoken by a large number of people. Christianity is the dominant religion.

ECONOMY

Most people work on the land, but the soils and climate do not favor agriculture. About half Namibia's food is imported, largely from South Africa. Livestock farming is the chief activity, producing mainly beef, mutton and karakul (Persian lamb) pelts. The once-prosperous offshore fishing industry is gradually recovering from overfishing in the 1970s.

Mineral resources include diamonds, gold and various metal ores, including uranium dioxide. Copper and lead are smelted at Tsumeb in the north of the country, but other minerals are exported unprocessed. Namibia depends on South Africa for fuel and manufactured goods. Its own industries are limited to food processing. The road and rail networks, though expanding, are still largely confined to the south, so that local air services are vital; there is an international airport at Windhoek.

Healthcare services are concentrated in a few large hospitals. Major diseases include tuberculosis, measles and malaria. Primary education is compulsory, but literacy is low outside the white community; skilled training must be sought in South Africa.

Dressed up to the nines The traditional costume of these Herero women was copied from 19th-century German Lutheran missionaries. The long skirts and puff-sleeved bodices have been colorfully adapted.

Botswana

REPUBLIC OF BOTSWANA

BOTSWANA IS A LANDLOCKED STATE THAT borders Namibia to the west and north, South Africa to the south, and Zimbabwe and Zambia to the northeast. It is in customs union with South Africa, on which its economy depends.

GEOGRAPHY

The country is part of the southern African plateau. It is covered in the west by the dry scrub of the Kalahari Desert, and in the east by savanna grassland. Surface water is confined to the north and east, in the river basins of the Okavango, the Chobe and the Limpopo. The Okavango river flows southeast from Angola via Namibia and into the Okavango Basin, from where there is an eastward seasonal flow toward Lake Dow and the Makgadikgadi Pan. The climate varies from semiarid to subtropical, with a hot season from October to March. Rainfall is scanty and unpredictable, especially in the south, and droughts can last for years.

Savanna grassland is the dominant vegetation type, while palms grow in the Okavango Delta and deciduous forests border the Chobe river on the northern border. Wildlife is rich and plentiful in the many game reserves and national parks. Typical animals include lions, elephants, giraffes and many species of poisonous snake.

SOCIETY

The original inhabitants of present-day Botswana were nomads called the San (Bushmen), of whom few remain. Eastern parts were mainly settled by Bantu peoples called the Tswana. The first European arrivals were British missionaries from the south in the early 19th century. From the 1820s onward there were raids by the Ndebele and other refugees escaping from the northward advance of the Zulus. In 1835 the Boers – descendants of the 17th-century Dutch settlers in South Africa – reached the southern borders of the country. The discovery of gold in the east led to several territorial conflicts. In 1884 the British occupied the area at the instigation of the British-born South African politician Cecil Rhodes (1853–1902), and in 1885 it became the British Bechuanaland Protectorate. It remained so until full independence was achieved in 1966.

Legislative power rests with the 40-member parliament, most of whom are directly elected for a five-year term; there are many different political parties. The 15-member House of Chiefs advises on ethnic and constitutional matters. There are eight major ethnic groups, but the Tswana account for three-fourths of the population. The two official languages are Tswana and English. Christianity is the dominant religion, though traditional beliefs are also widely held.

ECONOMY

Agriculture is dominated by cattle farming, which provides a major export. Most arable farming, in contrast, is for subsistence needs. Cash crops include groundnuts (peanuts) and sunflower seeds. Diamonds, copper and nickel are the main mineral resources, and a further major export; their mining has been extensively subsidized by South Africa. Sizable coal deposits are used to generate electricity. There have recently been attempts to develop the country's small manufacturing base, and to exploit its possibilities for tourism.

The main railroad was formerly operated by Zimbabwe, but is now under Botswanan management. Most roads are of dirt or gravel, but there is an all-weather north–south artery. Domestic and international flights are provided by Air Botswana. Health care is widely available through mobile clinics. Major health problems include tuberculosis, gastroenteritis and malnutrition. There is free primary education, and literacy levels are relatively high.

NATIONAL DATA – BOTSWANA

Land area	585,370 sq km (226,013 sq mi)			
Climate			Temperatures	Annual
	Altitude m (ft)	January °C(°F)	July °C(°F)	precipitation mm (in)
Francistown	983 (3,225)	25 (77)	15 (59)	448 (17.6)

Major physical features highest point: Tsodilo Hills 1,375 m (4,511 ft)

Population (2006 est.) 1,639,833

Form of government multiparty republic with one legislative house

Armed forces army 8,500; air force 500

Capital city Gaborone (220,558)

Official language English

Ethnic composition Tswana (or Setswana) 79%; Kalanga 11%; Basarwa 3%; other, including Bushmen, Kgalagadi, and White 7%

Religious affiliations Christian 71.6%; Badimo 6%; other 1.4%; unspecified 0.4%; none 20.6%

Currency 1 pula (BWP) = 100 thebe

Gross domestic product (2006 est.) U.S. $18.72 billion

Gross domestic product per capita (2006 est.) U.S. $11,400

Life expectancy at birth male 33.9 yr; female 33.56 yr

Major resources diamonds, copper, nickel, salt, soda ash, potash, coal, iron ore, silver, asbestos, beans, cattle, meat processing, sorghum, manganese, maize/corn, tourism

Zimbabwe

REPUBLIC OF ZIMBABWE

Members of the ZANU PF District Choir (*above*) at the national Heroes' Day celebrations in the Zimbabwean capital, Harare. The image of Robert Mugabe on their costumes proclaims the leading role that he played in their liberation struggle.

ZIMBABWE, FORMERLY SOUTHERN RHODESIA, lies in southeastern Africa between the Zambezi river, which forms its northern border with Zambia, and the Limpopo river, which forms the border with South Africa. Botswana lies to the west and Mozambique to the east.

GEOGRAPHY

Zimbabwe is dominated by a broad mountainous ridge known as the High Veld, which spans the country from the southwest to the northeast. To the southeast of the capital, Harare (formerly Salisbury), is mount Inyangani, the highest point of the escarpment that borders Mozambique. The mountains are flanked to the north and south by the rolling plateaus of the Middle Veld. These in turn fall away to the Low Veld of the two great river basins. The Zambezi river on the northern border plunges over the Victoria Falls into a series of gorges. Farther downstream the valley is overlooked by steep escarpments, and has been dammed and then flooded to create Lake Kariba. The Limpopo river in the far south flows through much gentler country. Both rivers flow out through Mozambique into the Indian Ocean.

Zimbabwe lies within the tropics, but its climate is modified by altitude and aspect. Winters are dry and often cool, and rain is mostly confined to the hotter summer months. The rains are heaviest in the eastern mountains, which support areas of evergreen forest. The land to the south around the Limpopo river receives the least rain, and is arid in places. Over most of the country the natural vegetation is savanna grassland with extensive woodlands, where native trees include mohobohobo and baobab.

Large areas have been cleared for cultivation, limiting the habitat of the local wildlife: baboons, elephants, aardvarks and gorillas, with hippopotamuses and crocodiles in the rivers. Some conservation measures have been adopted, mostly for the benefit of tourism.

SOCIETY

The name Zimbabwe (meaning "house of the chief" in the language of the Shona people) was taken from the Great Ruins, massive fortifications in the south of the country. Once thought to be the legendary King Solomon's Mines, the oldest parts date from about the 9th century, and were built by Bantu invaders who had driven out the original inhabitants.

Some of the modern Bantu peoples arrived as late as the mid-19th century, about the same time as the Europeans, fleeing the expanding Zulu kingdom of Shaka to the south. One such group, the Ndebele, settled in the southeast (now called Matabeleland), dominating the more peaceable Shona people. In the 1890s the British South Africa Company, founded by the British-born South African politician Cecil Rhodes (1853–1902), began the settlement of the country. Both ethnic groups rebelled, but to little effect. The country was run by the British South Africa Company until 1923, when it rejected union with South Africa to become the autonomous British colony of Southern Rhodesia.

In 1957 limited political rights were extended to the majority African population. Their National Democratic Party (NDP) opposed the federation with Northern Rhodesia and Nyasaland (now Zambia and Malawi respectively), which had been established in 1953. The NDP was banned following a number of riots, and the African movement split into broadly ethnic divisions. The Ndebele people supported the Zimbabwe African People's Union (ZAPU), led by Joshua Nkomo (b. 1917), while the majority Shona peoples formed the Zimbabwe African National Union (ZANU) under Ndabaningi Sithole (b. 1920) and later Robert Gabriel Mugabe (b. 1925).

When the federation collapsed in 1963, the European-dominated Rhodesian Front party, led by the white Rhodesian politician Ian Douglas Smith (b. 1919), pressed for immediate independence under their own minority rule. When Britain failed to grant this, the Front made a unilateral declaration of independence (UDI) in 1965. Few countries recognized Rhodesia, tion. Agriculture, for example, became more diversified. The larger farms, most of which are still European-owned, produce most of the cash crops, including tobacco, cotton and sugar cane. An accelerated land reclamation program in 2000 added to the economic destabilization. African farming is mainly small scale, producing staples such as maize and cassava, to which wheat has now been added. Beef cattle are the main livestock, and the country is self-sufficient in dairy products.

Manufacturing was greatly encouraged by sanctions pressure, and is now a major part of the economy. Industries are diverse, including textiles, light machinery and the processing of various agricultural products. Major exports include tobacco and gold, while imports include machinery, manufac-tured goods and fuel. Fuel is needed for electricity generation, although some hydroelectricity is available.

The rail system is excellent, with links to neighboring countries, including the nearest ports in Mozambique. The poor road system is being improved. The main airport is at Harare in the northwest. Health care and education were well provided for in the 1990s until foreign debt crippled government budgets.

The mighty stone walls of the Great Zimbabwe ruins, part of a vast complex of buildings that were occupied from about the 9th to the 17th century. At its height it was the center of an extensive trading empire, whose wealth was based on gold.

and the United Nations imposed economic sanctions. However, South Africa and other states continued to allow vital supplies into the country. ZANU and ZAPU launched a guerrilla campaign against Ian Smith and his party from inside Zambia and later Mozambique. Eventually the parties united to form the Patriotic Front.

The war continued, and mounting international pressure forced the prime minister, Ian Smith, to make concessions toward black majority rule, allowing the formation of the more moderate United African National Council (UANC) under Bishop Abel Muzorewa (b. 1925). An internal settlement in 1979 led to multiracial elections and a UANC government. Eventually, a new constitution led to elections in 1980 which the Patriotic Front won, and Robert Mugabe became president of a fully independent Zimbabwe.

The Patriotic Front split, and Nkomo was ousted from government. However, in 1987 ZANU and ZAPU were again reunited. In 1990 the 1979 constitution lapsed, and a new constitution created a

single-chamber parliament in what had effectively become a one-party republic. In the late 1990s as Zimbabwe plunged into economic crisis, an opposition Movement for Democratic Change (MDC) arose. The MDC won significant support in national elections in 2000 despite intimidation at the polls. There were calls for Mugabe to resign. After the March 2002 presidential elections, Zimbabwe was suspended from the Commonwealth for one year as observers declared the elections flawed by violence and intimidation. Mugabe then charged MDC leader, Morgan Tsvangirai, with treason.

The official language is English, although a majority speak Chishona as their first language. Most indigenous Africans hold traditional beliefs, but a large number are Christian.

ECONOMY

The economy suffered during the war and sanctions of the 1970s, but some sectors benefited from the lack of outside competi-

NATIONAL DATA – ZIMBABWE

Land area	386,670 sq km (149,294 sq mi)			
Climate	Altitude m (ft)	Temperatures January °C(°F)	July °C(°F)	Annual precipitation mm (in)
Harare	1,479 (4,852)	21 (70)	14 (57)	841 (33.1)

Major physical features	highest point: Inyangani 2,592 m (8,504 ft)

Population (2006 est.) 12,236,805

Form of government one-party republic with one legislative house

Armed forces army 25,000; navy 4,000; air force 21,800

Largest cities Harare (capital – 1,607,022); Bulawayo (713,340); Chitungwiza (352,204); Mutare (193,629); Gweru (148,935)

Official language English

Ethnic composition Shona 82%; Ndebele 14%; other Africans 2.0%; Asian 1.0%;

Religious affiliations syncretic (Christianity and indigenous beliefs) 50%; Christian 25%; indigenous 24%; others 1%

Currency 1 Zimbabwean dollar (ZWD) = 100 cents

Gross domestic product (2006 est.) U.S. $25.05 billion

Gross domestic product per capita (2006 est.) U.S. $2,000

Life expectancy at birth male 40.39 yr; female 38.16 yr

Major resources coal, chromium ore, asbestos, gold, nickel, iron ore, vanadium, lithium, tin, copper, platinum group metals, maize/corn, millet, sugarcane, wheat, tobacco

South Africa

A Ngoma Zulu dance team performs in the traditional dress of their warrior elite. The Zulus are proud of their warrior traditions and their distinctive spear and shield, but performances like this are usually put on for the entertainment of tourists.

SOUTH AFRICA IS A LARGE COUNTRY THAT occupies the southern tip of the African continent, bordered by the Indian Ocean to the southeast and by the Atlantic Ocean to the southwest. In the north it borders (from west to east) Namibia, Botswana, Zimbabwe, Mozambique and Swaziland, while in the east it totally surrounds the independent state of Lesotho. Walvis Bay, the main port on the coast of Namibia, was South African territory until 1994.

From 1948 to 1991 South African politics were dominated by its policy of apartheid which involved the almost complete segregation of the white minority community from the rest of the population. Despite strong international reaction against this policy, the country remained very prosperous in a hostile economic climate, thanks to its enormous wealth of natural resources, notably its minerals, and its strategic position.

GEOGRAPHY

South Africa can be divided into two main topographic areas: the vast central interior plateau, and the lands that surround it. These areas are separated by the Great Escarpment.

The land

The Great Escarpment or plateau edge varies in altitude, structure and steepness of slope. It forms a rim around the central plateau from the Transvaal Drakensberg in the northeast to Namibia in the northwest. The highest peaks are found along the Lesotho border in the Natal Drakensberg.

The central plateau is also varied in altitude, being highest along the edges of the Natal Drakensberg and lowest in the northwest on the sandy plateau of the

Kalahari Desert. The judicial capital, Bloemfontein, lies in the center, just to the west of Lesotho. In the northeast the Witwatersrand ridge divides the basins of the Vaal and the Limpopo rivers. The Witwatersrand area is the site of South Africa's largest conurbations, including the administrative capital of Pretoria, and Johannesburg, the largest city. The Witwatersrand gold deposits occur in some of the world's oldest rocks. The central plateau is largely drained by the Orange river, of which the Vaal is a tributary.

The lands surrounding the central plateau include the dissected slopes below the plateau edge, and in the south the lower-lying plateau of the Great Karoo, which is bounded by the Great Escarpment and by the ranges of the Cape fold mountains, such as the Great Swartberge. Table Mountain, part of the fold-mountain belt, overlooks the legislative capital, Cape Town, on the southwest coast. The coastal plain is generally quite narrow except immediately south of Mozambique, and contains no good natural harbors.

Climate

Most of South Africa lies within the warm temperate zone, and the climate is predominantly dry. The higher altitudes across the vast inland plateau tend to cancel out the effects of changing latitude, so that temperatures are generally moderate over most of the country.

Off the west coast the cold Benguela Current flows northward from the southern oceans, restricting rainfall in the west of the country. Off the southeast coast is the warm Agulhas Current, which brings greater rainfall to the eastern side of the country. On the east coast most of the rain falls during the summer months of November to April. The winter months, May to October, are generally fairly cool, bringing rain or even snow to the far southwest. However, on the east coast even the winters are warm.

Uneven distribution of rainfall means that more than half the country is arid or semiarid. Only about one-twentieth of the land area receives more than 1,000 mm (40 in) of rain per year.

Vegetation and wildlife

Higher rainfall in the east creates a characteristic landscape of savanna grassland. In the drier country to the west this gives way to the semiarid scrubland of the Karoo. In the north there is dry savanna bushland. Along the west coast the effects of the cold Benguela Current

have helped to create the sand desert of the Namib. In the far southwest the vegetation is Mediterranean in character. Along the warmer east coast, small forested areas can be found in the ravines that line the Great Escarpment.

South Africa boasts a great variety of wildlife, but much of it is under threat of extinction, and the majority is confined to game reserves. Lions, leopards, rhinoceroses, elephants, hippopotamuses, antelopes and giraffes are all found here. Altogether there are eight game reserves, including the famous Kruger National Park, the Kalahari Gemsbok National Park and the Addo Elephant National Park near Port Elizabeth on the south coast. All contribute to animal conservation.

SOCIETY

South Africa's policy of apartheid produced international condemnation, economic sanctions, arms-sale embargoes and exclusion from international sporting events. There had also been violent internal opposition from black African movements such as the African National Congress (ANC). Only in recent years have there been any significant changes within the system.

History

The Portuguese explorer Bartolomeo Diaz (c. 1450–1500) rounded the Cape of Good Hope in 1488, but it was not until 1652 that the Dutch East India Company founded a post there for supplying provisions. Dutch settlers, later called Afrikaners or Boers ("farmers"), moved inland from the Cape, encountering Bantu agriculturalists (Xhosa) who were extending their territories westward. Disputes with the Bantus developed into war, while the indigenous San (Bushmen) and Hottentot (Khoi) populations were either subjugated or driven away.

At the end of the 18th century the British twice captured the Cape Colony before purchasing it (for £6 million) in 1814. They systematically began to conquer Bantu territories to the northeast. In the 1830s disgruntled Boers left the colony to begin the Great Trek, which took them across the Orange and Vaal rivers to found the South African Republic (later known as the Transvaal) in 1838 and the Orange Free State in 1854. Britain acknowledged the independence of both states in the 1850s, but after 1870 the discovery of diamonds and gold there brought new waves of settlers. The British fought two wars with the Boers, eventually defeating them in 1902. Afterward they forged South Africa into a single colony, establishing the Union of South Africa in 1910. In the 1920s South Africa's dominion status was recognized.

During World War I South African forces drove the Germans out of South West Africa (present-day Namibia). In World War II, however, the Afrikaners

Voting for the first time, this South African woman embraces the opportunity to have a say in her own future. The huge distances that many had to travel did not deter black citizens from coming to the polling stations in their millions.

were critical of the government's decision to join what they saw as Britain's war against Germany. In 1948 the Afrikaner-backed National Party came to power, and began to implement the policy of apartheid or separate development.

Black South Africans were segregated in ten different "homelands" or Black States, which occupied only about one-eighth of the land area; no similar provision was made for other nonwhite South Africans (mainly Asians brought by the British as indentured laborers on the sugar plantations of Natal). Despite the limited resources of the homelands and their effective dependence on South Africa, the government announced that they would eventually become separate states. By 1981 four had become so, but none was recognized outside South Africa. In practice racial separation was never complete. Black laborers frequently worked alongside whites, especially in towns and in the mineral industries, but were denied skilled employment and were paid minimal wages. In cities such as Johannesburg, white residential areas are close to the center, while black areas such as Soweto lie much farther out.

In 1961 South Africa became a republic and seceded from the British Commonwealth. Anti-apartheid protest was repressed with increasing force, and in 1976 the response to a riot by unemployed black youths in Soweto brought death and injury on an unprecedented scale.

In 1990 the African National Congress (ANC) leader Nelson Mandela (b. 1918) was released after 27 years of imprisonment, signaling a change of heart on the government's part. The ANC suspended its 30-year guerrilla war against the government, while president Frederik Willem de Klerk (b. 1936) began to dismantle apartheid. A 1992 referendum approved a new constitution for multiracial majority rule despite the fact that only whites were allowed to vote in the referendum. The country's first multiracial elections were held in 1994. The new government announced plans to establish a "Truth and Reconciliation Commission"(TRC) to investigate violations of human rights perpetrated under the apartheid regime and to award reparations to the families of those who had suffered abuse. The TRC concluded its report in 1998 stating that the apartheid regime had been a crime against humanity. It also reported abuses by the ANC and Inkatha Freedom Party (IFP). Many public figures including Winnie Madikizela-Mandela and former president Pieter Willem Botha caused controversy.

Government

Nelson Mandela became the first president of a newly democratic South Africa in May 1994. In 1996 a new constitution was approved. One of its stated aims being to heal the divisions of the past and establish a society based on democratic

NATIONAL DATA - SOUTH AFRICA

Land area	1,219,912 sq km (471,011 sq mi)			
Climate		Temperatures		Annual
	Altitude m (ft)	January °C(°F)	July °C(°F)	precipitation mm (in)
Cape Town	12 (39)	21 (70)	12 (59)	515 (20.2)

Population (2006 est.) 44,187,637

Form of government multiparty republic with two legislative houses

Armed forces army 36,000; navy 4.500; air force 9,250

Capital cities Cape Town (legislative - 3,660,197); Pretoria (administrative - 1,757,505); Bloemfontein (judicial - 498,404)

Official languages Afrikaans, Ndebele, Northern Sotho, Southern Sotho, Swati, Tsonga, Tswana, Venda, Xhosa, Zulu, English

Ethnic composition Black African 79%; White 9.6%; Mixed race 8.9%; Indian/Asian 2.5%

Religious affiliations Zion Christian 11.1%; Pentecostal/Charismatic 8.2%; Catholic 7.1%; Methodist 6.8%; Dutch Reformed 6.7%; Anglican 3.8%; other Christian 36%; Islam 1.5%; other 2.3%; unspecified 1.4%; none 15.1%

Currency 1 South African rand (ZAR) = 100 cents

Gross domestic product (2006 est) U.S. $576.4 billion

Gross domestic product per capita (2006 est.) U.S. $13,000

Life expectancy at birth male 43.25 yr; female 42.19 yr

Major resources gold, chromium, antimony, coal, iron ore, manganese, nickel, phosphates, tin, uranium, gem diamonds, platinum, copper, vanadium, salt, natural gas, fruits, maize/corn, cotton, potatoes, tobacco, livestock, tourism values, social justice and fundamental human rights. The new constitution provides for an executive president elected by a 400-seat democratically elected National Assembly and a National Council of Provinces (NCOP) which has 54 permanent members and 35 special delegates. The NCOP represents provincial interests and considers legislation passed by the National Assembly. The nine new provinces which were established in 1994, and in which the former homelands were reintegrated, have their own legislatures.

Until the early 1990s, South Africa had troubled relations with its neighbors and funded dissenting groups in order to destabilize their regimes. Under Mandela, South Africa has pursued an independent foreign policy, establishing diplomatic relations with China but also strengthening ties with the United States. South Africa showed its commitment to future regional integration by joining the Southern African Development Community (SADC) in 1994.

People
Pre-1994, the South African government distinguished four racial groups, but the true picture is far more complex. Almost three-fourths of the population are blacks, generally of Bantu ancestry, the largest groups being the Zulu, the Xhosa and the Sotho. The whites make up less than one-fifth of the population; they are the descendants of European settlers, and the majority of them are Afrikaners. The so-called coloreds are of mixed racial origin, and form about one-tenth of the population. There is also a small Asian minority, mostly of Indian descent.

Afrikaans and English are the official languages, but at least eight others are widely spoken. More than three-quarters of the people are Christian (religious practice remains very strong among rural Afrikaners, members of the Dutch Reformed Church). Some traditional beliefs survive among black Africans, and there are small Hindu and Muslim minorities.

ECONOMY

Although South Africa is rich in mineral resources, it has no petroleum deposits, and in the light of the world reaction to apartheid, this has had a profound effect on the organization of its economy.

Agriculture
Although almost half the country's population lives in rural areas, agriculture employs only a small portion of the work force, and provides an even smaller percentage of gross domestic product. Maize is the chief staple, while export crops include sugar cane, citrus fruits, grapes, cotton and tobacco. Livestock is varied, and produces a number of export products. Land shortage among the traditional subsistence farmers has forced many black South Africans to become migrant laborers.

There is good coastal and trawler fishing, mainly off the west coast, where the rich plankton in the Benguela Current attract pilchards, mackerel, sardines and snoek. Although forestry potential is relatively small, the existing stands produce the greater part of South Africa's domestic needs.

Industry and resources
It was South Africa's gold and diamond resources that helped to make the country so attractive to European settlers. Gold and precious stones still account for up to half the country's total exports. Other important minerals include asbestos, uranium and other valuable metal ores. Ore-based industries include three iron-and-steel plants and an aluminum smelter.

Energy conservation is a necessary policy, as all petroleum except that produced from coal has to be imported, and many oil-producing countries have imposed an embargo on exports to South Africa. Low rainfall reduces the potential for hydroelectricity, but several schemes have been initiated, including a large installation on the Orange river. The country's substantial coal output has been increased to supply the electricity industry, and a nuclear power station in the Western Cape has been operating since 1982. The result is that South Africa generates about half the electricity used in the African continent, and the government's energy-conservation policy allows much of this to be exported. Mozambique and Namibia are among the many countries that are linked to the South African national grid.

The development of South Africa's manufacturing industries has been exceptionally rapid, and they now provide almost one-fourth of export earnings. The principal manufactured goods include beverages, chemicals, textiles, transportation equipment and electrical goods.

Trade and commerce
South Africa's chief trading partners include the United States, Germany, Japan, Switzerland and the United Kingdom.

Foreign investment declined after the Soweto riots of 1976, and most industrial finance now comes from within the country, helped by South Africa's wealth in gold. Tariff barriers protect local manufacturing industries and reduce imports.

Transportation and communications
The backbone of South Africa's transportation system is the state-owned railroad network. At least one-third of it has been electrified, and diesel locomotives service the rest of the system. The same adminis-

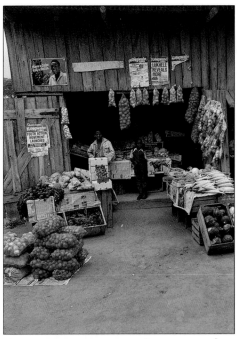

A well-stocked vegetable stall (*above*). The country's subsistence farmers grow a wide variety of fruit and vegetables for their own use and for sale. Maize is the staple crop.

The imposing facade of City Hall, Cape Town (*left*) is dwarfed by the timeless bulk of Table Mountain behind. In the 1980s the ANC rally in front of it would have seemed almost unthinkable, but now the police look on without interfering.

Health and welfare

Health care is better in urban areas than rural and while the majority of white South Africans contribute to private medical schemes black people are still largely dependent on state health care. Facilities are poor in the areas which were previously segregated, the hospitals are overcrowded while poverty and malnutrition make people susceptible to infection and disease. In the late 1990s, an estimated 20 percent of the South African population was HIV positive with the highest rates among young women aged 15-25 and young men aged 20-29. In 1994, the government introduced a new national health plan which included free health care for children under 6 years and pregnant women as well as the establishment of additional health clinics.

The government is also taking action against poverty and the chronic housing shortage, providing additional grants and pensions to those in need.

A new multi-racial education system was established in 1996 providing free, compulsory education for children aged 7-15. In the 1980s the universities were formally segregated but later began to admit students of all races. By 2000, the number of African students at former white universities was around 60 percent.

tration is also responsible for bus services on the extensive road network. However, few of the small, local roads in rural areas have been surfaced.

Water transportation is confined to the coast, as there are no navigable rivers. The harbors are state-owned; the main passenger port is at Cape Town in the southwest, while freight ports include Durban and the new complex at Richard's Bay, both on the east coast. Internal air services are excellent, and there are international airports at Cape Town, Johannesburg and Durban. There are many private airline companies in addition to the state-owned South African Airlines.

An independent broadcasting authority was set up in 1994 to establish a system free from state control. There are three television channels broadcasting in 11 languages as well as cable and satellite services. Newspapers are published in English and Afrikaans along with other black languages and black empowerment of the press has been encouraged. Black readership is subsequently increasing.

Lesotho

L ESOTHO, FORMERLY BASUTOLAND, IS A small kingdom near the southern tip of Africa. It is completely surrounded by South Africa.

GEOGRAPHY

Lesotho is the only country in the world where all the land is above 1,000 m (3,300 ft). Along its southeastern border the Drakensberg range rises to Thabana Ntlenyana, the highest point in southern Africa. The Orange, South Africa's longest river, rises in the Maloti Mountains in central-northern Lesotho. To the west the land descends to a more fertile plateau. The most populous area is the valley of the Caledon river, which forms the northwestern border.

The climate is warm-temperate, with

NATIONAL DATA – LESOTHO

Land area	30,355 sq km (11,720 sq mi)			
Climate		**Temperatures**		**Annual**
	Altitude m (ft)	January °C(°F)	July °C(°F)	precipitation mm (in)
Maseru	1,528 (5,013)	21 (70)	8 (46)	725 (28.5)
Major physical features highest point: Thabana Ntlenyana 3,482 m (11,425 ft)				
Population (2006 est.) 2,022,331				
Form of government constitutional monarchy				
Armed forces army 2,000				
Capital city Maseru (115,559)				
Official languages Sesotho, English				
Ethnic composition Sotho 99.7%; Europeans, Asians, and other 0.3%				
Religious affiliations Christian 80%; traditional beliefs 20%				
Currency 1 loti (LSL) = 100 lisente; 1 South African rand (ZAR) = 100 cents				
Gross domestic product (2006 est.) U.S. $5.195 billion				
Gross domestic product per capita (2006 est.) U.S. $2,600				
Life expectancy at birth male 35.55 yr; female 33.21 yr				
Major resources water, diamonds, sand, clay, building stone, clothing, asparagus, beans, livestock, maize/corn, pulses, wheat, mohair, wool, leather, jute				

more precipitation and greater ranges of temperature at higher altitudes. The low-lying areas have warm, wet summers, while the highlands have bitterly cold winters with snow on the peaks. The dominant vegetation is grassland scrub, giving way to better grassland on the mountain slopes and to alpine vegetation on high ground. Scattered woodland of wild olive and willow trees grows in the remoter valleys. Antelopes, hares and reptiles are common.

SOCIETY

In the early 19th century a Sotho chief, Mshweshwe (c. 1786–1870), secured his people against the double threat of the advancing Zulus and the colonizing European nations. After many disappointments he eventually won British protection. In 1871 the country was annexed to the Cape Colony, but became a separate crown colony in 1884.

Lesotho became an independent kingdom in 1966. In 1970, when the opposition Congress Party won the election, the outgoing prime minister, Chief Leabua Jonathan (1914–87) declared a state of emergency, during which the king, Mshweshwe II (1938–96) was deposed. The king was later allowed to resume a figurehead role. In 1986 a military coup restored the full monarchy, and political activity was banned. However, in 1990 the king's powers were again curtailed, as Major-General Justin Metsing Lekhanya, chairman of the 14-member Military Council, placed the country under military rule. Later the same year the king was once more deposed and replaced by his son (Letsie III). Free elections in 1993 ended military rule, but the government was dismissed in 1994. There is a National Assembly made up of an 80-member lower house and a Senate of 22 chiefs and 11 nominated members.

Almost all Lesotho's population belong to the Sotho peoples, who are Bantu in origin. Sotho and English are the official languages. Most people are Christian.

ECONOMY

Economically, Lesotho is mostly dependent on South Africa, with which it is in customs union. However, it is also a member of the Southern African Development Coordination Conference (SADCC), created in 1980 with the aim of lessening its member states' dependence on South Africa's dominant economy.

Agricultural land is very limited; sub-

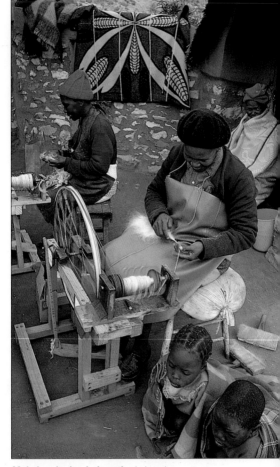

Mohair spinning in Lesotho is largely a cottage industry using treadle wheels. Mohair is one of the country's most prestigious products, both as yarn and in finely woven materials such as those proudly displayed on the wall behind.

sistence crops include maize, sorghum, wheat and beans, but yields are low. Cattle-rearing is uncommercial, but wool and mohair from sheep and goats are a vital export. Mineral resources are scanty; diamond production is controlled by De Beers Consolidated Mines of South Africa. Industry is limited to small-scale manufacturing. However, water is exported to South Africa, and major hydroelectric schemes are planned. Earnings from migrant laborers in South Africa help to balance the economy, and a growing tourist industry brings welcome hard currency.

Roads are few except in the west, and only a minority of them are surfaced. A short railroad links Maseru to the South African network. There is a commercial airport at Maseru, but no international services. There are several hospitals and clinics, mostly in towns, but health conditions are poor; the most common ailments are those associated with diet deficiencies. Education is free but not compulsory. Enrolment is high in primary schools and literacy is increasing.

Swaziland

KINGDOM OF SWAZILAND

THE KINGDOM OF SWAZILAND BORDERS Mozambique to the east, and is otherwise entirely surrounded by South Africa, with which it is in customs union.

GEOGRAPHY

The landscape descends in a series of steps from the mountainous High Veld in the west to the Lebombo Escarpment in the east. Numerous rivers, including the Umbeluzi and the Great Usutu, have cut deep gorges as they flow eastward. Mbabane, the administrative capital, and Lobamba, the royal and legislative capital, are in the western half of the country.

The climate is subtropical with a wet summer season, but becomes cooler and wetter with increasing height. Grasslands flourish on the High Veld, as do plantations of pine and eucalyptus. On lower land woodland savanna is typical. The wildlife includes lions, leopards, hippopotamuses, antelopes and giraffes.

SOCIETY

In 1893 Swaziland was effectively placed under South African administration, having been granted almost entirely to concessionary companies. However, after the Boer War (1899–1902) between the Dutch settlers and the British, it came under British control. Limited self-government was introduced in 1963; four years later the Kingdom of Swaziland was proclaimed, and full independence from Britain followed in 1968. In 1973 and 1977 the king dismissed parliament and a new constitution was adopted.

The reigning King is Mswati III (b. 1968). The House Assembly has 55 elected members and 10 appointed by the King. There is also a House of Senators consisting of 30 members. Ten of these are elected by the House Assembly while 20 are appointed by the King who retains much of his personal power. There are no political parties. The Swazi national Council is a consultancy body for Swazi men.

Swazis form the vast majority of the population, but there are also other ethnic groups, and a small European community. There are some 70 clans, including the royal Dlamini clan. The official languages are English and the royal (Siswati) dialect of Swazi.

ECONOMY

The great majority of the labor force work on the land, mostly as subsistence farmers. Cash crops such as sugar cane, citrus fruits and tobacco are produced by European-controlled companies and by Swazis on resettlement schemes. The Low Veld is particularly suitable for livestock, a traditional sign of wealth. Forestry is a major export earner, and there is a wood-pulp factory near Bunya to the southwest of Lobamba.

Production of local mineral resources, principally asbestos and iron ore, has declined. Power is generated from coal and a series of hydroelectric schemes. Manufacturing is varied but small scale. The economy depends heavily on the wages of migrant Swazi laborers in South Africa, and on the income from tourists from neighboring countries.

Good roads link the principal towns, but most of the system is still unsurfaced. There is one railroad, linking Swaziland to Mozambique. Aircraft from South Africa and Mozambique fly into an airport at Matsapa near Lobamba, and there are many small local airstrips.

Health facilities to combat malnutrition and endemic diseases are provided by the government. The welfare system provides disability, retirement and survivor pensions. Education is free but not compulsory; literacy levels are rising.

A Swazi woman (*above*) leads a colorful processional dance, waving a sheaf of long grass. Her particular traditional costume indicates that she is a married woman.

A Swazi craft stall (*left*) in the open-air market at Mbabane, the administrative capital. Tourists flock here to choose from the wide selection of traditional mats, baskets and bead curtains that are on offer. If customers cannot find exactly what they want, the local craftworkers may make goods on the spot to their requirements.

NATIONAL DATA - SWAZILAND

Land area	17,203 sq km (6,642 sq mi)			
Climate		**Temperatures**		**Annual**
	Altitude m (ft)	January °C(°F)	July °C(°F)	precipitation mm (in)
Mbabane	1,145 (3,757)	20 (68)	12 (54)	1,442 (56.7)

Major physical features	highest point: Emlembe 1,862 m (6,109 ft)

Population	(2006 est.) 1,136,334

Form of government	monarchy with two legislative houses

Armed forces	no armed forces

Largest cities	Mbabane (administrative - 81,312); Lobamba (legislative - 4,819)

Official languages	Swazi, English

Ethnic composition	Swazi 84.3%; Zulu 9.9%; Tsonga 2.5%; Indian 1.6%; others 1.7%

Religious affiliations	Zionist 40% (a combination of Christianity and traditional ancestor worship), Roman Catholic 20%; Muslim 10%; Anglican, Bahai, Methodist, Mormon, Jewish and other 30%

Currency	1 lilangeni (SZL) = 100 cents

Gross domestic product	(2006 est.) U.S. $5.91 billion

Gross domestic product per capita	(2006 est.) U.S. $5,200

Life expectancy at birth	male 32.1 yr; female 33.17 yr

Major resources	asbestos, coal, clay, cassiterite, hydropower, forests, small gold and diamond deposits, quarry stone, citrus fruits, maize/corn, sugarcane, timber, tobacco

Madagascar

M ADAGASCAR IS A LARGE ISLAND STATE lying off the southeast coast of Africa, opposite Mozambique. It remains very distinct from the African mainland, not least in its people and their culture, and in its unique and diverse wildlife.

GEOGRAPHY

Madagascar is the fourth largest island in the world. Geologically, the island was formed when it broke away from the landmass of Africa approximately 50 million years ago.

The east coast is lined with coral beaches and lagoons, and from here the land rises sharply to the great central plateau that covers most of the island. This is ridged and crossed by many rivers, and is interrupted by mountainous massifs. The highest of these is the Massif du Tsaratanana in the north, which culminates in the peak of Maromokotro. There are volcanic features such as crater lakes, and steep sea cliffs in the far north and south. On the western side the mountains fall away more gradually. While the eastern rivers plunge steeply to the Indian Ocean, those to the west such as the Ikopa and Mangoky meander through gentler landscapes, washing down rich sediments to the broad coastal plains.

The climate is greatly varied, from temperate in the highlands to tropical along the coasts; there are heavy seasonal rains in the north and east, but conditions are relatively dry in the central highlands and semiarid in the south. Until recently most of the island was covered in forests, ranging from tropical rainforest to cooler evergreen and deciduous woodlands. However, these forests have been devastated by slash-and-burn cultivation so

that only small areas are left. The exposed soil is now suffering drastically from severe erosion.

Madagascar's isolation has allowed a unique and rich diversity of wildlife to evolve. This includes the small, spiny, insectivorous tenrec, and many remarkable species of lemur. There are also crocodiles, chameleons, butterflies and great numbers of birds. Sea fish native to the area include the spiny globefish and a famous "living fossil" called the coelacanth. The wildlife has, however, also been devastated, both by deliberate killing and by the destruction of habitats. Several lemur species have become extinct, and those remaining are endangered. Conservation efforts are small-scale, and though the government is attempting to limit the destruction, it may already be too late.

SOCIETY

Madagascar was probably first settled some time before the 10th century by seafarers from Indonesia. Then in about the 14th century settlers arrived from the African mainland, together with Arab and Asian traders, who introduced Muslim influences. The people of Madagascar, the Malagasy, thus show considerable ethnic mixing, but their language is still clearly Indonesian in origin.

In the 16th century a French trading post was established, but the coast became a haunt of Indian Ocean pirates, and so the Malagasy began to settle the island's interior. They established powerful kingdoms, often with the help of European weapons. The central kingdom of the Merina people eventually emerged as the most dominant, founded by King Andrianampoinimerina (reigned 1787–1810) with guns obtained from the French in exchange for slaves. After 1810 his son Radama I (reigned 1810–28) joined with the British rulers of the neighboring island of Mauritius to end the slave trade and to expel the French. However, Radama's successors invited foreigners back in, and French Roman Catholics vied for influence with British Protestants. Matters came to a head with the Franco-Merina Wars of the 1880s.

In 1890 Britain recognized Madagascar as a French protectorate, and by 1895 the Merina had been subjugated by the French. Malagasy resistance was suppressed, and French culture came to dominate the prevailing mix of indigenous and Protestant custom. New economic development and health care

programs were introduced. During World War II Madagascar was occupied briefly by British and South African troops, but became a French overseas territory in 1945. In 1947 there was a strong nationalist revolt, and in 1958 the people voted in a referendum for an autonomous Malagasy Republic. Full independence followed in 1960 under President Philibert Tsiranana (b. 1912).

In 1975 his appointed successor, Major-General Gabriel Ramanantsoa (b. 1906), broke ties with France and sought to align the country with the Soviet bloc. In 1976, following a series of coups, Lieutenant-

Commander Didier Ratsiraka (b. 1936) became the president of the newly constituted Democratic Republic of Madagascar, in which the Supreme Revolutionary Council was the most powerful body. A new constitution was approved in 1992 and multi-party elections were held the following year. The prime minister is appointed by the president.

ECONOMY

The Madagascan economy suffered greatly from the abrupt transition to state planning that followed the 1975 revolution. Anti-government strikes and demonstrations have hindered economic reform. Foreign debt was over $4 billion in 1996. France remains the country's chief trading partner.

Madagascar still relies largely on traditional agriculture. The majority of its exports are of cash crops such as coffee, cloves, vanilla, sisal and sugar. Cattle are the main livestock, and various fruits are grown, largely for local consumption. The main staple crops are rice and cassava, but the island is not self-sufficient in either, because production cannot keep pace with the fast-growing population; the introduction of collective farming after 1975 caused much resentment without improving efficiency. Fisheries have so far remained undeveloped. There has been considerable reforestation, usually with non-native species.

Chromite and graphite are mined for export, together with small amounts of gold and precious stones. There are no energy resources except hydroelectricity, which fulfills half of the island's requirements; the rest is generated with imported fuels. Manufacturing industry is based mainly on agriculture; typical products include soap, textiles, sisal rope, sugar and other processed foods. There is a printing works and a vehicle-assembly plant in the capital, Antananarivo, roughly in the center of the country.

Only a few major roads are surfaced, but the few railroad and port facilities are good, and there are excellent internal and international air services. Primary education is free and compulsory, and is provided both by the state and by Protestant and Roman Catholic churches. However, literacy levels are still relatively low. Health services are available mostly to more affluent city-dwellers, while the majority live in poor conditions. There has, however, been considerable success in the eradication of endemic diseases.

Outrigger canoes glide back to shore after a day's fishing at Nosy Komba, Madagascar, watched by the village children. Many of the smaller communities live by fishing, which is mainly a subsistence occupation.

NATIONAL DATA – MADAGASCAR				
Land area 581,540 sq km (224,534 sq mi)				
Climate		Temperatures		Annual
	Altitude m (ft)	January °C(°F)	July °C(°F)	precipitation mm (in)
Antananarivo	1,310 (4,297)	22 (72)	15 (59)	1,365 (25.9)
Major physical features highest point: Tsaratanana Massif 2,876 m (9,436 ft); longest river: Mangoky 560 km (350 mi)				
Population (2006 est.) 18,595,469				
Form of government multiparty republic with one legislative house				
Armed forces army 12,500; navy 500; air force 500				
Capital city Antananarivo (1,538,349)				
Official languages French, Malagasy				
Ethnic composition Malagasy 98.9%; Comorian 0.3%; Indian 0.2%; French 0.2%; others 0.4%				
Religious affiliations traditional beliefs 52%; Christian 41%; Muslim 7%				
Currency 1 Madagascar ariary (MGA) = 5 iraimbilanja				
Gross domestic product (2006 est.) U.S. $17.27 billion				
Gross domestic product per capita (2006 est.) U.S. $900				
Life expectancy at birth male 54.93 yr; female 59.82 yr				
Major resources graphite, chromite, gold, coal, bauxite, salt, quartz, tar sands, precious stones, mica, fish, hydropower, tourism, bananas, cassava, cloves, coffee, maize/corn, peppers, potatoes, rice, sisal, sugarcane, vanilla, zirconia				

Mauritius

REPUBLIC OF MAURITIUS

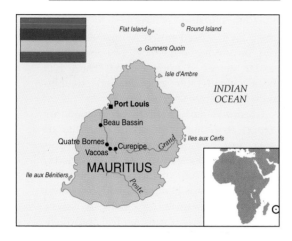

M AURITIUS, ONE OF THE MASCARENE Islands, is an independent island state in the Indian Ocean some 800 km (500 mi) east of Madagascar. Its territory also includes Rodrigues to the east and the tiny dependency of the Agalegals to the north.

GEOGRAPHY

Mauritius is a rugged volcanic outcrop fringed with coral reefs. From the broad coastal plains in the north, it rises to a plateau surrounded by mountain peaks – the remnants of a giant volcano. The highest point is Piton de la Petite Rivière Noire in the south. Rodrigues is similarly volcanic and rugged in character. The tropical climate is moderated by the southeasterly trade winds, which bring ample rainfall and the occasional cyclone.

NATIONAL DATA – MAURITIUS

Land area	2,030 sq km (784 sq mi)			
Climate		**Temperatures**		**Annual**
	Altitude m (ft)	January °C(°F)	July °C(°F)	precipitation mm (in)
Port Louis	55 (180)	28 (72)	23 (59)	711 (27.9)

Major physical features highest point: Piton de la Petite Rivière Noire 826 m (2,711 ft); largest island: Mauritius 1,865 sq km (720 sq mi)

Population (2006 est.) 1,240,827

Form of government multiparty republic with one legislative house

Armed forces Paramilitary 2,000

Capital city Port Louis (158,196)

Official language English

Ethnic composition Indo-Mauritian 68%; Creole 27%; Sino-Mauritian 3%; Franco-Mauritian 2%

Religious affiliations Hindu 48%; Roman Catholic 23.6%; other Christian 8.6%; Muslim 16.6%, other 2.5%; unspecified 0.3%; none 0.4%

Currency 1 Mauritian rupee (MUR) = 100 cents

Gross domestic product (2006 est.) U.S. $16.72 billion

Gross domestic product per capita (2006 est.) U.S. $13,500

Life expectancy at birth male 68.66 yr; female 76.66 yr

Major resources sugarcane, tea, bananas, fish, tourism, potatoes, tobacco

The dense forest that once covered Mauritius has largely been cleared, to the detriment of local wildlife. This includes some remarkable species, notably the introduced hedgehog-like tenrec, and a fruit bat. Mauritius was also once the home of the now extinct dodo. Conservation projects have been set up with British assistance.

SOCIETY

Mauritius was named in 1598 by Dutch navigators after Prince Maurice of Nassau (1567–1625), but remained uninhabited until it was colonized by the French in the 18th century. They called it Ile de France, and imported African slaves to work on their sugar plantations, both here and on neighboring Rodrigues. In 1810 both islands were ceded to Britain, but the French and mixed-race Creole communities each retained their own language and culture. Following the abolition of slavery, large numbers of Indian laborers were brought in, and Chinese traders also settled on Mauritius.

After World War II economic reforms and increasing local self-government were introduced. In 1968 Mauritius was granted independence within the British Commonwealth. Its parliamentary government endured despite political turbulence until the declaration of a republic in 1992. English is the official language, but French, Creole, Hindi and Chinese are in everyday use. The main religious groups are Hindus, Christians (mainly Roman Catholics) and Muslims, along with a smaller number of Buddhists.

ECONOMY

Mauritius still remains heavily dependent on sugar cane, although great efforts are being made to diversify. Other crops include tea, bananas, tobacco and various food crops, and fishing methods are also being modernized.

The manufacturing sector is developing fast, thanks to the establishment of free-trade zones, within which imported raw materials are processed and reexported; successful products include textiles and clothing, plastics and electronics. The island's natural beauty and lively cultural mix have encouraged a flourishing tourist industry, which is now the country's second-largest revenue earner. Exports are still heavily outweighed by imports of food, fuel, machinery and manufactured goods. Hydroelectricity and even sugar-derived fuels are being developed to reduce the country's reliance on imported energy sources.

The roads on Mauritius are good and mostly surfaced, and buses have replaced the old railroad system. The capital, Port Louis, on the northwest coast, has very good dock facilities, and there is an international airport at Plaisance in the southwest of the island. The Mauritius Broadcasting Corporation operates radio and television services. Health and welfare services are good despite problems of overpopulation. The vast majority of children receive primary education, and secondary education is readily available. The University of Mauritius provides mainly technical subjects.

Sugar cane fields in Mauritius spread out beneath jagged peaks of volcanic stone – an image that unites the island's two great assets: the exotic beauty of its landscape – a draw for tourists – and the crop on which the country depends.

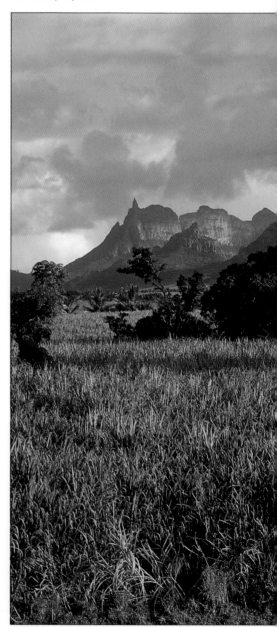

Comoros

FEDERAL ISLAMIC REPUBLIC OF THE COMOROS

Fishing boats lie in the harbor of Moroni, capital of the Comoros. Behind is the ornate white facade of the Great Mosque, witness to the islands' long Arab connections. Although the population is today very mixed, Islam is the dominant religion and culture.

THE COMOROS ISLANDS LIE IN THE INDIAN Ocean between northern Madagascar and the African coast. Three out of the four islands form the independent republic. The southeasternmost island, Mayotte, has remained a French dependency.

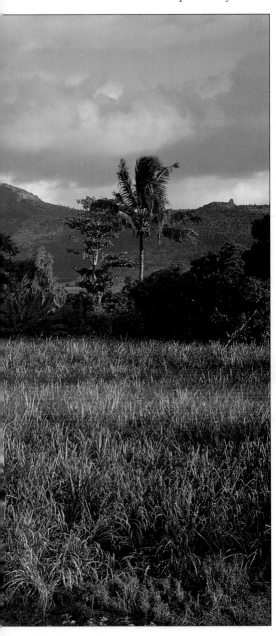

GEOGRAPHY

The archipelago is volcanic in origin. Njazidja (formerly Great Comoro) in the northwest is mostly a rough plateau that rises southward to the active crater of Mount Kartala. Moroni, the national capital, lies on the west coast. The climate is hot with seasonal rains, but the ground is porous, and only the upper slopes support rainforests. The island of Nzwani (formerly Anjouan) to the southeast is very similar in character, but although its soils are better, much of the topsoil has been eroded after land clearance. Mwali (formerly Mohéli), the smallest island, has well-watered river valleys and densely forested hillsides. Wildlife includes several species of land bird and a lemur that are endemic to the islands.

SOCIETY

The Comoro Islands were settled by many different peoples at various times, including mainland Africans, Malagasy, and Arab and Iranian traders. Arabs were the dominant group when the first Europeans arrived in the 16th century. In 1843 Mayotte was annexed by France, and in 1886 the other islands became French protectorates. Arabic and French are the official languages in this predominantly Muslim state, but the lingua franca is Comorian, which resembles Swahili.

After World War II the islands became a French overseas territory, and in 1961 they were granted internal autonomy. In 1975 President Ahmed Abdallah (1919–89) declared independence unilaterally. Abdallah was deposed later that year, but in 1978 he was restored after a coup, and became president of the newly constituted Federal Islamic Republic. In 1989 Abdallah was assassinated but 1990 saw the first multiparty elections since independence and Said Mohamed Djohar became president. Further coups in 1995 and 1999 and attempts at secession by the islands of Nzwani and Mwali left Comoros in a troubled state as the century closed.

ECONOMY

The Comorian economy has traditionally depended on subsistence agriculture. Goats and cattle are the main livestock, and food crops include cassava, sweet potatoes and mountain rice, but much food is imported. Exports consist mainly of cash crops such as vanilla and cloves.

There are a few hydroelectric plants, but most fuel has to be imported. Manufacturing industry consists mostly of small–scale processing of vanilla, cloves and other spices. Tourism also provides an important source of revenue. France remains the chief trading partner and the main source of foreign aid.

The roads are often impassable in the rainy season, but good air and ferry services operate between the islands and to Madagascar and East Africa. There is an international airport on Njazidja. The three French-built hospitals suffer from a lack of medical staff. Education is provided both by Islamic schools and by French-style lycées.

NATIONAL DATA - COMOROS

Land area 2,170 sq km (837 sq mi)

Climate	Altitude m (ft)	Temperatures January °C(°F)	July °C(°F)	Annual precipitation mm (in)
Moroni	12 (39)	27 (81)	23 (73)	2,542 (100)

Major physical features highest point: Mount Karthala (Njazidja) 2,361 m (7,746 ft); largest island: Njazidja 1,148 sq km (443 sq mi)

Population (2006 est.) 690,948

Form of government federal multiparty republic with one legislative house

Armed forces army 700 (reorganized in 1995); navy 200

Capital city Moroni (44,518)

Official languages Arabic, French

Ethnic composition Comorian (Bantu/Arab/Malagasy) 96.9%; Makua 1.6%; French 0.4%; others 1.1%

Official religion Islam

Religious affiliations Sunni Muslim 98%; Roman Catholic 2%

Currency 1 Comoran franc (KMF) = 100 centimes

Gross domestic product (2002) U.S. $441 million

Gross domestic product per capita (2005) U.S. $600

Life expectancy at birth male 60 yr; female 64.72 yr

Major resources negligible apart from cash crops such as vanilla, copra, cloves, ylang-ylang, coffee, sweet potatoes, rice, coconut, maize/corn, yams

Dependencies in the region

MAYOTTE
FRANCE

The southeasternmost island of the Comoros archipelago is a French dependency with the official status of a *collectivité territoriale*. It is mountainous, largely volcanic and fringed by coral reefs. The capital and port, Dzaoudzi, is perched on a rocky outcrop. The wet, tropical climate supports lush forest vegetation.

The people are mainly Malagasy in origin and speak Comorian, which is closely related to Swahili. Most are Muslim (Islam was introduced by Arab traders), but there is a strong Roman Catholic minority. In 1975, when Comoros declared its independence from France, Mayotte chose by plebiscite to remain French. Comoros still claims the island, supported by a 1979 United Nations resolution. However, in 1984 the French abandoned plans to return Mayotte to Comoros on account of local opposition. The island has its own elected assembly, but is also represented in the French parliament.

The economy is largely agricultural; the main cash crops are vanilla, ylang-ylang (used in perfumes), coconuts and coffee. Manufacturing is mainly limited to food processing. The education system includes both Islamic and French schools.

REUNION
FRANCE

Réunion, the most southerly of the Mascarene Islands, lies in the Indian Ocean off the east coast of Madagascar and southwest of Mauritius. It is an overseas department of France.

The island is mountainous and volcanic in origin; the highest peak, Piton de la Fournaise, is still active. The tropical climate is cooled by southeasterly trade winds, which bring ample rainfall and occasional severe cyclones. The rich rainforest that once covered the island is all but gone, except in uninhabitable upland areas, where the tamarind tree is still found. Much of the native wildlife has now disappeared, but that which remains includes several species of bat, and among the birds the Bourbon flycatcher and the Mascarene bulbul.

The island, claimed by France in 1638, was originally known as Bourbon. It was first settled in the 17th century by French colonists, who brought African slaves to work their sugar plantations. These were

later followed by Malay, Chinese and Indian laborers. The present population includes all these peoples, but is predominantly mixed race, or Creole. The everyday speech is Creole, but French is the official language, and most people are Roman Catholic. The islanders are French citizens, and return five deputies and three senators to the French parliament. Internal government is by an elected council of 44 members, headed by a French-appointed prefect.

Réunion's economy is predominantly agricultural, and is largely dependent on sugar cane. Other cash crops include tobacco, vanilla, ornamental and perfume plants, and various fruits and vegetables. Sugar, together with its derivatives, rum and molasses, accounts for the great majority of exports, but tourism is an increasing source of revenue. Imports include food, fuel and manufactured goods. Manufacturing at home is mostly confined to sugar refining and other food-processing industries. Health facilities are being developed to combat indigenous diseases, and the French-based education program is contributing to the increase in the literacy figures.

ST HELENA
UNITED KINGDOM

St Helena – a remote volcanic island lying some 1,930 km (1,200 mi) off the southwest coast of Africa – is a British colony which itself has two dependencies: Ascension Island to the northwest and Tristan da Cunha to the southwest.

St Helena covers some 120 sq km (47 sq mi) of mostly rugged and mountainous

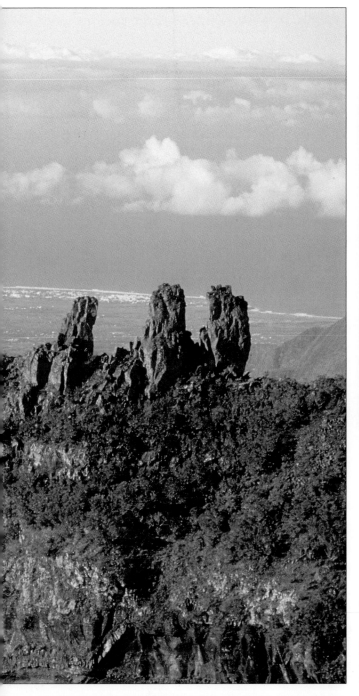

A striking aerial view of Dzaoudzi (*above*) The capital of Mayotte, it occupies a rocky outcrop connected by a causeway to the islet of Pamanzi, part of the coral reef that surrounds the main island of Mayotte. Water to supply the town is drawn from reservoirs that are filled in the rainy season.

Towering mountain peaks (*left*) on Réunion, eroded into spectacular shapes, betray the island's volcanic origins. Volcanic activity continues in the eastern part of the island, where there is one active volcano and nine dormant cones.

terrain, but with enough fertile land to support a small farming economy for its 6,000–7,000 inhabitants. The main port and capital is Jamestown, which also acts as an administrative center for Ascension Island and Tristan Da Cunha.

The island was discovered on 21 May 1502 by the Portuguese navigator Jõao da Nova (d.1509) and named after St Helena (c. 248–329), the mother of Constantine the Great, whose feast day it was. It was subsequently settled by the East India Company and became a British colony in the mid-19th century. Following his abdication as emperor, Napoleon Bonaparte (1769–1821) was exiled there by the British from 1815 until his death in 1821.

Ascension Island is a barren, rocky peak, almost totally lacking in vegetation and covering only 88 sq km (34 sq mi). It was uninhabited until Bonaparte was

exiled to St Helena, when a small British garrison was established on Ascension as a precautionary measure. Today it has a tiny agricultural community, most of whom are St Helenians. Ascension Island is perhaps most famous for its green turtles, which come to the island between December and May each year to lay their eggs in the sand.

Strategically, however, Ascension Island has become a vital link across the South Atlantic, and many of its 1,000 inhabitants are service personnel or communications workers. It came into the public spotlight briefly during the British recapture of the Falkland Islands from Argentina in 1982, when it acted as a critical air–sea staging post for British and US armed forces. It also provided a base for US tracking equipment during the Apollo space program. In addition, the

island is a relay station for telecommunications between South Africa and Europe, and links international broadcasting via satellite and submarine cables. It was given its name by a Portuguese expedition – led by Jõao da Nova – which landed there on Ascension Day, 1501.

The Tristan da Cunha group of islands – Tristan da Cunha, Inaccessible Island, Gough Island and the three Nightingale Islands (Nightingale, Stoltenhoff and Middle) – lies about 2,550 km (1,600 mi) off the coast of Cape Town. Tristan da Cunha itself is a volcanic cone, 100 sq km (40 sq mi) in area, which last erupted in 1961 and was settled again in 1963. It has only a few hundred residents. An administrator who represents the British government heads the local council on Tristan da Cunha and acts as the magistrate for the whole group. The other islands in the group are largely uninhabited, but the low cliffs of Nightingale Island and its two rocky islets support millions of nesting seabirds.

Tristan da Cunha first came to prominence as a garrison in the early 19th century; now three main activities support the islands' economy. The group's main source of revenue comes from selling postage stamps to philatelists. It also lies at the center of a small fishing industry run from Cape Town, which exports goods to Japan, the United States and Britain. In addition, Gough Island is the home of a small weather station.

The Nordic Countries

COUNTRIES IN THE REGION

NORWAY · SWEDEN · FINLAND

DENMARK · ICELAND

A tranquil fishing village (*left*) in the Lofoten islands, off the northwest coast of Norway. The sea has always played an important role in the region's economic and cultural life, and most of the population is found along the coasts. Fishing and sailing are popular leisure pursuits even with city dwellers.

ICELAND

North Cape · Grimsey · Arctic Circle
Straumnes · Fontur
Ísafjördur
Akureyri
Breidafjördur
Langjökull · Hofsjökull · Ódádahraun · Vatnajökull
Faxaflói
Reykjavík · Höfn
Thjórsá · Hvannadalshnúkur 2119
Heimaey · Vestmannaeyjar
Surtsey

- ■ capital city
- ● major town

height of land (meters)
2000
1000
500
200

□ permanent ice
▲ mountain peak

ARCTIC OCEAN

North Cape

Senja · Tromsø · Haltiatunturi 1328 · Inari
Vesterålen
Lofoten · Hinney · Narvik · Kebnekaise 2111 · Kiruna · Lapland · Lokka Reservoir
Vest Fjord
Svartisen 1699▲ · Kemijärvi · Arctic Circle
Hornavan · Kemi
Luleå · Kemi
Skellefteå · Oulu
Umeå · Lake Oulu · Pielinen
Trondheim Fjord · Jakobstad · **FINLAND** · Kalla
Trondheim · Vaasa · Kuopio
Östersund · Ori
Ålesund · Storsjön · Jyväskylä · Hauki
NORWAY · Näsi · Puula · Saimaa
Snøhetta 2286 · Sundsvall · Pori · Tampere
Galdhøpiggen 2469 · Siljan · Päijänne
Jotunheimen · Turku
Lillehammer · Gävle · Helsinki
Bergen · **SWEDEN**
Hardanger Fjord · Dal · Åland
Sogne Fjord · Klar · Drammen · **Oslo** · Uppsala · Gulf of Finland
Bokna Fjord · Stavanger · Vänern · Örebro · **Stockholm**
Mälaren
Kristiansand · Norrköping
Skagerrak · Göta · Vättern · Gotland
North Sea · Ålborg · Göteborg · Jönköping · Bolmen
DENMARK · Kattegat · Kalmar · Öland
Århus · Helsingborg
Jutland · Baltic Sea
Esbjerg · Copenhagen · Malmö
Odense · Sjaelland
Fyn · Bornholm
Lolland · Falster

Gulf of Bothnia

Ume · Angerman · Indal · Ljusnan · Göta
Kjølen Mountains · Törne · Kemi · Lule

211

Norway

KINGDOM OF NORWAY

NORWAY LIES ALONG THE WESTERN COAST of the Scandinavian peninsula. Its spectacular rugged land supports a relatively small population.

GEOGRAPHY

Long and thin – only 450 km (280 mi) across at its widest point – Norway narrows in places to 80km (50 mi).

The land

Norway's ancient granite bedrock was eroded by ice age glaciers and shaped into high mountains and deep valleys. There are more than 160,000 lakes; the largest is lake Mjøsa in the southeast. Fjords (highwalled sea lakes in glacial valleys) turn the coastline into a maze flanked by some 50,000 islands.

Oslo and the southeast are in the lowlands. The Glama river and others run down from the mountains, creating fertile valleys and lakes in the main agricultural area. Other important lowland farming is in the Trøndelag in central Norway, and near Stavanger in the southwest.

Glacial erosion has flattened some mountains, creating immense plateaus such as the Hardangervidda in southern central Norway. Farther north lies the Jotunheimen mountain range. Toward the coast is the Jostedalsbreen icefield with its flowing glaciers. Northern Norway, lying mostly above the Arctic Circle, ends in the Finnmark Plateau. Norway also includes the bleak Arctic islands of Svalbard and Jan Mayen.

Climate, plants and animals

Below the Arctic Circle, the North Atlantic

A peaceful scene in summer Vetle Fjord is a northern arm of southern Norway's Sogne Fjord, the longest – at 184 km (115 mi) – of the country's great coastal inlets. It was carved out by the grinding action of glaciers during the last ice age, and flooded by the sea.

Drift gives western Norway a relatively temperate climate. Eastern Norway has colder winters, warmer summers and lower rainfall. About one-quarter of the country is forested. The richly varied vegetation includes blueberries, cranberries and cloudberries – the fruits of which are Scandinavian delicacies. Recently, large areas have been damaged by acid rain from industrial air pollution originating in neighboring countries. Large predatory animals such as wolves, bears and lynx have been hunted almost to extinction in the southern forests.

SOCIETY

Norway's immense mountains and deep fjords have always made land travel and communications extremely difficult, so it is not surprising that the country has a long maritime and seafaring history

History

Norway's earliest settlers, Central European hunter–gatherers, gradually moved north as the last ice age ended. By the 9th century AD settled farming was well established. As the population expanded, settlers began trading and raiding in Scotland, Ireland, northern England and parts of northern France (Normandy). They also settled in Iceland and Greenland, and crossed the Atlantic to North America. The name Viking comes from Skagerrak, the strait between Norway and Denmark, then called the Vik.

King Harald Fairhair (c. 860–940) united Norway about 900. Christianity and the eventual decline of the Viking era strengthened the monarchy, but the country was devastated by the bubonic plague, which reached Norway in 1350.

The Kalmar Union of 1397 combined the royal lines of Norway, Sweden and

Filigree of ice Snow lies on the ground on many days, and winters are long and dark. But the moderating influence of the warm North Atlantic Drift and westerly winds means that coastal waters are rarely frozen over even when the land is snowbound.

Denmark. In 1814 Norway was forced into a union with Sweden but did gain limited internal self-government. After a series of popular protests, Sweden recognized Norway's independence in 1905.

Norway sought to remain neutral in World War II but was occupied by the German army from 1940 to 1945. After the war, Norway joined the North Atlantic Treaty Organization (NATO) and the European Free Trade Association (EFTA).

Government

The constitutional monarchy adopted in 1814 remains today. The parliament, or Storting, is elected by universal adult suffrage for a 4-year term and in turn selects a prime minister and cabinet. The king acts as head of state but his functions are mainly ceremonial.

People

Most Norwegians share common ethnic and linguistic origins with their Scandinavian neighbors. Bokmal, the urban educated form of Norwegian, and Nynorsk (New Norsk), based on rural western dialect, are gradually being combined into Samnorsk. Most Norwegians belong to the national Lutheran Church.

NATIONAL DATA - NORWAY

Land area	307,442 sq km (118,673 sq mi)			
Climate		Temperatures		Annual
	Altitude m (ft)	January °C(°F)	July °C(°F)	precipitation mm (in)
Oslo	96 (314)	-4 (24)	17 (62)	655 (26)
Bergen	39 (127)	1 (34)	14 (58)	2,074 (82)

Major physical features highest point: Galdhøpiggen 2,469 m (8,100 ft); longest river: Glåma 611 km (380 mi)

Population (2006 est.) 4,610,820

Form of government multiparty constitutional monarchy with one legislative house

Armed forces army 14,700; navy 5,310; air force 5,000

Largest cities Oslo (capital - 521,886); Bergen (237,430); Trondheim (154,351)

Official language Norwegian

Ethnic composition Norwegian 95.8%; other Scandinavians 1.0%; others 3.2%

Religious affiliations Church of Norway 85.7%; Pentecostal 1%; Roman Catholic 1%; other Christian 2.4%; Muslim 1.8%; other 8.1%

Currency 1 Norwegian krone (NOK) = 100 øre

Gross domestic product (2006) U.S. $207.3 billion

Gross domestic product per capita (2006) U.S. $47,800

Life expectancy at birth male 76.91 yr; female 82.31 yr

Major resources petroleum, natural gas, iron ore, copper, lead, zinc, pyrites, nickel, fish, timber, hydropower, apples, barley, coal, livestock, oats, potatoes, silicon, titanium with Russia (1741–43) left Finland, then a duchy of Sweden, occupied by Russian troops, while war with Prussia (1757–62), the rising power in northern Germany, put a heavy strain on the country's finances. Gustav III (1746–92) restored absolute rule and attempted major social reforms. His efforts to regain the Finnish provinces were foiled by treason in 1788, and four years later he was assassinated.

During the Napoleonic wars at the beginning of the 19th century, Sweden aligned itself with Britain, but its ally could offer little support against the combined forces of France, Russia and Denmark. In 1809 Finland and the Åland Islands at the southern end of the Gulf of Bothnia were ceded to Russia, and in 1810 one of the emperor Napoleon's marshals, Jean-Baptiste Bernadotte (1763–1844), was chosen as crown prince, taking the name of Charles John. He failed to negotiate a return of the ceded territories, but in 1813 took a combined Prussian, Russian and Swedish army into Denmark, forcing its king, Frederik VI (1768–1839), to surrender Norway under the Treaty of Kiel in 1814. Shortly afterward, following a brief rebellion, Norway was forced into political union with Sweden.

Liberal reforms in the 19th century included free trade (1846) and a two-chamber parliament (1865–66). In 1905 Oscar II (1829–1907) dissolved the union of the Swedish and Norwegian crowns.

Sweden remained neutral throughout both World Wars. It later joined the United Nations (UN), but stayed outside the North Atlantic Treaty Organization (NATO). From 1946 onward the Social Democratic Party developed a comprehensive welfare system. The murder of prime minister Olaf Palme (1927–86) deprived the Social Democrats of their most charismatic and persuasive leader.

Government

Sweden is a constitutional monarchy, though today the monarch's role is almost entirely ceremonial. The Riksdag or parliament consists of a single chamber with 349 members elected for a four-year term. Everyone aged 18 or over can vote. Some 310 seats are allocated directly, and the rest are divided between parties in order to achieve proportionality with the accumulated votes throughout the country. This proportional representational system is designed to avoid having an overabundance of small or fringe parties.

The principal organs of government lie within 14 committees, and committee

Seasonal festivities (*far right*) Communities in Sweden still celebrate Midsummer's Eve – the longest day – in the traditional way with children dancing around a cross of flowers and leaves, and with all-night parties. Candles on St Lucy's day (13 December) light up the winter gloom, and spring is greeted on Walpurgis night (the last day of April) with the lighting of bonfires.

Symbol of Nordic political union (*right*) The castle stronghold of the historic city of Kalmar overlooks the narrow straits lying between the mainland and Öland, a long narrow island off the southeast coast. It was here that the Union of Kalmar was sealed in 1397 with the coronation of Eric of Pomerania as king of Denmark, Norway and Sweden. Queen Margrethe I (1353–1412) who was regent of Denmark (from 1375), of Norway (from 1380) and of Sweden (from 1389) had been instrumental in making her nephew Erik undisputed king of Scandinavia. After his coronation Margrethe remained regent and effectively Scandinavia's ruler until her death. Her diplomacy that had led to the Union of Kalmar united Denmark, Norway and Sweden until 1523 and Denmark and Norway until 1814.

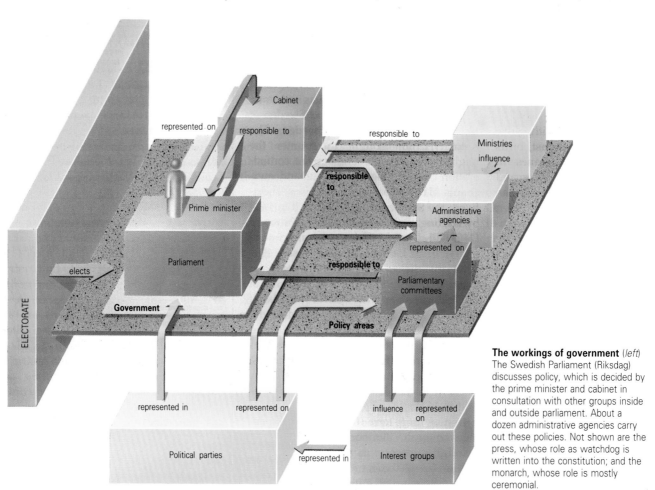

The workings of government (*left*) The Swedish Parliament (Riksdag) discusses policy, which is decided by the prime minister and cabinet in consultation with other groups inside and outside parliament. About a dozen administrative agencies carry out these policies. Not shown are the press, whose role as watchdog is written into the constitution; and the monarch, whose role is mostly ceremonial.

members are elected by proportional representation. Executive control belongs to an 18-member cabinet led by the prime minister, who is appointed by the speaker of the Riksdag and must be approved by its members. The cabinet, or an individual minister, can be dismissed by a simple majority vote in the Riksdag. At local level there are 24 counties (each headed by a governor) and 23 county councils (the island county of Gotland in the Baltic does not have one). There are in addition 284 town councils.

Environmental awareness brought the Green Ecology Party enough support to make its first appearance in the Riksdag in 1988. The victory of the conservative Moderate Party (1991) turned Sweden from socialism towards a free market economy. The Social Democratic Labor Party was returned to power in 1994 but had to continue with unpopular levels of taxation and welfare cuts until the economy improved in 1999.

People

The people of Sweden share common ethnic and cultural origins. The Swedish language is closely related to Danish, Norwegian, Icelandic and Faeroese, and all belong to the same branch of the Indo-European family of languages as German and English. The most sizable ethnic minority are the Sami (Lapps), who have a distinctive culture and language. However, most Samis also speak Swedish, and in the north both languages are taught in schools. Recent immigration has created other minority groups including Finns, Greeks and Turks.

Most Swedes belong to the Lutheran Church whose organization and finances are closely linked with the state, but it retains religious independence. Swedish culture is lively and productive, particularly in literature, art and film-making.

ECONOMY

The prime natural resources of Sweden are lumber, iron ore and water, which is harnessed for hydroelectric power. All three are largely confined to Norrland. Forestry and engineering are the chief export earners since the decline of the shipbuilding industry.

Agriculture

The lumber industry is carefully regulated by the government. Some 60 percent of Sweden's total land area is devoted to productive forestry, which yields turpentine, resins, dyes, rayon and plastics as well as traditional timber products such as pulp for paper. The rivers of Norrland supply timber and hydroelectric power to sawmills and pulp mills on the Bothnian

The Vikings: founders of Russia?

In the 8th century a strong kingdom based around Uppsala in Sweden made the first of many forays into the eastern Baltic. Within 50 years two Swedish settlements had developed in what is now Latvia, with another at Apulia in modern Lithuania. Although these towns were probably established as trading posts, they became the starting point for Viking expansion and conquest farther south.

In the early 9th century dense forest separated the Baltic coast from the Black Sea and Central Asia, which at that time were dominated by two great trading empires. The Arab Caliphate, with its capital at Baghdad, controlled everything south of the Caucasus, including the caravan routes to China and the Far East. The Byzantine empire, with its capital at Constantinople (modern Istanbul), controlled the Black Sea and southeastern Europe. For enterprising and adventurous traders there were two possible routes southward from the Baltic coasts – down the Dnieper river to Constantinople, and down the Volga river to the Caspian Sea and Baghdad. Archaeological and written evidence indicates Swedish Vikings were using both routes by the end of the 9th century.

The Russian Primary Chronicle, also known as the Nestor Chronicle, describes the origins and growth of Russia up to the 12th century. It cites a tradition that the royal families of Novgorod and Kiev were founded by Scandinavians – and there is tangible evidence to support this idea. Archeologists digging at Staraya Ladoga in northern Russia have found the remains of a Swedish colony. Their excavations have shown that Swedes

lived there from the early 9th century to the middle of the 11th. The town was ideally placed for access to both major routes south, and another written source proves the existence of a northern kingdom of "Rus" as early as 839. It also records that these "Rus", although not Swedish, were "of Swedish origin". Inscriptions and archeological finds from Sweden confirm strong links with Kiev and Constantinople, and Scandinavian names began to appear in the Slavic regions throughout the 9th century: Helgi changed to Oleg, Valdemar became Vladimir, Ingvar became Igor and so on.

So was Russia founded by Swedish Vikings? Probably not as such – but the Vikings almost certainly did found a great number of towns in Russia and the Ukraine, including Novgorod, Izborsk, Polotsk, Smolensk, Murom, Rostov, Chernigov and Kiev.

Viking hoard A cache, found in an 11th-century burial site in Sweden, consists principally of Byzantine and Islamic coins, and provides striking evidence of the great distances traveled by the Viking adventurers.

coast. From there the finished products are shipped abroad or sent to the industrial south for further processing.

Less than 7 percent of Sweden's land area is farmed, and agriculture is protected by government levies; however, it does meet four-fifths of the domestic requirement. Most cultivated land is in Skåne and the central lowlands, where arable crops include sugar beet, barley and wheat. Elsewhere farms are smaller and livestock (cattle, sheep and pigs) is more important; only hardy varieties of grass for fodder, potatoes, barley and rye can be grown successfully in the north's shorter growing season. Fur-bearing animals such as mink and fox are also raised for their pelts. Increasing mechanization and rationalization of farms is causing a rural exodus. Some 83 percent of the population now live in towns and cities, and farming employs less than 3 percent of the workforce.

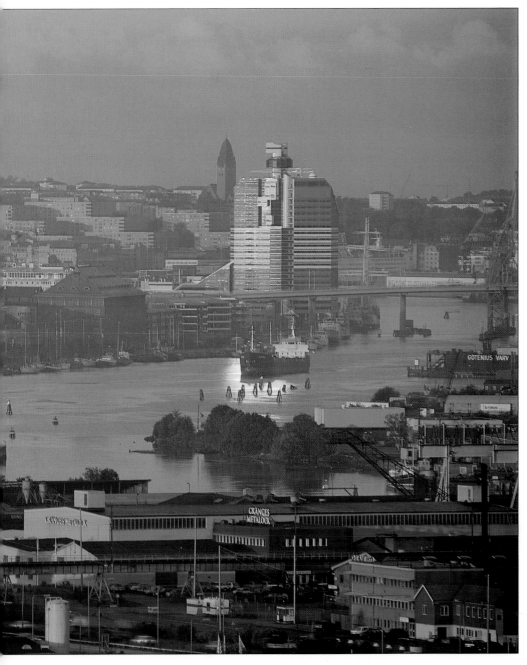

Transportation and communications

State involvement in transportation and communications is quite high. The trucking industry, dominated by two government-run companies, is the chief goods carrier. Buses are the backbone of most local public transportation services. Air services are dominated by the Scandinavian Airlines System (SAS), a joint venture of the Danish, Norwegian and Swedish governments.

Since the railroads were built in the 19th century internal shipping services have declined in importance, though international shipping traffic remains very active. Recently the state-owned railroads have also suffered low profits and shutdowns. Most people now travel in cars, increasing demand for improved roads.

There are 168 daily newspapers, but many of them are state-subsidized and the number has been falling steadily since World War II. Sveriges Radio operates a comprehensive radio network and two television channels. There are four commercial television channels.

Health and welfare

Health insurance is compulsory and comprehensive. The government provides virtually everyone with child allowances, pensions for old age and disability, and benefits for sickness, maternity and unemployment. Medicines, doctors' fees and hospital charges are largely paid for by the state. As a result, both direct and indirect taxation is very high. Postwar governments have always been committed to full employment. In the 1990s unemployment figures averaged 7.5 percent.

Education

Sweden has a completely unified education system that coordinates university and vocational studies with secondary-school programs. Everyone has to attend comprehensive schools between the ages of 7 and 16, and special provision is made for the disabled. English is taught from the third year on, and German or French can also be selected in later years. In the seventh year pupils choose their own subject groups.

Everyone is entitled to continue their studies after leaving school, to prepare for university or for vocational training. There are 13 universities, of which the oldest is Uppsala (founded 1477). Adult education is available through daytime and evening classes, or at the folk high schools. Distance or home learning is available with support from radio, television or correspondence courses.

Unlike Norway, fishing – mostly from harbors on the west coast – makes a marginal contribution to the economy. The catch is dominated by fish such as herring, mackerel and cod.

Industry and trade

Kiruna in northern Norrland has one of the world's richest deposits of iron ore, and a processing plant has been built at the port of Luleå on the Bothnian coast. Other mineral resources include a range of precious metals as well as copper, zinc, lead and uranium. There are low-quality coal deposits in the south.

Sweden is one of the largest users of hydroelectric power in the world. A series of plants, mostly along the Norrland rivers, provides a large percentage of the country's electricity requirements. Coal and nuclear fuel are imported, along with all of Sweden's oil. In 1988, after the nuclear disaster at Chernobyl in the

Seaport city Göteborg, overlooking the straits between the North and Baltic seas, is Sweden's largest port and second largest city. Its industries include Volvo, textiles and the SKF steel giant. It is also a large freight and passenger ferry center.

Ukraine (then part of the Soviet Union), the decision was taken to phase out the country's 12 nuclear power stations, which supplied half of Sweden's energy requirements, by 2010.

The manufacturing industry in Sweden benefits from the country's rich resources. Products include electronic and telecommunications hardware as well as transportation equipment and machinery.

Close to half of the country's heavy industrial production is sold abroad, making the economy increasingly dependent on foreign trade. Sweden had been a full member of the European Free Trade Association (EFTA) since 1960, but, after a referendum in favor, became a member of the European Union (EU) in 1995.

Finland

REPUBLIC OF FINLAND

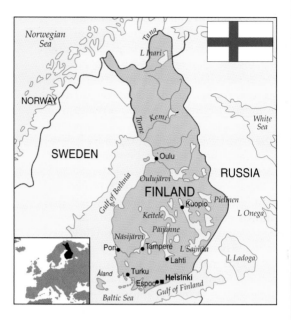

FINLAND'S CULTURE AND LANGUAGE ARE quite distinct from those of the other Nordic Countries. For long periods of its history it was ruled by Sweden and Russia, but neither was able to subdue the Finnish spirit.

GEOGRAPHY

Finland is the sixth largest country in Europe and one of the most northerly. Its climate and topography reflect the fact that one-third of the country's area lies within the Arctic Circle.

NATIONAL DATA – FINLAND

Land area	338,145 sq km (130,559 sq mi)			

Climate	Altitude m (ft)	Temperatures January °C(°F)	July °C(°F)	Annual precipitation mm (in)
Helsinki	51 (167)	–6 (21)	17 (63)	635.4 (25)

Major physical features highest point: Haltiatunturi (northern Finland) 1,328 m (4,357 ft); longest river: Kemi 548 km (343 mi); largest lake: Lake Saimaa 1,147 sq km (443 sq mi)

Population (2006 est.) 5,231,372

Form of government multiparty republic with one legislative house

Armed forces army 1,920; navy 5,000; air force 2,800

Largest cities Helsinki (capital - 559,046); Espoo (231,704); Tampere (202,932)

Official language Finnish, Swedish

Ethnic composition Finn 93.4%; Swede 5.7%; Russian 0.4%; Estonian 0.2%; Roma 0.2%; Sami 0.1%

Religious affiliations Lutheran National Church 84.2%; Greek Orthodox in Finland 1.1%; other Christian 1.1%; other 0.1%; none 13.5%

Currency 1 euro (EUR) = 100 euro cents

Gross domestic product (2006) U.S. $171.7 billion

Gross domestic product per capita (2006) U.S. $32,800

Life expectancy at birth male 74.99 yr; female 82.17 yr

Major resources timber, iron ore, copper, lead, zinc, chromite, nickel, gold, silver, limestone, cereals, dairy products, fish, livestock, potatoes

"Land of lakes and marshes" is the literal meaning of Suomi, the name the Finns give their country, and this view of the Lapland landscape in summer shows how well it fits the description. Birch trees flourish on the peaty soils of this part of the country.

The land

Finland is landlocked to the north and east; to the south lies the Gulf of Finland and to the west the Gulf of Bothnia. The Åland Islands, lying at the entrance to the Gulf a few miles off the southwestern coast of Finland, are an extension of the coastal lowland. Some 80 of the 3,000 islands in the archipelago are inhabited.

Apart from a small hilly area in the northwest, Finland is lowlying, its ancient granite rocks ground down by the glacial icesheets. The ice remained until only a few thousand years ago, leaving an undulating landscape dotted with lakes. The Finnish name for the country, Suomi, means "land of lakes and marshes". There are over 60,000 lakes covering at least one-tenth of its area, and they are linked by extensive river systems. Lake Saimaa, the largest, covers 4,400 sq km (1,700 sq mi). Most of the lakes, however, extend no more than 25 sq km (10 sq mi). The longest rivers are the Kemi and the Oulu, both of which drain west into the northern end of the Gulf of Bothnia.

Climate

Separated from the Atlantic by the Scandinavian peninsula, Finland's climate is influenced by the continental weather system. Above the Arctic Circle the weather is typically extreme; very cold winters are matched by short summers in which temperatures often soar during the long days. Farther south the climate is moderated by the proximity of the sea. The mild summers allow Finland's agricultural area to extend farther north than in the other Nordic countries. However, from mid November until April the whole country is snow-covered and the sea is frozen over.

Plants and animals

Most of the country is heavily forested, chiefly with pine, spruce and other conifers. Hardy broadleaf trees grow in the south. In the extreme north the forests give way to swampy tundra, a landscape of few trees but rich in lichens and berry-producing plants, especially cloudberries.

The woodlands are home to many mammal species, including bears, wolves and lynx; elk and reindeer are relatively common. The rivers are rich in salmon and trout, and they also support ducks and other waterfowl.

As elsewhere in the Nordic Countries, the forests, lakes and rivers show signs of environmental damage from acid rain, caused by windblown air pollution originating outside the region. In addition, radiation from the Chernobyl nuclear disaster in the Ukraine in 1986 still affects Sami (Lapp) reindeer herds.

SOCIETY

Although Finland has been independent only since the early 20th century, its people retain a proud and enduring sense of national identity.

History

Finland was first inhabited by neolithic nomadic hunter–gatherers, probably ancestors of the present-day Sami. Later settlers came from around the Baltic area, speaking a Finno-Ugaric language that is closely related to both Estonian and Lithuanian but is completely unrelated to the Scandinavian languages, which are Indo-European in origin.

Finland remained outside the cultural orbit of Europe until the 12th century, when the Swedish kings launched a crusade to subdue and convert its pagan inhabitants. The Swedes' English-born bishop, Henry, murdered by a Finnish peasant, later became Finland's patron saint. Between the 13th and 19th centuries Finland was effectively a territory of Sweden, and its culture and education were entirely Swedish-dominated. Swedish became the language of government and the aristocracy.

Russia, however, continually eroded parts of Finnish territory, and in 1809, when Sweden failed to defend Finland, the Finns accepted an offer by Tsar Alexander I (1777–1825) to make the country an autonomous grand duchy. Finland enjoyed considerable political freedom, and prospered under the favorable rule of Alexander II (1818–81). He encouraged wider use of Finnish, which had survived only as the language of the working people; this was partly to weaken links with Sweden, and partly to foster a sense of nationalism and new belief in a Finnish cultural identity. Harsh moves by his successor, Nicholas II (1868–1918), to Russify the country only encouraged a growing nationalist movement that successfully united all levels of society.

At the outbreak of the Russian revolution (1917–18) Finland seized the opportunity to declare its independence; a communist coup was defeated and a republic established in 1919. The Åland islanders in 1917 sought to become part of Sweden, with which they had long economic and cultural links dating back to the 12th century. Finland granted them autonomy in 1920, but refused to allow them to secede; today they remain an autonomous part of the state of Finland.

In 1939 the Soviet Union took advantage of its pact with Nazi Germany to seize much of Karelia in eastern Finland. With the Allies unable to help, Finland had to turn to Germany, for which the Soviet Union imposed heavy postwar reparations. During the Cold War era Finland managed to steer a delicate course, maintaining its democratic tradition and mixed economy without offending the Soviet Union. The term "Finlandization" came to describe the situation of a small, neutral country coexisting beside a communist superpower.

Government

The president is elected for six years by popular vote with a second round of voting if there is no absolute majority. The president appoints the administration, the prime minister and cabinet. Legislative authority is vested in a single-chamber parliament, the Eduskunta with 200 directly elected members. The autonomous province of the Åland Islands has its own elected congress and land councillor, or prime minister, but the provincial governor, appointed by the Finnish government, has the right to veto the decisions made by congress.

Proportional representation in Finland has allowed a mass of major and minor parties to develop, and most governments are center-left coalitions. The largest parties are the Social Democrats, the Center Party and the National Coalition Party; the once-popular Communist Party has suffered from declining membership.

People

Finland is more linguistically complex than the other Nordic Countries. In most of the country the Finnish-speaking majority is dominant, including regional groups with their own dialects, such as the Tavastlanders. Most of the country's Sami inhabitants, who speak their own distinct language and now number only about 2,500, live as seminomadic reindeer herders in the north. In the coastal south and the Åland Islands there is a Swedish-speaking minority. Swedish is still an official language alongside Finnish, but is in declining use.

The Lutheran Church is the national church, to which about 90 percent of the population belong. A tiny proportion (about 1 percent) belong to the Finnish Orthodox church.

Finland's international cultural contribution belies its size and its linguistic isolation. The arts, particularly the music of Jean Sibelius (1865–1957) and the folk-epic *Kalevala* compiled by Elias Lönnrot (1802–84), played an important part in establishing Finland's distinctive identity and, latterly, its nationhood. In the field of town-planning, the architects Alvar Aalto (1898–1976) and Eliel and Eero Saarinen (1873–1950 and 1910–61) have been influential in shaping the modern cityscape. Finnish design skills have become world-famous, particularly in glasswork and the bright Marimekko fabrics. Also internationally famous, particularly with younger readers, are the novels of Tove Jansson (b. 1914) about the Moomintroll family and their adventures.

ECONOMY

Finland enjoys its share of the Nordic Countries' prosperity, and its national income is currently growing at a much faster rate than its population.

Agriculture and forestry

Timber is Finland's main natural resource, and the country's forestry and lumber industries are among the most important in Europe. Agriculture is confined to the south by climate; only about 9 percent of the land is cultivated. Most farmers own smallholdings and combine agriculture with forestry. Cereals are grown, mainly as fodder, with potatoes and sugar beet and, in the extreme south, rapeseed and some fruit. Livestock production traditionally concentrated on the raising of dairy cattle, but this has declined significantly; there is also some pig and poultry farming.

Industry

Industry – primarily manufacturing – accounts for one-quarter of the country's income. Mining industries are small but significant, including important European supplies of copper, nickel and vanadium, and other rare metals such as zinc, chromium and titanium. Iron ore is no longer mined.

Manufacturing growth after World War II was spurred on both by the Soviet Union's demands for reparations and by favorable trade agreements soon after, whereby Russian oil was exchanged for Finnish manufactured goods. This was a considerable advantage for Finland, as the country was largely dependent on peat for its energy. One-third of the country's energy requirements is derived from hydroelectric plants, and more than one-third from nuclear plants.

Heavy industries include specialized shipbuilding, particularly of icebreakers, and the manufacture of machinery and cement, together with building and construction, chemicals and petroleum refining. Important lighter industries include electronics, textiles, printing and publishing, and all types of timber products, including wood and paper. Tourism is increasingly important. Most major industries have some degree of state ownership or government investment.

Floating assets Forestry is the mainstay of the Finnish economy. The cut logs are floated down the rivers to huge timber mills on the coast. Wood products, including paper, wood pulp, board and plywoods, account for over one-third of exports.

Trade and commerce

Following a referendum in favor, Finland became a member of the EU in 1995.

Timber related products, such as paper, dominate exports, while petroleum, machinery and transport equipment are the major imports.

Transportation and communications

The road network is extensive, though only about half the roads are surfaced. The rail network was privatized in 1995. It is extensively developed in the southwest, and is undergoing modernization and electrification elsewhere. Air services operate between various domestic airports all around the country, and internationally via Malmi airport at Helsinki. The inland waterway network of lakes and rivers is an important transportation route for the timber industry, and its scenic beauty also attracts tourists.

Health and welfare

Finland's state welfare system offers a wide range of benefits, including some subsidized housing in the towns to counter an acute national shortage. The population is aging as life expectancy increases under Finland's excellent health service. Elderly people enjoy good pensions and generous care provisions. The nine-year comprehensive education system is free and compulsory.

Finland Awakes!

Perhaps more than any other single person, it was the composer Jean (or Jan) Sibelius who helped to put his struggling new nation on the international map.

Sibelius was born in 1865 in the small town of Hämeenlinna in the south of Finland. Like most educated Finns at that time, his family were Swedish-speaking, but their forward-looking views led them to Finnicize their Swedish name, Sibbe, and Jean was sent to the Finnish-speaking grammar school. It was there that he first showed unusual talent for the violin, and composed his first piece at the age of 10. He was sent to university in Helsinki to study law, but soon abandoned this for music, going to Berlin and Vienna to study.

His time abroad reinforced his love of Finnish, as did his secret engagement to Aino Järnefelt, daughter of a fervently nationalist aristocratic family. Finnish influences became pervasive in his music. His first major work, the Kullervo Symphony, was a massive choral symphony based on a tragic legend from the *Kalevala* folk epic. First performed in 1892, it won him immediate recognition.

In 1899 Sibelius, by now beginning to establish an international reputation and devoting himself fulltime to composing, wrote the first of his seven symphonies. However, the political upheavals of that year were having an effect on Finnish society. Under Tsar Alexander II (1818–81), Finland had enjoyed freedoms unknown in Russia itself, but following the accession of Nicholas II (1868–1918) a Moscow-appointed governorship was imposed, and Russian was made the legal state language. Major Finnish newspapers were closed down, and protest meetings and political plays were banned. To find a way around the ban a series of allegorical tableaux was put on. Sibelius wrote the music for these tableaux and some of them, including the *Karelia Suite*, became the basis of his most popular works. For the last tableau he wrote *Finland Awakes!* – a strikingly dramatic finale.

Renaming the piece *Finlandia*, Sibelius redrafted the finale as a concert overture for the Helsinki Orchestra to play at the Paris World Fair in 1900, where Finland was given a pavilion of its own on a par with other countries. It is from this World Fair that Finland usually dates its acceptance as a nation among nations. The pavilion, designed by Finnish architects including the young Eliel Saarinen, and filled with original Finnish works of art and design, made an immense impression, as did *Finlandia*. Soon the piece was being played all over the world. Even today *Finlandia* is considered the country's unofficial national anthem.

Monument to a national hero The Sibelius monument in Helsinki honors the composer whose music captured the growing sense of Finnish identity. His *Finlandia* overture was performed in 1900, 17 years before Finnish independence.

Denmark

KINGDOM OF DENMARK

Dᴇɴᴍᴀʀᴋ, ᴛʜᴇ ѕᴍᴀʟʟᴇѕᴛ Sᴄᴀɴᴅɪɴᴀᴠɪᴀɴ country, enjoys possibly the highest standard of living in the European Union. Until recently it also had the highest income tax and foreign debt.

GEOGRAPHY

Mainland Denmark, the Jutland peninsula, extends 300km (200mi) north from the border with Germany. Dunes, lagoons

NATIONAL DATA - DENMARK

Land area	42,394 sq km (16,368 sq mi)			
Climate		**Temperatures**		**Annual**
	Altitude m (ft)	January °C(°F)	July °C(°F)	precipitation mm (in)
Copenhagen	9 (30)	0 (32)	17 (63)	587 (23)

Major physical features highest point: Yding Skovhøj (central Jutland) 173 m (568 ft); largest island: Sjaelland 7,016 sq km (2,709 sq mi)

Population (2006 est.) 5,450,661

Form of government multiparty constitutional monarchy with one legislative house

Armed forces army 12,500; navy 3,800; air force 4,200

Largest cities Copenhagen (capital – 1,086,762); Arhus (228,547); Odense (145,554); Alborg (121,549)

Official language Danish

Ethnic composition Danish 95.1%; other European 1.5%; Asian 1.3%; other 2.1%

Religious affiliations Evangelical Lutheran 95%; other Protestant and Roman Catholic 3%; Muslim 2%

Currency 1 Danish krone (DKK) = 100 øre

Gross domestic product (2006) U.S. $198.5 billion

Gross domestic product per capita (2006) U.S. $37,000

Life expectancy at birth male 75.49 yr; female 80.22 yr

Major resources petroleum, natural gas, fish, salt, limestone, chalk, stone, gravel, sand, electronics, textiles, cattle, cereals, coal, cryolite, fodder, iron ore, lead, molybdenum, natural gas, pigs, potatoes, poultry, uranium, zinc

and sandbars shelter the west coast from North Sea storms; the eastern plains face the Baltic sea and an archipelago of 483 islands, of which 97 are inhabited.

The land
Denmark averages only about 30m (100 ft) above sea level. Its landscapes are the product of ice age glaciation; glacial debris divides eastern and western Jutland. Denmark's largest lagoon, the Lim Fjord, cuts north across the country. In the extreme north, the long sandspit known as the Skaw points toward Sweden.

East of Jutland, across the Little Belt strait, is the island of Fyn and, to the southeast, the smaller island of Langeland. Across the Great Belt is the island of Sjaelland. Copenhagen, the capital city, is on its eastern coast. Suburbs now cover most of the coast from Copenhagen to Helsingor (Elsinore). Far out in the Baltic lies the granite island of Bornholm.

Climate, plants and animals
Denmark's cool and temperate climate is warmed by the North Atlantic Drift. Summers are warm, with high rainfall.

Conifer plantations have replaced the natural deciduous woods, and trees have been planted as windbreaks on the North Sea coast. Few large mammals have survived in the wild, but there are red deer in Jutland's forests and plantations, and a profusion of bird life. Pollution has affected marine and freshwater animal life.

SOCIETY

Progressive, but deeply interested in their past, the Danes have achieved a healthy and equitable society.

History
Reliable historical accounts begin in the 9th century, when the Danes resisted the northward expansion of the Franks. King

Early morning in Copenhagen Dawn brushes the bronze statue of Hans Christian Andersen's *Little Mermaid* in Copenhagen harbor, a symbol of Denmark's close association with the sea, as are the wharves and dockyards in the background.

Harald Bluetooth of Jutland (c.910–c.985) brought Christianity to his people. His son Sweyn Forkbeard (d.1014) conquered England, creating an empire that lasted for two generations.

Norway and Sweden were united under the crown of Denmark by the Kalmar Union (1396–1523), which tried to counter the commercial power of the German-based Hanseatic League. Civil war in 1534–36 resulted in the formation of the new Lutheran Church of Denmark, and many Catholic properties were seized. King Christian IV (1577–1648) tried and failed to reestablish Danish authority over the Baltic, and modern national borders were established by 1645. Norway, still subject to the Danish crown,

The Faroes – islands of sheep

The Faeroes are a group of 22 islands lying in the Atlantic between Scotland and Iceland, formed by basaltic lavas similar to those found in Iceland. Some 18 of the islands are inhabited. The characteristic landscape is harsh and rugged with high coastal cliffs and treacherous straits between islands.

The climate, with mild winters and cool summers, benefits from the warm North Atlantic Drift. Trees survive only in plantations that are sheltered from the frequent strong winds. Native wildlife consists mainly of seals and a rich variety of seabirds, including puffins and eider, valued for their feathers. There are no native land mammals: rats and mice were introduced by early settlers, and Viking colonizers in the 9th century brought sheep to the islands. An old proverb says that "sheep's wool is Faeroe gold", and the islands' name even means "sheep islands".

In about 1000 AD the Faeroese were converted to Christianity. The islands came under Norwegian rule in 1035, and passed to Denmark in 1380. From 1709 to 1856 a Danish royal trade monopoly held back economic development and impoverished the population. In the 19th century a growing nationalist movement led to the creation of the written Faeroese language, based on Old Norse. Demand for home rule strengthened, and in 1948 the Faeroes became a self-governing dependency of Denmark. Today the Faeroes have their own 32-member parliament, the Lagting, which elects the Landsstyri, an administrative committee. The Faeroese government has been pressing hard for independence since the early 1990s and wants to phase out Danish subsidies by 2012.

Nearly one-third of the inhabitants live in or around the main harbor and capital, Thorshavn. Faeroese is the main language, though Danish is still taught in all schools. The islands have their own radio broadcasting system and there is also a limited television service. Most people are members of the Danish Lutheran Church.

The islanders enjoy many distinct cultural traditions. Ballad-dances, less common than they used to be, take place between Christmas and Lent in the Christian calendar. Ballads concerning legend, folklore or Faeroese history are sung aloud to provide the music for a ring-dance. The wording of the ballads has remained unchanged from one generation to the next.

Woolen yarns and knitted garments are produced by the islanders, but today fishing and associated industries are far more important, providing some 95 percent of exports. Whaling, for meat and oil, has always been part of the Faeroese economy and continues today despite international concern over the depletion of whale stocks.

Less than one-twentieth of the land is cultivated, mostly producing grass for fodder and vegetables, especially potatoes. Coal is mined on Sudhuroy, but hydroelectric power meets all the islands' needs. There are regular interisland ferries and shipping communications with Denmark, Britain, Norway and Iceland throughout the year. In summer, passenger and vehicle ferries serve a small tourist industry. The airport on Vágar offers regular flights to Iceland and Norway and daily flights to Copenhagen.

Living in isolation A turf-covered wooden cabin on Strømø, the largest of the Faeroe Islands, is silhouetted against a silver, wind-ruffled sea. Fishing and farming communities are scattered around the islands' deeply indented coasts.

was relinquished in 1813 after Swedish forces defeated Denmark following the Napoleonic wars.

In 1864 war broke out with Prussia and Austria over long-disputed Schleswig and Holstein at the southern tip of Jutland. Denmark lost, but in 1920 a plebiscite returned Schleswig to Danish rule.

Although neutral in World War II, Denmark was occupied by German troops from 1940 and had an active resistance movement. As a member of the North Atlantic Treaty Organization (NATO) from 1949, Denmark refused to allow United States' air bases on Danish soil.

In 1973 Denmark joined the European Community (EC). In 1999 Denmark ratified the Treaty of Amsterdam, the EU's new constitution, setting the agenda for enlargement of the Union including relaxation of border controls.

Government
Denmark has been a constitutional monarchy since 1849. The monarch appoints the prime minister and the cabinet, who are accountable to the 179-member single-chamber parliament or Folketing.

Parliamentary elections based on proportional representation are held every four years. All citizens aged 18 and over may vote. Greenland and the Faeroes each have two representatives in the Folketing but follow Danish foreign affairs, defense, financial and legal policies.

Denmark's non-nuclear policy has led to debates over the presence of nuclear weapons on visiting warships.

People
The Danes are closely related to the other Scandinavian peoples, and there is a significant German minority. Both German and English are taught in Danish schools.

More than 90 percent of Danes belong to the national Lutheran church. The country has produced internationally renowned writers including the philosopher Soren Kierkegaard (1815–55) and Hans Christian Andersen (1805–75). The Royal Danish Ballet preserves 19th-century choreographic traditions.

ECONOMY

Denmark today is an industrial society, but agriculture still plays a vital role.

Agriculture
Farming accounts for some two-thirds of the land area, but occupies only one-sixteenth of the labor force. Small farms and agricultural cooperatives form an efficient system of intensive farming, emphasizing livestock and dairy products. The major crop is barley. About half the country's farm produce is exported, mostly to the EU. The rest supplies Denmark's food-processing industry, its second-largest employer. Denmark's fishing industry is among the world's top 15; herring, cod and flatfish account for half of the total catch.

ing in the 1980s caused a severe balance-of-payments problem. The government responded with tax cuts and reductions in welfare expenditure.

Transportation and communications

The sea has always been central to Denmark's main lines of communication. In addition to sea routes there are about 420km (260 mi) of inland waterways. Two bridges connect Jutland with Fyn, and in 1989 a massive project was launched to make a permanent link between Fyn and Sjaelland, formerly accessible only by ferry. The Øresund bridge connecting Copenhagen with Malmö, Sweden opened in July 2000. Road and railroad networks are extensive and well maintained.

Denmark has 13 airports. Kastrup, near Copenhagen, serves Jutland and Bornholm as well as international traffic, and is a major base for the co-owned SAS (Scandinavian Airlines System).

Until the 1980s Danish broadcasting was controlled by Danmarks Radio, an independent public body funded by license fees. A second television channel with commercial advertising was launch-ed in 1988, and Danes can receive broadcasts from Sweden, Norway and Germany. There are about 60 newspapers.

Health and welfare

Denmark has one of the oldest and most extensive welfare programs in the world. It offers universal benefits including pensions and free health checks for children. Compulsory health insurance provides free hospital treatment and home nursing. Unemployment insurance is voluntary but widespread.

Education is compulsory from age 7 to 16. Most schools are free; private schools receive state support. Youth schools offer both vocational and other training. The traditional folk high schools, widely imitated abroad, provide a broad-based curriculum of adult education. There are five universities and a number of professional institutions, commercial and technical colleges.

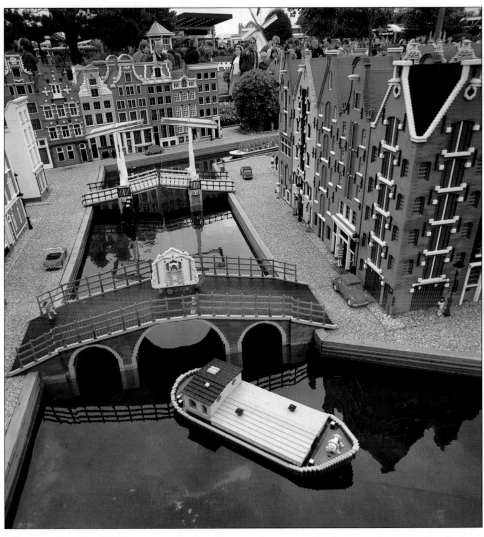

Model attractions A model of Copenhagen harbor on display at Legoland, a purpose-built park at Billund in central Jutland – home of Lego building blocks. The displays, made from 30 million blocks, attract tourists from all over Europe and beyond.

The Øresund Bridge (*above*) opened in July 2000 and links Scandinavia with the European continent via a 10-mile railway and freeway that runs between the Danish capital, Copenhagen, and the Swedish city, Malmö. A combined bridge and tunnel, its budget was $2 billion.

Industry

The country's mineral resources are heavily exploited. They include granite, boulder clay, white chalk and kaolin.

North Sea gas and oil supply about half of Denmark's energy. Surplus energy is sold to Germany and Sweden. Wind turbines are increasingly common in flat, windy Jutland, and newer houses may have solar or geothermal heating.

Industrial products include cement, crude steel, machinery and transportation equipment, processed foods, chemicals, paper and furniture. Shipping and tourism are also important.

Trade and commerce

Denmark trades almost entirely within Europe, particularly with Germany, Britain and Sweden. High consumer spend-

Iceland

REPUBLIC OF ICELAND

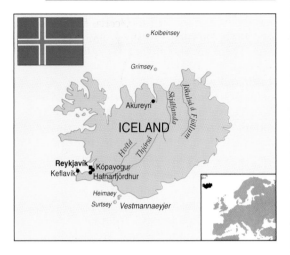

ICELAND IS AN ISLAND REPUBLIC JUST SOUTH OF the Arctic Circle. Its culture derives from the original 9th-century Viking settlers. The country and its people have grown up in a harsh, volcanically active environment unmatched anywhere else in Europe.

GEOGRAPHY

Geologically Iceland is very young. It has been built up by a series of volcanic eruptions from the mid-Atlantic ridge – the boundary of two of the Earth's tectonic plates. In 1973, the island of Heimaey off the south coast had to be evacuated after the eruption of the volcano Helgafell. Earthtremors are frequent, and hot springs are common in volcanic areas, most notably at Geysir in the southwest. Similar springs gave the capital its name – Reykjavik, or "Smoky Bay" – and today they provide most of the country's hot water and heating.

At the heart of the island is a high plateau partly covered by a glacier and surrounded by mountains. Glaciers account for over one-tenth of Iceland's total land area. The largest is Vatnajökull in the southeast, covering an area of 8,300 sq km (3,200 sq mi), which, in late 1996, split open due to volcanic activity beneath. Most of the population live on the narrow coastal plain around Reykjavik in the southwest. In the north steep valleys penetrate deep into the interior.

The northwest of the island tends to be cooler than the southeast, and snowfall there is more frequent. Temperatures are well above the average for this latitude, due to the influence of warm southwesterly winds from the tropics associated with the North Atlantic Drift, but gales are common, especially in the mild, humid winters. Summers are generally cool, and rainfall is high throughout the year. Aurora borealis, or the northern lights, can often be seen, especially in fall and early winter.

Woodland once covered large areas of the island, but has long since been cleared by settlers. The characteristic vegetation is mosses and lichens with a few shrubs. Sparse grassland is limited to valleys and the coastal plain.

Foxes are Iceland's only native mammal. Mice and rats came with the Vikings, as did domestic animals such as the sturdy Icelandic pony. Reindeer were introduced in the 18th century. Bird life is rich and varied, especially on sea cliffs and at Lake Mývatn in the north. Many migratory birds visit the island.

Water and rock A waterfall plummets down a cliff of basalt rock at Seljalands Foss in southern Iceland. The island's dramatic scenery includes lunar-like lava fields, vast glaciers and snowfields, active volcanoes and geysers spouting steam high into the air.

SOCIETY

The first settlers in Iceland were probably Irish monks. They were followed in about 870 AD by Vikings from Norway, Ireland and Scotland. The chieftains in each district set up a "thing" or local assembly to settle quarrels and feuds common among landholders, and about 930 a national assembly, the Althing, was established – the basis of Europe's oldest parliament. It was through a decision reached in the Althing that the population accepted Christianity in 999–1000.

During the 13th century Iceland's chieftains accepted Norwegian rule in order to preserve domestic peace, but in 1380 Norway itself fell under Danish sovereignty, and Iceland entered a long period of decline, exacerbated by the Danish crown's commercial monopoly, which crippled trade.

In the 19th century the demand for independence and constitutional reform grew. Trading restrictions were lifted, and the Althing was granted a degree of legislative authority. Conditions remained harsh, however, and there was massive emigration to Canada until the development of the fishing industry brought increased prosperity at the beginning of the 20th century; the island achieved full self-government in 1918.

During World War II British and American forces used Iceland as a base. The German invasion of Denmark in 1940 had severed communications with the island, and in 1944 the Althing cut the last ties with Denmark by declaring Iceland an independent republic.

Since 1951 there has been an American military base at Keflavík; this, and Iceland's membership of the North Atlantic Treaty Organization (NATO), have been hotly debated political issues. Fishery protection is another major concern. In 1975 the government declared a 370-km (200-mi) fishing limit – a decision that led to one of several "cod wars" with the United Kingdom. However, in 1976 the British government recognized the new international limit.

Icelanders are proud of their Viking roots, and many can trace their ancestry back to the earliest settlers, whose names are recorded in the *Landnama Bok*, or Book of Settlement. The modern Icelandic language is directly descended from Old Norse, the language of the Vikings. The medieval sagas, which mingle imagination and folklore with historical facts about the early settlers, can be easily read by modern Icelanders. Most of the islanders belong to the Lutheran Church, which was established in Iceland in the 16th century.

ECONOMY

Fishing is the cornerstone of Iceland's economy, accounting for three-quarters of all exports. Much of the fishing fleet is government-owned. Attempts are being made to tackle the problems of over-

The sun at midnight illuminates the harbor of Höfn, in southeastern Iceland. Fishing, once a smallscale family activity, is now big business and the huge trawlers and processing plants are owned by a few large companies or cooperatives.

fishing and declining fish stocks of cod, capelin and herring. The Marine Research Institute regulates operations with a system of quotas and bans. Fish farming is proving increasingly successful. Concern over loss of fishing rights is a major reason for Iceland's reluctance to join the European Union.

Until the beginning of this century sheep farming was the main occupation of all Icelanders. There are still some 4,500 farms, but only about 1 percent of the total land area is cultivated, producing fodder crops, potatoes and turnips. Grazing for livestock (sheep, cattle and ponies) occupies about 20 percent of the land. Fresh vegetables and flowers are available all year round from greenhouses heated by geothermal power, a cheap and abundant energy source.

Several major industrial installations are also run on geothermal power. Minor industries include fish processing, skins and furs. In the 1990s Iceland has diversified into software production, biotechnology and financial services. Tourism is also growing steadily, especially in whale-watching, ecotourism and cycle touring.

Iceland has about 11,000 km (7,000 mi) of roads, but many of them are impassable in winter. There are no railroads on the island and in many places Icelandic ponies still provide an important means of local transportation. Icelandair operates domestic and international flights from Reykjavik. The merchant fleet carries up

to half of Iceland's imports and more than half of its exports.

Icelanders are among the healthiest people in the world. Compulsory health insurance allows many medical services to be offered at minimal cost, and hospital care is free. All education is also free, and compulsory from the ages of 7 to 16. Further education is available and there is a university at Reykjavik.

NATIONAL DATA – ICELAND				
Land area 100,250 sq km (38,707 sq mi)				
Climate		Temperatures		Annual
	Altitude m (ft)	January °C(°F)	July °C(°F)	precipitation mm (in)
Reykjavik	52 (170)	0 (32)	11 (52)	817 (32)
Major physical features highest point: Hvannadalshnúkur 2,119 m (6,952 ft); longest river: Thjórsá 230 km (143 mi)				
Population (2006 est.) 299,388				
Form of government multiparty republic with one legislative house				
Armed forces paramilitary 130				
Capital city Reykjavik (184,244)				
Official language Icelandic				
Ethnic composition Icelandic 96.4%; other Europeans 2.5%; Asian 0.7%; others 0.4%				
Religious affiliations Lutheran 85.5%; Reykjavik Free Church 2.1%; Roman Catholic 2%; Hafnarfjorour Free Church 1.5%; other Christian 2.7%; other or unspecified 3.8%; unaffiliated 2.4%				
Currency 1 Icelandic króna (ISK) = 100 aurar				
Gross domestic product (2006) U.S. $11.4 billion				
Gross domestic product per capita (2006) U.S. $38,100				
Life expectancy at birth male 78.23 yr; female 82.48 yr				
Major resources cattle, dairy products, diatomite, fish, fodder, hay, hydroelectric and geothermal power, potatoes, poultry, sheep				

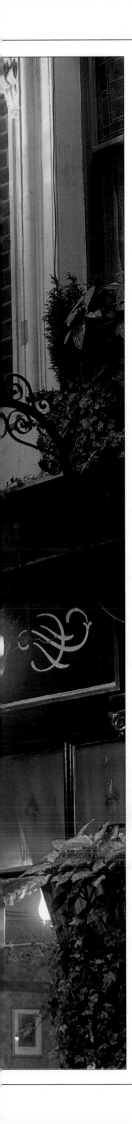

The British Isles

COUNTRIES IN THE REGION

UNITED KINGDOM · IRELAND

The traditional English pub (*left*), or public house, is found on countless street corners in towns and villages throughout the region – the meeting place of people from all walks of life, both young and old. *The Sherlock Holmes*, in London, named after the celebrated 19th-century fictional detective, is a typical example.

capital city
major town

height of land (meters)

1000
500
200
0 (sea level)

mountain peak

Unst
Yell
Shetland Islands
Mainland
Lerwick

Sanday
Mainland
Kirkwall
Hoy
Orkney Islands
John O'Groats

Butt of Lewis
Lewis
Outer Hebrides
Harris
N Uist
Skye
S Uist
Barra
Rhum
Inner Hebrides
Mull
Jura
Islay

North West Highlands
Moray Firth
Inverness
Loch Ness
Cairngorm Mts
Ben Nevis ▲ 1344
Grampian Mountains
Perth
Aberdeen
Dundee
Firth of Tay
Firth of Forth
Edinburgh
Glasgow
Clyde
Irvine
Arran
Southern Uplands
Tweed
Cheviot Hills

ATLANTIC OCEAN

North Sea

Malin Head
Londonderry
Mts of Antrim
Donegal Bay
Lough Erne
Lough Neagh
Belfast
Sligo
Mourne Mts
Achill I
Lough Mask
Lough Corrib
Lough Ree
Galway Bay
Galway
Shannon
Aran Is
Burren
IRELAND
Lough Derg
Wicklow Mts
Barrow
Limerick
Suir
Wexford
Waterford
Macgillycuddy's Reeks
Cork

Dublin

Dumfries
Newcastle upon Tyne
Cumbrian Mts
▲ Scafell Pike 977
North York Moors
Middlesbrough
UNITED KINGDOM
Pennines
Ouse
York
Hull
Leeds
Douglas
Isle of Man
Irish Sea
Anglesey
Manchester
Liverpool
Sheffield
Trent
The Wash
Snowdon ▲ 1085
Cambrian Mountains
Norwich
The Fens
Aberystwyth
Birmingham
Coventry
Severn
Gt Ouse
Ipswich
St George's Channel
Cotswolds
Oxford
Chiltern Hills
Brecon Beacons
Swansea
Newport
Cardiff
Bristol
Thames
London
North Downs
Bristol Channel
Mendip Hills
Exmoor
South Downs
Southampton
Brighton
Strait of Dover
Dartmoor
Exeter
Isle of Wight
Plymouth
English Channel
Land's End
Isles of Scilly

United Kingdom

THE UNITED KINGDOM OF GREAT BRITAIN AND NORTHERN IRELAND

T HE UNITED KINGDOM LIES OFF THE ATLANTIC northwest coast of Europe and consists of the union of England, Scotland and Wales (Great Britain) with the six counties of Northern Ireland.

GEOGRAPHY

A fairly small land area contains great diversity of landscape and geology. Much of southern and central England has gently rolling countryside, but northern England, Scotland, Wales and Northern Ireland have a high proportion of rugged terrain and mountains.

The land

Scotland and its offshore islands – the Outer Hebrides, the Orkneys and the Shetlands – are the northernmost part of Britain. The Scottish mainland divides into the Highlands in the north, the Lowlands in central Scotland, and the Southern Uplands to the south. The highest peak in the British isles is Ben Nevis in the western Scottish Highlands at 1,344 m (4,408 ft). The Cheviot Hills to the far south form a natural border with England.

In England, the Pennine chain rises south of the Cheviots and extends into central England. In the northwest are the Cumbrian mountains and the Lake District; in the northeast are the North Yorkshire moors. South of the Pennines, much of central England has rolling hills, while the east tends to be flat. Britain's longest river, the Thames, rises in the Cotswold hills in the south central region

The Channel Islands

The Channel Islands, which lie in the English Channel off the northwest coast of France, are a self-governing group of islands. They have been dependencies of the British crown since the Norman invasion of 1066, when they belonged to the Duchy of Normandy. There are four main islands: Jersey, Guernsey, Alderney and Sark. Jersey has its own constitution and a separate system of government.

During World War II the Channel Islands were the only British territory to be occupied by Nazi Germany. English is now the dominant language, though many people speak French, the official language of government on

Fertile arable land (*above*) is common in southern England. Five percent of Britain is now forested, with enclosed fields such as these being the rule. The use of traditional hedgerows is being revived to preserve wildlife habitat.

Natural harbors and rich farmland (*left*) characterize the county of Devon in the extreme southwest of England. Tourism and fishing are the major industries along the coast, and the area's mild winters make it a popular retirement area.

Jersey. The distinctive Jersey cattle are a main export, together with potatoes, flowers, fruit and a much diminished crop of greenhouse tomatoes. The Battle of Flowers is an annual tourist event that also helps to promote the island's successful cut-flower industry. Their southerly location has made the islands a popular resort area, while low taxes have attracted a number of very wealthy businesses to register there.

The vexed question of sovereignty over some of the very small islands was revived in the 1970s, when France and the United Kingdom disputed the right to explore for oil nearby.

Albert Harbor marina, St Helier, Jersey Lying only 19 km (12 mi) west of France, Jersey is a popular port of call for both British and French tourists, particularly yacht owners. Most place-names on the island are French

and flows southeast into the heart of London (the capital city) and out into the North Sea.

Wales, on the western border of England, is bounded by the Irish Sea to the north, St George's Channel to the west, and the Bristol Channel to the south. The Cambrian Mountains fill most of the center of the country. The second longest river in Britain, the Severn, rises in the Cambrian mountains and flows into the Bristol Channel.

Across the Irish Sea is the island of Ireland. The Republic of Ireland occupies some five-sixths of the land area, with the British province of Northern Ireland in the northeastern corner. North of Belfast, the provincial capital, are the Mountains of Antrim; one slope forms spectacular chalk and basalt cliffs facing Scotland's southwestern coast.

Climate

Despite the northerly latitude, the climate is generally mild. Warm air derived from

NATIONAL DATA - UNITED KINGDOM

Land area 241,590 sq km (93,278 sq mi)

Climate	Altitude m (ft)	Temperatures January °C(°F)	July °C(°F)	Annual precipitation mm (in)
London	7 (22)	4 (39)	17 (63)	611 (24)

Major physical features highest point: Ben Nevis 1,343 m (4,406 ft); lowest point: Holme Fen (The Fens) -4 m (-13 ft); longest river: Severn 290 km (180 mi)

Population (2006 est.) 60,609,153

Form of government multiparty constitutional monarchy with two legislative houses

Armed forces army 117,300; navy 40,700; air force 48,585

Largest cities London (capital - 7,172,091); Birmingham (989,141); Leeds (715,402); Glasgow (629,501); Sheffield (513,234); Edinburgh (452,194); Liverpool (439,473); Manchester (395,323); Bristol (380,615); Cardiff (305,353); Belfast (277,391)

Official language English

Ethnic composition White (English 83.6%; Scottish 8.6%; Welsh 4.9%; Northern Irish 2.9%) 92.1%; Black 2%; Indian 1.8%; Pakistani 1.3%; Mixed 1.2%; other 1.6%

Religious affiliations Christian (Anglican, Roman Catholic, Presbyterian, Methodist) 71.6%; Muslim 2.7%; Hindu 1%; other 1.1%; unspecified or none 23.6%

Currency 1 pound sterling (GBP) = 100 pence

Gross domestic product (2006) U.S. $1.903 trillion

Gross domestic product per capita (2006) U.S. $31,400

Life expectancy at birth male 76.09 yr; female 81.13 yr

Major resources coal, petroleum, natural gas, iron ore, lead, zinc, tin, limestone, salt, clay, chalk, gypsum, potash, silica sand, slate, barley, fruit, vegetables, fish, potatoes, sugar beet, wheat, tourism the North Atlantic Drift and from western Africa often combines with cold, wet air from Iceland and the Arctic, producing high winds and heavy rainfall.

There are strong regional variations. Western Scotland has heavy snow in the mountains. Northern Ireland is humid, but there is little snow in winter. Wales has cool summers and mild winters, with high rainfall in the uplands. Southern England has the highest summer temperatures and the widest annual range; winters are often cold and frosty.

Plants and animals

Over 5,000 years ago much of Britain was covered in dense deciduous woodland, but today hardly any survives. Moorland is common to Scotland, Northern Ireland and the uplands of northern England and Wales. Most valleys and woodlands have been cleared to make way for farmland. Commercial tree plantations are common.

Large mammals such as wolves were long ago hunted to extinction or died out as their habitats were destroyed. The largest surviving mammal is the Red deer, found in the Scottish Highlands. Smaller species include the Red fox, the badger, the otter and the stoat. Bird life is

The rugged mountains of Snowdonia (*right*) in northwestern Wales support slate quarrying, hydroelectricity generation and aluminum smelting, as well as attracting outdoor sports enthusiasts.

The Isle of Man

The Isle of Man lies in the Irish Sea midway between the northwest coast of England, the south coast of Scotland and the east coast of Northern Ireland. It is not part of the United Kingdom, but a crown possession that is virtually self-governing.

The island is about 48 km (30 mi) long by 16 km (10 mi) wide, with a rocky coastline enclosing a central highland area rising to 620 m (2,034 ft) at Snaefell. Low-lying land on either side produces cereals and potatoes.

A number of Neolithic monuments on the island indicates that there were settlers as early as 2000 to 1500 BC. The next wave of settlers were the Celts, and their language, Gaelic – known on the island as Manx – remained in daily use until the 19th century. In the 5th century AD the island was settled by Irish missionaries, and in the 9th cen-

tury it was invaded by Vikings from Norway. It remained a Norwegian dependency until it was sold to Scotland in 1266. It came under English control in the early 15th century.

Since 1828 the island has been a crown possession with a governor appointed by the monarch. It has its own two-chamber parliament, which dates from the 9th century. The legislative council, or upper house, and the elected lower house, the House of Keys, sometimes assemble together as the Court of Tynwald, but they normally operate as separate bodies.

The Isle's chief industry is tourism, and its Tourist Trophy (TT) motorcycle races attract many visitors. It is also famous for its tailless Manx cats.

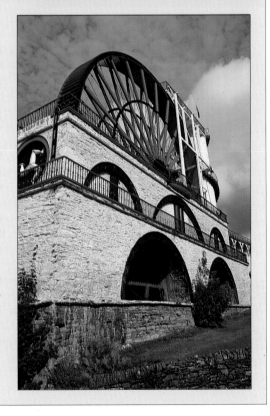

The Laxey Wheel (or Lady Isabella Wheel), on the Isle of Man, is the largest of its kind in the world. Built in 1854, it stands 22 m (72 ft) high, and was used to pump water from the Laxey iron ore mines in the Agneash Valley. Now it is a tourist attraction.

Angles, Saxons and Jutes (Germanic peoples from northern Europe). By the 7th century, when missionaries reintroduced Christianity, Anglo-Saxon kingdoms had emerged.

During the 8th and 9th centuries the northeast and east of England came under Viking rule from Norway and Denmark. Regained by the mid 10th century, England, now a single kingdom, was again invaded and ruled by the Danes from 1016–42.

In 1066 the English king Harold II (b. 1022) was killed in battle by William, Duke of Normandy, who claimed the English throne and established a royal dynasty. The new feudal structure of English society gave his barons considerable power. Royal power was limited in 1215 by the Magna Carta or Great Charter, which led to the establishment of Great Councils, later known as parliaments. The first recognizable parliament met during the reign of Edward I (1239–1307).

Ireland had been conquered in the late 12th century. In 1284 Wales was made subject to the English crown, but only became fully united with England in 1536. In 1314 Edward II (1284–1327) was forced to recognize Robert the Bruce (1274–1329) as king of Scotland. Territorial quarrels with France led to the Hundred Years War (1337–1453), and English forces were eventually expelled from all French land.

The Wars of the Roses (1455–85) saw a bloody internal struggle for the throne, ending with the accession of Henry VII (1457–1509), who founded the Tudor dynasty. Henry VIII (1491–1547) broke away from the Roman Catholic Church in 1533 and proclaimed himself Supreme Head of the Church of England. Under his daughter Elizabeth I (1533–1603), England enjoyed a golden age of prosperity, literature and exploration. The navy defeated a Spanish invasion by sea, and colonies were founded in North America.

Elizabeth's successor, James VI of Scotland (1566–1625), became James I of England and united the two kingdoms. During his reign thousands of Scottish and English Protestants settled on Catholic owned lands in northern Ireland, thereby sowing the seeds of conflict between Catholics and Protestants that has lasted to this day. James's son Charles I (1600–49) quarrelled with parliament over the extent of the monarch's power, leading to civil war and his execution in 1649. The parliamentary general Oliver Cromwell (1599–1658) was appointed Lord Protector of the Commonwealth, but two years after his death, Charles II (1630–85) was invited to return to the throne, though with reduced powers.

James II (1633–1701), a Catholic, alienated large numbers of his subjects and was forced into exile in 1688 (the

The huge stone circle at Stonehenge in central southern England is the country's most fascinating prehistoric puzzle. Built between 1800 and 1300 BC, it was possibly used for astronomical purposes and worship, but the religious practices of the people it served are not known.

rich and varied, and freshwater fish are plentiful.

SOCIETY

The British Isles were the birthplace of modern parliamentary democracy, of the Industrial Revolution and of the welfare state. An equally great contribution is the English language, spoken by some 350 million native speakers worldwide, and by another billion as a second language.

History

Britain's first settlers came from mainland Europe across an Ice Age land bridge that later disappeared. There is evidence of Neolithic settlement as early as 7000 BC and Bronze Age metalwork in about 1800 BC. Celtic peoples from Central Europe arrived in about 450 BC.

Rome occupied most of what is now Wales and England from the 1st to the 5th centuries AD, and introduced Christianity and the Latin language. As the empire collapsed, the Celtic Britons were driven into the far north and southwest of Wales and England by waves of settlers – the

Henry VIII (*above*) by the German artist Hans Holbein the Younger (1497–1543), English court painter from 1536. His astute portraits of Henry and his wives provide an excellent pictorial record of the Tudor court.

The Spanish Armada (*top right*) In 1588 Philip II of Spain attempted to invade England with an armada of 130 ships against England's 197 ships with more advanced firepower. The battle took only 100 English lives but cost Spain over half of its 27,000 troops.

The Great Screen at Winchester Cathedral (*left*) a magnificent Norman church built in the late 11th century in central southern England on the site of the Saxon church of St Swithun. The cathedral is the home of one of the most famous choral schools in Britain and is a center of excellence for church music.

"Glorious Revolution"). His Protestant daughter Mary II (1662–94) ascended the throne with her husband, the Dutch prince William of Orange (1650–1702). England, Wales and Scotland were formally united into the kingdom of Great Britain in 1707. Under a new dynasty, the Hanovers, a Scottish rebellion to restore the exiled Stuart line was defeated in 1715 and a second one bloodily suppressed at the Battle of Culloden in 1746.

In the 1760s Britain defeated France over rival claims in India and North America, becoming a leading colonial power. Explorations of Australia and New Zealand in 1770 by Captain Cook (1728–79) were the basis for later settlements. But in 1775, angered by taxes, the American colonies rebelled. Winning the resulting war, they became independent in 1783. Britain had better results against the emperor Napoleon I of France (1769–1821): in 1805 with a naval victory at Trafalgar off the Atlantic coast of Spain, and on land at the Battle of Waterloo in 1815 which finally defeated Napoleon.

Britain's rise as an overseas trading nation coincided with the Industrial Revolution, which, based on local iron

William Shakespeare

William Shakespeare (1564–1616) is regarded by many literary experts as the greatest writer of all time, and certainly the greatest dramatist and poet to have written in the English language. A genius with words and images, his poetry and plays explore the complete range of human emotions and conflicts. His versatility was extraordinary, with works ranging from histories to comedies, tragedies and romances.

Shakespeare was born and educated at Stratford-upon-Avon in the English Midlands, where he married Anne Hathaway (1556–1623) at the age of 18. Later he moved to London, where he became an actor with the city's leading theater company, and by 1584 he had emerged as a rising playwright.

The first play attributed to his authorship, based on the life of the English king Henry VI (1421–71), was produced in 1590–92. The precise sequence in which the rest of the plays were written and performed is hard to establish, though the early historical plays were written as a tribute to Queen Elizabeth I to support the Tudor claim to the throne. By 1597 he had earned enough money from his writings to buy New Place, one of the largest houses in Stratford, still standing today. In 1598 he became involved in the building and management of a new London theater, the Globe, whose foundations were rediscovered in 1989. Shakespeare's famous love sonnets, 154 in all, were mostly written between 1593 and 1596, but were not published until 1609.

In the early years of the 17th century, under the patronage of King James I, Shakespeare produced his great tragedies: *Hamlet*, *Othello*, *King Lear* and *Macbeth*, all profound explorations of human character.

Publication in 17th-century England was a chancy business. Copyright laws were virtually nonexistent, and copyists would often write down the text of a play during its performance and then publish it for their own profit. Shakespeare's name was well established, and some unscrupulous publishers applied it to the writings of others to increase their sales. The first collected edition of Shakespeare's plays, the *First Folio*, was published in 1623, just seven years after the playwright's death, and contained all but one of what we now take to be his plays.

In the 18th century scholars began to suggest that Shakespeare, the product of a provincial schooling, could not have written all the plays published under his name. A number of apparent cyphers and cryptograms in the plays were presented as evidence that the philosopher and essayist Francis Bacon (1561–1626) had written them. Modern

Shakespeare's original Globe Theater, built in Southwark, London, in 1599. It burned down in 1613 during a performance of "Henry VIII", was rebuilt in 1614, and was demolished in 1644 during the English Civil War. It has now been rebuilt.

scholars, however, have demonstrated that this theory cannot be supported. Although Shakespeare's surname occurs in a number of different spellings in various documents, this was not at all unusual at the time. In fact, the documentary evidence for Shakespeare's life is remarkably comprehensive for a man of his station in life at this period.

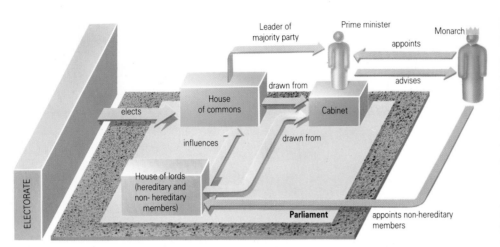

ore and coal deposits, established a new economic base in manufacturing. During the reign (1837–1901) of Queen Victoria (1819–1901) the British empire expanded to cover one-quarter of the world's land surface, and on its behalf the British army fought in various regions almost continuously from 1851 to 1902.

By the end of the 19th century reforms improved the welfare of the working class, while in Ireland there was a growing demand for Home Rule. In 1914 the United Kingdom entered World War I in alliance with France and Russia against Germany and Austria, suffering enormous casualties. Southern Ireland (later the Republic of Ireland) won independence in 1922 as the Irish Free State, but the province of Northern Ireland elected to remain part of the United Kingdom.

In 1931 the United Kingdom's self-governing overseas dominions of Australia, Canada, Ireland, New Zealand and South Africa became known as the Commonwealth of Nations. At home there was rising unemployment and economic depression well into the 1930s. London withstood heavy bombing by German aircraft in World War II; British casualties in the war were again very high. Rationing lasted into the 1950s.

A socialist Labor government elected immediately after the war created the first welfare state and placed major industries such as transportation, coal mining, electricity and gas under state ownership or control. Later Conservative governments, notably in the 1980s and 1990s, returned most of these to the private sector.

By the 1970s the British Commonwealth was almost entirely an association of independent states. The 1960s saw an increase in the conflict in Northern Ireland between the Protestant majority, who wished to remain in the United Kingdom, and the Catholic minority seeking union with the Republic of Ireland.

Britain's constitutional monarchy (*above*) Britain has no written constitution: the relationship between the monarch as head of state and the prime minister as head of government is based on rigid convention. The 650 members of the Commons are elected locally for a maximum five-year term by voters aged 18 or over. Executive power rests with the prime minister (leader of the majority party in the Commons) and a cabinet of ministers.

The Houses of Parliament (*right*) are built on the site of the 14th-century royal Palace of Westminster, which was destroyed by fire in 1834. The present buildings are the seat of government for the United Kingdom and house both the upper and lower chambers. Big Ben, one of London's most famous features, is the name of the clocktower's 13-tonne bell. It was named after Sir Benjamin Hall, the commissioner of works at the time of its installation.

Violence and terrorism escalated in the 1970s and 1980s, but a ceasefire came into force in 1994. In February 1996 the Irish Republican Army (IRA) renewed its bombing campaign. A fragile peace was brokered with multi-party talks held at Stormont in 1998. It was agreed that a new Northern Ireland Assembly would be set up, but the disarmament of the paramilitary forces is a major stumbling block.

The United Kingdom has a permanent seat on the Security Council of the United Nations, and is a founder member of the North Atlantic Treaty Organization (NATO). In 1973 it joined the European Community (EC). In 1982 it recaptured the Falkland Islands in the South Atlantic after an Argentinian invasion.

Government

The United Kingdom is a monarchy governed by parliamentary democracy but without a written constitution. The law is based on parliamentary statutes, on the interpretation of statutes in the law courts, and (except in Scotland) on common law. The parliament consists of the House of Commons and the House of Lords; members of the former are elected while the latter inherit their positions or are appointed. The monarch's right to

veto legislation has not been used since the 18th century. London became the first city to have an elected mayor in 2000.

At the local level, counties (regions in Scotland) and districts have their own elected councils. After referendums in 1998 the government agreed to set up a locally elected parliament in Scotland and assembly in Wales, with administration powers being transfered from London.

People

The English, Irish, Scottish and Welsh are all centuries-old mixes of peoples. The most numerous are the English. English is the main language, though Gaelic is still spoken in parts of Scotland, while in Wales the use of Welsh has increased since its introduction into local schools.

Since 1945 immigration from Commonwealth countries, notably the West Indies, India and Pakistan, created a multiracial and multicultural society. There are growing numbers of Muslims and other faiths. Among the 90 percent of the population that is nominally Christian, church attendance is low. Less than 10 percent of people now live in rural areas.

Animal pyres (*above*) such as this one in Cumbria, England were a familiar site in 2001. Drastic measures were taken to eradicate an epidemic of hoof and mouth which brought terrible losses to farming and tourism.

Lloyd's of London (*right*) A giant in the world of international insurance, the Corporation of Lloyd's moved into its dramatic modern premises in 1986 in the heart of London's financial district. All around it the major banking and financial services of the city are packed into the region known as the "Square Mile".

ECONOMY

Undermined by increasing competition in the 20th century, Britain's early industrial superiority was diminished by the breakup of its empire and the economic destruction of two world wars. Since 1945 the most significant developments have been the exploitation of North Sea oil and gas reserves; the changing patterns of trade within the European Union (EU); and the shift to a services economy.

Agriculture and fisheries
Agriculture supplies about half the demand for food but employs only 1 percent of the workforce. Eastern areas tend to be arable, western areas pastoral. Sheep, cattle, pigs and poultry are the main livestock, while wheat, barley, potatoes and sugar beet are the main crops. Farming is highly mechanized.

Fishing is also highly mechanized but is declining as a result of overfishing and foreign competition. There have been disputes with Iceland and Denmark over territorial waters. Fish farming, especially of salmon, is increasing. Productive woodland covers a little more than 5 percent of the country. Forest management is divided between the government-run Forestry Commission and private landowners.

Industry
Britain's limited mineral resources have been heavily depleted. By the 1990s nearly all coal mines in southern Wales –

once the mainstay of the economy – had been closed. Most of the coal consumed in Britain now comes from cheaper imports.

Natural gas first came ashore in 1962. North Sea oil followed in 1975, making Britain a net oil exporter. It is now self-sufficient in energy production. Nuclear power provides up to a quarter of the energy supply. Smallscale hydroelectricity is generated in Scotland.

The chief manufacturing products include steel, motor vehicles, machinery, electrical goods, aerospace equipment, leather, textiles and clothing. New indus-

tries such as chemicals and electronics have replaced declining traditional industries such as shipbuilding. Britain's wealth of historic sites, diverse landscapes, and numerous cultural events have encouraged the growth of the tourist industry. Britain remains one of the largest trading nations, but its trade imbalance has grown steadily.

Transportation
Canals and railroads were built in the 19th century to meet the urgent transportation needs of the Industrial Revolution.

Oxford and Cambridge

In termtime, the narrow cobbled lanes of Oxford and Cambridge – England's two most ancient universities – are crowded with students hurrying between colleges and lecture halls and libraries as they have done for the last 800 years. The modern universities of today evolved from corporations, or private societies, of students and masters that developed in Europe in the 11th and 12th centuries to provide a place of education and learning outside the narrow confines of the monastic and cathedral schools. The two oldest were at Bologna, in Italy, and Paris. Oxford, England's first such school of learning, was set up at the beginning of the 12th century after English students were banned from the University of Paris.

At first the university had no buildings of its own. Students and masters lived in lodgings in the town, and met in any suitable room or open space they could find. Disputes between "town" and "gown" were frequent and often very violent. Quarrels broke out between rival groups of scholars, and in 1209 a dissident group left to set up a sister university in Cambridge.

In the 13th century the first colleges in both Oxford and Cambridge were built, founded by wealthy patrons as places where masters and scholars could live and work as a self-contained community, in sets of rooms surrounding a quadrangle or court. Each college had a chapel for common worship, and a large hall where members could dine together every day.

About 1500 the practice began of admitting boys (often as young as 12 or 13) to courses of study. Then they were trained for careers in the church, the law, or the civil service. Today students are taught in every branch of learning from computer technology to Egyptology. The first women's colleges were established in 1871 (in Cambridge) and 1878–79 (in Oxford), though in Oxford they had to wait till 1920, and in Cambridge till 1947, to be admitted to full membership of the university. In the 1970s and 1980s most colleges became coeducational.

At both Oxford and Cambridge, college buildings of all periods from the 13th to the 20th centuries are scattered through and around a basically medieval town center. As a result

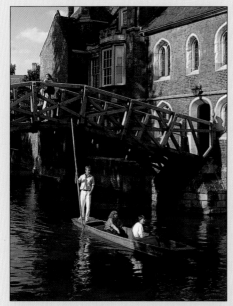

Punting on the river Cam, tourists float past historic Queens' College, Cambridge. Members of the university have included the internationally renowned scientists Isaac Newton (1642–1727) and Charles Darwin (1809–82).

the visitor can explore a superb display of architecture from every period, with an almost unparalleled richness of historical association. But anyone expecting to find a "university campus" among the cluster of streets and buildings will be disappointed.

Gateway to Europe A high speed train leaves Waterloo Station, London, bound for France via the Channel Tunnel. Although the Tunnel has improved links between Britain and Europe, it has also opened up another route for illegal immigrants looking for a better life in Britain.

The entire rail network was nationalized in 1947. Investment has been concentrated on mainline routes between cities, along with containerized freight traffic. An extensive network links the suburbs with central London. In the late 1990s the railroad system was privatized.

Since 1960 the motorway or expressway system has been extended and has become overburdened with traffic, especially in the heavily populated southeast. Yet government attempts to reduce car use have met with resistance. The bus network is well developed.

British Airways, a privatized company, has one of the world's most comprehensive international air networks. Other airlines operate from London's two major airports, Heathrow and Gatwick, and from regional airports.

A road and rail tunnel under the English Channel, under construction since the late 1980s, opened in 1994 linking England to France and the main European transportation network.

Communications and media

The telecommunications industry is also privatized. All British newspapers are privately owned, mostly by two big groups. There is no formal censorship. The British Broadcasting Corporation (BBC) operates as an independent public corporation controlling a substantial share of local and national radio and television. Wales has a number of Welsh language channels. The Independent Television Commission (ITC) licences and regulates commercial television and the Radio Authority, commercial radio. Television broadcasting is now set to provide up to 500 channels along with high speed Internet access.

Health and welfare

The national health service, introduced in 1948, offers free healthcare and hospital treatment. Eye tests, dental treatment and prescriptions are free for children, the elderly and those with low incomes. There are 2,500 national health hospitals and many private feepaying hospitals.

Education is both free and compulsory between the ages of 5 and 16. A national curriculum was introduced in the late 1980s. There are also many hundreds of fee-paying or public schools outside the state system. Almost two-thirds of students go on to some form of higher education.

Ireland

REPUBLIC OF IRELAND (EIRE)

THE REPUBLIC OF IRELAND OCCUPIES SOME five-sixths of the island of Ireland, about 80 km (50 mi) off the west coast of Great Britain. Ireland is renowned for natural beauty and turbulent history.

GEOGRAPHY

Ireland's lush green landscapes have earned it the title "the Emerald Isle."

The land
Ireland is surrounded by a broken rim of mountains. The central plain is low-lying – less than one-sixth of the country rises above 200 m (700 ft) – with hills and ridges protruding from the limestone layers beneath. Ice-age glaciers deposited fertile, rock-scattered soils. Small lakes, known as loughs, and bogs are common. The main rivers – the Liffey, the Blackwater, the Clare and the Shannon (the longest river in the British Isles) – are broad and slow moving.

Shorter, fast-flowing rivers fall to the coasts from the highland fringes. The largest ranges are the granite Wicklow Mountains in the east and Macgillicuddy's Reeks in the extreme southwest.

The coastline is most dramatic in the west, where it is rocky and deeply indented, with many spectacular cliffs. In the far southwest the mountains strike out into the Atlantic along a ragged row of broad peninsulas separated by deep inlets such as Dingle Bay and Bantry Bay.

Climate
The warm waters of the North Atlantic Drift create moderate temperatures, and mild Atlantic winds bring rain on two days out of three. Annual average rainfall ranges from 2,540 mm (100 in) in the mountains of the southwest to 762 mm (30 in) in the east. Summers are cool and

Lough Leane, Co. Kerry is the largest of a series of picturesque lakes in the far southwest. Lough Leane (or Lower Lake) covers 2,024 ha (5,001 acres) and encompasses 30 islands, one of which – Inisfallen – has a 9th-century abbey.

winters mild, with very little snow except in the mountains.

Plants and animals
Flourishing grasslands cover much of the country, but plant life is less rich than Britain's and there are fewer trees. Under one-twentieth of the country is forested; little forest has been restored even for commercial purposes. Peat bogs – habitats for many rare plants – carpet large areas of central Ireland. Long exploited for domestic fuel, the bogs are now threatened by the use of peat in commercial power stations and for export for garden and horticultural use. Along the west coast, where soils are poorer, heath and scrub vegetation predominate.

At the end of the last ice age Ireland was colonized by plants and animals migrating overland from Europe. However, many common European and British animals such as weasels, moles and snakes never reached the island, which

was cut off from Britain some 11,000 years ago when the glaciers melted and the sea level rose. According to folklore, St Patrick – the patron saint – banished the snakes.

SOCIETY

Ireland's history is one of turmoil, bloodshed, religious conflict and the fight for independence. Today it is a stable democracy with a strong literary tradition.

History
The earliest known occupants of Ireland date back to about 7000 BC, and Bronze Age artifacts from Ireland were traded all over Europe. In about 300 BC, Celtic ironworkers arrived from Europe and gradually became organized into five tribal groups or kingdoms.

The Irish Celts were unaffected by the Roman conquest of Britain in the 1st century AD. Converted to Christianity in the 5th century, they remained a center of faith despite the collapse of the Roman empire and several centuries of raids by Vikings, who established a fortified base at Dublin before they were finally de-

(1633–1701) was defeated at the Battle of the Boyne by the Protestant forces of King William III (1650–1702). The English government imposed civil disabilities and land confiscations (known as the penal laws) on Irish Catholics, but uprisings continued. In 1801 an Act of Union abolished Ireland's parliament, integrating it with that of Britain.

Under English rule, Ireland was divided into large estates owned by absentee landlords and farmed by local Irish peasants. From 1846 to 1851 the potato crop – the Irish staple food – was destroyed by a fungal disease, leading to the Great Famine. More than a million people died; many more emigrated to the United States, halving the population in Ireland.

From about 1870, the movement for Irish Home Rule gathered momentum, led in the British House of Commons by the Irish Member of Parliament Charles Stewart Parnell (1846–91). Home rule was finally granted in 1914, though violently opposed by Ulster's majority Protestant population. World War I delayed its implementation, fueling discontent. In 1916 an armed insurrection was suppressed by the British army; 14 leaders were executed. Eamon de Valera (1882–1975), a survivor, formed the Sinn Fein party and in 1919 set up a provisional Irish government. After a bitter civil war, Britain agreed to partition the island. The six northeastern counties became Northern Ireland, a province of the United Kingdom. The self-governing Irish Free State was created in the south.

In 1937 the Irish Free State renamed itself Eire and adopted a new constitution with an elected president, severing its links with the British crown. In 1949, on leaving the British Commonwealth, it declared itself an independent republic. Economic development was slow, and large numbers continued to emigrate in search of work, particularly to Britain. In 1973 Ireland joined the European Community (EC) now the European Union (EU).

In the 1960s the fragile stability in Northern Ireland began to break down as the minority Catholic population began campaigning for civil rights. In the 1970s the situation became increasingly violent, British troops were sent onto the streets and direct rule from Britain was imposed. Various attempts at reconciliation floundered during the 1980s and it was not until 1998 that the Good Friday Agreement set up a new constitution for Northern Ireland that was ratified by referendum held in both southern and Northern Ireland. In 1999 the Republic of

A Celtic cross at Drumcliffe in the northwest of Ireland, erected by early Christians in the 10th century on the site of a former monastery. The Anglo–Irish poet and mythologist W. B. Yeats (1865–1939) is buried in Drumcliffe churchyard.

feated by Brian Boru (941–1014), the first king of all Ireland, in 1014.

In 1172 Ireland was invaded by Anglo-Norman knights sent by England's king Henry II (1133–89). His sovereignty was recognized by the Church and many of the Irish knights. Over the next 300 years the Anglo–Norman barons became assimilated and Ireland effectively came to be ruled by the earls of Kildare.

England's break with the Roman Catholic Church in 1533 under King Henry VIII (1491–1547) increased the resentment of the Irish, who remained firmly Catholic. Repeated uprisings took place, often aided by Catholic Spanish or French troops. Many Irish Catholics emigrated to Europe. When the earls of Ulster in the north chose this course, James I of England (1566–1625) encouraged Scottish Protestants to settle on the vacated lands. In 1641 an uprising among the Irish of Ulster led to the massacre of thousands of colonists. An equally savage massacre of the Irish was at the hands of Parliamentary forces fighting in Ireland under Oliver Cromwell (1599–1658).

In 1690 an uprising led by England's deposed Roman Catholic king James II

NATIONAL DATA – IRELAND

Land area	68,890 sq km (26,599 sq mi)			

Climate		Temperatures	Annual	
	Altitude m (ft)	January °C(°F)	July °C(°F)	precipitation mm (in)
Dublin	81 (266)	5 (41)	15 (59)	733 (28.8)

Major physical features highest point: Carrauntoohill (Macgillicuddy's Reeks) 1,041 m (3,414 ft); longest river: Shannon 370 km (230 mi)

Population (2006 est.) 4,062,235

Form of government multiparty republic with two legislative houses

Armed forces army 8,500; navy 1,100; air force 860

Largest cities Dublin (capital - 1,186,159); Cork (119,143); Galway (65,832); Limerick (54,023); Waterford (44,594)

Official language Irish, English

Ethnic composition Irish 95%; British 2.7%; others 2.3%

Religious affiliations Roman Catholic 88.4%; Church of Ireland 3%; other Christian 1.6%; other 1.5%; unspecified 2%; none 3.5%

Currency 1 Euro (EUR) = 100 euro cents

Gross domestic product (2006) U.S. $177.2 billion

Gross domestic product per capita (2006) U.S. $43,600

Life expectancy at birth male 75.11 yr; female 80.52 yr

Major resources natural gas, peat, copper, lead, zinc, silver, barite, gypsum, limestone, dolomite, barytes, cereals, fish, livestock, potatoes, sugar beet, vegetables, tourism

Ireland gave up its territorial claims to the whole island and in 2000 the Northern Ireland Assembly was established.

Government

By its constitution, adopted in 1937, Eire is a parliamentary democracy. The head of state is the president, who is elected for up to two seven-year terms. The National Parliament or Oireachtas consists of the Dail, whose 166 members are appointed by the Taoiseach (prime minister) or elected by various public bodies. The Dail initiates all legislation, the Seanad can revise or delay its enactment and the President may refer proposed legislation to the Supreme Court if it might infringe the Constitution. Voting in parliamentary elections is through a system of proportional representation.

People

The distinction between English-speaking and Celtic-speaking Irish largely vanished as a result of the mass emigrations of the 19th century. Although Irish Gaelic is taught in all schools, it is now the first language only in some rural areas. The Irish are noted for their love of good talk, and Ireland has produced many famous writers, such as Jonathan Swift (1667–1745), Oliver Goldsmith (1730–74), Oscar Wilde (1854–1900), George Bernard Shaw (1856–1950), William Butler Yeats (1865–1939), James Joyce (1882–1841) and Samuel Beckett (1906–1991).

The story of the Irish people has been one of continued emigration. However, Ireland's economic renaissance in the late 20th century led to net immigration for the first time and in 2000 Ireland found itself in the strange position of needing to attract thousands of immigrants to deal with skills shortages.

The vast majority – over 90 percent – are Roman Catholic; other groups include Church of Ireland Episcopalians, Presbyterians, Methodists and Jews. The influence of the church remains strong over social issues such as divorce, abortion and birth control, though the 1990s have seen a demand for greater flexibility on these moral issues. Freedom of conscience and religious practice are guaranteed under the Irish constitution. In 1995 a referendum was narrowly in favor of divorce.

ECONOMY

Ireland both encourages private initiative and undertakes the administration of essential services and development.

Agriculture

Agriculture still accounts for employment among ten percent of the workforce. Most Irish farms are owner-occupied smallholdings. The limited area of arable land is highly productive. Cereals, sugar beet and potatoes are the main crops. Large areas of rich grassland support livestock almost all the year round. Cattle and pigs are kept in great numbers, while sheep graze on the upland pastures. Meat, milk and dairy products account for nearly one-sixth of Irish exports.

Forestry is limited, and concentrates mostly on fast-growing conifers. Sea fisheries were modernized with government aid in the 1970s. Although potentially rich, they are subject to overfishing controls. Salmon farming in the country's rivers, a potential source of earnings, has recently been endangered by pollution.

A son of Dublin

Every year on "Bloomsday" (June 16th) visitors and residents in Dublin retrace the footsteps of the city's best-known fictional character, Leopold Bloom. He was the creation of another Dubliner, James Joyce (1882–1941), a man who dropped out of school for two years at the age of nine due to his family's financial difficulties, but went on to university and published his first critical essay at the age of 18. Subsequently he became one of the most influential writers in the history of modern English literature.

Joyce was a quiet, mild-mannered, but deeply unconventional man who was plagued by poor health, including extremely poor sight, and a precarious financial situation. Given a scholarship to a Jesuit grammar school as a teenager, he performed well academically but lost favor with his teachers as he questioned his Roman Catholic faith. In 1904 he met Nora Barnacle, with whom he lived for the rest of his life. He refused to marry her until 1931, when he agreed to his daughter's wish that her parents should get married.

Joyce spent his adult life abroad, chiefly in France and Switzerland, but wrote exclusively about the people of Dublin, whose hospitality and poetic speech he loved as much as he hated their rigid, oppressed and oppressive social institutions. *Dubliners*, published in 1914, was the title of his first work of fiction; the publication date was delayed because the original publisher feared a libel suit and destroyed the printed galleys.

Ulysses, his best-known novel, first appeared in serial form in New York in 1917, but several episodes were confiscated and burned by the New York post office in 1919–20 on the grounds of obscenity. In 1921 the New York Society for the Prevention of Vice took Joyce's publishers to court and won a ban on further publication. The decision was finally reversed by a higher court in 1936, and a complete edition appeared that year. But Ulysses had long since assured James Joyce's international reputation. The novel records a day in Dublin – 16 June 1904 – with incidents based on scenes from Homer's ancient Greek epic, the *Odyssey*, and from Joyce's own life.

James Joyce is probably Dublin's most famous citizen, yet he wrote much of his work in exile in Zurich or Trieste, and was banned from publication at home. Ireland is now a writer's tax haven and home to an international literary set.

The date was deliberately chosen by Joyce in remembrance of the day he first met Nora Barnacle. Now it is celebrated by Joyce's admirers all over the world.

A **horsefair** (*above*) in a village southwest of Cork. The Irish thoroughbred is prized throughout the racing world and horse breeding contributes significantly to agricultural earnings, but trading in more humble horses remains an Irish passion.

Adjoining pubs in the same street of Allihies, Co. Cork. The pub is so much a part of Irish culture and social life that many small towns boast a dozen or more licensed establishments. Many ordinary retailers provide "bar" facilities for thirsty shoppers.

Industry and commerce

Ireland's mineral resources are not great, and only gypsum, zinc, and lead ores are mined in any quantity. There is little coal but large reserves of peat which are being developed for home and commercial use. Irish offshore fields supply natural gas meeting 20 percent of national requirements. Oil and gas exploration continues.

The 1980s and 1990s saw changes in output and employment. The industrial sector now accounts for 39 percent of GDP and manufacturing has expanded strongly, especially in engineering, pharmaceuticals, computers and computer software, telemarketing and financial services. In 1998 Ireland was the second biggest exporter of software in the world. Foreign investment has been important in this expansion, in 1998 there were over 1,000 foreign-owned companies. The services sector accounts for slightly over half of GDP and for almost 61 percent of employment. In the late 1990s, Ireland had become one of the most trade-dependent countries in the world. Exports include machinery, transport equipment, chemicals, and food products. A quarter of exports were to the UK. At the end of the 1990s the economy had an average annual growth of over 8 percent.

Transportation and communications

Ireland's road system is extensive, even in the most rural areas, and its major roads have generally been modernized in recent years. By contrast, the railroad network has shrunk, with many branch lines closed in the 1960s and 1970s. The railway and bus services are state-supported and the extensive inland waterways are maintained by the government as a public amenity. Ireland has several seaports, with the most important for trade being Dublin and Cork. Air traffic is provided through Aer Lingus with competition from eighteen independent operators. There are international airports at Dublin, Shannon and Cork.

Radio Telefis Eireann is the state-owned broadcasting company. The Independent Radio and Television Commission controls an independent television channel and the many commercial radio stations. There are Irish language radio stations and a TV channel. Cable television and British television channels are also available. Publishing and printing are major industries, there are eight national dailies and five Sunday newspapers. Freedom of the press is guaranteed by the constitution, but the publication of matter deemed to undermine Roman Catholic moral standards is forbidden.

The telecommunications industry is privatized. There are three mobile phone operators and expansion of this sector is a high priority. Ireland intends to become a leader in the field of electronic commerce and internet-based industry.

Health and welfare

Eire operates a social insurance scheme providing all major pensions, unemployment, and health benefits, with added charges that are calculated according to income. Health conditions and facilities are generally high, and hospitals raise extra income through the internationally famous Irish Hospitals' Sweepstake.

Education is free and compulsory between the ages of 6 and 15, and schools are almost entirely run by religious organizations. Further education is provided by secondary schools, secular vocational schools, and at a higher level by technical colleges and the major universities. The most famous of these is Trinity College, Dublin, founded in 1591. The National University, was founded in 1908.

France and its neighbors

COUNTRIES IN THE REGION

FRANCE · ANDORRA · MONACO

The Arc de Triomphe de l'Etoile (*left*), hung with the tricolored flag of France, at the Place Charles de Gaulle. Commissioned in 1806 by Napoleon, it looks down the Avenue Champs-Elysees and across the Place de la Concorde to its twin, the Arc de Triomphe du Carrousel at the Louvre.

■ capital city
● major town

height of land (meters)

3000
2000
1000
500
200
0

▲ mountain peak

English Channel

Lille
Amiens
C de la Hague
Le Havre
Rouen
Oise
Ardennes
Reims
Metz
Marne
Strasbourg
Nancy
Caen
Seine
Paris
Vosges
Brest
Rhine
Rennes
Aube
Orléans
Langres
Plateau
Mulhouse
Belle Ile
Loire
Angers
Tours
Sologne
Yonne
Dijon
Doubs
Jura Mts
Nantes
FRANCE
Saône
Besançon
Vienne
Indre
Morvan
L Geneva
Creuse
Allier
Loire
Ile de Ré
Ile d'Oléron
Mont Blanc
4807
Charente
Limoges
Puy de Dôme
1463
Clermont-
Ferrand
Lyon
A L P S
Puy de Sancy
1885
St Etienne
Isère
Bordeaux
Dordogne
Grenoble
Massif du
Pelvoux
M a s s i f
C e n t r a l
Rhône
Bay of Biscay
Garonne
Lot
Maritime
Alps
Tarn
Cévennes
Durance
Nice
MONACO
Monaco
Gulf
of
Gascony
Les Landes
Adour
Toulouse
Camargue
Nimes
Montpellier
Marseille
Toulon
P Y R E N E E S
Perpignan
Mediterranean Sea
Vignemale Peak
3298
Andorra la Vella
ANDORRA

C Corse
Bastia
Mt Cinto
2710
Ajaccio
Corsica

France

FRENCH REPUBLIC

FRANCE, IS THE LARGEST COUNTRY IN Western Europe, and has a long history of cultural influence both within Europe and across the world as a result of its former colonial power.

GEOGRAPHY

France is bordered on the northwest and west by the English Channel and Atlantic respectively, on the southwest by the Pyrenees, on the south by the Mediterranean, and in the southeast and east by the Alps. In the north and northeast are low-lying plains and hills.

The land

The steep mountain range of the Pyrenees divides France from Spain in the southwest. At the eastern end of the Pyrenees mountains the Mediterranean coast curves down to the Rhône delta. From the interior the mountains of the Massif des Maures, the Massif d'Esterel and the Maritime Alps stretch to meet the sea. Here on the Côte d'Azure lie the wealthy resort towns of the French Riviera.

The Alps form a barrier along the Italian border. Farther north, the high limestone ridges of the Jura form the border with Switzerland. North of these the Vosges mountains overlook the Rhine valley and mark the boundary with Germany. The northeastern border meets Germany, Luxembourg and Belgium. Cliffs and beaches on the north coast give way to the English Channel. West of the peninsula of Cotentin in Normandy, Brittany juts out into the Atlantic. The low plateaus of the Armorican Massif stretch inland, and the sandy plains run along the west coast to the Pyrenees.

Climate

The climate of France is temperate, with three regional variations. In the north and west, Atlantic winds bring high humidity and unpredictable weather, especially in Brittany. Summers are cool and winters mild; farther south, summers are warmer and winters colder.

The Paris Basin has a more continental climate, with considerable rain in the spring and fall, and thunderstorms in the summer. In eastern France and the Massif Central, winters are colder, with heavy snow in the mountains. Strasbourg, near the German border, has the biggest temperature range in the country.

The Mediterranean climate of Provence in the southeast has only a few days of frost each year. Summers are hot and dry. The mistral, a cold dry northerly wind, is funneled down from the Alps mainly in the winter months.

Plants and animals

The fertile plains of the north and west rise to hills covered in gorse and heather, with oak forests growing in sheltered

Volcanic wonderland (*above*) The Massif Central, seen from the Puy Mary, is a huge granite plateau covering much of south-central France. It is dotted with extinct volcanoes and hot volcanic springs, around which spa towns such as Vichy have grown up.

Wedded to the sea (*left*) The deeply indented south Brittany coast in the northwest of France provides excellent natural harbors for the traditional fishing fleets of the area and the pleasure boats of vacationers from Paris and abroad.

The four regions of the mountainous Massif Central dominate central and southern France. In the south are the Causses – a high, arid limestone plateau dissected by deep river canyons. To the east, steep granite massifs stretch north toward Burgundy, dropping sharply into the Rhône and Saône valleys.

Around Limousin, in the northwest of the Massif, are stepped plateaus cut by more river gorges. The higher central plain of the Auvergne rises to the Puy de Sancy, the highest peak in the Massif Central. Many local features here were formed by volcanoes.

The lowlands are defined by river systems. The largest rivers of the northeast are the Meuse and the Moselle, both tributaries of the Rhine. The Saône river rises in the Vosges and meets the Rhône at Lyon; the Rhône flows through Lake Geneva between the Jura and the Alps.

In the northwest, the Seine drains the Ile-de-France – the low-lying region around Paris – and flows northwest to the English Channel. Beyond the plains to the southwest is the longest river, the Loire, extending some 1,020 km (630 mi) and draining much of western France. In the far southwest the river Garonne rises in the Pyrenees. It is fed by the Tarn and Lot rivers, and meets the Dordogne north of Bordeaux, where the two rivers flow into the broad Gironde estuary.

NATIONAL DATA – FRANCE

Land area	545,630 sq km (210,669 sq mi)			
Climate		Temperatures		Annual
	Altitude m (ft)	January °C(°F)	July °C(°F)	precipitation mm (in)
Brest	103 (338)	7 (45)	17 (63)	1,145 (45)
Paris	53 (174)	5 (41)	20 (68)	650 (25.5)
Marseille	8 (26)	7 (45)	24 (75)	555 (21.8)

Major physical features highest point: Mont Blanc 4,807 m (15,771 ft); longest river: Loire 1,020 km (634 mi)

Population (2006 est.) 60,876,136

Form of government multiparty republic with two legislative houses

Armed forces army 137,000; navy 44,300; air force 64,000

Largest cities Paris (capital - 2,142,800); Marseille (795,600); Lyon (468,300); Toulouse (426,700); Nice (339,000); Nantes (276,200); Strasbourg (273,100); Montpellier (244,700); Bordeaux (229,500); Lille (222,400)

Official language French

Ethnic composition French 90.6% (including Occitan 2.7%; Alsatian 2.3%; Breton 1%; Catalan 0.4%); Algerian 1.5%; Portuguese 1.4%; Moroccan 0.8%; Spanish 0.6%; Italian 0.6%; others 4.5%

Religious affiliations Roman Catholic 83%–88%; Protestant 2%; Jewish 1%; Muslim 5%–10%; unaffiliated 4%

Currency 1 euro (EUR) = 100 euro cents

Gross domestic product (2006) U.S. $1.871 trillion

Gross domestic product per capita (2006) U.S. $30,100

Life expectancy at birth male 76.1 yr; female 83.54 yr

Major resources coal, iron ore, bauxite, zinc, uranium, antimony, arsenic, potash, feldspar, fluorospar, gypsum, timber, fish, barley, cattle, fruit, vegetables, grapes/wine, maize, oats, oil, natural gas, pigs, potatoes, poultry, salt, sheep, sugar beet, timber, uranium, wheat, tourism

Corsica

The island of Corsica, which lies off the northwest coast of Italy and is separated by a narrow strait from Sardinia, has been part of France for 200 years. Since 1975 it has been a single French region divided into two *départements*. It is a popular holiday resort, and many French people have second homes there.

Corsica's coastline has been compared to a fist with one finger extended toward Italy. The east coast is flat and marshy, while the west coast rises steeply to a mountainous interior. A chain of peaks runs like a backbone down the center of the island. The climate is Mediterranean, cooled by altitude over much of the island. Substantial rainfall feeds the island's many rivers and lush vegetation. More than one-fifth of Corsica is forested; chestnut trees grow in abundance, with tall pines on higher slopes. Much of the rest is covered with maquis, whose aromatic plants give the island a distinctive fragrance. Wildlife includes mouflon, wild goats and large numbers of feral pigs.

Corsica was conquered by the Romans in the 3rd century BC (there are many Roman ruins on the island). As the western Roman empire collapsed during the 5th century AD, Corsica was overrun many times by invading Germanic peoples. The island was eventually claimed by the Pope, but in 1077 it was ceded to Pisa in Italy. After a long struggle the Pisans were finally ousted by the Genoese in 1299. The Genoese ruled the island for about four centuries, putting down rebellions by Corsican nationalists. The island was sold to France in 1769 – the year in which its most famous native, Napoleon Bonaparte, was born. After brief periods of British occupation during the Napoleonic Wars, it finally became a *département* of France. In recent decades a nationalist movement has been active in demanding separation from France, often using terrorism.

The Corsican people, over one-third of whom live in the main towns of Bastia and Ajaccio, are full French citizens. They speak French as their official language, but most also speak one of the island's many dialects – some of which are similar to the Tuscan dialect of Italian. The Corsicans also resemble Italians in appearance, traditional dress and way of life. The island's population is declining and aging, largely because so many young people emigrate. However, it has also seen a huge influx of immigrants, especially from the former colony of Algeria after it gained its independence in 1963.

The main reason for the high level of emigration is the lack of opportunity in Corsica's mainly agricultural economy. Wheat is the main cereal, and sheep the main livestock animal. Major exports include citrus fruits, wines and cheeses. Fuel and machinery have to be imported, and there is little industry apart from tourism – a major source of income. The road network has been improved as tourism has increased, and the main towns are linked by rail. Regular sea and air services operate to the Italian mainland.

The citadel of Calvi in the northwest of Corsica was fortified by the Genoese in the 16th century. Corsica itself is a natural fortress of 20 mountain peaks. It was the first part of France to free itself from German occupation in World War II.

areas. The northeastern regions and much of the Paris Basin are covered in loess soils. There are lush pastures and slopes on the western and central slopes of the Pyrenees and the Massif Central. The vegetation in the southeast is typically Mediterranean, with olive, box, thyme, rosemary and cork oak. Maquis scrubland occurs in drier parts. Unique wildlife includes feral white horses, bulls and flamingoes in the Camargue at the Rhône delta on the southern coast.

The river Lot (*right*) at Estaing, near Espalion. The Lot rises west of the Massif Central and flows in a deep valley to join the Garonne river south of Bordeaux. Along its banks can be seen many ancient castles, quiet villages and terraced vineyards.

Jagged peaks and icy pools (*below*) in the French Alps, part of a massive mountain range that extends across Switzerland, northern Italy and Austria. The range is between 65 and 2 million years old. Mont Blanc in France is the highest peak at 4,807 m (15,770 ft).

SOCIETY

After centuries of cultural and military power, France remains very influential as a founding member of the European Community.

History

The 30,000-year-old remains of the first modern human, Cro-Magnon man, were discovered in France and named for the caves in the Dordogne where they were found. The history of France as a country began with settlement by the Gauls, a Celtic people who had spread west from the Rhineland about 800 BC. By the 5th century BC they had conquered large areas of Europe and were rivals to Rome, which they failed to conquer in 390 BC.

Rome conquered the whole of France between 58 and 50 BC, and the region remained stable for nearly 500 years. The Christian Frankish kings (the Merovingian dynasty) helped keep Gaul's Roman culture alive after the empire disintegrated, but were gradually displaced by the Pepins. One of these, Charles Martel (c. 688–741), defeated Muslim raiders from Spain (the Moors) at Poitiers in 732 and founded the Carolingian dynasty. His grandson, Charlemagne (742–814), spread Frankish rule back into Germany and to the borders of Scandinavia and Spain, thus establishing the major Christian kingdom of western Europe with his capital at Aix-la-Chapelle (now Aachen in Germany).

The Capetian kings from 987 began as figureheads but gradually extended their authority. Philip VI (1293–1350), the first of the Valois kings, had much of present-day France under his control. But large areas came under English control in the Hundred Years War (1337–1453), while the whole country suffered from high taxes and the Black Death (1347–51). Charles VII (1403–61), aided by Joan of Arc (c. 1412–31), drove out the English and had reconquered his realm by 1453.

France in the 16th century was overshadowed by religious conflict between Catholics and French Protestants or Huguenots. The Wars of Religion ended in 1589 when the Protestant prince Henry of Navarre (1553–1610) became a Catholic in order to inherit the throne as Henry IV, founding the Bourbon dynasty and granting tolerance to the Huguenots. After his death, however, the old divisions reappeared. Under Louis XIII (1601–43) the ruthless chief minister Cardinal Richelieu (1585–1642) built up French power abroad and suppressed all dissent at home. A series of revolts known as the Frondes followed under his successor, Cardinal Mazarin (1602–66).

Louis XIV (1638–1715), the "Sun King," built the most opulent court in Europe to distract the nobility from politics. Government became even more centralized and the Huguenots were persecuted. Louis sought to dominate Europe and strengthen France's territorial possessions. France colonized parts of North America, India and the Caribbean, where plantations used African slaves to produce vast wealth in coffee and sugar.

During the 18th century, expensive wars lost France both Canada and India, and ruined its financial system. Under Louis XV (1710–74) the monarchy's popularity waned. The economy approached breaking point under the well-meaning but inept Louis XVI (1754–93) and his hated queen Marie Antoinette (1755–93). The French Revolution, which began in 1789, finally toppled the monarchy. Seizure of the government by fanatics such as Danton (1759–94) and Robespierre (1758–94) led to the Terror, a period of mass murder and anarchy. Louis XVI and Marie Antoinette were both executed.

From 1799 a brilliant young Corsican general, Napoleon Bonaparte (1769–1821), assumed power, proclaiming himself Emperor in 1804. He instituted many domestic reforms, among them the Napoleonic Code, which is still central to French law. But his chief concern was expanding his empire by foreign conquest, and he overextended himself and his army. Defeated in 1814, he was exiled from France but returned in 1815 to face final defeat at Waterloo in Belgium.

The unpopular restored Bourbon kings were followed by the unstable Second Republic, and Napoleon's charismatic nephew, Louis-Napoleon Bonaparte (1808–73), was elected president. In 1852 he staged a coup and declared himself Napoleon III of the Second Empire. During his reign the Industrial Revolution spread throughout France, bringing peaceful social change and unprecedented prosperity – along with political and financial corruption.

In 1870 Otto von Bismarck (1815–96), architect of the rising German empire, goaded France into war with Prussia. The French armed forces collapsed and Napoleon III was defeated and captured. The new Third Republic sued for peace, and the Germans annexed Alsace-Lorraine. A radical uprising established a Commune in Paris, which was savagely defeated: more than 20,000 Communards died.

Under the Third Republic France regained domestic stability, consolidated its presence in Algeria, and established more colonies in West Africa and Indochina. Much of World War I took place on French soil, and France lost almost 1.4

Symbolic victory On 14 July 1789 a Paris mob stormed the Bastille – the state prison – and freed its 7 prisoners in an enraged gesture against Bourbon despotism. The Bastille was an 8-tower fortress built in the 14th century to protect Paris from English attack.

million men. After the war France recovered Alsace-Lorraine and occupied the German Rhineland from 1923 to 1930.

Northern France came under German rule in the invasion of 1940, while the south was governed by the collaborationist Vichy regime under Marshal Philippe Pétain (1856–1951). Both were opposed by the British-based Free France movement led by General Charles de Gaulle (1890–1970). In 1947 the Fourth Republic was established, but it lost public confidence during disastrous wars in French Indochina (Vietnam) and Algeria.

In 1958 General de Gaulle returned to serve 10 years as president of the Fifth Republic. French prosperity revived in the newly established European Economic Community (EEC), but in 1968 the country was shaken by violent student riots in Paris. At the same time de Gaulle was forced to grant independence to Algeria and other colonies, but he pursued nationalist policies, resisting US influence in Europe and repeatedly vetoing British entrance into the EEC.

De Gaulle's successors, Georges Pompidou (1911–74) and Valéry Giscard d'Estaing (b. 1926), both encouraged industrial development but faced growing economic problems. In 1981 de Gaulle's old rival, François-Maurice Mitterrand (1916–96), was elected France's first socialist president and was re-elected in 1988. Jacques Chirac (b. 1932) was elected president in 1995. In January 1996 he announced the end of French nuclear testing in the Pacific, and revealed plans to phase out military conscription by 2001.

Government

The constitution of the Fifth Republic, adopted in 1958, made France a multi-party democracy with the president as head of state elected by popular vote for a 7-year term. Appointed ministers, led by the prime minister, carry out administrative business. The president can also summon or dissolve parliament, hold a referendum and assume emergency powers.

The two parliamentary chambers are the Senate and the National Assembly. The Senate's 321 members have limited legislative powers. The National Assembly's 577 deputies are the main legislative body, but finance, defense and other national issues lie in the hands of the president, who has more power than is usual in Western democracies – a legacy of the de Gaulle era and the political instability that preceded it.

Local government wields little power apart from administration and public

Hero or tyrant?

Nearly 200 years after his death, Napoleon (1769–1821) is a controversial figure. The impact he had is unquestioned: within a few years he changed the history of France and of Europe. What is still debated is whether or not his regime was beneficial to France.

Napoleon was born at Ajaccio on Corsica in 1769, only shortly after the Italian city-state of Genoa had sold the island to France. His name Napoleone Buonaparte – is an indication of his Italian ancestry. Corsicans were then Italian in both language and culture, a feeling that Napoleon himself never lost, although he was educated in France and graduated from the military academy in Paris before becoming an artillery lieutenant. Briefly involved in the Corsican nationalist movement, he then joined the Jacobin revolutionaries in France. He distinguished himself against the British at the siege of Toulon becoming a brigadier general.

His career prospered in spite of the political turmoil of the early 1790s and by 1795 he was given the command of the army of the interior. In 1796 he married his mistress, Josephine de Beauharnais (1763–1814), widow of a general who had been guillotined during the Revolution. That same year he led a successful invasion of Italy, and forced Austria to yield valuable territory. He also organized an antiroyalist coup in the French government. At the zenith of his power he declared himself Emperor of France and King of Italy; married the daughter of the Emperor of Austria; and controlled Austria, Prussia, Spain and Portugal.

His defeats were equally spectacular. Clashing with Britain, he invaded Egypt in 1798 to cut off wealth derived from colonial India. He occupied the country successfully and introduced many European political institutions, administration and technical skills there. But his fleet was destroyed by the British under the command of Admiral Horatio Nelson (1758–1805) in the Battle of the Nile (1798). An attempted invasion of Britain ended in another naval defeat at Nelson's hands at Trafalgar (1805). In 1812 Napoleon turned on Russia, the other great power not yet under his control, which had become hostile to French ambitions in the Baltic. He defeated the

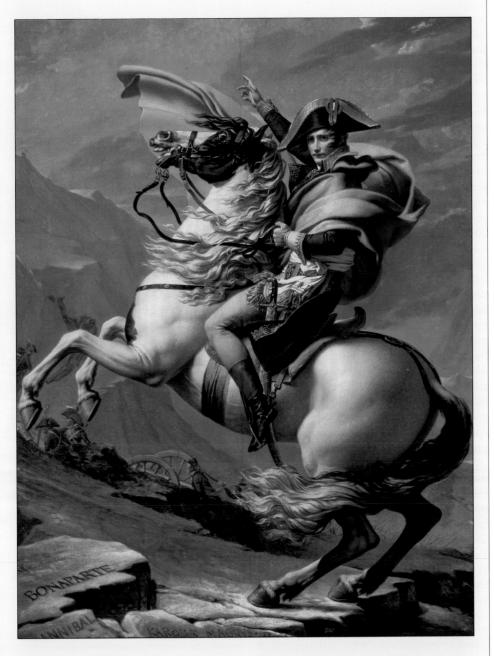

Russian army at Borodino and entered Moscow. But the Russians' scorched-earth tactics deprived his army of essential supplies, and the savage Russian winter forced him into a disastrous retreat. Defeat followed defeat, forcing the French army back within the borders of France. Invaded by combined Russian, Prussian, Austrian and British forces, Napoleon was forced to abdicate in 1814, but returned within a year from exile on the Mediterranean island of Elba. His final defeat by British and Prussian forces at Waterloo in Belgium came in 1815. By then he was so unpopular in France that he was forced to put himself under British protection. He died after six years in exile on the remote Atlantic island of St Helena.

Napoleon crossing the Alps An idealized portrait by Jacques-Louis David (1748–1825). Born in Corsica and educated in France, Napoleon had a spectacular career: brigadier-general at the age of 24, military dictator at 30, emperor at 35.

Was he a far-seeing statesman or merely a ruthless tyrant? Much that he did for France was good – he restored its prestige abroad and laid the foundation of the modern French state by establishing the prefecture system of administration, the Code Napoleon, the Bank of France and other financial institutions. But – like so many dictators before and after him – he was unable to distinguish between self-interest and the national interest, and his ambitions led to his ruin. However, he was a hero to many of his countrymen.

services. There are 22 regions, 96 *dé-partements* and 4 overseas *départements*; the smallest local government body is the commune, responsible for municipal services and establishments such as schools, hospitals and fire stations. The commune is headed by a mayor and deputies, who are important figures in local life.

People and language

In addition to physical differences and regional dialects, linguistic minorities reflect the mixed ancestries of the French people. The northern accents are close to standard French; those of the south have highly distinctive nasal vowel sounds. The Occitan dialect of the south is derived from the Provençal language of the Middle Ages, and a small written tradition has been preserved. The most marked language differences are in outlying areas: Catalan, spoken in the eastern Pyrenees; Breton, a Celtic language of Brittany; the Germanic dialect of Alsace; and Flemish (similar to Dutch) along the border with Belgium. The Basque language of the western Pyrenees (also spoken in northern Spain) is linguistically unique among European languages.

The Académie Française – established in 1635 to uphold the highest standards in French language and literature – insists on the use of French-derived terms, newly coined if necessary, to replace adopted English language phrases. "Le computer" has become "le ordinateur."

Culture and religion

French culture has enriched the world from the time of Charlemagne onward. Playwrights, painters, philosophers, writers, composers, film directors and sculptors are included in the long list of influential French cultural figures. But perhaps France's most lasting contribution has been its cooking. From traditional peasant dishes to elegant nouvelle cuisine, it has come to dominate the tables of the world.

There is no state church in France. The majority of the people belong to the Roman Catholic Church. There are large Protestant and Jewish minorities, and a large number of Muslims among the immigrant-based communities from Northern Africa.

ECONOMY

One of the most highly developed in the European Union (EU), the French economy has enjoyed a high growth rate since the 1950s.

Strategic landmark (*above*) The Eiffel Tower, during the Universal Exposition of 1900. Named after its French designer, it was derided by the critics, but its importance in World War I as a communications tower saved it from demolition.

Impressions of light and shadow (*right*) Edouard Manet's 1874 painting "Monet Working on his Boat in Argenteuil". These two French masters shared a passion for painting outdoors – *en plein-air* – and for exploring the transient effects of light on color.

Agriculture, forestry and fisheries

France, the largest agricultural producer in western Europe, has benefited enormously from EU subsidies, though the agricultural sector is shrinking. Small farmers, who form the majority, are offered incentives to consolidate farms, modernize techniques and form agricultural cooperatives. Overproduction of wheat, dairy produce and wine is causing problems in the EU and overseas.

Wheat, barley, oats, rye, maize, rice, sunflower seeds and root crops, chiefly sugar beet and potatoes, are the main crops. Normandy and Brittany are famous for apples and the making of cider and calvados (apple brandy). Aquitaine in the southwest is equally known for plums. Peaches and apricots are grown in the warm Rhône valley.

France is one of the world's largest producers of wine. Languedoc in the south produces the most wine, but the largest share of high-quality wines comes from Bordeaux, followed by Burgundy and some of the regions in the southeast.

Livestock farming accounts for an increasing proportion of agricultural production; the government has encouraged farmers to shift to meat production to reduce surpluses of cereals and butter. Cattle are the main stock animals throughout the country. Pigs are also important. Sheep and goats are raised principally in the less fertile highlands of central and southern France.

About one-fifth of France is forested, and it is Europe's third largest producer of raw timber. Large areas of forest are state-managed and there are extensive reforestation programs.

Fisheries are confined mainly to the Atlantic coast, as fish stocks are poor along the Mediterranean coast. La Rochelle, Concarneau, Boulogne and Lorient are the major fishing ports.

Industry, trade and commerce

Industry was devastated in World War II. Its recovery was initially financed by American aid under the Marshall Plan, which sought to improve Europe's war-torn economies with cash input, and coordinated by a government-led recovery plan. France is now the world's fifth greatest industrial power, after the United States, Japan, Germany and China.

Compared with other European countries, there are few large companies. There is much public and relatively little private investment. Industry suffered during the oil-based recessions of the 1970s and 1980s, but there has since been some recovery. France's mineral resources are now limited. Lorraine's supplies of iron ore and coal have dwindled, and most is now imported. Bauxite, potash, rock salt and uranium ore are mined in smaller quantities.

Almost three-quarters of electrical power is now generated by state-owned nuclear plants. Many hydroelectric plants have been built along the extensive river systems. Petroleum and natural gas reserves are found chiefly in Les Landes and in the central part of the Paris Basin, but most has to be imported. France nevertheless has one of Europe's largest refining and petrochemical industries.

France is also a leading steel and automobile manufacturer, as well as a major producer of both civil and military aircraft. It is involved in joint European aircraft projects such as Concorde and the Airbus. Other major industries are armaments, rubber and chemicals. The textile industry still thrives, and food processing is a quickly growing sector. Prestige industries such as perfume and fashion are world market leaders.

Manufactured goods, and machinery in particular, constitute France's largest category of exports. Other important exports include automobiles, agricultural products and chemicals. The country's major trade is mostly within the EU.

The state is heavily involved in banking and finance. The major clearing banks are state-owned, while the main stock market, the Paris Bourse, plays a smaller role than in other Western countries.

Transportation
France has developed a comprehensive transportation system, headed by Europe's most extensive road network, but there are fewer urban expressways than in other European countries. The streets of provincial cities tend to be choked with traffic.

The state-owned railroad system has its heart in Paris. In recent years it has been modernized and considerably improved. New tracks accommodate the new high-speed trains known as TGVs (*trains à grande vitesse*), which run at speeds of up to 300 km/h (186 mph) on the main routes between Paris and Lyon, the French Riviera and Switzerland. A newly opened route links Paris with Nantes and Bordeaux. The network extended to Britain via the Channel Tunnel, which opened for business in 1994.

Paris, the hub of air transport, has three main airports: Charles de Gaulle and Le Bourget to the north, and Orly to the south. Internal air services have grown rapidly and are mostly operated by Air France's subsidiary, Air Inter. France boasts the longest inland waterway system in Europe. Many of its rivers have

High fashion France has long been the world leader in women's fashion, its most exclusive designers based in Paris's Faubourg Saint Honoré. Chanel, Dior, Cardin and Saint Laurent are all household names, although few can afford their prices.

Champagne – king of all wines

Wine has been grown on the chalky soils of the Champagne region, southeast of Paris, at least since Roman times. But when exactly the *méthode champenoise*, the process that adds the sparkle to the world famous white wine, was developed is uncertain. Legend ascribes its invention to a certain Dom Pierre Pérignon, a local cellarer and Benedictine monk, in the 17th century.

The method consists basically of bottling the wine after its first fermentation and allowing it to ferment a second time within the bottle. It is not easy to achieve this. The bottles have to be securely corked and placed in sloping racks, then turned and shaken a little every day. This process, known as *rémuage*, can last for anything from six months to several years. It was originally done by men walking along the rows – one of the labor-intensive aspects of production that made champagne so expensive. These days huge, specially engineered racks are used, which can be shaken as a unit and will turn the bottles automatically. Some companies have introduced completely automated systems controlled by computer that carry out *rémuage* continually – all day, every day – speeding up the maturing process without, apparently, affecting the quality of the wine produced.

The *rémuage* finishes with the bottles upended, so that the deposit created by the second fermentation collects on the cork. The bottles are then passed through a bath of freezing brine. This encases the deposit in a plug of solid ice that is removed with the cork. The bottle is then topped up with a solution of cane-sugar syrup in old champagne, and restoppered. If little or no syrup is added, the resulting champagne becomes *brut*, or extra dry; more, and it becomes *sec* (dry), or *demi-sec* (medium dry – acutally quite sweet). There are rarer extremes: sugary *doux* is reminiscent of cough syrup, and sugarless *brut sauvage* is positively arid.

The wine is produced in three areas to the south of the ancient city of Reims. The northernmost area is the Montagne de Reims, where one of the principal varieties of champagne

grape, the black Pinot Noir, is grown. To the south are the vineyards along the Marne Valley from Château-Thierry to Epernay – home of another black grape variety, the Meunier. South of Epernay is the district known as the Côte des Blancs, where the white Chardonnay grape is grown.

Champagne is almost always blended by its producers from the products of particular vineyards. The less common varieties known as *blanc des blancs* and *blanc des noirs* are made exclusively from white and black grapes respectively. The best champagne will be blended only from the wines of a particularly good year, and hence known as vintage champagne. Otherwise different years will be blended to produce the best flavor; from skillful

Ten percent of the French labor force earn their living from the automobile industry. The firm PSA Peugeot-Citroën has operations in 140 countries and sold 1.5 million automobiles in 1999.

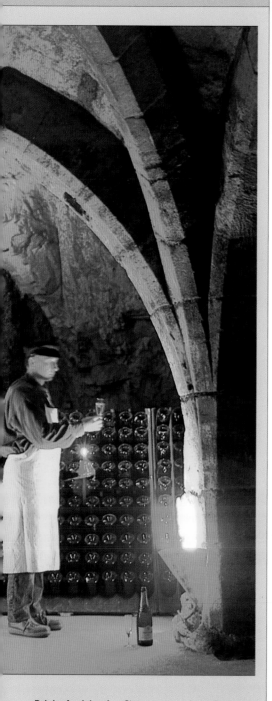

Drink of celebration Champagne, perhaps the world's most copied wine, comes from the region of chalky soil around Reims, where summers are hot and winters cold. No other region may legally call its wine champagne

hands these nonvintage champagnes may be as good or better than the vintage varieties.

Many other regions in France make their own sparkling wine by the *méthode champenoise*; much of it is excellent. Other sparkling wines, including most of the equally good German varieties, are produced by the much cheaper *cuve close* technique. But only the true champagne can boast the distinctive crisp, uncloying flavor – derived, apparently, from the chalky soil of its native slopes – that has earned it its truly regal title: king of all the wines.

been canalized, though only the most modern parts of the system in the industrial northeast can carry large international freight barges. Pleasure boating is also popular.

Media and communications

France has a vigorous publishing tradition. Many of its newspapers and magazines – among them *Le Monde*, *Le Figaro*, *Paris-Match* and the leading women's magazines *Elle* and *Marie Claire* – are read internationally.

Broadcasting, until recently, was much less diverse, partly because it was a state monopoly. In 1975 the central controlling body, the Office de Radiodiffusion-Télévision Française (ORTF), was broken up into separate companies controlled by representatives of the state, parliament, the press and the broadcasters themselves. The number of television channels increased to seven. Since the advent of cable and satellite broadcasting, the viewers' choice has become even wider.

Health and welfare

The French social-security system was founded in the 1930s and expanded after World War II. It is mostly financed by the state, with compulsory worker and employer contributions.

Medical expenses are dealt with by reimbursement, using a standard fee

scale to allow private medicine to coexist with public services. Medical facilities are excellent by world standards, but there are shortages in the number of hospital beds, and severely limited special care facilities such as mental institutions and nursing homes for the aged.

Education

France has an excellent tradition of public education, which is free at all levels and compulsory between the ages of 6 and 16. The Ministry of National Education draws up curricula and administers the system. After primary education up to age 11, *lycées* or grammar schools prepare some students for the national *baccalauréat* examinations; others go on to secondary colleges providing general or technical education.

Higher education was reformed after the student unrest of 1968 and now consists of multidisciplinary universities, polytechnics and professional colleges such as medical schools. The quality of public education makes private education less popular; most private schools are Roman Catholic, but some receive limited government assistance.

Andorra

PRINCIPALITY OF ANDORRA

THE TINY AND ISOLATED PRINCIPALITY OF Andorra lies in the heart of the eastern Pyrenees, bounded by France to the north and Spain to the south. It is centered around the triple valley complex of the Valira river, which flows south into Spain and forms the only breach in the surrounding mountains. The landscape is extremely rugged, with high peaks towering above deep valleys and gorges that were originally carved out by glaciers. No part of the country lies below an altitude of 900 m (3,000 ft).

The climate is typically alpine, but with variable rainfall. Winters are cold, with heavy snow usually blocking the one pass into France. Spring and summer are mild and sunny. The plants and animals are also alpine in character. Hardy trees such as pine, fir and birch have been largely cleared, giving way to extensive mountain pastures on the slopes, where the chamois and Spanish goat roam wild.

Andorra first gained its independence from Spanish Muslim rule in 803 AD when Charlemagne (742–814) recovered the area from the Muslims. In 819 his son Louis the Pious (778–840) gave the region to the Spanish Bishop of Urgel in northeastern Spain. In 1278 an agreement was drawn up to divide dominion of Andorra between Spain and the French Counts of Foix, and apart from a short period following the French Revolution, it has remained as a co-principality up to the present day.

The heads of state are the President of France and the Bishop of Urgel, but since the new constitution of 1993 they only have power to veto matters that affect the borders of Spain and France. The Andorran legislature (General Council of Andorran Valleys) has 28 elected members.

Strong cultural links are maintained with the neighboring Spanish province of Catalonia. Catalan is the main language, though French and Spanish are also spoken. Most of the population are Roman Catholic, and Andorra belongs to the diocese of Urgel across the border in Spain. The only large town is the capital, Andorra la Vella. In recent decades the population has been greatly swelled by immigrants, mostly from Catalonia, lured by the growing tourist industry and the promise of jobs. Immigrants now make up some two-thirds of the population.

The economy is traditionally based on agriculture. In the winter cattle and sheep are kept in the villages or in lower-lying pastures in Spain, and in summer they are moved up the mountains to graze on the pastures; many of the villagers move with their livestock to temporary summer settlements. There is some forestry, and the tiny area of cultivable land is devoted to tobacco growing. Industry is mostly limited to timber and tobacco processing and hydroelectric power generation.

Tourism is now the main source of revenue. Visitors are attracted by the magnificent scenery, the local architecture and folklore, and the good winter skiing. Roads have been greatly improved, but there is no railroad and there is only one through road from Spain into France.

Beneficial tax and customs laws have encouraged the development of Andorra as a place of international commerce. Both French and Spanish currencies are used. Education, television, postal and telephone services are provided jointly by the French and Spanish governments. The Council General is responsible for other public services.

Mountain hideaway Andorra, high in the eastern Pyrenees, encompasses a cluster of mountain valleys that provides tourists with excellent skiing. They also come for the scenic beauty and duty free goods. The official language is Catalan.

NATIONAL DATA - ANDORRA

Land area	464 sq km (179 sq mi)			
Climate		Temperatures		Annual
	Altitude m (ft)	January °C(°F)	July °C(°F)	precipitation mm (in)
Andorra la Vella	1,080 (3,543)	2 (36)	19 (67)	808 (31.8)

Major physical features	highest point: Coma Pedrosa 2,946 m (9,665 ft)

Population (2006 est.) 71,201

Form of government parliamentary democracy with one legislative house

Armed forces no armed forces

Largest cities Andorra la Vella (22,035)

Official language Catalan

Ethnic composition Spanish 38%; Andorran 37%; Portuguese 11.5; French 6%; others 13.5%

Religious affiliations Roman Catholic 89.1%; other Christian 4.3%; other 1.6%; nonreligious 5%

Currency 1 euro (EUR) = 100 euro cents

Gross domestic product (2004) U.S. $1.84 billion

Gross domestic product per capita (2004) U.S. $24,000

Life expectancy at birth male 80.61 yr; female 86.61 yr

Major resources hydropower, mineral water, timber, iron ore, lead, barley, cattle, potatoes, rye, sheep, tobacco, vegetables, tourism

Monaco

PRINCIPALITY OF MONACO

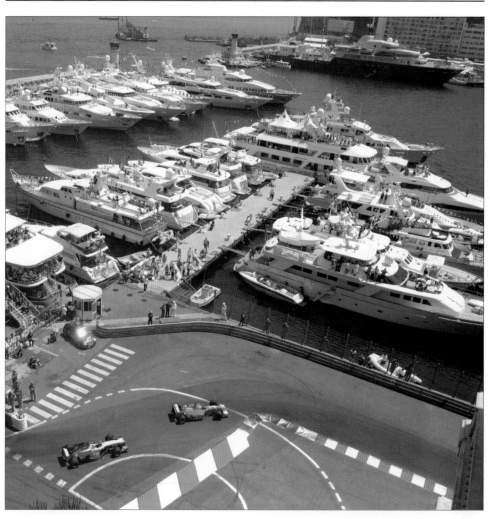

Grand Monaco The Monaco Grand Prix, held each May, is one of the most glamorous venues on the circuit. Monaco is a tiny principality with some big drawcards, from motor racing and gambling at the famous casino to its splendid Oceanographic Museum and old harbor.

THE TINY INDEPENDENT ENCLAVE OF Monaco is situated on the French Mediterranean coast just to the east of Nice and very close to the Italian border. It enjoys a Mediterranean climate, with hot, dry summers and mild winters; and its sheltered location on the lower slopes of the rugged, maquis-covered Maritime Alps affords protection against the cold northerly wind known as the mistral.

The whole area – only 1.95 sq km (0.75 sq mi) – is densely populated, and there are four main sectors or quartiers. To the northeast is fashionable Monte Carlo, with its hotels, villas and high-rise apartments climbing up the hillside. The commercial center of La Condamine looks across the modern harbor to the old town of Monaco proper, which is perched on a rocky headland. Across the old harbor (now a marina for luxury yachts) to the southwest is the newer industrial quarter of Fontvieille, part of which has recently been reclaimed from the sea.

From 1297 until 1792, Monaco was ruled by the princes of the Grimaldi family from Genoa. In 1792, however, the local people declared the principality a republic, and in 1793 revolutionary France took control of Monaco. In 1815, at the Congress of Vienna, Monaco regained its independence, but placed itself under the protection of the kings of Sardinia. The Grimaldis were restored to power in 1861, but by then the territory had been considerably reduced. A treaty with France in 1918 ensured Monaco's continuing autonomy even if the Grimaldi line were to become extinct. In 1962, following a disagreement with France over taxes, Prince Rainier III (b. 1923) introduced a new constitution, whereby legislative power is shared between the monarch and the 18-member elected National Assembly.

Only a small minority of the population are native Monégasques – more than half are French citizens. There are also a number of Italians and a significant international community. The majority of the population are Roman Catholic, and the official language is French, though the local Monegasque dialect is also spoken.

The principality's chief industry is tourism. Its location and excellent climate have made Monaco one of Europe's most prestigious resorts. Monte Carlo contains a world-famous casino and a theater that is renowned for its opera and ballet. Monaco is also home to two international motor sport events: the Monte Carlo Rally and the Monaco Grand Prix round the streets of the principality.

Monaco is in customs and monetary union with France. Up until 1962 its tax-exempt status provided a haven for the world's rich, fueling the town's prosperity. Taxes have since been introduced, though native Monegasques are exempt from income tax. Monaco is a major international business center, drawing revenue from taxes, from radio and television franchises, and from state-operated businesses such as tobacco, postage stamps and the casino. Some light industries have developed in Fontvieille.

NATIONAL DATA – MONACO

Land area	1.95 sq km (0.75 sq mi)			
Climate		Temperatures		Annual
	Altitude m (ft)	January °C(°F)	July °C(°F)	precipitation mm (in)
Monaco	55 (180)	10 (50)	24 (75)	816 (32.1)

Major physical features	Mont Agel 140m (459ft)
Population	(2006 est.) 32,543
Form of government	nonparty constitutional monarchy with two legislative houses
Armed forces	no armed forces
Capital city	Monaco (15,500)
Official language	French
Ethnic composition	French 47%; Monegasque 16%; Italian 16%; other 21%
Religious affiliations	Roman Catholic 90%; others 10%
Currency	1 euro (EUR) = 100 euro cents
Gross domestic product	(2000) U.S. $870 million
Gross domestic product per capita	(2000) U.S. $27,000
Life expectancy at birth	male 75.85 yr; female 83.74 yr
Major resources	tourism, financial services, property revenue, ceramics, metal works, textiles, instruments, plastics

The Low Countries

COUNTRIES IN THE REGION

NETHERLANDS · BELGIUM · LUXEMBOURG

Mist lies in the valley (*left*) in Luxembourg's most fertile region, the Gutland; throughout the Low Countries mixed farming predominates. But in this densely populated region intense agricultural production is never far from highly developed manufacturing, agricultural and commercial industries.

West Frisian Islands
Schiermonnikoog
Ameland
Terschelling
Delfzijl
Vlieland
Waddenzee
Groningen
Texel
IJsselmeer
NE Polder
North Sea
E Flevoland
Haarlem
S Flevoland
IJssel
Amsterdam
Veluwe
Oude
NETHERLANDS
The Hague
Utrecht
Arnhem
Lek
Waal
Maas
▼ ~7
Rotterdam
Rhine
Overflakkee
Schouwen
E Schelde
Walcheren
W Schelde
Eindhoven
Vlissingen
Kempenland
Zeebrugge
Antwerp
Ostende
Bruges
Schelde
Ghent
Brussels
Maastricht
BELGIUM
Hesbaye
Liège
Meuse
Botrange ▲
Namur
694
Charleroi
Hautes Fagnes
Sambre
Meuse
Ourthe
Our
Ardennes
Clervaux
Sûre
LUXEMBOURG
Luxembourg
Moselle
Esch

■ capital city
● major town

height of land (meters)
500
200
0 (sea level)

▲ high point
▼ depression

263

Netherlands

KINGDOM OF THE NETHERLANDS

THE SMALL, DENSELY POPULATED NETHER-lands is remarkable not only for its predominantly flat landscape – one-third of which lies below sea level – but also for the vast tracts of land that have been reclaimed from the sea. The country is bordered by the North Sea to the north and west, Germany to the east and Belgium to the south. It controls access to the North Sea from three major European rivers – the Schelde, Meuse and Rhine – making the Netherlands an important international commercial center.

NATIONAL DATA - NETHERLANDS

Land area	33,883 sq km (13,082 sq mi)			
Climate		Temperatures		Annual
	Altitude m (ft)	January °C(°F)	July °C(°F)	precipitation mm (in)
Amsterdam	-4 (-13)	3 (37)	17 (63)	831 (33)

Major physical features highest point: Vaalserberg (in far southeast) 321 m (1,053 ft); lowest point: (in the west) -7 m (-22 ft); longest rivers: Rhine (part) 1,319 km (820 mi) and Meuse (part) 933 km (580 mi)

Population (2006 est.) 16,491,461

Form of government multiparty constitutional monarchy with two legislative houses

Armed forces army 23,150; navy 12,130; air force 11,050

Largest cities Amsterdam (capital - 1,002,868); Rotterdam (989,956); The Hague (610,245); Utrecht (366,186); Eindhoven (302,274)

Official language Dutch

Ethnic composition Dutch 83%; other 17% (mainly Germans, Turks, Moroccans, Antilleans, Surinamese, and Indonesians)

Religious affiliations Roman Catholic 31%; Dutch Reformed 13%; Calvinist 7%; Muslim 5.5%; other 2.5%; none 41%

Currency 1 euro (EUR) = 100 euro cents

Gross domestic product (2006) U.S. $512 billion

Gross domestic product per capita (2006) U.S. $31,7000

Life expectancy at birth male 76.39 yr; female 81.67 yr

Major resources natural gas, petroleum, peat, limestone, salt, sand, gravel, arable land, cereals, fish, fruit, vegetables, livestock, potatoes, poultry, sugar beet, horticultural plants

GEOGRAPHY

As early as 1000 AD, simple sluices were being used in low-lying areas of the Netherlands to improve drainage. Four hundred years later the clay flats of the west and north were crisscrossed by an intricate network of dikes and drains. From the 15th to the 17th century, windmills were built on a large scale to pump water from low-lying land behind the dikes, creating "polders" – areas where the water level can be controlled – for farming. Steam pumps were introduced in the 19th century, making it possible to reclaim the Haarlemmermeer in 1852. Schiphol airport lies in the northeastern corner of the Haarlemmermeer Polder 4 m (13 ft) below sea level.

The land
The north coast of the Netherlands is fringed by the long arc of the West Frisian Islands. These are formed from dunes and sandbars at the outer limits of the Waddenzee – a stretch of shallow saltwater lying between the islands and the mainland. The Waddenzee is separated from the freshwater IJsselmeer by a broad 32 km (20 mi) long dike.

The dike was the first stage of a massive scheme, conceived as early as 1891, to reclaim land from what was then the Zuider Zee, and to link the province of North Holland with the northwestern province of Friesland. While the dike was under construction the shallow Wieringermeer, to the southwest, was drained. This became the first of four large polders to be reclaimed, and today they cover an area of more than 2,000 sq km (800 sq mi).

The western provinces of North and South Holland are older reclaimed lands.

In the southwestern province of Zeeland, the rivers Rhine, Schelde and Meuse flow out into the North Sea by way of a delta made up of estuaries and islands. After disastrous flooding in 1953 most of the estuaries and sea-arms were sealed off by 32 km (20 mi) of new dikes, and one inlet was spanned by a tidal barrage. Two major shipping lanes remain open, however: the New Waterway linking Rotterdam with the sea, and the Western Schelde, which gives access to the Belgian port of Antwerp. In early 1995, the government declared a state of emergency during severe flooding.

Inland, the flat, clay lowlands of the north give way to sandy soils in the eastern, central and southern Netherlands. In the extreme southernmost province of Limburg, a chalk plateau rises toward Vaalserberg, which is the highest point in the Netherlands.

Climate
Summers are relatively cool, though there can be very warm days, and winters are generally mild and wet. Winds from the north may bring cold weather in the spring, especially to the north and east. Rainfall is frequent, but often light and

Low-lying coastal marshes (right) are part of the distinctive landscape of the Netherlands. Flat as well as low, they offer little resistance to sea winds, and from the 15th century windmills were used extensively to drain the land for farming.

Behind the protective coastal sand dunes (below) a scrubland habitat supports a range of plants adapted to tolerate strong winds and high levels of salt. Lichens, mosses, low shrubs and deep-rooted grasses bind and stabilize the sands.

evenly distributed thoughout the year. In the fall, heavy cloud cover, fog and mist are common.

Plants and animals

Compared with prehistoric times, today both forest and marshland are much reduced. Most of the oak and birch woodlands in eastern areas have been planted, as have the large coniferous and deciduous forests on the central hills of the Veluwe. Large areas of agricultural land are now occupied by commercial nurseries.

In the north the Waddenzee, with its sand dunes, mudflats and marshes, is the most important wetland in western Europe, supporting many species of native and migrating birds, as well as fishes and crustaceans. Marram and other coarse grasses cover the coastal dunes, while the tidal mudflats attract salt-loving plants such as glasswort; otters are found on the polders.

SOCIETY

The histories of the Netherlands, Belgium and Luxembourg are all intimately connected. They are known collectively as the Low Countries, and many attempts have been made to unite them.

History

In the 1st century BC the Celtic and Germanic peoples of the Low Countries were invaded by Rome's legions. Roman occupation came to an end some 450 years later with the rise to power of the Germanic Franks in the early 5th century AD.

The Franks founded a kingdom that covered the area between the Rhine and the Loire rivers. Later they extended it to the coastal area of what is now the Netherlands, which was occupied by another Germanic people, the Frisians.

After the death of the Carolingian emperor Charlemagne (742–814), the greatest of the Frankish rulers, there was a gradual decline in Frankish power. In 888 the Carolingian empire finally broke up and from its ruins the kingdoms of France and Germany emerged.

From the 10th century the Low Countries were effectively ruled by the vassals of the German Holy Roman Emperor. In the late 14th and early 15th centuries, marriages, inheritance and annexation brought many of these territories under the control of the French dukes of Bur-

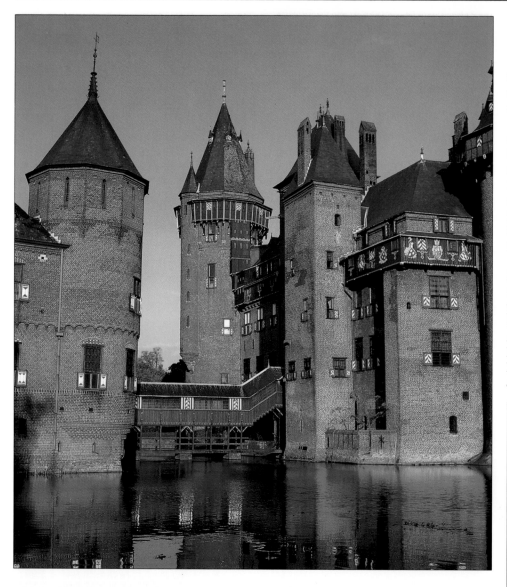

finally recognized the provinces' independence by the Peace of Westphalia agreement in 1648.

The 17th century was a golden age of scientific and cultural achievements for the Dutch Republic. It produced among others, Antonie van Leeuwenhoek (1632–1723), the first person to study microscopic organisms; the greatest Dutch painter Rembrandt van Rijn (1606–69) and the Jewish philosopher Baruch Spinoza (1632–77).

In 1794 the French invaded and when Napoleon I (1769–1821) became emperor of France in 1806 he appointed his younger brother, Louis Bonaparte (1778–1846), as king of the Netherlands. After Napoleon's defeat in 1814, hereditary stadtholder (governor), William VI of Orange (1772–1843), returned as King William I to create a constitu-

The Dutch East India Company

By the end of the 15th century Dutch commercial explorers from the newly independent United Provinces were keen to challenge Portugal for a share of the spice trade.

To divide the risks – as well as the profits – involved in trading with the East Indies, rich Amsterdam merchants invested in a joint-stock company called the Dutch East India Company. The government granted the company a trading monopoly between the Cape of Good Hope and the Straits of Magellan, with the right to draw up treaties with neighboring provinces, maintain armed forces and appoint officials in new territories.

In 1603 the Dutch took the island of Amboina from the Portuguese, in the first of many Portuguese losses.

Malacca fell to the Dutch in 1641, and control of Ceylon, with its precious stones, pearls and cinnamon trees, was finally wrested from the Portuguese in 1656. By 1755, most of the island of Java was in Dutch hands.

In the 18th century the company changed from a commercial shipping venture to a territorial administration interested in agricultural produce. It began purchasing local produce. The produce went to Europe, and the profits remained in Dutch hands. But

gundy. The Low Countries then passed to the powerful Habsburg dynasty through the marriage of Mary (1457–82), daughter of the last Duke of Burgundy, to the Austrian archduke Maximilian (1459–1519), heir to the Holy Roman Empire. In 1504 their son, Philip the Handsome (1478–1506), and his wife inherited the Spanish crown.

The Spanish king who succeeded Philip, Charles V (1500–58), regarded the Spanish Netherlands (as the Low Countries were now called) as a single unit, and attempted to rule them as such. His fiercely Catholic son Philip II (1527–98), who became King of Spain in 1556, was determined to wipe out what he saw as the Protestant heresy in the Netherlands. A Burgundian was appointed as the Catholic Primate of the Netherlands, provoking outright resistance and fueling Protestant opposition movements. Alarmed, Philip II sent Fernando Álvarez de Toledo, third Duke of Alba (1508–82), with an army of 10,000 to the Netherlands to enforce order and deal with all heretics.

The Duke's repressive rule quickly brought open rebellion led by Prince William the Silent, Prince of Orange

De Haar castle, to the northwest of Utrecht. Built in medieval times, its strong defenses afforded protection over the centuries from the warring French, German and Spanish forces in the region.

(1533–84), a Germanborn nobleman living at the Burgundian court at Brussels who came to side with the Protestants. By 1573, most of the provinces of Holland and Zeeland were rebel-held. In 1576 William marched into Flanders, took the town of Ghent and signed a treaty – called The Pacification of Ghent – allowing freedom of worship in the Low Countries. Three years later the seven northern Dutch provinces of the Netherlands signed the Union of Utrecht, creating a loose federation known as the United Provinces. The 1581 Act of Abandonment declared the independence of all the Low Countries from Spain, but it was only truly effective in the north.

In the years that followed, the United Provinces forged and protected their independence with a well-equipped and efficient army. Commercially, the Dutch became Europe's carriers, and burgeoning trade – especially with the East and West Indies – ensured the success of their banking and financial institutions. Spain

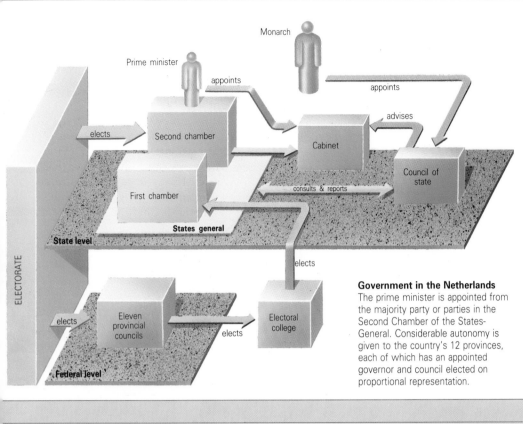

Government in the Netherlands
The prime minister is appointed from the majority party or parties in the Second Chamber of the States-General. Considerable autonomy is given to the country's 12 provinces, each of which has an appointed governor and council elected on proportional representation.

tional monarchy. His kingdom included modern Belgium and Luxembourg, but they seceded to become independent countries in 1831 and 1867 respectively.

The Netherlands remained neutral during World War I, but in May 1940, soon after the outbreak of World War II, the country was invaded by Nazi Germany.

After the war, American economic aid – part of the Marshall Plan – helped the Dutch to create a powerful industrial base and neutrality was abandoned in favor of full participation in Western European affairs. Overseas, a nationalist revolution in the Dutch East Indies (modern Indonesia) eventually led to full independence from the Netherlands in 1949. Independence came to Dutch New Guinea (Irian Jaya) in 1962, though it was annexed by Indonesia the next year, and to Suriname in South America in 1975.

Building an empire In the early 17th century the Dutch East India Company built its headquarters at Batavia (Djakarta) on the Indonesian island of Java. Fleets carrying precious spices returned each year to Holland, to the profit of the shareholders. In 1625 thousands of clove trees were destroyed in order to raise the price of cloves in Europe.

these measures could not halt the company's slow decline. Its monopoly was broken, first by smuggling, and then, after the Treaty of Paris in 1784, by its old rivals the British, who were granted the right to trade in the East

Indies. In 1799, while the Netherlands lay under the dominion of the French Napoleonic empire, the company was finally wound up by the Dutch government, which acquired all its debts and possessions.

Government

The monarch is the head of state, but most real power is vested in the democratically elected parliament, or States-General, which meets in The Hague. The constitution provides for two parliamentary chambers. The 150 members of the Second Chamber are directly elected by the people for four years and only they can propose and amend laws.

The 12 provinces are administered by elected councils and an appointed commissioner. Both national and local elections involve a proportional system in which a large number of parties and opinions are represented.

The Netherlands belongs to the North Atlantic Treaty Organization (NATO), and is a full member of the European Union (EU).

People

Most of the 15 million people speak standard Dutch, though West Frisian is still spoken in the northern province of Friesland. Some Saxon dialects can still be heard in the northeastern area bordering Germany. More than half the population is Christian – Roman Catholics comprise 31 percent and various Protestant groups 21 percent. The Netherlands also has many new immigrant communities, including people from former Dutch colonies, as well as small Turkish and Moroccan minorities, mostly made up of so-called "guest workers".

The country maintains a strong cultural tradition, particularly in music and art, boosted by state subsidy. Drama and literature – seldom translated for a global audience – also remain vigorously creative in the Netherlands.

Wheels of yellow cheese (*above*) are carried on traditional "sledges" at the cheese market at Alkmaar near Amsterdam. Highly productive dairy farming provides milk for processing into cheese and butter, much of which is exported.

Flowers under glass (*right*) A flower exhibition at Zoetermeer, near Rotterdam. Flowers have been an important Dutch industry since the 17th century: in the 1630s tulip speculation ran wild and many people were ruined with the crash of the market.

ECONOMY

Benefiting from its key location at the mouth of the Rhine, the Netherlands has become one of Europe's most prosperous nations.

Agriculture

Farms are large and highly mechanized, but they employ only a tiny percentage of the workforce. About one-fifth of the land is devoted to agriculture and almost two-thirds of this is pasture. The principal cattle-farming regions are Friesland and North Holland, the two coastal provinces on either side of the IJsselmeer. The country remains a major exporter of cheese and butter, and meat products, bacon and eggs are also significant exports. Agricultural produce represents nearly one-quarter of the value of the country's exports.

Crop farming (sugar beet, potatoes and vegetables) is often combined with cattle farming, but is generally less important. The Netherlands grows a considerable amount of salad vegetables – for example cucumbers, tomatoes and lettuces – under glass.

Another unique feature of the landscape is the bulbfields. In addition to the famous export trade in bulbs, cut flowers – tulips, roses, chrysanthemums and lilies – are exported to florists and distributors all over the world.

Forestry and fisheries

Forests account for a little under one-tenth of the land area of the Netherlands. Many smaller woods have been planted as recreational areas, or to prevent the movement of sand dunes.

The small fishing industry has tended to concentrate on shellfish, but stocks have been greatly reduced by the coastal barriers that were erected as flood control measures. The North Sea herring fisheries have also declined, largely due to falling herring populations and the extension of fishing limits by neighboring countries.

Industry

During the 1950s substantial natural gas reserves were discovered, both offshore and in the northeastern province of Groningen. Much of the gas is exported. The Netherlands produces its own crude petroleum from wells in the provinces of Drenthe (in the northeast) and South Holland. Output is too small to meet domestic demand, but much of the crude petroleum that is imported into the Netherlands is reexported after passing through Europoort – a large storage, refining and petrochemical complex outside Rotterdam. At the ports of Delfzijl in the far northeast and Vlissingen in the extreme southwest, smelting works fueled by natural gas have been set up to extract aluminum.

About one-seventh of the labor force works in manufacturing, chiefly located in a densely populated arc known as the Randstad, which runs from Dordrecht in the south, via The Hague, Leiden, Haarlem, Amsterdam and Hilversum, to Utrecht. Important manufacturing industries include food processing, chemicals and petrochemicals, motor vehicles, commercial aircraft, electrical equipment and computers and microelectronics.

Trade, commerce and transportation

Exports from the Netherlands amount to about half its income. More than half the country's external trade is within the EU, in particular with the other Benelux countries of Belgium and Luxembourg and with Germany. The dense road system is extensive and well maintained, with Dutch trucking companies handling a large share of the EU road freight. Utrecht is the hub of the state-owned rail network.

Rotterdam-Europoort, the largest port in Europe, handles the biggest tonnage of any harbor in the world. Nearly 5,000 km (3,000 mi) of canals provide the main arteries for freight transportation, linking the major coastal ports with inland areas and main European rivers. The national carrier Royal Dutch Airlines (KLM) provides passenger services to most major international destinations.

Media and communications

Post, Telegraph and Telephone (PTT), the national mail and telephone service, was privatized in 1989 and divided into PTT-Post and PTT-Telecom. The press includes more than 40 national and local daily newspapers. Broadcasting, which is controlled by the Netherlands Broadcasting Foundation (NOS), is free of government interference.

Health, welfare and education

All citizens are entitled to retirement, disability and survivors' pensions, and there are also children's allowances. For people in the labor market, incapacity and unemployment insurance are compulsory, and most have health insurance.

School attendance is compulsory for children between the age of 6 and 15. Education is provided in both state and private schools, all of them government funded. Universities, colleges and technical institutions are of a high standard.

The industrial complex of Europoort is an extension of the inland port of Rotterdam some 28 km (17 mi) away. Building started on the complex in 1957, and the port can now accommodate the world's largest bulk carriers of oil, ore and grain.

Portugal

REPUBLIC OF PORTUGAL

PORTUGAL OCCUPIES THE WESTERN SIDE OF the Iberian Peninsula in southwest Europe, between the Atlantic Ocean and Spain. Its territory includes the Atlantic islands of Madeira and the Azores.

GEOGRAPHY

The country falls into two main geographical regions, roughly divided by the river Tagus. The lusher, more mountainous north contrasts with the drier lowland areas of the south.

NATIONAL DATA - PORTUGAL

Land area	91,951 sq km (35,493 sq mi)			
Climate		Temperatures		Annual
	Altitude m (ft)	January °C(°F)	July °C(°F)	precipitation mm (in)
Lisbon	77 (252)	11 (51)	22 (71)	702 (27)

Major physical features highest point: (mainland) Estrela (Serra de Estrela) 1,991 m (6,532 ft); (Azores) Ponta do Pico 2,351 m (7,713 ft); longest river: Tagus (part) 1,007 km (626 mi)

Population (2006 est.) 10,605,870

Form of government multiparty republic with one legislative house

Armed forces army 26,700; navy 10,950; air force 7,250

Largest cities Lisbon (capital – 1,977,000); Porto (1,303,000)

Official language Portuguese

Ethnic composition Portuguese 91.9%; African 1.6%; Brazilian 1.4%; others 5.1%

Religious affiliations Roman Catholic 87.4%; other Christian 3.7%; Protestant 1.3%; others 1.1%; nonreligious 6.5%

Currency 1 euro (EUR) = 1 euro cent

Gross domestic product (2006) U.S. $203.1 billion

Gross domestic product per capita (2006) U.S. $19,100

Life expectancy at birth male 74.43 yr; female 81.2 yr

Major resources fish, cork, iron ore, copper, zinc, tin, tungsten, silver, gold, uranium, marble, clay, gypsum, salt, hydropower, cereals, citrus fruits, figs, vegetables, fish, grapes/wine, olives, rice, timber, tourism

The land

Portugal's northeast plateaus are an extension of the Spanish Meseta, divided by mountains, river valleys and deep gorges. Farther west, the land drops away toward the hills bordering the Atlantic coast. The inland plateaus extend south beyond the valley of the river Douro, falling away westward to a broad coastal plain.

At the southeast border of the plateau region, mountain chains form a barrier between northern and southern Portugal, becoming lower as they approach the coast. To their south the river Tagus runs from east to west, entering the Atlantic at the capital, Lisbon.

South of the Tagus are undulating hills and plains. The coastal lowlands spread inland into the uplands of Alto Alentejo in the east, which dwindle southward to form the low rolling hills of Baixo Alentejo. In the far south, the hills and scarps of the Algarve run from the Spanish border across to Cabo de São Vicente at Portugal's southwestern tip.

Climate, vegetation and wildlife

The Mediterranean climate is heavily influenced by sea and landscape. Winters are cool and wet, and summers on the coast are humid. Summers in the interior are hot and dry; drought is common in the south and east. The mountains of the northwest are barriers to clouds from the Atlantic and receive more than 1,000 mm (40 in) of rain a year. The plateaus of the northeast are drier and cold in winter.

The higher rainfall in the north supports forests. The south is less wooded, but the southern Algarve region is famous for its groves of carob, fig and almond trees. Olive trees are found on lower slopes throughout the country, and Alpine vegetation in the higher mountains. Portugal has a rich variety of mammal life, including wolves, wild boar and lynxes. Reptiles are also plentiful, especially lizards and chameleons in the south. Bird life, which includes some species found only on the Iberian Peninsula, is enriched by seasonal migrants en route to and from Africa.

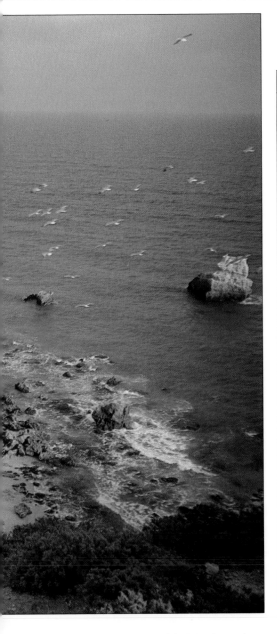

Gulls circle the cliffs near Albufeira (*above*) one of many resort towns in the Algarve, Portugal's busiest tourist area. Once the Moorish kingdom of Al Gharb, the province offers holidaymakers a wonderful climate and good beaches.

SOCIETY

Portugal's history includes an overseas empire in which the Portuguese language and culture have left a significant mark. Macao, its last colony, was returned to China in 1999.

History

Little is known about Portugal's earliest inhabitants, the Iberians. The Celts began to enter through the Pyrenees about 1000 BC. The Lusitani were a Celticized people who gave their name to the Roman province in central Portugal. The modern name comes from the Roman fortress Portus Cale (now Oporto).

Following the breakup of the Roman empire in the 5th century AD, Portugal was contested between the Suebi, invading from the north, and the Visigoths

The Azores

The Azores are a scattered achipelago in the mid Atlantic made up of nine islands and many smaller islets. There are three main groups. The largest, São Miguel, is also the nearest to Portugal, lying about 1,200 km (750 mi) due west of Lisbon. Most of the population lives here and on the neighboring island of Santa Maria. Other islands include Terceira and Pico in the central group, with Flores and Corvo in the far west.

All the islands are volcanic in origin, and the archipelago remains volcanically active. The resulting landscape is varied and often spectacularly beautiful, with many crater lakes. The mild Atlantic climate produces almost no rain during the summer months. The vegetation is rich, but there are almost

An island archipelago of volcanic origin in the mid Atlantic, the Azores were first settled by the Portuguese in 1431. Two crater lakes – Verde and Azur – on the largest island of São Miguel typify the spectacular landscape.

no native trees, and wildlife consists mostly of introduced species.

The uninhabited Azores became a Portuguese colony in 1431. A part of metropolitan Portugal, the islands constitute three of its 22 administrative districts, with a greater degree of autonomy than those on the mainland.

The islands' main crops include tea, tobacco and a variety of tropical fruits. Wine, fish and handicrafts are the main exports. Tourism is an increasingly important source of revenue. There is little industry, but the main airport on Santa Maria is an important transatlantic refueling station.

already present as Roman auxiliaries. The country remained fragmented until it was overrun in 711 by Muslim Moors from northern Africa. Only the north remained in Christian hands. Reconquest was begun by Afonso Henriques (1109–85), who became king. His successors recaptured Portugal from the Moors, until in 1270 the present Portuguese frontier was defined; it is the oldest surviving border in Europe.

The new kingdom soon became involved in power struggles with the Spanish kingdom of Castile. In 1385 John

I of Castile (1358–90) invaded, but was routed at Aljubarotta. Under their own king John I, Portugal established political and trading links with England and Flanders, and conquered Ceuta in northern Africa.

From the 15th century, under the patronage of Prince Henry the Navigator (1394–1460), Portuguese sailors explored the Atlantic coast of Africa and sailed east to India, Indonesia and China. They also crossed the Atlantic to the Americas, where in 1500 Pedro Alvares Cabral (1467–1520) claimed Brazil for Portugal.

Madeira

Madeira is a small archipelago lying in the Atlantic about 1,000 km (600 mi) southwest of Portugal off the coast of Northern Africa. Formed by volcanoes, it consists of two islands and several uninhabited islets.

The island of Madeira is steep and mountainous, with deep valleys. There is little flat land except on the south coast around the capital, Funchal. Porto Santo to the northeast is smaller and flatter, and has only a tiny population. Madeira has a mild climate, with moderate rainfall in the winter, supporting lush vegetation. There are no native mammals, though the rare Mediterranean monk seal may be seen off the coast. There are 40 bird species and numerous insects.

Madeira was uninhabited until it was colonized by the Portuguese in the 15th century. The islands were discovered by João Gonçalves Zarco on an expedition sponsored by Prince Henry the Navigator. The name Madeira, meaning timber, refers to the dense forest that covered the island until it was cleared to provide fertile farmland for settlers.

The islands today constitute one of Portugal's 18 national administrative districts, but with a greater degree of autonomy. The economy depends mostly on the export of handicrafts, bananas, sugar and the famous Madeira wine, which is fortified with cane sugar and has a distinctive smoky taste. Tourism is increasingly important. Visitors are attracted by the spectacular scenery and relaxed lifestyle.

Traditional farming methods are practiced throughout much of northern Portugal. Maize, seen growing here, has been one of Portugal's chief crops since the late 15th century when it was brought back from the Americas by Christopher Columbus (1451–1506)

Due to intermarriage with the Castilian royal family, the throne passed to the Spanish king Philip II (1512–80) in 1580. Spain treated Portugal like a minor colony, fit only for exploitation. Portuguese South American possessions, threatened by the Dutch, were poorly defended. In 1640 the Portuguese revolted, supported by the French. John, Duke of Bragança (1603–56), became King John IV, but Portugal's independence was only recognized in 1668. Portugal's recovery was speeded by a new influx of wealth from its colonies, especially by Brazil's gold and diamond reserves.

The French emperor Napoleon I invaded Portugal in 1807. Portugal's ally Britain evacuated the Portuguese royal family to Brazil, and after a three-year campaign, British troops finally drove the French out of Portugal in 1811. The royal family remained in Brazil, which became a joint kingdom with Portugal. After a democratic revolt in 1820 a liberal constitution was introduced.

Two years later Brazil seceded from Portugal. An inconclusive civil war, known as the War of the Two Brothers (1826–34), marked the beginning of a long period of turmoil and decline. By the end of the 19th century, republicanism was gaining the upper hand. In 1908 King Charles (1863–1908) and his heir were assassinated. His successor Manuel II (1889–1932) was overthrown by a revolution in 1910. Portugal became a republic, initially under the presidency of the writer Téofilo Braga (1843–1924), then in 1911, after an election, under Manuel José de Arriaga (1842–1917).

After continuing unrest, a right-wing army coup in 1926 brought Antonio de Oliveira Salazar (1889–1970), a professor of economics, to prominence. He restored the country's devastated finances, turning Portugal into a semi-fascist "corporative republic" under his personal, if discreet, dictatorship. Salazar refused to allow any progress toward independence in any of Portugal's African colonies. In 1968 Dr Marcelo Caetano (1906–80) replaced Salazar.

The African independence movements became increasingly hostile to Portugal, leading to guerrilla war. In 1974 General António de Spinola (b. 1910) was sacked for criticizing Portugal's colonial policy. His dismissal led to a left-wing coup. Spinola was made president, acceptable to all sides, with an equally broadbased cabinet. Within a year all the Portuguese colonies (excluding only tiny Macao in Asia) had gained independence. The Marxist-led Armed Forces Movement (MFA) became increasingly dominant. The minority Communist Party won control of labor organizations and began expropriating private land. Spinola was ousted in 1974. In the following year the MFA mounted a military coup, but this was suppressed, and a transitional constituent assembly was elected in 1975.

A new democratic constitution in 1976 led to national elections. General António Ramalho Eanes (b. 1935) was chosen as president with Dr Mario Soares (b. 1924) as prime minister. He began to repair the economy, badly affected by the revolution, and Portugal gradually achieved greater stability. In January 1986 Portugal joined the European Community, now the European Union (EU), and later that year Dr Soares was elected Portugal's first civilian president since the 1926 coup.

Government

Portugal has been a republic since 1910, but has been truly democratic only since 1976. The president, as head of state, is chief of the armed forces. He is advised by the Council of the Revolution, which he leads, and by the Constitutional Commission; both bodies are less powerful under civilian rule. The president also appoints a prime minister to form the Council of Ministers, or cabinet. The legislative parliament, the Assembly of the Republic, is elected every four years by proportional representation.

Portugal's traditional provinces have no official status. The 18 administrative districts each have their own governor. The Azores and Madeira districts have a greater degree of autonomy.

People
The Portuguese people are a homogeneous group of typically Mediterranean appearance. Immigrants from the former colonies live mainly in Lisbon and Oporto. Recently many Portuguese have emigrated, particularly from rural areas, to other western European countries.

ECONOMY

Portugal's economy suffered badly from the 1974 revolution, the African guerrilla wars and the loss of income from the colonies. Recovery, now underway, has been further encouraged by Portugal's membership of the EU.

Agriculture and fisheries
Portugal's predominantly agricultural economy has a relatively low output. In the north, small farm holdings limit the possibilities for modern methods. The large estates (*latifundia*) of the south were nationalized in 1975, but have not yet been transformed by modernization.

Large quantities of wheat and maize, the chief crops, still have to be imported to meet domestic needs. Fruit crops include figs and olives. Wine is the major agricultural export, along with tomatoes and cork. Timber and woodpulp are also exported. A commercial fishing industry has developed along the Atlantic coast.

Industry and trade
Portugal has few energy reserves. Gas and petroleum are imported; about one-third of electricity comes from hydroelectric plants. Small amounts of tungsten, copper pyrites and kaolin are exported.

Manufacturing is increasingly important. Heavy industry is concentrated in Lisbon's shipyards and has suffered from falling demand. Fish canning, tobacco processing, cork products, footwear and textiles, and paper are the main light industries and exports. Portugal's chief trading partners are within the EU. Revenue from tourism is growing fast, particularly in the Algarve.

Transportation and communications
The road network covers most of the country and is now being improved. The railroad system, linking all the main towns, has been extensively electrified.

Airlines serve the main airports at Portela de Savacem near Lisbon, Oporto, and Faro in the Algarve. Portugal has a large merchant fleet. Radio and television were nationalized in 1975; satellite television is received from elsewhere in Europe. Most daily newspapers are either owned or subsidized by the state. Censorship ended with the 1976 constitution.

Welfare and education
The state provides an extensive social welfare system. Further schemes are run by trade unions, professional associations and major private employers. There are both public and private hospitals some run by charitable and religious organizations. Free education is available in state schools, and there are some private schools. There are five universities.

Dependencies in the region

GIBRALTAR
UNITED KINGDOM

Gibraltar is a small, self-governing British colony, 6.5 sq km (2.5 sq mi) in extent, occupying the southern tip of Spain. Facing it across the Straits of Gibraltar is the Spanish enclave of Ceuta on the coast of northern Africa.

Gibraltar consists of a single steep limestone ridge of rock shaped like an arrowhead that rises to 425 m (1,394 ft). From the Mediterranean side it is almost unassailable. A low, sandy isthmus joins it to Spain. The Rock has many caves and fortifications from all periods of its history. There are no natural springs or rivers, so rainwater catchment reservoirs occupy a large area of the summit. There is almost no rainfall in the hot summer.

The name Gibraltar derives from the Arabic Jabr al-Tariq, or "Mount Tariq," after the Moorish emir Tariq ibn Ziyad, who captured it in 711. It remained under Muslim rule until 1462. Captured by the British in 1704, during the War of the Spanish Succession, in the 19th century it was developed as a major British naval base and dockyard. Gibraltar played an important strategic role during World Wars I and II, and today it remains an airbase and communications center.

Spain has never ceased to lay claim to it, but in 1967 Gibraltarians voted overwhelmingly to remain British. In early 2002 talks were held between Britain and Spain to resolve the issue and decided to conclude a deal by summer 2002. Joint sovereignty is one of the issues under discussion but any decision made will be determined by a Gibraltarian referendum.

The government of the colony is headed by a governor, as representative of the British sovereign, who appoints a council of ministers from the majority party in the 18-member assembly. Gibraltarians are a mixture of Moorish, Maltese, Spanish, Genoese, British and Jewish descent. Most speak Spanish and English, and most are Roman Catholic. They possess full British citizenship. About one-third of the population are resident foreigners, including Asians, Moroccans and British military personnel. Spaniards cross daily to work on the Rock.

The colony's economy depends largely on the extensive port facilities and on light manufacturing, banking and finance. All food is imported. Tourism is becoming increasingly important, and is served by the airport close to the Spanish border. There is good education and healthcare.

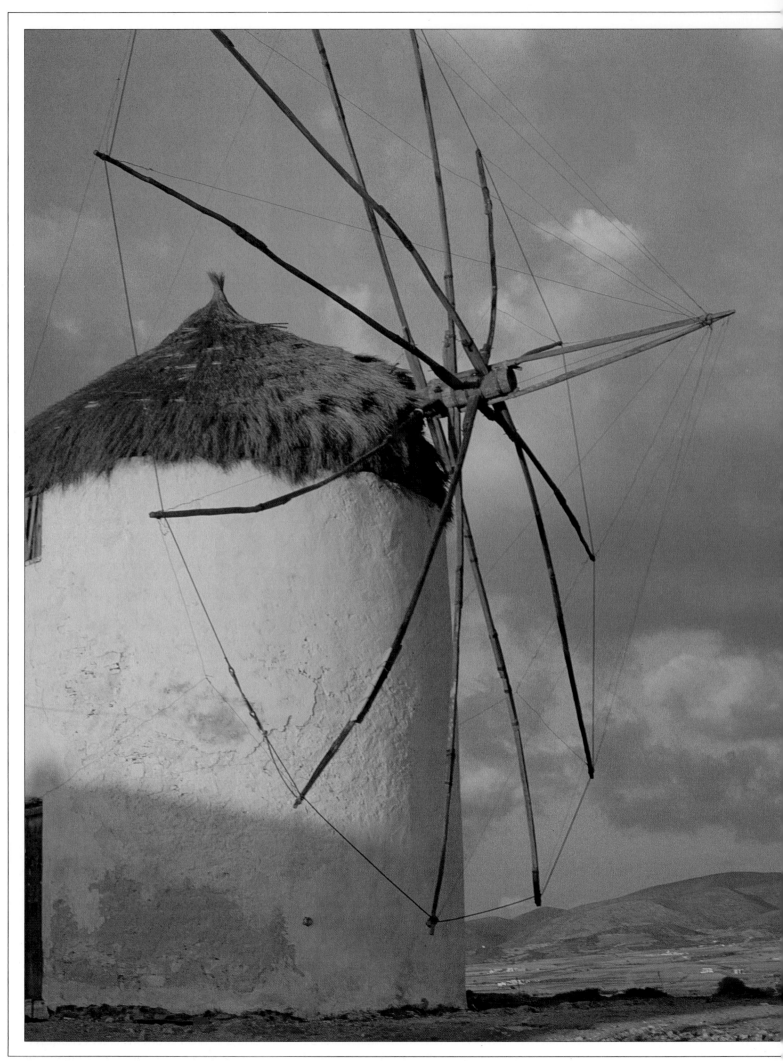

Italy and Greece

COUNTRIES IN THE REGION

ITALY · VATICAN CITY · SAN MARINO

GREECE · MALTA · CYPRUS

Bathed in Mediterranean sun (*left*), a Greek windmill on Antiparos in the Cyclades, a group of islands off the southeast coast of Greece. The harsh, arid, sun-baked landscape and whitewashed buildings are common to both Italy and Greece, where traditional ways of life are still followed in many rural areas.

- ■ capital city
- ● major town

height of land (meters)

- 3000
- 2000
- 1000
- 500
- 200
- 0

- ▲ mountain peak

Italy

ITALIAN REPUBLIC

ITALY HAS ONE OF THE OLDEST HISTORIES in Europe, second only to Greece. It is also one of Europe's youngest unified states. The rural landscape is spectacular, and the country's long cultural history has left a legacy of ruined temples and classical sculptures, medieval castles and fine Renaissance churches and painting.

NATIONAL DATA - ITALY

Land area	294,020 sq km (181,252 sq mi)			
Climate		**Temperatures**		**Annual**
	Altitude m (ft)	January °C(°F)	July °C(°F)	precipitation mm (in)
Rome	17 (56)	8 (46)	23 (73)	724 (28.5)

Major physical features highest point: Monte Rosa 3,505 m (11,500 ft); longest river: Po 652 km (405 mi); largest lake: Lake Garda 370 sq km (143 sq mi)

Population (2006 est.) 58,133,509

Form of government multiparty republic with two legislative houses

Armed forces army 112,000; navy 34,000; air force 45,152

Largest cities Rome (capital - 2,547,677); Milan (1,308,735); Naples (984,242); Turin (900,608); Palermo (670,820); Genoa (620,316); Bologna (373,473); Florence (366,901)

Official language Italian

Ethnic composition Italian 94.1%; Sardinian 2.7%; Rhaetian 1.3%; others 1.9%

Religious affiliations Roman Catholic 90%; Protestant and Jewish communities and growing Muslim immigrant community

Currency 1 euro (EUR) = 100 euro cents

Gross domestic product (2006) U.S. $1.727 trillion

Gross domestic product per capita (2006) U.S. $29,700

Life expectancy at birth male 76.88 yr; female 82.94 yr

Major resources coal, hydroelectric power, mercury, zinc, potash, marble, barite, asbestos, pumice, fluorospar, feldspar, pyrite (sulfur), natural gas and crude oil reserves, fish, cereals, citrus fruits, grapes/wine, iron ore, livestock, olives, potatoes, sugar, vegetables, tourism

GEOGRAPHY

To the north Italy is separated from France, Switzerland, Austria and Slovenia by the formidable barrier of the Alps. Southward, beyond the fertile expanse of the great northern plain, are the Apennines, which form a rocky backbone running from the Alps in the northwest down to Calabria in the toe of the peninsula. The islands of Sicily to the south and Sardinia to the west each have a distinct character of their own.

The land

The western and central Alps reach their highest points in Italy at Mont Blanc – more than 4,800 m (15,700 ft) – on the border with France, and at Monte Rosa – 4,634 m (15,203 ft) – on the Swiss border.

The northern Italian lakes – Maggiore, Como, Iseo and Garda, noted for their beauty – are water-filled valleys scooped out by ice age glaciers. Northeast of Lake Garda, the Dolomites, a breath-taking series of limestone peaks that are part of the eastern Alps, rise to the Marmolada – 3,342 m (10,965 ft).

The fertile soils of the north Italian plain were washed down by the melting of ice age glaciers. Over 320 km (200 mi) long and up to 200 km (125 mi) wide, more than half of its 70,000 sq km (27,000 sq mi) area is drained by the Po, Italy's longest river.

The long, rocky spine of the Apennines, south of the plain, begins in the Ligurian Alps. The mountains hug the northwest coast of the Italian Riviera, and then sweep in a bow shape down the

The Amalfi coast (*left*) as seen from the Villa Rufolo at Ravello, south of Naples. Many small resort towns such as Ravello shelter between the steep cliffs of this spectacular peninsula, where earth tremors are still a frequent occurrence.

The towering cliffs of Volterra (*above*), to the southeast of Pisa, form part of the Colline Metallifere, or Ore Mountains, rich in minerals and rising to a height of 1,050 m (3,400 ft). Farther north, the Apuan Alps are a famous source of high-quality marble.

length of the peninsula, reaching their highest point at Monte Corno in the Gran Sasso d'Italia northeast of Rome. On the western side, rivers such as the Arno and the Tiber run through a complex landscape of hills and plateaus. To the east, dozens of rivers create a landscape of parallel ridges and valleys running down to a narrow coastal plain. In the southeast a limestone platform forms the promontory of Gargano, and in the extreme south the Gulf of Taranto divides the "heel" (Apulia) from the "toe" (Calabria) of Italy.

Across the Strait of Messina lies the island of Sicily, the largest island in the Mediterranean. Almost entirely mountainous, with the highest ranges in the north, the eastern end is dominated by the active volcano of Etna. Italy lies on the boundary between the Eurasian and the African tectonic plates where earth tremors occur frequently. Other volcanoes in the region include Stromboli in the Lipari islands to the north of Sicily, and Vesuvius on the Bay of Naples, made famous by the eruption of 79 AD, which destroyed the city of Pompeii. The nearby Phlegraean Fields bubble with volcanic gases. Near Rome, the uplands between Monte Amiata and the Alban Hills are clearly volcanic: there are mud springs at Viterbo, and the area is dotted with crater lakes such as Bolsena and Nemi. Thermal springs rise as far north as the Euganean Hills near Padua. The whole region is subject to occasional severe earthquakes.

Off the western coast of Italy, the rugged mountains of the island of Sardinia are generally formed of older igneous and metamorphic rocks than those of the mainland.

Climate

In the Alps, the winter months are usually cold and can bring heavy snow, while summers are short and cool; but there is considerable variation. Lower temperatures on the higher mountain slopes tend to be offset by intense sunshine.

The northern plain, which is shielded from the Ligurian Sea by the Ligurian Alps, enjoys long, warm summers ideal for cultivation, though winters are short and often bitter. Farther south the climate is characteristically Mediterranean, with warm and sunny summers and mild winters. The high Apennines receive up to three times as much rainfall as the lowlands, where there may be prolonged summer drought.

Strong winter winds blow off the Ligurian Sea. The island of Sardinia endures the cold northerly maestrale, while the Italian Riviera is swept by the westerly *libeccio*. In spring the hot *sirocco* from Northern Africa brings hazy, stifling conditions to southern Italy.

Plants

Like much of Europe, the Italian lowlands, were once clothed in dense woodland, but little remains in the northern plains after 22 centuries of intensive farming. Cultivated crops dominate the landscape; even the poplars on the river margins have been planted. Farther south, cleared and abandoned woodland has reverted to a dense scrubland of aromatic plants (macchia) or, in stony areas, to garigue – open ground scattered with low shrubs and flowering annuals. In some areas, such as Calabria in southern Italy, the environment has been reduced to near desert.

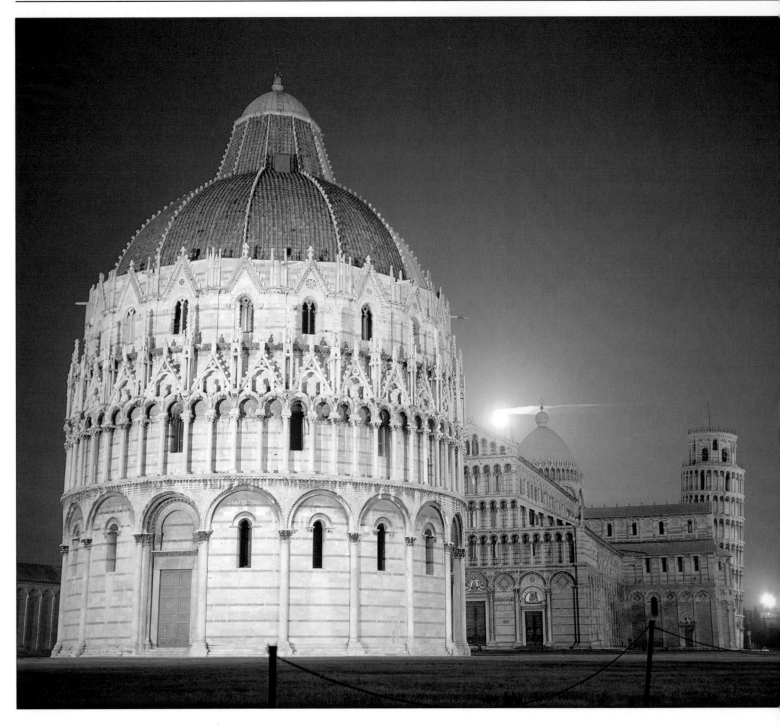

Rainfall in the Apennines encourages deciduous oak, chestnut and beech trees. On the western side and in Sardinia, evergreen oaks cling to lower mountain slopes up to about 600 m (2,000 ft). In the Alps very few plants can survive above 2,250 m (7,500 ft). Forest management has preserved stands of oak and sweet chestnut, which have also survived in the foothills. Around the lakes olive, cypress and laurel flourish.

Animals
Centuries of hunting and other human activities have reduced the number of animal species to be found in Italy, though the Alps shelter animals such as ermine, Alpine rabbits and mountain partridge; larger Alpine mammals include ibex and the increasingly rare Brown bear

(both protected species) as well as chamois and roe deer. In the south of the country foxes and even wolves live in the mountains; on Sardinia there are fallow deer, mouflon sheep and wild boar. The variety of fish around the coasts include bluefin tuna, swordfish and white shark.

Pollution is a massive and growing problem. Along the Adriatic coast seasonal blooms of algae, nourished by the nutrients in the industrial and agricultural chemicals washed down by the rivers of the northern plain into the sea, have deprived other marine life of oxygen, killing thousands of fish, and covering tourist beaches in a stinking slime. However, new water purification plants, brought into operation in the 1990s, have begun to improve conditions in the sea.

Distinctive buildings of Pisa. The leaning tower, one of the world's most famous buildings, now tilts more than 5.2 m (17 ft) in contrast to the cathedral baptistery (foreground) and the Duomo. Signs of subsidence began appearing soon after construction began in 1174.

SOCIETY

In the course of its long history Italy was at one time the power base of the mighty Roman empire, the birthplace of the Roman Catholic Church and the cradle of the Renaissance. It has a literary tradition going back almost unbroken to Roman times, and a long history of political disunity that continues to this day.

History
As early as the 8th century BC Italy was dominated by two major civilizations: the Greeks in Sicily and the south, and the

Etruscans in the center. Under Etruscan rule the small city of Rome became increasingly important as a power base and trading center, but in the late 6th century BC the native Latin-speaking people of the city – the Romans – overthrew the Etruscans and established Rome as a republic. Over the next five centuries the Romans waged war almost continuously on the neighboring peoples, imposing centralized power across the whole peninsula. Under Rome's first emperor, Augustus, they were able to gain control of the entire Mediterranean area.

At its height, in the 1st and 2nd centuries AD, the Roman empire stretched from Britain in the northwest to Egypt in the southeast, creating the cultural boundaries of modern Europe. However, by the 3rd century the western half of the empire had begun to decline economically, and the emperor Constantine established a new capital, Constantinople, in the east. The western empire became progressively weaker as waves of invaders from central Europe swept into Italy, and in 476 the last Roman emperor was deposed.

During the mid 6th century the country was invaded again, this time by the Eastern empire of Byzantium, which claimed Italy as an outlying province. Further Germanic invaders, the Lombards, conquered much of Italy in the late 5th and early 6th centuries, but were themselves overthrown by the Frankish king, Pepin. He gave the lands around Rome and Ravenna to the pope, the head of the Roman Catholic Church, marking the beginning of the Papal States.

In the 9th and 10th centuries Italy was nominally unified within the Holy Roman Empire, but the country remained fragmented into powerful city-states. Among the greatest in late medieval times were Florence and the maritime states of Genoa and Venice. From the 15th century the north came increasingly under the rule of Spain, and later of the Austrian empire. The south, particularly Sicily and Naples, also came under Spanish rule, but then changed hands several times between Spain, Austria and France. The country did not break free from foreign domination until the bloody wars of the Risorgimento, a cultural and political move-

ment for unification led by figures such as Giuseppe Garibaldi (1807–82) and the Piedmontese statesman Conte Camillo Benso di Cavour (1810–61). Victor Emmanuel II (1820–78), king of Sardinia and Piedmont, was installed (1861) as the first king of a reunited Italy. Venice joined in 1866, and Rome in 1870. However, the pope, having been granted sovereignty over the Vatican City, refused to recognize the new order. San Marino also remains an independent enclave.

In the late 19th and early 20th centuries Italy became a colonial power in its own right, occupying Libya, Italian Somaliland and Eritrea. In 1922 Benito Mussolini (1883–1945) seized power. Under his fascist dictatorship Italy expanded its empire to include the whole of Ethiopia between 1936 and 1945, and became an ally of Nazi Germany during World War II. In 1943 Mussolini lost support of the Fascist Council and was arrested by royal command. Though rescued and reinstated by the Germans as a puppet, he was murdered in 1945 trying to escape. The monarchy was discredited, and in 1946 it was abolished and a republic declared.

Florence – city of the Renaissance

During the Middle Ages, Florence – a town on the river Arno at the foot of the Apennines – emerged as one of the most powerful city-states in Italy, its wealth based on woolen cloth and international banking. Its merchant aristocracy supported social improvements ranging from street paving to schools, hospitals and one of Europe's foremost universities. This turbulent but exciting time produced many talents, notably the painter and architect Giotto (1266–1337) and two of the world's greatest writers, Dante Alighieri (1265–1321) and Giovanni Boccaccio (1313–75). The city entered a period of financial eclipse in the mid 14th century, but recovered under the powerful merchant princes of the Medici family, Cosimo (1389–1464), Piero (1416–69) and Lorenzo the Magnificent (1449–92), who ruled Florence by dominating its senate, the Signoria.

Directly or indirectly, the Medici were the patrons of a remarkable generation of artists, writers and

thinkers who have come to epitomize "Renaissance man" – widely talented creators who were deeply interested in all aspects of humanity and human knowledge. Leonardo da Vinci (1452–1519) – artist, inventor, writer – exemplified Renaissance man, as did Lorenzo de' Medici himself, a notable poet and naturalist. Other Renaissance artists included the architect Filippo Brunelleschi (1377–1446) and innumerable painters, among them the Della Robbia family; Tommaso Masaccio (1401–28); Sandro Botticelli (1445–1510); Raffaello Sanzio, known as Raphael (1483–1520); and Michelangelo Buonarroti (1475–1564).

Under the later Medici, Florentine arts declined, and by the 17th century they had stagnated altogether. Their influence, however, had spread far beyond Italy, and stagnation preserved many Renaissance treasures intact – to the benefit of the thriving tourist trade, which began in the 18th century. Modern Florence still owes its rich character and its distinctive skyline to the Renaissance churches, palaces and public buildings that are its living inheritance from the Medici.

Detail from "La Primavera (Spring)" by Sandro Botticelli (1445–1510), tempera on wood. Botticelli used line to depict both contour and movement, showing typical Renaissance passion for sculpturally solid classical figures bathed in light.

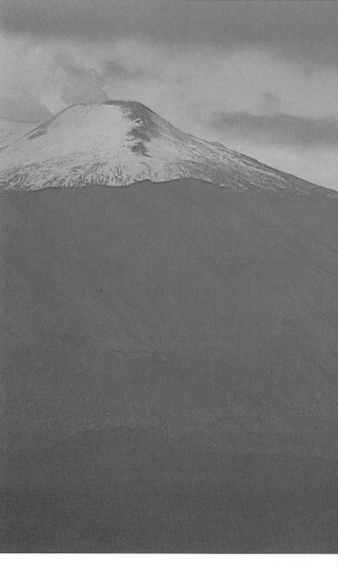

People

Centuries of foreign domination have left a distinctive mark on the Italian people and their dialects. In the south and Sicily, Greek and Arab influences are strong, while in the north people tend to be taller and fair-skinned. In the postwar period these distinctions became more blurred as many people from the rural areas of the south migrated to the industrialized cities of the north, such as Milan and Turin.

In the Valle d'Aosta and Trentino-Alto Adige (South Tyrol) regions – isolated enclaves in the high Alps on the northern borders – there are long-established French- and German-speaking communities. The Alto Adige only became part of Italy after World War I, as a reward for victory over Austria, and many of its inhabitants resented the transfer of authority. There are also significant ethnic minorities from neighboring states such as Albania and Slovenia.

Each of the former independent states boasts a distinctive character and dialect, but the Tuscan form of the Italian language, which was used by the 14th-century poet Dante Alighieri in his masterpiece the *Divine Comedy*, has become the literary standard. The Roman Catholic Church has played a major part in shaping the country. Until 1870 part of the country was under direct papal rule. More than 90 percent of Italians are Catholic, though in the postwar period the authority of the church has declined, particularly over controversial moral issues such as divorce, contraception and abortion. Religious regard for the sanctity of family life is still important, especially in rural communities.

The Italian Renaissance of the 14th to 16th centuries created an astonishing

Italy's government (*above*) The president, elected by parliament and the regional councils, has little actual power. A 1997 parliamentary commission recommended great changes to the constitution including a directly elected president.

Mount Etna (*right*) Europe's highest active volcano at 3,323 m (10,902 ft) looms over eastern Sicily. It remains snow-capped for six months each year and at night the glow of molten lava from its crater can be seen for many miles. It has been active for at least the last 2,500 years.

wealth of painting and sculpture, as well as a school of music that dominated Europe throughout this time. The artists of Florence in particular turned to the art of classical Greece and Rome for inspiration. It was also a period that saw the growth of skepticism and freethinking, belief in the individual and new discoveries in science and mathematics.

Opera is an Italian invention: it is still a national passion today and, like the rest of the country's magnificent heritage, is a potent tourist attraction. The richness of Italian cooking has been equally influential in more recent years. Widespread emigration has established Italian cooking throughout the world. In the growing fast-food culture the Italian pizza, introduced into the United States by Italian immigrants, has been the hamburger's greatest rival.

Government

The present structure of Italian government was adopted in the Republican constitution of 1947. The country is divided into 20 autonomous regions, each with its own legislature and executive council headed by a regional president.

The national parliament consists of a lower chamber of deputies with 630 members elected for a five-year term, and an upper house of 315 senators elected by region for five-year terms. There are also 11 life senators appointed by the president, who is elected by both chambers for a seven-year term.

There were many changes among political parties in the 1990s: the Christian Democrats were devastated by scandal and reformed into the Italian Popular Party (PPI); the Communists split into the Democratic Party of the Left (DS) and the Reconstructed Communism Party (RC); and the centrist Democratic Union for the Republic (UDR) was formed in 1994. Three new parties, Forza Italia, the Northern League and the National Alliance caused ripples of alarm across Europe when they took a majority of seats at the 1994 general election, espousing neo-fascist sentiments. Proportional representation and fierce regional self-interest have contributed to a splintered political scene with many coalition governments – 49 between 1945 and 1991. There were two attempts to abolish proportional representation in the 1990s by referendum, but low turnouts led to no mandate.

Activities of the Mafia have been a major challenge to governments. A determined effort was made to combat the Mafia in the 1990s, following the murders of some members of the judiciary.

ECONOMY

Italy's dependence on imports for food and energy has created a large and almost permanent balance-of-trade deficit and a weak currency. However, despite a huge public deficit, industrial unrest, high wage settlements and high inflation, the country had experienced a boom economy and a thriving manufacturing industry until the beginning of the 1990s, when it was hit hard by the world recession. Recovery was slow and tough economic policies were pursued by the government in order to meet European Union criteria that would allow Italy to achieve European monetary union in 1999.

Agriculture

Despite a long farming tradition, Italy is a massive net importer of agricultural pro-

Small family-run farms (*below*) are a feature of the Italian landscape, and they employ 6.6 percent of the workforce. Italy is one of Europe's major producers of fruit, vegetables, wine and olive oil. About 40 percent of the land is arable.

ducts. Although some 65 percent of available land is still farmed, much of it is in hilly or mountainous country. The fertile northern plain accounts for roughly 40 percent of total production, almost as much as the whole of the south. Of more than 3 million farms, 94 percent are very small; one family will own several holdings, and part-time farming is common. Typical livestock includes cattle, pigs, chickens, sheep and goats. Traditional exports (wine, olives and fruit) remain important despite growing competition within Europe.

Forests account for some 25 percent of usable land, but the quality of the timber is generally low and none is exported. The small fishing industry relies largely on individual enterprise. Pollution of the Adriatic from industrial agricultural chemicals is now a serious threat to the fishing industry.

Industry

There are few mineral resources in Italy; only mercury, sulfur, rock salt and marble are present in significant quantities. However, other minerals are also extracted, notably in Sardinia and Sicily. Oil deposits in Sicily provide some 6 percent of Italy's energy needs. Reserves of iron ore, coal and natural gas are too small to meet national requirements, and up to 80 percent of the country's energy is imported. However, Italy earns valuable revenue by processing imported metals and refining imported oils.

The 1981 National Energy Plan proposed energy conservation, a nuclear power program, increased use of gas and the conversion of oil-fired power stations to coal. Implementing these ideas has proved slow and laborious, particularly as the construction of nuclear power complexes (in Piedmont, Lombardy, Latium and Apulia) has been blocked by both local governments and environmental protection movements.

In this mountainous country hydroelectricity is an obvious potential source of power. More than half the power requirements of northern Italy are met by installations near lakes Como, Maggiore and Garda, and there are similar schemes in Sardinia. Geothermal power sources are also being tapped while a major natural gas pipeline runs from Algeria.

Manufacturing industry, now involving more than one-fifth of Italy's labor force, is largely concentrated in the north. FIAT (for Fabbrica Italiani Automobile Torino) in Turin is one of Europe's largest manufacturers of automobiles. Fashion items, especially shoes and clothes, are important exports, as are chemicals for agriculture and industry. Until the late 1990s state involvement in industry was the highest in the European Union, with three state holding companies. The Instituto per la Ricostruzione Industriale (IRI) is chiefly involved in mechanical products, steel, banking and airlines; the Ente Nazionale Idrocarburi (ENI) in hydrocarbons; and the Ente Finanzario per l'Industria Meccanica (EFIM) in light engineering.

Tourism is an important source of revenue, providing employment and an influx of foreign currency. Each year, Italy receives over 20 million tourists.

Trade and commerce

Trade and manufacturing industry are major contributors to the economy, closely followed by services. Italy's chief trading partners are Germany, France, the United States, the United Kingdom, the Netherlands, Switzerland and Belgium. Imports include machinery and transportation equipment, precision machinery, crude petroleum, chemicals and chemical products, food and live animals, and metal and semiprocessed metal.

Transportation and communications

Italy's road network, the origins of which date back to Roman times, is the heart of its communication system, covering some 300,000 km (186,000 mi). To date the IRI has built some 7,000 km (4,500 mi) of autostradas (toll expressways). Other roads are well maintained but cannot

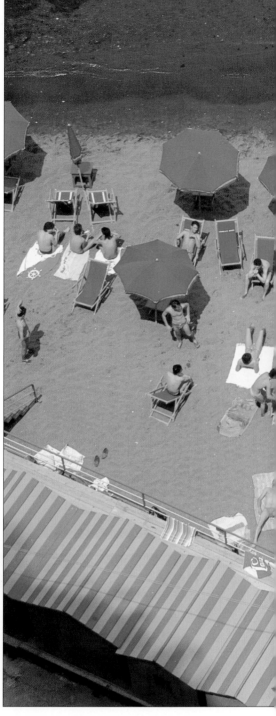

The Galleria in Milan (*below*) is a bustling arcade of shops and cafés. Spurred by a huge postwar influx of immigrants from the impoverished south, Milan developed into Italy's financial, manufacturing and commercial capital.

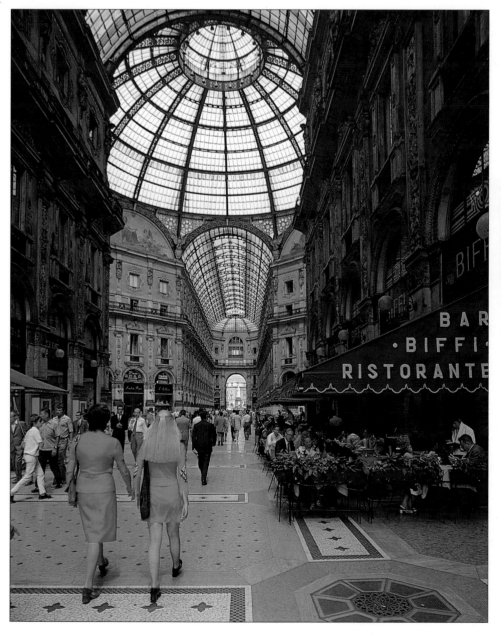

support fast-moving traffic. There are many road passes over the Alps, but most are closed during the winter; the all-weather St Bernard and Mont Blanc tunnels keep traffic flowing at these times.

The part state-run railroad, Ferrovie dello Stato, is being modernized. More than half the 16,000 km (10,000 mi) of track are now electrified, and more investment is being made in the system. Main routes run parallel to the autostradas, and in the Alps railroad traffic passes through the Mont Cenis, Simplon, St Gotthard and Brenner passes. Rome has a clean and efficient subway system.

Apart from the Po, most of Italy's rivers are unsuitable for navigation. The Italian merchant fleet carries less than one-third

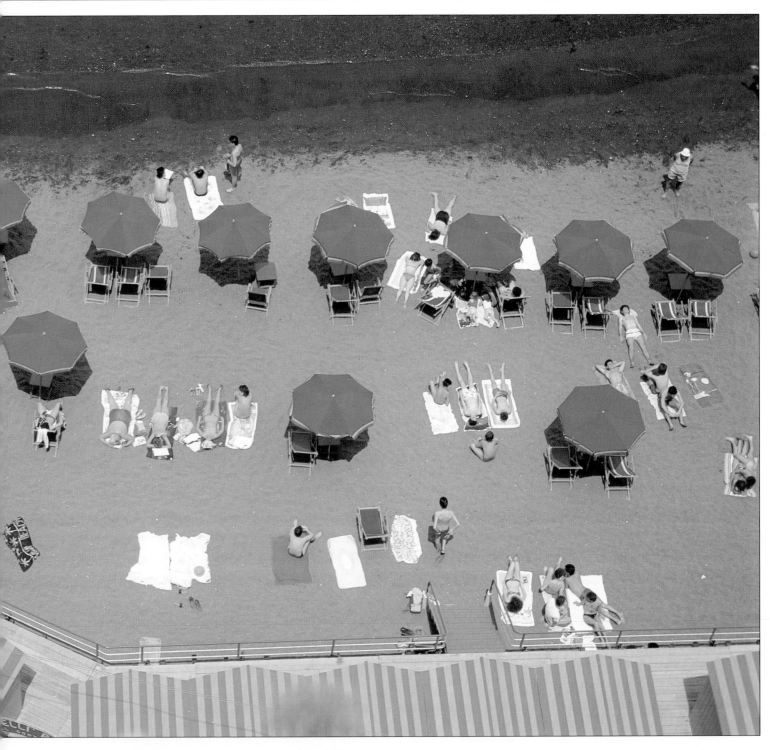

A mecca for sunlovers Rows of orderly sun umbrellas such as these at Sorrento, south of Naples, are a common sight on Italy's long, sandy beaches. Each year some 20 million tourists visit the country, providing a vital contribution to the economy.

of the country's external trade, creating a massive balance-of-payments deficit on maritime freight. The state, a major shareholder in the merchant fleet, has responded with a restructuring program that includes the purchase of new vessels. Principal services are within the Mediterranean and to North America and Australia. Genoa, Naples, Venice, Trieste and Leghorn are the chief ports.

The national airline, Alitalia, flies to most major cities worldwide. Italy has over 30 airports, the largest being Leonardo da Vinci, 24 km (15 mi) southwest of Rome.

There are nearly 100 daily newspapers; the most influential are *Corriere della Sera*, published in Milan, and *Repubblica*, pub-lished in Rome. In addition to the two state TV channels there are several hundred private radio and television channels appealing to local audiences.

Health and welfare

Since 1980 Italians have enjoyed a comprehensive national health service that aims to provide free medical care for all citizens. However, minimum charges are made for essential medicines, medical examinations and hospital treatment. A unified medical insurance plan is also being put into operation. Welfare includes unemployment and disability benefits, as well as retirement pensions and family allowances, and there is a scheme for industrial injuries.

Education is compulsory between the ages of 6 and 14, and there is a standardized national curriculum. Private schools, often run by the church, flourish alongside the public elementary schools (ages 6–11) and secondary schools (ages 11–14). Preschool facilities (ages 3–6) are excellent. Further education and preparation for university is available at a wide variety of institutions, and although scholarships or grants are hard to obtain, recent legislation aims to improve this situation. Standards of literacy are high.

Vatican City

ALTHOUGH COMPLETELY SURROUNDED BY the city of Rome, the Vatican City is an independent state – the smallest in the world. It lies on the west bank of the river Tiber, its medieval walls cutting it off from the city beyond – except at St Peter's Square on its southeastern flank.

The Vatican City is the last remnant of the Papal States, the great temporal domain that successive popes had built up around Rome over the centuries. The papacy refused to recognize the incorporation of these lands into the newly independent Italian state in 1860 and 1870. In 1929, however, it finally con-cluded the Lateran Treaty with the government of Mussolini, ceding all claims to the papal territories in return for sovereignty over the Vatican City.

Within its walls the pope has absolute legislative, executive and judicial powers, exercised in practice through a commission of five cardinals and an appointed governor and council. The Vatican also has its own army, the Swiss Guard – about a hundred soldiers, whose 16th-century parade uniforms belie their responsibility for papal security.

The Vatican operates its own radio station, telephone and postal systems independently of the Italian state. It has its own banking system, issuing its own coinage and having no exchange controls. This is a powerful advantage in the Vatican's complex involvements in international finance, carried out chiefly through its merchant bank, L'Istituto per le Opere di Religione.

All essential supplies and energy are imported and the Vatican's extensive publishing operations are the nearest thing to an export industry. Its products are sold all over the world, and its daily newspaper, *L'Osservatore Romano*, has influence far beyond the city boundaries.

Tourism is also important. The city attracts great numbers of visitors from all over the world, both for its religious significance and for its rich cultural legacy. Pilgrims flock to St Peter's basilica, until recently the world's largest church, to celebrate the great festivals of Christmas and Easter. The church dominates the main public entrance to the city. A church stood on this site in the 4th century, but the building took its present shape in the 16th and 17th centuries when it was embellished by some of the greatest artists and sculptors of the Renaissance, among them Raphael, Michelangelo and Gian Lorenzo Bernini (1598–1680). The most famous works of art in the Vatican City are probably the frescoes on the walls and ceiling of the Sistine Chapel, the personal chapel of the popes in the Vatican Palace. These include priceless works by both Michelangelo and Raphael.

The Vatican also has jurisdiction over a number of buildings elsewhere in Rome, including the churches of St John Lateran and St Paul Outside the Walls, the largest church in the city after St Peter's; and the pope's summer palace at Castel Gandolfo, about 20 km (12 mi) southeast of Rome.

St Peter's Square, the magnificent keyhole-shaped colonnade in the Vatican, seen from the dome of St Peter's basilica. Many great artists contributed to rebuilding St Peter's in the 16th century, a grandiose project designed to be a marvel of Christendom. The cost was so great that it led to abuses in the sale of indulgences to raise funds, precipitating an attack on the papal system by the German reformer Martin Luther (1488–1546) and leading to the Reformation.

NATIONAL DATA – VATICAN CITY

Land area	0.44 sq km (0.17 sq mi)			
Climate		Temperatures		Annual
		January °C(°F)	July °C(°F)	precipitation mm (in)
		8 (46)	23 (73)	724 (28.5)

Major physical features none

Population (2004 est.) 780

Form of government city-state with one appointed absolute ruler - the pope

Armed forces Swiss Guard 100

Official language Italian

Official religion Roman Catholicism

Religious affiliations Roman Catholic 100%

Currency 1 Vatican lira (VL) = 1 euro (EUR) = 100 euro cents

Major resources tourism

San Marino

REPUBLIC OF SAN MARINO

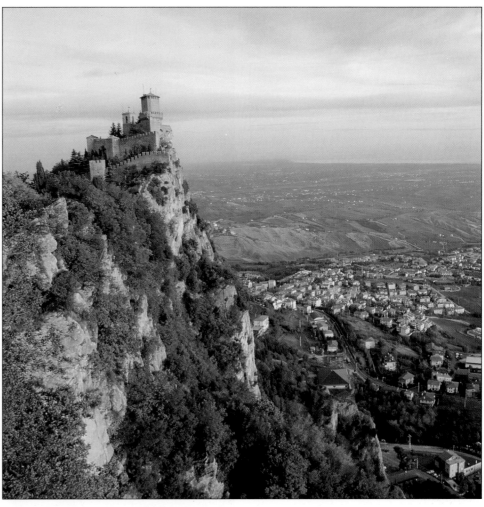

THE MOST SERENE REPUBLIC OF SAN MARINO, smallest republic in Europe, second smallest in the world, is an enclave in the Apennine mountains of northern Italy. The towers and battlements of the Rocca Fortress in San Marino, the capital, are perched on the rugged western slope of Mount Titano, while woods and pastureland cover the foothills. This ancient state largely owes its survival to its rugged and isolated terrain, which helped preserve it from predatory neighbors throughout the turbulent Renaissance and after.

Legends trace San Marino's origins to the 4th century AD, when St Marinus, a Dalmatian stonemason, was thought to have settled there with his Christian followers. In recorded history, it first appears in the 12th century as a self-governing commune.

In 1600 a constitution was adopted that remains broadly in force today (although in the 18th century government was shared between a few leading families). The Congress of Vienna (1815), which met to decide the future shape of Europe after the Napoleonic wars, agreed to preserve San Marino's independence. In the mid 19th century the leaders of the Italian unification movement, among them Giuseppe Garibaldi (1807–82) took refuge in San Marino, and this helped it to conclude favorable treaties with the new Kingdom of Italy after 1861. This newfound security, along with the development of modern roads and communications, helped reduce San Marino's isolation. Although many of its people have emigrated, chiefly to Italy and the United States, a large number of Italians have also moved to San Marino and settled there. Italian is the official language, though a much older dialect is also still spoken.

The mountainous landscape, ancient fortifications and unspoilt countryside now attract increasing numbers of tourists, which enables the republic to survive as an independent state. Until this century agriculture, well suited to the mild Mediterranean climate, was the main occupation. Cattle are reared on mountain pastures, and the lower slopes are planted with vines and cereals. Until the 1940s there was no industry except stone quarries (now exhausted) on Mount Titano. Since World War II various light manufacturing industries have been introduced, though tourism is quickly overtaking them as a source of revenue. San Marino is now investing in its own resort hotel facilities. Related industries, ranging from traditional handicrafts to the republic's own finely printed postage stamps, have also benefited.

An extensive welfare system is provided; many San Marino citizens are employed on state concerns, and assisted with home ownership. Free education is provided to the age of 14, and there are grants for further study outside the republic. Government is headed by a two-house parliament. The Great and General Council are elected by universal suffrage at five-year intervals; every six months the Council elects the State Congress, a council of executive ministers, from among its own members, and two captains-regent to head it. The largest political party is traditionally communist, but moderate in the Italian tradition.

The Rocca Tower fortress dominates San Marino, the triple-walled capital of Europe's smallest republic, self-governed since the 12th century. Its first known settlers were a group of Christians led by St Marinus seeking refuge from persecution in the 4th century.

NATIONAL DATA - SAN MARINO

Land area	61 sq km (24 sq mi)
Climate	mediterranean
Major physical features	highest point: Monte Titano (capital on its slopes) 739 m (2,424 ft)
Population	(2006 est.) 29,251
Form of government	multiparty republic with one legislative house
Capital city	San Marino (2,294)
Official language	Italian
Ethnic composition	Sammarinesi 84.9%; Italian 14.6%; others 0.5%
Religious affiliations	Roman Catholic 95.2%; nonreligious 3%; others 1.8%
Currency	1 euro (EUR) = 100 euro cents
Gross domestic product	(2001) U.S. $940 million
Gross domestic product per capita	(2001) U.S. $34,600
Life expectancy at birth	male 78.23 yr; female 85.5 yr
Major resources	building stone, cattle, dairy products, olives, vines, wheat, tourism

Greece

HELLENIC REPUBLIC

T HE GREAT FLOWERING OF ART, LITERATURE and ideas that took place in the city-states of Greece between the 8th and 5th centuries BC has been a major influence on the development of European culture and civilization. Today the country is renowned throughout the world for its rich cultural heritage – particularly its architecture and sculpture – as well as its great natural beauty.

NATIONAL DATA – GREECE

Land area	130,800 sq km (50,502 sq mi)			
Climate		Temperatures		Annual
	Altitude m (ft)	January °C(°F)	July °C(°F)	precipitation mm (in)
Athens	107 (351)	9 (48)	28 (82)	424 (16.6)

Major physical features highest point: Olympus 2,917 m (9,570 ft); longest rivers: Vardar (part) 388 km (241 mi) and Aliàkmon 314 km (195 mi)

Population (2006 est.) 10,688,058

Form of government multiparty republic with one legislative house

Armed forces army 110,000; navy 19,250; air force 23,000

Largest cities Athens (capital – 745,514); Salonika (363,987); Piraeus (175,697); Pátrai (161,114)

Official language Greek

Ethnic composition Greek 98%; other 2%

Religious affiliations Greek Orthodox 98%; Muslim 1.3%; other 0.7%

Currency 1 euro (EUR) = 100 euro cents

Gross domestic product (2005) U.S. $236.8 billion

Gross domestic product per capita (2005) U.S. $22,200

Life expectancy at birth male 76.72 yr; female 81.91 yr

Major resources lignite, petroleum, iron ore, bauxite, lead, zinc, nickel, magnesite, marble, salt, hydropower potential, barytes, chrome, cotton, fruit, vegetables, grapes/wine, livestock, olives, sugar beet, tobacco, wheat, tourism

Stripped to the bone The thin, rocky soils of the Eurotas valley and Taygetus mountains in the southern Peloponnese are typical of much of Greece. Clearance of the natural vegetation and overgrazing by sheep and goats have caused widespread soil erosion

GEOGRAPHY

Greece is a land of peninsulas and island chains formed by parallel mountain ranges that were flooded by the rising levels of the Mediterranean. Its coastline is deeply indented with bays and inlets. The whole region is geologically unstable and experiences frequent tremors.

The land

Mainland Greece forms the southern end of the Balkan peninsula. The Pindus mountains in the northwest extend from the Dinaric Alps of the Former Yugoslav Republic of Macedonia and Albania. In the far north, Greek Macedonia and Thrace are bounded by the Rhodope mountains on the Bulgarian border and the Maritsa river on the Turkish border.

East of the Pindus the lowland plain of Thessaly is bounded on the north by mountains. The Othrys range runs southeast into the Aegean Sea and reappears to form the northern part of Euboea, a long, narrow island parallel to the coast.

To the south is the Peloponnese, separated from the northern mainland by the Corinth Canal, which links the Aegean and Ionian seas. The mountainous plateau of the Peloponnese ends

Peaceful scene on the tip of a volcano Thira, the southernmost island of the Cyclades in the Aegean, is the remnant of a huge volcanic eruption in about 1500 BC. It consists largely of lava and pumice, the latter being the island's major export.

in four rocky peninsulas, the longest of which, the Taygetus, arcs southeast to form the island chain of Kithira, Crete and Rhodes.

More than 2,000 islands (less than 200 of which are inhabited) account for one-fifth of the total land area. Crete, in the southern Aegean, is the largest. The Ionian islands, west of the mainland, include Corfu, Levkas, Cephalonia and Zakinthos. The two largest clusters are the Cyclades in the south Aegean and, to their east, the 12 Dodecanese.

Climate
The Mediterranean climate has marked regional variations. In the northeast, cold winter air is drawn in from the eastern Balkans. An icy wind, the *boreas*, brings snow to the highest mountains. In the south, warm winds make mild winters. Rainfall is most frequent during fall and winter, and heaviest on the western coastal ranges. Summers are hot and dry with heatwaves. Sea breezes may fan the coast, but hot dry winds called *meltemi* intensify the heat in Crete and the southern islands.

Plants and animals
Thousands of years of clearing and grazing by sheep and goats have reduced the original forest cover to one-fifth of the land area, mostly in the north. Maquis scrubland predominates in cleared lowland areas and throughout the Peloponnese. Flowering plants, including crocuses and anemones, make a brilliant display in spring. The Ionian islands of Corfu and Levkas receive more rain and have lush vegetation.

The northern forests shelter some large mammals that have now virtually vanished from the rest of Europe, such as wolves, brown bears and lynxes. There is a wide variety of reptiles in the south including chameleons. Turtles are endangered by tourism and pollution on the beaches where they lay their eggs.

SOCIETY

Greeks have a strong national pride and identity derived from their classical heritage. Influenced by later Mediterranean civilizations, present-day Greek life is particularly rich and vital.

History
Between 2000 and 1500 BC Crete was the center of the Minoan civilization, usually considered Europe's earliest. By the 16th century BC its influence passed to Mycenae in the Peloponnese, which in turn was overrun about 1200 BC by Dorian warriors from the north, who had developed the use of iron weapons. The Greek city-states may have originated at this time as villages joined for self-defense within the walls of a citadel (*acropolis*).

These city-states became more prosperous; among them Athens established trading links and colonies around the Mediterranean. Literature, architecture, politics, art and philosophy were vigorously practiced. This rich civilization reached its zenith in the 5th century BC, but rivalry between the leading city-states eventually weakened them, and they were taken over by Philip II of Macedonia (382–336 BC). His son, Alexander the Great (356–323 BC), spread Greek culture across Asia to northern India, but after his death his empire was divided. Internal struggles continued to weaken the Greek city-states and by the end of the 3rd century BC Greece had succumbed to the rising power of Rome.

The Romans admired Greek art and scholarship but allowed Greece to fall into economic decline. After the 4th century AD, Greece was part of the eastern Byzantine empire ruled from Constantinople. The Greek language and Eastern Orthodox Church remained unifying forces in spite of successive invasions. From the 11th century the Byzantine emperors were unable to retain control of the Peloponnese, where Norman and French adventurers established feudal kingdoms. Venice controlled many Ionian and Aegean islands. By 1460, following the fall of Constantinople, the Turkish Ottoman empire absorbed nearly all of Greece. The Ionian islands remained in Venetian hands until 1797. During the 18th and 19th centuries, the Orthodox Church became a focus for growing nationalism. In 1830, after a bitter struggle, an independent Greek state was established in the Peloponnese, central Greece and the Aegean. Added to these were the Ionian islands in 1863, Thessaly in 1881, Epirus and Macedonia in 1913, and Thrace in 1918. After an unsuccessful attempt to annex Anatolia from Turkey after World War I, the Greek community in Turkey was expelled.

Occupation by German forces between 1941 and 1944 was followed by civil war. By the time monarchist forces defeated the communists in 1949, 10 percent of the Greek population had died in battle or of starvation. A military coup in 1967 overthrew the monarchy; a new republic proclaimed in 1973 lasted only a year. The ruling military clique was almost certainly involved in the overthrow of the government of Cyprus in 1974. Under the constitution of 1975, democratic elections were restored and Andreas Papandreou (1919–96) was elected in 1981 as Greece's first socialist prime minister. His popularity waned during the 1980s, and he fell from office in 1989. The socialist PASOK party won the 1993 general elections and were re-elected in 1996 and 2000.

People

The Greek people form a uniform majority. There are also significant Macedonian, Turkish and Albanian minorities. Almost all Greeks are active members of the Greek Orthodox Church. Although Greece is rapidly becoming an urban society, most Greeks feel strong attachment to their rural roots. Strict marriage and divorce laws and the position of women have been improved by recent legislation.

Government

Greece is governed by a single-chamber parliament of 300 members. The leader of the majority party becomes prime minister and head of government. The president is elected by parliament for a five-year term and is the supreme commander of the armed forces. Governments are elected for a four-year term, and all Greeks who are over the age of 18 are required to vote.

Greece's 13 administrative regions are divided into 51 departments. Mayors and councils are elected locally, but are funded by the national government.

Greece is a member of the North Atlantic Treaty Organization (NATO) and also a member of the European Union (EU). Greece's continuing involvement in Cyprus continues to cause difficult relations with neighboring Turkey.

ECONOMY

Greece is the poorest country in the EU. Services (especially tourism, banking and shipping) are the most profitable sector and do much to relieve a continuing balance-of-payments problem.

Agriculture

Agriculture employs 18 percent of the workforce, but productivity is poor. Only about one-quarter of the land can be cultivated. Many farmers still rely on draft animals. A wide variety of crops includes cotton, tobacco, olives and citrus fruits. Sheep and goats are raised throughout the country. Fishing, though modernized, remains a minor industry. Greece has benefited from EU subsidies to agriculture.

Industry

Greece has some mineral resources, but most raw materials must be imported, including 75 percent of its fuel. New plans for increased energy production focus on hydroelectricity and converting oil-burning power stations to locally

available lignite and peat. Geothermal, solar and wind power are also utilized. Most of the country's industry is based in or near Athens; Salonika in the northeast is the only other significant industrial city. Manufactured products include cigarettes, textiles, finished clothes, footwear, cement and fertilizer. The Greek shipping industry is one of the largest in the world and accounts for more than 30 percent of the income derived from services. However, rising fuel prices and demands for higher wages are affecting profits. Tourism provides foreign exchange as well as jobs in the service and construction industries.

Trade and commerce

Greek trade has expanded within the EU, but low-value exports and high-value imports have created a steadily growing

A hard living (*above*) Some 90 percent of Greece's farms are small to medium sized, typically 5 ha (12 acres), due to fragmentation by inheritance laws, poor farming methods and the loss of male workers to the cities. In addition, soils are rocky and rainfall is sparse.

trade deficit. Apart from crude oil, major imports include meat, milk, iron, steel, automobiles and pharmaceutical products. Important exports include textiles and handicrafts, tobacco, olive oil, tomato puree, petroleum products and minerals.

Transportation and communications

Greece's geography accounts for the importance of shipping and for the lack of surfaced roads well into the 20th century. Modern highways now link Athens with other centers. There is an extensive network of rural bus routes on the mainland and in Crete, and ferries between the mainland and the islands. The railroad system, though modernized, is confined to the mainland. The state-owned Olympic Airways operates both domestic and international flights. There are nine international airports, including Heraklion (Crete) and Rhodes, mainly servicing the major tourist industry.

Broadcasting is state-controlled and, like newspapers, subject to some censorship. There are 130 newspapers.

Health and welfare

Malaria, typhoid, dysentery and tuberculosis have been brought under control by improved medical care and water supplies. Life expectancy is now comparable with most European countries. Most hospitals are in the larger cities, but there are local health centers in rural areas and the islands. Care is free for those unable to afford it. A social security system is supplemented by the close-knit family. The state school system offers 9 years of free compulsory education between the ages of 6 and 15. Entry to the national universities (private universities are banned by the constitution) is controlled in order to limit unemployment among graduates.

A tourist haven (*above*) Cruise ships lie at anchor offshore from Mykonos ("white island" in Greek), one of the smallest islands in the eastern Cyclades. With poor soil and little fresh water, Mykonos depends heavily on tourism for income. As a result, it is well connected to mainland Greece by air from Athens and to many islands by boat. It supports a population of 5,500 and hosts a branch of the Athens School of Fine Art.

Table talk in Athens (*left*) One of the most striking features of modern Greek society is its great uniformity. Most of the population are Greek and belong to the Greek Orthodox Church. They also share a magnificent heritage, a strong sense of equality and a passion for politics. Thanks to a rise in the birthrate after World War II, Greece is now a young and increasingly urbanized country – one-third of its people live in Greater Athens.

305

Malta

REPUBLIC OF MALTA

THE REPUBLIC OF MALTA CONSISTS OF THE islands of Malta, Gozo and Comino in the Mediterranean between Sicily and north Africa. The commanding position they hold in the relatively narrow straits between Sicily and Tunisia makes them strategically important, and they have changed hands many times throughout their history.

GEOGRAPHY

The islands are the peaks of a massive limestone outcrop stretching south from Sicily, dissected by deep bays and valleys. With no fresh water apart from low rainfall, desalinization plants are necessary for a stable water supply.

The natural vegetation is sparse due to centuries of overgrazing and erosion. The remains of prehistoric animal species have been found in extensive caves, but today only a few small mammals (such as hedgehogs, shrews and bats) are found. Birds are more numerous, Malta lying on an important migratory route between Africa and Europe.

SOCIETY

Malta, possibly a former colony of Carthage (in north Africa), was part of first the Roman and then the Byzantine empires. Seized by the Arabs in 870, it was reconquered and ruled by Christian feudal lords until 1530 when it was ceded to the Knights of St John, a religious and military order. Malta was seized by Napoleon in 1798 but returned to the Knights in 1802. However, after a rebellion the island became a British colony in 1814. It became an important staging post on the sea route to India via the new Suez Canal (1869) and developed a dockyard industry.

In World War II Malta withstood repeated attacks by Germany and its allies, earning the entire population the British George Cross for bravery.

Full independence was granted in 1946. Good relations with Britain were maintained until the Maltese Labor Party came to power in 1971, swinging toward the Eastern Bloc, ousting NATO forces and sharply increasing Britain's rental fees for the dockyards. In 1979 the dockyards were nationalized and all British forces expelled. An attempt to replace Britain with Libya as mentor and partner proved stormy. In 1987 the Labor Party was ousted by the Nationalist Party, which sought to restore relations with the West while maintaining links with Libya.

Prehistoric Malta was inhabited by people from Sicily or Italy, and there has been a strong Italian and north African influence. The Maltese language is derived from Arabic with Sicilian elements. English is the second language, and Italian is widely spoken.

Christianity survived the Muslim occupation, and most Maltese are Roman Catholics. The church is a major property owner, wielding considerable influence.

At independence Malta's constitution set up a two-chamber parliamentary system. In 1974 this was amended to make Malta a republic, with a single chamber that is elected by proportional representation. This House of Represen-

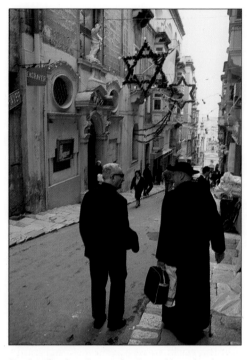

Passing the time A passerby chats with a priest in Valletta, capital of Malta. Locals speak Maltese, a distinct language blended from North African Arab and Sicilian Italian, but written in Latin script. Schools, however, teach in English.

tatives (65 seats) elects the president for a five-year term. Malta is divided into six administrative regions.

ECONOMY

Agriculture is restricted to terraced hillsides and irrigated lowlands. Vegetables, cereals and citrus fruit are the chief crops; cattle and goats the main livestock. Intense cultivation has brought near self-sufficiency, but agriculture accounts for only 4 percent of gross national product. Fishing is local and small-scale.

The economy depends largely on shipping and port facilities. Light industry has become increasingly important. The textile and garment industries, which are among the major exports, have suffered from European Union (EU) import controls. Other products include machinery and electronics. The major source of foreign exchange is tourism. In 1995 it was agreed that EU accession negotiations would start in 1997.

Roads are extensive and well surfaced. Ferries run between Malta and Gozo, and Valletta's modern airport is served internationally. Publishing and the media are active and are generally bilingual, as is television. Communications systems are excellent, and strong social and health services reflect a high standard of living. Education is compulsory and free between the ages of 5 and 16.

NATIONAL DATA - MALTA

Land area	316 sq km (122 sq mi)			

Climate	Altitude m (ft)	Temperatures January °C(°F)	July °C(°F)	Annual precipitation mm (in)
Valleta	70 (230)	12 (54)	26 (79)	519 (20.4)

Major physical features largest island: Malta 246 sq km (95 sq mi); highest point: southwest cliffs 253 m (830 ft)

Population (2006 est.) 400,214

Form of government multiparty republic with one legislative house

Armed forces joint forces 2,237

Capital city Valletta (7,137)

Official language Maltese, English

Ethnic composition Maltese 95.7%; British 2.1%; others 2.2%

Religious affiliations Roman Catholic 98%; other 2%

Currency 1 Maltese lira (MTL) = 100 cents

Gross domestic product (2006) U.S. $8.122 billion

Gross domestic product per capita (2006) U.S. $20,300

Life expectancy at birth male 76.83 yr; female 81.31 yr

Major resources limestone, salt, tourism, port facilities, fruit, goats, pigs, poultry, tobacco, vegetables, wheat

Cyprus

REPUBLIC OF CYPRUS

CYPRUS, THE THIRD LARGEST ISLAND IN THE Mediterranean, is divided by a long-running conflict between the majority Greek and minority Turkish population.

GEOGRAPHY

On the north coast are the Kyrenia mountains; in the south, the Troodos, from which the Pedias and Kariyoti rivers flow toward the bays of Famagusta and Morphou. The fertile Mesaoria plain in the center is the site of Nicosia, the capital. Winter is the rainy season, and summers are hot and dry. Scrubland predominates; the formerly extensive woodlands now remain only in the mountains. Many animals have been overhunted, and fishing is poor. Pasture used by sheep, goats and pigs covers one-tenth of the island. Crops include wheat, barley, citrus fruits, potatoes, tobacco, grapes, olives, almonds and vegetables.

SOCIETY

Once a colony of the Greeks, Phoenicians and Romans, Cyprus spent more than 700 years as part of the Byzantine empire before passing to the Ottomans in 1573 and to Britain in 1878. It became an independent republic in 1960. By 1963 the Turkish minority had withdrawn from the government. Amid a terrorist campaign in favor of union with Greece, UN forces were sent to keep the peace.

In 1974 a coup ousted and exiled the president. Turkey invaded Cyprus five days later, seizing the north of the island. Some 200,000 Greek Cypriots fled south. The Turkish Republic of Northern Cyprus, proclaimed in 1983, is recognized only by Turkey. UN forces remain, but negotiations for Cyprus to join the EU in the 2000s have prompted the two sides to hold open-ended talks to resolve the "Cyprus problem" once and for all.

Four-fifths of Cypriots are of Greek descent. The Turkish-speaking minority arrived from 1573. Almost all Greek Cypriots belong to the Greek (Eastern) Orthodox Church while the Turkish Cypriot community is mostly Sunni Muslim. English is widely spoken.

The Republic of Cyprus still claims legitimacy under the constitution of 1960. Many constitutional provisions for Turkish Cypriot representation have been suspended including the vice-presidency and seats in the House of Representatives. The breakaway Turkish state introduced a new constitution in 1985 and now elect their own government.

ECONOMY

Cyprus' economic expansion – funded by foreign aid – was severely disrupted in the Turkish invasion of 1974. Although most factories came under Turkish control, the Turkish sector remains dependent on agriculture and support from Turkey. The Greek sector has enjoyed impressive growth in industry and tourism. The Republic offers six compulsory years of free primary education and three years of free secondary education.

Kykko monastery in the eastern Troodos mountains, decorated in classical Byzantine fashion. Greek Cypriots are members of the Church of Cyprus, an independent Eastern Orthodox Christian church, while Turkish Cypriots are mainly Sunni Muslim.

NATIONAL DATA – CYPRUS

Land area	9,240 sq km (3,568 sq mi)			

Climate		Temperatures		Annual
	Altitude m (ft)	January °C(°F)	July °C(°F)	precipitation mm (in)
Nicosia	175 (574)	10 (50)	29 (84)	355 (13.9)

Major physical features highest point: Mount Olympus 1,951 m (6,403 ft)

Population (2006 est.) 784,301

Form of government multiparty republic with one legislative house (Turkish republic not recognized internationally)

Armed forces national guard 10,000

Largest cities Nicosia (capital – 202,488); Limassol (156,286); Larnaka (70,000); Famagusta (43,975)

Official language Greek, Turkish

Ethnic composition Greek 77%; Turkish 18%; other 5%

Religious affiliations Greek Orthodox 78%; Muslim 18%; Maronite, Armenian Apostolic, and other 4%

Currency 1 Cypriot pound (CYP) = 100 cents; 1 Turkish New lira (YTL) = 100 kurus

Gross domestic product (2006) Republic of Cyprus: U.S. $17.79 billion; north Cyprus: U.S. $4.54 billion

Gross domestic product per capita (2005) Republic of Cyprus: U.S. $22,700; (2004) north Cyprus: U.S. $7,135

Life expectancy at birth male 75.44 yr; female 80.31 yr

Major resources copper, pyrites, asbestos, gypsum, timber, salt, marble, tourism, almonds, beans, carob, citrus fruits, chromium, clay, grapes/wine, iron ore, olives, potatoes, vegetables, barley, wheat cratic Republic (East Germany). In 1952, the Soviet Union established a heavily guarded cordon of land along the whole border. West Berlin – an isolated enclave 190 km (120 mi) within East Germany – became the favored destination for refugees. An uprising against the communist regime in June 1953 was crushed by Soviet tanks.

In 1955 the Warsaw Treaty Organization unified the military command structure of the Eastern bloc countries. West Germany achieved full sovereignty, and became a member of the North Atlantic Treaty Organization (NATO).

In order to stem the exodus of East Berliners to West Berlin, East Germany built the massive Berlin Wall in 1961. In 1968, a thaw in relations between East and West was interrupted by the Soviet invasion of Czechoslovakia, in which East German troops played a token role. Chilly relations lasted until 1970, when a treaty confirmed existing German frontiers and renounced the use of force for political purposes. Two years later, a pact between East and West Germany opened a door to cooperation.

The 1980s saw a slowly increasing wave of pressure for reform in the Soviet bloc. In 1989 a wave of demonstrations was followed by largescale resignations in both national and local governments. In November the border with West Germany was opened, releasing a flood of emigrants. East Germany's communist government finally collapsed in the spring of 1990. On 3 October 1990, East and West Germany were formally reunited.

Government

West Germany's constitution of 1949 created a federal government based on 10 *Länder*, or autonomous states. The 1990 Unification Treaty admitted the five reconstituted East German *Länder* to the federation; the reunited city of Berlin became the 16th *Länd* (state) and the reunited Germany's capital city.

The 669 members of the Federal Assembly (Bundestag) are elected every four years by popular vote; they in turn elect the chancellor, who heads the government. The 69 members of the Federal Council (Bundesrat) are appointed by the 16 *Länder*. The president, who is the formal head of state, is elected every five years by an electoral college made up of delegates from both the federal and state assemblies. The government structure within each *Länd* is generally similar to that of the federal government.

People

The German people have a common history going back many centuries, but a unified language began only in the 16th century. Martin Luther's Bible translation of 1522 laid the foundations of today's written language.

Ethnic Germans make up most of the country's population, and include many communities who fled former German territory east of the Oder river. Germany is also home to numerous minorities, notably Turks, Yugoslavs and Italians, who came as "guest workers" to West Germany in the postwar years. Since 1990 the reunited country has taken in great

numbers of refugees from Africa, Asia and the poorer parts of Eastern Europe, stoking violent right-wing opposition and a neo-Nazi resurgence, especially among the poorer youth of eastern Germany who resent the foreigners' claims on the heavily burdened resources of the unified government.

Religious practice reflects a long history of divided loyalties since the Reformation. Generally Lutherans are strongest in the north and east, Roman Catholics in the south and in the Rhineland.

Germany has a rich legacy of art and literature. Its renowned composers include Johann Sebastian Bach (1685–1750), Ludwig van Beethoven (1770–1827) and Richard Wagner (1813–83).

ECONOMY

Reunification has had a dramatic impact on the German economy. The infrastructure of former East Germany – its housing, transportation networks, telecommunications and industrial base – is very poor in comparison with the west. Its industries also caused largescale environmental damage. Addressing these problems is stretching state resources.

Agriculture and industry

Since 1990 increased mechanization has boosted output, but agriculture remains second to industry in economic importance. The former East German cooperatives have little place in the free-market economy. Just over one-third of Germany's land is under cultivation. Germany is a

The huge steel plants at Duisburg, on the junction of the Rhine and Ruhr rivers, are supplied with iron ore and coal from the canals of the Ruhr valley. This is the most heavily industrialized part of Germany, and pollution is a major problem.

leading producer of hops, which supply more than 5,000 varieties of beer. The Rhine and Moselle valleys produce grapes for wine-making, while dairy cattle supply milk, cheese and butter.

Germany's extensive forests are a well-developed resource, though East Germany imported lumber until 1990. A large modern fishing fleet operates in the North Sea, the Baltic, the Atlantic, the Pacific and even the Antarctic.

Mining and manufacturing account for about one-quarter of the nation's income, despite relatively poor mineral resources. Most electricity is generated through the burning of fossil fuels. There are 21 nuclear and hydroelectric plants.

German manufacturing is widely diversified, but the east is less developed, and uncompetitive products can no longer find a market in the new Germany. The main industrial area extends southwest from the Ruhr to Cologne. Manufactured goods include chemicals, machinery, iron and steel products, automobiles and trucks, textiles and clothing, and modern electronic and precision equipment. Construction is another important industry, especially since reunification.

In western Germany, tourism is a profitable industry – though German travelers abroad spend four times what the industry earns at home. Tourism is less developed in the east, though the many resorts and historic sites have great potential to attract visitors.

Transportation and communications

The contrast between eastern and western Germany is more marked in this sector than many others. The west has a comprehensive network of surfaced roads, including a sophisticated system of *autobahns* (expressways) linked to neighboring countries. In the east, by contrast, many roads were still unsurfaced in 1990, and the autobahns had changed surprisingly little since they were first built, just before World War II. The state of the railroads is very similar.

The German waterways form a complex network. The Rhine is at the center of a vast canal system from the heart of Europe to the North Sea. Further canals link the Rhine to the Elbe, the Oder and the Danube, creating a doorway to Eastern Europe. Germany's most important port, Hamburg, is easily accessible from the North Sea, while the Kiel Canal allows shipping through Schleswig–Holstein to the Baltic.

Lufthansa, the federal state airline, was an elite international carrier from the 1950s onward. Domestic air traffic is light in the west, largely because the road and rail networks are so efficient.

Telecommunications throughout Germany have needed vast capital investment. The media, too, faced problems of readjustment. West Germany's press had always been in private hands, operating without government interference; in the east, the press was tightly controlled. Radio and television were state-owned in both East and West Germany, but West German broadcasting preserved an independent voice in the divided country.

Welfare and education

Healthcare and medical provision throughout Germany are generally very good, though the west has significantly more resources. The west also has a well-developed private health network.

Welfare systems were well developed in both countries before reunification, with a comprehensive range of benefits. However, the East German system was strongly tied to employment, and included extra provisions such as free nursery care for working mothers. After 1990 many in the east resented the loss of these services. Unemployment benefits paid to the many citizens in the east who lost their jobs as a result of reunification have been a drain.

Education is free and compulsory, and standards are generally high. An examination known as the *Abitur* is the passport to higher education at college or university. However, the system is very different in east and west, and there are discrepancies in the west between individual *Länder*. Germany's universities have a history of excellence, notably in theology, philosophy, philology and chemistry. Among the most historic institutions are Heidelberg, founded in 1386, and Tübingen, 1477. Marburg, founded in 1527 in central Germany, is the oldest Protestant educational establishment.

River of wine A barge glides past vineyards reflected in the peaceful waters of the Moselle river in western Germany. The steeply terraced vineyards along the sheltered Rhine and Moselle river valleys produce Germany's world-famous white wines.

Switzerland

S WITZERLAND IS A LANDLOCKED COUNTRY IN the mountainous heart of Europe. Politically a federal state made up of 26 separate cantons – many of them tiny, and speaking as many as four different languages – the Swiss nation still preserves a remarkable sense of unity.

GEOGRAPHY

Switzerland is renowned for its spectacular Alpine scenery. In no other nation does it so dominate the landscape.

Topographically Switzerland comprises three distinct areas. The plateaus of the Jura Mountains in the northwest give way to long, grass-covered mountainous ridges divided by deep wooded valleys and gorges; the southern edge overlooks Lake Neuchâtel and the Aare valley.

The Mittelland plateau is the least mountainous part of Switzerland, and holds most of the population.

The southern margins of the Mittelland merge into the foothills of the Alps, which occupy well over half Switzerland's total land area. The two ranges of the Alps almost meet at the St Gotthard massif, where the Rhône and Rhine rivers rise; the rivers form parts of Switzerland's two largest lakes, Lake Geneva and Lake Constance respectively. The highest peaks are in the Pennine Alps in the southwest along the Italian frontier, and include Monte Rosa at 4,634 m (15,203 ft) and the Matterhorn at 4,477 m (14,688 ft). The Bernese Alps to the north of the Rhône form the northern rim of Europe's largest glacier, the 24 km (15 mi) long Aletsch. Alpine rivers have carved several gorges here, and there are a number of spectacular lakes, such as Lake Lucerne and Lake Lugano.

Switzerland's climate is temperate continental, but the weather varies depending on terrain and altitude. Midwinter temperatures normally fall below freezing everywhere. The highest Alps lie under perpetual snow and ice. Summers in the Mittelland, however, are warm, and in the southern valleys they are often quite hot. Rain falls throughout the year.

Over the centuries, grazing sheep and goats have stripped much of the country's original plant cover, although rare plants, and flowers such as the edelweiss, can still be found above 2,000 m (6,500 ft). The lower slopes support woodland; the southern valleys contain trees, such as cypresses and figs. As well as the typically Alpine animals – marmots, chamois and ibex (reintroduced) – Switzerland harbors many species familiar in other parts of Central Europe.

SOCIETY

In the 13th, 14th and 15th centuries the Swiss cantons struggled to achieve independence. Since then they have kept that freedom, forging disparate cultures, religions and languages into a strong national identity. In 58 BC the Romans placed permanent garrisons in the Alpine passes of Helvetia – modern Switzerland. In 259 AD, the Alemanni were the first of several successive waves of Germanic invaders to arrive. The area later came under Frankish rule, and by 1033 it had become a part of the Holy Roman Empire.

In 1291 three cantons around Lake Lucerne– Uri, Schwyz and Nidwalden –

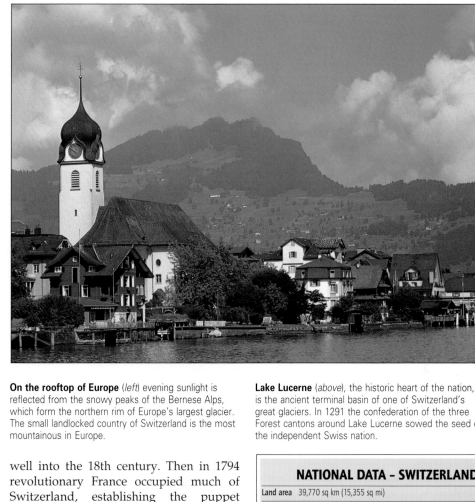

On the rooftop of Europe (*left*) evening sunlight is reflected from the snowy peaks of the Bernese Alps, which form the northern rim of Europe's largest glacier. The small landlocked country of Switzerland is the most mountainous in Europe.

Lake Lucerne (*above*), the historic heart of the nation, is the ancient terminal basin of one of Switzerland's great glaciers. In 1291 the confederation of the three Forest cantons around Lake Lucerne sowed the seed of the independent Swiss nation.

formed a confederation to break free of Habsburg rule. In 1332 Lucerne joined; Zurich joined in 1351, followed by neighboring Glarus, Zug and Bern. By 1388 the Swiss Confederation was firmly established. During the next 150 years the Confederation grew in power and influence, while continuing to struggle for its independence against the threat of warring neighboring states.

The rise of Protestantism led to open conflict between cantons. Yet, while religious wars ravaged Europe, the Swiss remained neutral, patching up their domestic differences. The Peace of Westphalia (1648) finally acknowledged the cantons' freedom from imperial rule. Switzerland maintained its neutrality

well into the 18th century. Then in 1794 revolutionary France occupied much of Switzerland, establishing the puppet Helvetic Republic four years later. In 1815 the last remaining cantons were admitted to the confederation, and Switzerland regained its neutrality.

In the mid 19th century, following the forced secularization of Jesuit monasteries, the Catholic cantons formed a protective union known as the Sonderbund. However, this alliance was nullified in 1847, and in 1848 a new constitution created the modern federal state. Bern became the capital of the confederation. Swiss neutrality made the country a haven for many international bodies, such as the International Red Cross and the short-lived League of Nations. The Geneva Convention of 1864 established laws for the conduct of war. Today many United Nations' and other international agencies have offices there. Even so, a referendum in 1986 rejected United Nations' membership as a threat to Swiss neutrality.

Switzerland is a federal republic made up of 20 full cantons and six half-cantons. National policy is decided by the Federal Assembly: the 200-member National Council, elected every four years; and the Council of States, whose 46 members are appointed by the cantons. Federal

NATIONAL DATA - SWITZERLAND

Land area	39,770 sq km (15,355 sq mi)			
Climate		**Temperatures**		**Annual**
	Altitude m (ft)	January °C(°F)	July °C(°F)	precipitation mm (in)
Zurich	540 (1,772)	0 (32)	18 (79)	1,086 (42.7)

Major physical features highest point: Monte Rosa 3,505 m (11,500 ft); longest river: Rhine (part) 1,319 km (820 mi), Rhone (part) 813 km (505 mi); largest lake: Lake Constance (part) 544 sq km (210 sq mi)

Population (2006 est.) 7,523,934

Form of government federal multiparty republic with two legislative houses

Armed forces joint forces 4,300

Largest cities Zurich (342,518); Geneva (177,535); Basel (165,051); Bern (capital - 122,707); Lausanne (116,332); Lucerne (116,332)

Official language German, French, Italian, Romansch

Ethnic composition German 65%; French 18%; Italian 10%; Romansch 1%; other 6%

Religious affiliations Roman Catholic 41.8%; Protestant 35.3%; Orthodox 1.8%; other Christian 0.4%; Muslim 4.3%; other 1%; unspecified 4.3%; none 11.1%

Currency 1 Swiss franc (CHF) = 100 centimes

Gross domestic product (2005) U.S. $241.8 billion

Gross domestic product per capita (2005) U.S. $32,300

Life expectancy at birth male 77.69 yr; female 83.48 yr

Major resources hydropower, timber, salt, apples, barley, building stone, grapes, livestock, potatoes, salt, wheat, tourism

Chocolate, a Swiss specialty, was first made in Zurich in 1845 by the master confectioner Rudolf Sprüngli (shown in the portrait). The company, now Lindt-Sprüngli, exports highest quality chocolates to over 80 countries around the world.

government is administered by the Federal Council, consisting of seven members, each from a different canton, whom the Federal Assembly elects for four-year terms. Each year a member of the council is elected federal president. Major issues, such as constitutional changes, are decided by popular referendum. Women were denied voting rights until 1971 (1989 in the two Appenzell half-cantons).

Switzerland has four different linguistic communities. About 65 percent of the population speak Swiss German (*Schwyzerdütsch*); they live in the central, northern and eastern cantons and the local dialects differ greatly. French speakers, who constitute about 18 percent, live in the west, and Italian-speakers live in the southern canton of Ticino. Romansh, a descendant of Latin, is spoken by only one percent of Swiss. Switzerland is evenly divided between Roman Catholics (49 percent) and Protestants (48 percent). Most Protestants are either Lutherans or Calvinists.

ECONOMY

Switzerland is one of the most prosperous nations in the world. Its strategic location at the crossroads of Europe, combined with its political stability and tradition of confidentiality, has boosted its development as a center of international finance. Switzerland's few flat areas are heavily populated, leaving little fertile land for farming. Nearly half the nation's food has to be imported. However, Swiss cheese and chocolates are exported all over the world. Mountain slopes provide pasture for beef and dairy cattle. Forests – protection against avalanches and floods – cover

many valley slopes, and clearance is forbidden by law.

Hydroelectric power meets almost all the country's needs; nuclear power stations supply the rest. A 10-year ban on expansion was imposed in 1990. High-grade manufactured goods for export are the mainstay of the Swiss economy. They include chemicals, heavy machinery and precision instruments such as clocks and watches. Trade tariffs provide the greater part of federal government revenue, while tourism is an all-year-round source of foreign income.

Several important trans-European routes pass through the Swiss Alps. The road system is very well developed; tunnels keep traffic flowing in winter. There are hundreds more tunnels on the country's extensive electrified railroads.

The northern city of Basel, on the Rhine, is Switzerland's only freight port.

The Matterhorn, one of the most spectacular mountain peaks in the world and one of the ultimate challenges to skilled mountaineers, seen from the terrace of an Alpine hostel. Tourists are attracted to Switzerland all year round and provide the main source of income for many Swiss communities.

Swissair is among the world's major commercial carriers, with principal airports at Zurich and Geneva.

Broadcasts and media are in three languages. The Swiss Broadcasting Corporation, a nongovernmental body, supervises all television and radio output.

Healthcare is free or subsidized. The welfare system provides child allowances and pensions; all salaried workers must take out unemployment insurance.

Education is free and compulsory from the age of 7 to 14, and secondary education is either academic or vocational. Higher education attracts an unusually large number of foreign students.

Liechtenstein

PRINCIPALITY OF LIECHTENSTEIN

Liechtenstein's decorative postage stamps, famous around the world, contribute about 10 percent of the national income.

LIECHTENSTEIN IS A VERY SMALL CENTRAL European principality on the east bank of the upper Rhine, sandwiched between Austria to the east and Switzerland to the west and south. Although it uses Swiss currency and belongs to the Swiss customs union, it is otherwise a fully independent state.

GEOGRAPHY

Liechtenstein measures about 25 km (16 mi) from north to south, and is 6 km (3.7 mi) wide. To the north and west the land is part of a broad floodplain bordering the Rhine. Once marshy, it was drained in the 1930s for agricultural purposes. To the east, the foothills of the Rhaetian Alps rise to snow-capped peaks along the southern border. The climate is surprisingly mild considering its continental location, and rainfall is plentiful. The valley suffers from sharp frosts, but the *föhn* – a warm, dry wind that blows from the south – brings an early spring.

These conditions allow a rich variety of plant life, with Alpine plants at higher altitudes, and forests of beech, maple, elm and ash on the lower slopes. The forests cover two-fifths of the land area, and are rigorously protected. Animal life includes several species of deer, as well as marmots, foxes, badgers and weasels.

SOCIETY

The area was first settled by the Alemanni, a Germanic people, in about 500 AD, and Liechtensteiners still speak an Alemannic dialect of German. Under Charlemagne (742–814) it became part of the Carolingian empire, and for centuries the land was divided between two independent counties based at Vaduz (the present-day capital) and Schellenberg in the north. In 1719 these two counties were amalgamated to form the principality of Liechtenstein, which remained a part of the empire. In 1806 the Austrian king abandoned the title of Holy Roman Emperor, and Liechtenstein became part of the Rhine Confederation. In 1815 it joined the German Confederation, and achieved full independence in 1866 when the Confederation was dissolved.

Liechtenstein is a constitutional monarchy with a single-chamber parliament, whose 25 members are elected every four years. There are 11 local authorities, each controlled by an elected mayor and council. Women may vote only in national elections. The official language is German, and most people are Roman Catholics, but fewer than two-thirds of the population are native Liechtensteiners; most of the rest are foreign industrial workers. No town in the country has more than 5,000 inhabitants.

ECONOMY

Since 1921 Swiss currency has been used in Liechtenstein, and its advantageous tax and company legislation have made the principality an important banking center. Revenue also comes from the registration of foreign firms, and from a thriving government-sponsored tourist industry. Liechtenstein is particularly famous for its decorative postage stamps, which add to the government revenue. Good roads provide excellent access from neighboring countries, and the Paris–Vienna railroad crosses its territory.

In 1930 the great majority of Liechtensteiners were farmers. Today, fewer than one in 40 of the population works on the land, and farms are being enlarged to improve efficiency, notably in stock breeding and dairy farming. The principal crops are corn and potatoes.

Liechtenstein has no raw materials or fuels, so all these items have to be imported. There is no heavy industry either, and consequently little pollution. Smallscale manufacturing produces a variety of goods, including optical, electronic and precision equipment, and various consumer goods.

Healthcare in the principality is excellent, and the Liechtenstein government contributes to facilities in neighboring Switzerland, so that they are made available to Liechtensteiners too. Compulsory insurance schemes financed by workers, employers and the government provide social security. Educational standards and literacy levels are both extremely high: Liechtensteiners receive 11 years of compulsory education, which may be followed by study at a technical, vocational, academic or other institution.

NATIONAL DATA - LIECHTENSTEIN	
Land area	160 sq km (62 sq mi)
Climate	temperate continental
Major physical features	highest point: Grauspitz 2,599 m (8,527 ft)
Population	(2006 est.) 33,987
Form of government	Hereditary constitutional monarchy on a democratic and parliamentary basis
Capital city	Vaduz (5,038)
Official language	German
Ethnic composition	Alemannic 86%; Italian, Turkish, and other 14%
Religious affiliations	Roman Catholic 76.2%; Protestant 7%; unknown 10.6%; other 6.2%
Currency	1 Swiss franc (CHF) = 100 centimes
Gross domestic product	(2001) U.S. $1.786 billion
Gross domestic product per capita	(1999) U.S. $25,000
Life expectancy at birth	male 76.1 yr; female 83.28 yr
Major resources	hydroelectric potential, grapes, livestock, maize (corn), potatoes, timber, vegetables, wheat, tourism

Slovakia

REPUBLIC OF SLOVAKIA

POLAND

CZECH REPUBLIC

Zilina

Martin • Hornad • Presov

Banská Bystrica • Hron • Kosice

SLOVAKIA

Trnava • Nitra

Bratislava

AUSTRIA • Danube

HUNGARY • Tisza

UKRAINE

Tisza

Vah

Morava

S LOVAKIA – UNTIL 1993 PART OF FORMER Czechoslovakia – is a newly created state at the center of Europe. The "Velvet Divorce" from the Czech Republic in January 1993 gave Slovakia complete independence for the first time in over a thousand years.

GEOGRAPHY

Slovakia is mostly mountainous, with lowland plains around the basin of the Danube on the southern border and on the eastern border with the Ukraine. To the west, the western Carpathian Mountains rise sharply from the border with the Czech Republic. Central Slovakia contains the parallel ranges of the Tatra and Slovakian Ore Mountains.

Slovakia's climate is continental, with cold winters, and warm summers, often with thunderstorms. In the far east and in the mountains winters can be very severe with heavy snow or rain. Precipitation is lower on the sheltered plains.

SOCIETY

Illyrian, Celtic and German peoples were the first settlers in modern Slovakia, and in the 6th century AD Slavic peoples arrived from Russia and the east. After the collapse in the 10th century of the Moravian empire, Slovakia began a long period under Hungarian control. Following the breakup of the Austro-Hungarian empire in 1918, the republic of Czechoslovakia was created, with Tomaš Masaryk (1850–1937) as its first president.

In the early 1990s, following the dissolution of the Soviet Union and the subsequent political upheaval in Eastern Europe, the Slovaks and the Czechs agreed to separate. From 1 January 1993 Slovakia gained full independence as a republic in its own right.

Czechoslovakia 1918–92

When the Austro-Hungarian empire collapsed at the end of World War I, the reorganization of its former territories and disparate peoples within new borders took no account of the cultural and nationalist aspirations of the peoples involved. Although obvious linguistic and geographic links between Czechs and Slovaks made the formation of the Republic of Czechoslovakia in 1918 seem a logical step, the union of these two peoples within one state was to be an uneasy one.

In 1918, Tomaš Masaryk (1850–1937) was elected president of the Republic of Czechoslovakia. He was faced with the task of uniting the powerful Czech kingdom of Bohemia and Moravia with Slovak provinces that had been under Hungarian rule for nearly a thousand years. He also faced the additional problem of appeasing several other minority groups including ethnic Germans, Poles and Ukrainians.

Under Masaryk's influence Czechoslovakia managed to sustain a parliamentary democracy while other democracies in Eastern Europe were falling

Alexander Dubček, in the "Prague Spring" of 1968, the year he was removed from office by the Soviets. In 1989 he returned to guide Czechoslovakia to democracy, but by 1993 ethnic conflicts had led to Slovakia's independence.

apart. But growing unrest between two of Czechoslovakia's major ethnic groups soon led to political tension. The Slovaks resented Czech domination of political and economic life, finding it difficult to maintain their national identity within the new republic. A great deal more threatening, however, was the disaffection of the Sudeten Germans in the Sudetic Mountains on the border with Poland, who looked to the emergent Nazi party in Germany for support in their nationalist aspirations. In 1938, under the notorious Munich Agreement between Britain, France, Italy and Germany, the Sudeten territories were ceded to Germany. Hitler subsequently used this as a pretext for invading and seizing large areas of Bohemia and Moravia. In 1939 both provinces were incorporated into a German Protectorate within the Third Reich, while a puppet government under the pro-fascist Josef Tiso (1887–1947) was set up in Slovakia.

After World War II the Soviet Union capitalized on its liberation of most of Czechoslovakia from German occupation and rapidly established a power base there. In 1948 the communists declared Czechoslovakia a People's Democracy. In the years that followed, political repression manifested itself in a series of Stalinist show trials, in which the communist leadership turned on its own ranks. Thirteen prominent communists, all of them Jews, and including the vice premier Rudolf Slansky (1901–52) were arrested, and all but three executed. The hardline regime in Czechoslovakia became ever more entrenched and by the time Antonin Novotny (1904–75) became president in 1957, Czechoslovakia's status had been reduced to nothing more than that of a Soviet satellite.

The brutalities of the old Stalinist regime were slow to recede in Czechoslovakia in spite of the political "thaw" in the Soviet Union under the premier Nikita Khrushchev (1894–1971). But eventually, in the 1960s, growing pressure for political and economic reforms resulted in Novotny's dismissal from office. A Slovak, Alexander Dubček (1921–92), took over as Czechoslovakia's First Secretary in 1967. The period of political tolerance and reform that followed in 1968 became known as the "Prague Spring".

Alarmed that the Czech example might be followed by other Eastern European satellite states, the Soviet Union launched an invasion of Czechoslovakia by Warsaw Pact troops in August 1968, on the pretext of "defending socialism". Resistance was put down, Dubček was removed from office and a puppet government installed under Gustav Husak (1913–91).

During the late 1980s – the Soviet Union's years of *perestroika* under the leadership of Mikhail Gorbachev (b.1931) – Husak scorned the reforms that were sweeping across the Soviet Union. But nothing could suppress the continuing calls for reform in Czechoslovakia and, after massive demonstrations, the communists were finally driven from power at the end of 1989. Alexander Dubček returned from obscurity as chairman of a new Federal Assembly with the former dissident playwright Vaclav Havel (b. 1936) as president. However, the transition to a parliamentary democracy was dogged by the age-old problem of Czech–Slovak relations, as each nation tried to look after its own best economic and political interests. A series of political crises followed in late 1990 and 1991 as divergences between Czechs and Slovaks deepened. On 1 January 1993 the two countries formally separated, and the states of Slovakia and the Czech Republic came into being.

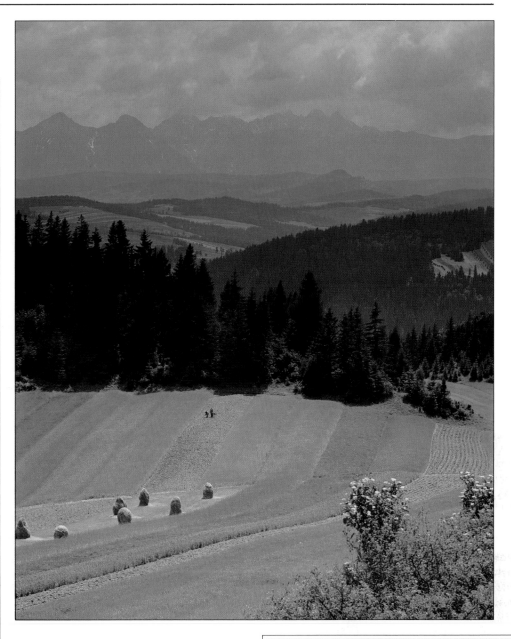

An elaborate green patchwork Strip farming and coniferous forests cover the rolling foothills of the Tatra Mountains – a range of the Carpathians – in central Slovakia. The extensive forests supply the country's lumber and paper industries.

ECONOMY

Although only one third of Slovakia is cultivated, agriculture produced 5 percent of GDP in 1996. Major crops include wheat, sugar beet, barley and potatoes. Sheep, pigs and cattle are the main livestock animals. Forestry is also important. Iron, copper, lead, manganese and zinc are mined from various sites, but in small quantities. Coal is mainly poor-quality lignite, used mostly for power generation, though the main source of power is hydroelectricity.

Postwar industrialization was concentrated in the Slovakian cities of Kosice and Bratislava; manufacturing in these areas is still the biggest sector of the economy. By 1995 extensive privatization had taken place and the economy had stabilized.

NATIONAL DATA – SLOVAKIA

Land area	48,800 sq km (18,842 sq mi)			
Climate		**Temperatures**		**Annual**
	Altitude m (ft)	January °C(°F)	July °C(°F)	precipitation mm (in)
Bratislava	132 (433)	-1 (30)	21 (70)	557 (21.9)

Major physical features highest point: Gerlach Peak 2,655 m (8,711 ft); longest rivers: Danube (part) 2,850 km (1,771 mi), Vah 394 km (245 mi)

Population (2006 est.) 5,439,448

Form of government multiparty parliamentary democracy with one legislative house

Armed forces army 12,860; navy 5,160; joint forces 2,175

Largest cities Bratislava (capital - 425,533); Kosice (235,281); Presov (92,147); Zilina (85,278); Nitra (86,138)

Official language Slovak

Ethnic composition Slovak 85.8%; Hungarian 9.7%; Roma 1.7%; Ruthenian/Ukrainian 1%; other and unspecified 1.8%

Religious affiliations Roman Catholic 68.9%; Protestant 10.8%; Greek Catholic 4.1%; other or unspecified 3.2%; none 13%

Currency 1 Slovak koruna (SKK) = 100 hellers

Gross domestic product (2006) U.S. $96.35 billion

Gross domestic product per capita (2006) U.S. $17,700

Life expectancy at birth male 70.76 yr; female 78.89 yr

Major resources brown coal and lignite; small amounts of iron ore, copper, manganese ore, zinc, salt, antimony, cereals, livestock, magnesium, mercury, potatoes, sugar beet, barley, timber, uranium

Hungary

REPUBLIC OF HUNGARY

Hungary is one of the oldest states in Eastern Europe, with an ancient Magyar culture. Landlocked, it is bordered by Slovakia and the Czech Republic to the north, Austria to the west, Slovenia to the southwest, Croatia and Serbia to the south and Romania to the east; in the far northeast it shares a short frontier with the Ukraine.

GEOGRAPHY

Hungary is mainly low lying, the broad fertile plains of the central Danube basin extending across much of the country. Low hills and mountains dominate the north and west.

A rich lowland area – the Great Hungarian Plain – extends eastward from the river Danube, which flows south across the center of the country. To the north it is rimmed by mountains. West of the river Danube lies hilly Transdanubia rising to several mountain chains. To the northwest of these uplands is the Little Hungarian Plain, a lowland area bounded to the north by the Danube.

Hungary's climate is continental, with cold winters and warm summers. Droughts are common in the summer months, though thunderstorms can also produce heavy rainfall. High winter snowfalls in the mountains of neighboring countries (including the Alps to the west) cause a spring thaw that swells Hungary's rivers.

Hungary's once extensive natural forests were cleared long ago. Some small broadleaf forests have survived, and there has also been some reafforestation in mountain areas. Here deer and Wild boar are still abundant. Some natural marshlands remain. Intensive industrialization has caused pollution; effluent controls and conservation programs have yet to make much impact. The marshlands are a

Vineyards in the Mecsek hills (*above*) near Pecs in the far south of Hungary, where summer temperatures are warm and rainfall can be heavy. Local wines such as Tokaj and "Bull's Blood" are extremely popular in Hungary, and are also exported to the West.

paradise for waterfowl, including exotic migratory species such as flamingoes and spoonbills.

SOCIETY

Hungary has been settled by many different peoples. However, native Hungarians – descendants of the "Magyars", originally nomadic horsemen – have maintained a strong national identity.

Hungary was part of the extensive Bulgarian empire before it fell to the Byzantines and was incorporated into the eastern Roman empire. During the 9th century the area was overrun by the Magyars from the Volga steppes. Under their supreme chieftain Árpád (d. 907), the Magyars became a menace to the German kingdom. Árpád's descendant, King Stephen I (977–1038), established Hungary as a kingdom in 1001 and began

to build it into a feudal monarchy in line with much of Europe.

Extending its dominion, Hungary became one of Europe's largest kingdoms, but in 1241 a Mongol horde overran the country. From 1301 Hungary was ruled by foreign monarchs. Sigismund (1368–1437) neglected Hungary in favor of his other kingdoms in Germany and Bohemia (now the Czech Republic). The Turkish rulers of the nearby Ottoman Empire quickly seized this opportunity to make serious encroachments on Hungarian territory.

In the 16th century much of southeast Hungary was annexed to the Ottoman empire; the rest became part of the Holy Roman empire in 1558. But Turkish attempts at gaining supremacy failed, and in 1699 the Ottoman empire ceded Hungary to the Holy Roman Emperor.

Magyar nationalism grew under leaders such as Lajos Kossuth (1802–94), uniting Hungarian society against Austrian rule. The 1848 uprising in Hungary forced the Austrian Emperor Ferdinand (1793–1875) to abdicate in favor of his nephew Franz the Hungarian economy, and in 1989 Kádár was ousted.

In 1989 the new constitution paved the way for multiparty democracy. The 386 seats of the one-chamber assembly are now freely elected for a five-year term. After the first multiparty elections in 1990, the writer and former dissident Árpád Goncz (b. 1922) was elected president. The election of 1994 was won by the former ruling Communist Party, as voters showed their dissatisfaction at the fall in living standards. Goncz was re-elected president in 1995.

ECONOMY

During the 1990s, the former communist controlled economy is slowly gave way to free-market capitalism. By 2000 over 85 percent of the economy was privatized.

As much as three-fifths of the country is rich arable land, and Hungary is practically self-sufficient in food. Since 1990 land has been restored to precollectivation owners.

The main crops include cereals, sugar beet and sunflowers; pigs and poultry are the main livestock. Vineyards, mostly in upland areas, yield some excellent wines. Commercial forestry along the Tisza river supplies the cellulose industry.

Hungary's main mineral resources include low-grade coal (lignite), its main fuel for energy. Only small deposits of petroleum and natural gas have been tapped. Manufacturing industries are extensive, particularly around Budapest. Major products include chemicals and engineering, automobiles and transportation equipment. Lighter manufacturing has grown in importance.

A massive program of privatization is underway. Trade with the European Union (EU) has greatly increased, with Germany, Austria and Italy as the major trading partners. Hungary is hoping to gain EU membership in the early 2000s.

Hungary's road and rail networks, both centered on Budapest, are highly developed but increasingly crowded. The Danube carries a vast amount of freight by water, especially from abroad. The national airline, Malév, is a significant international carrier.

Political changes have brought restructuring of the comprehensive welfare benefits provided by the communist state, and there is economic hardship for some Hungarians. Primary and secondary education is free and compulsory, and further education is available at a variety of colleges and universities.

Baroque splendor The Fertöd Palace, built in the mid 18th century by the aristocratic Esterhazy family, leading Magyar landowners of the period. Between 1761 and 1790 the Austrian composer Joseph Haydn (1732–1809) was court musician to the Esterhazys.

Joseph (1830–1916). The rebellion was only put down with Russian help. In 1867, compromises on both sides won Hungary self-government as one half of the dual monarchy of the Austro-Hungarian empire. World War I shattered the Austro-Hungarian empire, and in the chaos that followed more than two-thirds of Hungary was annexed by surrounding countries and the rest fell to a communist coup. However, in 1920 Admiral Miklos Horthy (1868–1957) restored Hungary as a monarchy with himself as regent.

During World War II Hungary sided with Nazi Germany. As the Soviet army pushed westward in 1945 Hungary was occupied and the inevitable communist takeover followed in 1949. In 1956 Imre Nagy (1896–1958) led a nationalist rebellion which was savagely suppressed by the Soviet Union. A pro-Moscow regime under the leadership of János Kádár (1912–89) was then imposed.

While adhering closely to Soviet policies abroad, Kádár allowed a degree of political and economic liberalization at home. But the 1980s saw the collapse of

NATIONAL DATA – HUNGARY

Land area	92,340 sq km (35,653 sq mi)			

Climate		Temperatures		Annual
	Altitude m (ft)	January °C(°F)	July °C(°F)	precipitation mm (in)
Budapest	139 (456)	-1 (30)	21 (70)	516 (20.3)

Major physical features highest point: Kékes 1,014 m (3,327 ft); longest river: Danube (part) 2,850 km (1,771 mi); largest lake: Lake Balaton 601 sq km (232 sq mi)

Population (2006 est.) 9,981,334

Form of government multiparty republic with one legislative house

Armed forces army 23,950; navy 7,500; joint forces 850

Largest cities Budapest (capital - 1,697,343); Debrecen (204,297); Miskolc (175,701); Szeged (162,889); Pécs (156,567)

Official language Hungarian

Ethnic composition Hungarian 92.3%; Roma 1.9%; other or unknown 5.8%

Religious affiliations Roman Catholic 51.9%; Calvinist 15.9%; Lutheran 3%; Greek Catholic 2.6%; other Christian 1%; other or unspecified 11.1%; unaffiliated 14.5%

Currency 1 forint (HUF) = 100 filler

Gross domestic product (2006) U.S. $172.7 billion

Gross domestic product per capita (2006) U.S. $17,300

Life expectancy at birth male 68.45 yr; female 77.14 yr

Major resources bauxite, lignite, natural gas, tourism, cereals, fruit, vegetables, grapes/wine, livestock, oil, potatoes, sugar beet, sunflowers

Romania

REPUBLIC OF ROMANIA

ROMANIA – THE "LAND OF THE ROMANS" – was founded as part of the ancient Roman empire. In its southeastern corner it has a short coastline along the Black Sea. Apart from this port access Romania is landlocked. The river Danube forms its southern border with Bulgaria. Serbia lies to the southwest, Hungary to the west, and the Ukraine to the north and east, sandwiching Moldova to the northeast.

GEOGRAPHY

Romania is dominated by the great mountain barrier of the Carpathians that extends from the Ukraine in the north and sweeps through Romania in a broad arc, first running southeastward before turning sharply west toward Serbia. The highest and most spectacular peaks are in the southern part of the main range, known as the Transylvanian Alps. Here the mountains are separated into massifs by deep defiles created by rivers such as the Olt and the Buzǎu.

North and west of the Carpathians, the upland basin of Transylvania is bounded by the Bihor Mountains, which are also cut by major rivers: the Somes running northwest into Hungary, and the Mures flowing due west into the Banat lowlands bordering Hungary. The northeast of Romania is occupied by the low plateau of Moldova, which continues into the newly independent republic of Moldova, formerly part of the Soviet Union.

South of the Transylvanian Alps, the river Danube forms the southern edge of the flat, fertile plains of Walachia, in the center of which lies the capital, Bucharest. To the east, the uplands of Dobruja give way to a series of marshes and a broad, swampy delta leading to the Black Sea.

The climate of Romania is largely continental, with cold, relatively dry winters and warm summers, which can be prone to thunderstorms. The mountains modify the climate considerably, receiving most of the country's rain and snow, and protecting western and southern areas from the bitterly cold winter winds that blow from the northeast.

SOCIETY

In 106 AD Romania was conquered by the Romans, becoming the Roman province of Dacia. The native Thracian peoples were quickly Romanized, and their culture survived wave after wave of incursions by the Germanic Goths, the Huns, the Slavs, the Avars, the Bulgars and finally, in the 10th century, the Magyars.

In 1365, two independent Romanian principalities emerged: Walachia in the Danube plains of the south, and Moldova to the northeast of the Carpathian Mountains. Their independence was brief: in 1396 Walachia became a vassal state of the burgeoning empire of the Ottoman Turks, and Moldova followed in 1455.

With the wane of Ottoman power at the end of the 18th century, Romania came increasingly under Russian influence. This period fueled an intense and long-lasting dislike of the Russians. At the end of the Crimean War in 1856, the principalities of Walachia and Moldova were united, and the Treaty of Berlin of 1878 recognized the new Kingdom of Romania as an independent state.

Romania gained much new territory as a result of the peace treaty ending World War I; Bessarabia, Bukovina and Transylvania were all incorporated into the country. It also inherited new problems and tensions with the absorption of a diverse ethnic mix of peoples: Magyars, Jews, Germans and Ukrainians.

In 1940 Soviet troops occupied Bessarabia (the modern Republic of Moldova). Endeavoring to recover the lost territory, Romania allied with Nazi Germany and in 1941 joined the German invasion of the Soviet Union – with disastrous results. In 1944 the country was occupied by the Soviets who installed a puppet government. In 1947 the monarchy was abolished and Romania was proclaimed a republic.

In 1965 Romania entered a dark and repressive period of its history as Nicolae Ceauşescu (1918–89) became secretary general of the Romanian Communist Party. He assumed the presidency in 1967, and under his dictatorial rule corruption at high level became rife. Resentment erupted into popular revolt in December 1989. Ceauşescu's attempt to reassert his authority failed when the army turned against him and the dissident National Salvation Front (NSF) proclaimed a provisional government. Ceauşescu and his wife were captured, tried and executed on 25 December 1989.

Ethnic Romanians (*above*) make up 89 percent of the population. They can be distinguished from the Hungarian, German, Turkish and Gypsy minorities by their costume and language. Ceaușescu's regime strongly emphasized mass culture.

A typical Romanian Orthodox church (*left*) in the village of Humor. Villages and the Church both suffered under Ceaușescu, who suppressed worship and destroyed some 7,000 villages to make way for agro-industrial complexes.

tion own a car. Medical treatment is free and more hospitals are being built and more doctors trained to cope with the high demand. But the costly transition to a market economy has left little funding to support the social welfare system. Ceausescu's policy of denying Romanian women access to contraception in order to raise the birthrate has left the country with the problem of providing for large impoverished families that cannot support themselves.

Ion Iliescu (b. 1930) remained president through a period of political and economic troubles in the early 1990s. In 1996 the Democrats were voted into power, and new president Emil Constantinescu promised to deal with the corruption and organized crime which had hampered economic development. In the late 1990s reform was slow as a result of disagreements between the ruling coalition.

ECONOMY

Repeated attempts have been made to transform Romania into a free-market economy: state subsidies have been cut; price controls removed; and a program of mass privatization has been undertaken. Efforts at reform have been hampered by political and popular dissent, especially in areas where the closure of unprofitable heavy industries and coal mines would have an adverse effect on the local population. The IMF granted a large loan to Romania to assist with restructuring in the late 1990s. In 1998 Romania began negotiating to join the EU but it must first control inflation and unemployment.

Farming, collectivized under communist rule, is still a cornerstone of the Romanian economy. Ceaușescu's policy of destroying Romanian villages and replacing them with huge industrial complexes has done much to damage the

centuries-old way of life of the rural population. But the traditional crops grown in these areas, such as cereals and sugar beet, fruit and vegetables, are essential to the economy.

Sheep rearing is widespread, and there are many mixed farms in Transylvania and Banat. The Romanian forests are heavily exploited, less for sawn timber than for export products such as furniture and chipboard.

Although Romania is well supplied with coal, petroleum and natural gas, some fuel still has to be imported, together with the majority of raw materials for industry. A number of hydroelectric projects supply some of Romania's electricity, but most comes from thermal stations, many burning imported fuel. One nuclear power plant began operation in 1996. Since the 1960s much emphasis has been placed on the development of the manufacturing sector, dominated by heavy industry. The two main activities – machine building and metal processing – cause pollution. Much of the wildlife of the Danube is threatened by industrial effluent. The tourist industry is as yet underdeveloped. Romania has much to offer: ski resorts, historic towns and castles and stunning mountain scenery.

Romania's standard of living is lower than that of most other European countries; less than one percent of the popula-

NATIONAL DATA - ROMANIA

Land area	230,340 sq km (88,935 sq mi)		

Climate		Temperatures		Annual
	Altitude m (ft)	January °C(°F)	July °C(°F)	precipitation mm (in)
Bucharest	82 (269)	-2 (28)	22 (72)	595 (23.4)

Major physical features highest point: Mount Negoiu 2,535 m (8,317 ft); longest rivers: Danube (part) 2,850 km (1,771 mi) and Prut (part) 909 km (565 mi)

Population (2006 est.) 22,303,552

Form of government multiparty republic with two legislative houses

Armed forces army 66,000; navy 7,200; air force 14,000

Largest cities Bucharest (capital - 1,921,751); Iasi (321,580); Cluj-Napoca (318,027); Timisoara (317,651); Constanta (310,526); Craiova (302,622); Galati (298,584)

Official language Romanian

Ethnic composition Romanian 89.5%; Hungarian 6.6%; Roma 2.5%; Ukrainian 0.3%; German 0.3%; Russian 0.2%; Turkish 0.2%; other 0.4%

Religious affiliations Eastern Orthodox 86.8%; Protestant (denominations including Reformate and Pentecostal) 7.5%; Roman Catholic 4.7%; other (mainly Muslim) and unspecified 0.9%; none 0.1%

Currency 1 leu, plural lei (RON) = 100 bani

Gross domestic product (2006) U.S. $197.3 billion

Gross domestic product per capita (2006) U.S. $8,800

Life expectancy at birth male 68.14 yr; female 75.34 yr

Major resources petroleum, timber, natural gas, coal, lignite, iron ore, salt, hydropower, cereals, copper, fish, fruit, vegetables, grapes, gold, livestock, oil, potatoes, silver, sugar, uranium

Bulgaria

REPUBLIC OF BULGARIA

BULGARIA LIES ON THE EASTERN SIDE OF THE
Balkan peninsula bordering the Black
Sea. It is bounded by Romania to the
north (across the river Danube), Serbia
and Macedonia to the west and Greece
and Turkey to the south. The country was
one of the most acquiescent of the Soviet
satellite states. Since the quiet revolution
of 1989 it has been undergoing rapid
political and economic changes.

GEOGRAPHY

Two large mountain ranges divide the
country into four distinct areas: the
northern plateau, the east–west range of
the Balkan Mountains, the central and
coastal lowlands and the southwestern
Rhodope Mountains.

Across the north of the country the
steep banks of the Danube (Dunar) rise to
the low, fertile Danubian plain. South of
this the land rises gently to the fragmented ridge of the Balkan Mountains,
and then falls sharply to a complex series
of river basins in the central lowlands.
Here the river Maritsa drains southward
into Greece. Rivers in the eastern lowlands drain to the Black Sea coast. Southern and western Bulgaria is dominated by
the great massifs of the Vitocha, Rila, Rhodope and Pirin Mountains, which include
Musala, the highest peak in the Balkans at
2,925 m (9,594 ft). The capital city, Sofia,
lies in a high basin at the conjunction of
the Balkan and Vitocha Mountains.

The influence of the Mediterranean Sea
creates a temperate climate in the southeast with warm dry summers and mild
wet winters. In the north and west the
country is subject to greater extremes of
climate with colder winters and hotter
summers. The mountains are covered by
heavy snowfalls in winter.

SOCIETY

The Bulgars were originally a Turkic
people, who first crossed the river Danube from central Asia in the 7th century
AD. Gradually assimilated by the Slavs
that had settled there a century or so
earlier, the Bulgarians soon created a
large empire in the Balkans. This first
empire fell to its southern neighbor
Byzantium in 1018. A second empire,
wrested from Byzantium in 1185, survived until 1396, when the country was

Alexander Nevsky Cathedral (*above*) in Sofia was built
as a symbol of gratitude to the Russian troops who
liberated Bulgaria in 1878 from nearly 500 years of
Ottoman rule. The following year the city was
designated Bulgaria's capital.

taken by the Ottoman Turks. For the
next 500 years Bulgaria was subjected
to Turkish rule.

In the 19th century, growing nationalist
movements throughout the Ottoman empire fueled the decline of Turkish power.
The Congress of Berlin, called in 1878
to address the territorial disputes in the

Balkans, gave Bulgaria the status of an autonomous principality. Continuing upheavals in the Ottoman empire led Ferdinand of Saxe-Coburg-Gotha (1861–1948), Bulgaria's elected prince, to proclaim himself king in 1908. Bulgaria at this time was fully independent of Turkey. In 1912 Bulgaria joined forces with its neighbors Serbia, Greece and Montenegro to extract more territorial concessions from the declining Ottoman empire. However, in 1913 it turned on these allies and lost further territory to Greece and Serbia after a humiliating defeat. Balkan politics encouraged Bulgaria to align with Germany in World War I, and again in World War II. It was unable to realign before Soviet troops "liberated" Bulgaria from German occupation in 1944. Two years later the monarchy was abolished and so were all opposition parties. A Soviet-style communist republic was established in 1947, with the government in the hands of the Fatherland Front.

Bulgaria, like other Eastern European countries, was eventually caught up in the wave of change that swept through the communist bloc in the late 1980s. Following a series of strikes and demonstrations, Bulgaria's president of 27 years, Todor Zhivkov (b. 1911), was ousted in 1989. The constitution was amended and the Communist Party's sole right to govern was abolished. The next two years saw the decentralization of parts of the economy, the recognition of Turkish minority rights, and government liberalization on an unprecedented scale. A democratic coalition, led by Zhelyu Zhelev (b.1935) as president, came to power in 1992 and altered the process of election to Bulgaria's 240-member National Assembly whereby all are elected by proportional representation. American-style primary elections were introduced in 1996 and in June 2001 former king Simeon II became prime minister. He is the first exiled monarch in post-communist eastern Europe to return to power.

ECONOMY

The reforms of 1989 effectively abolished Bulgaria's centralized economy, allowing existing businesses to transform themselves into independent companies with the right to issue shares. The process of transition to a market economy has been difficult. Hyperinflation in the mid 1990s and the near collapse of the banking system prompted a restructuring program aided by a US$850 million IMF loan.

Cereals and vegetables are the main staples, while cash crops include sunflower seeds, tobacco and grapes; Bulgaria is famous for its fine wines. Livestock and forestry are less significant, though of growing importance. Fishing is a major industry; besides local fisheries in the Black Sea and inland waters, Bulgarian trawlers travel as far as the Mediterranean and the Atlantic.

Bulgaria has deposits of copper and some iron ore. There are also fossil fuel reserves of coal, but little petroleum or natural gas. Hydroelectric plants generate up to one-tenth of the country's power; one-half comes from thermal plants, and the rest from Soviet-designed nuclear stations. Bulgaria has become a major distribution center for energy in the Black Sea region. The communist regime invested heavily in manufacturing industry, and its products – notably machinery and equipment, metals and petrochemicals – earn the bulk of Bulgaria's export revenue. Tourism, particularly to the Black Sea coast, also brings useful foreign exchange.

Plovdiv, Bulgaria's second city, has a charming old quarter with well-preserved 19th-century houses, as well as some impressive Roman remains. The city saw a sixfold population increase between 1946 and 1985, and is now a major industrial center and market town.

NATIONAL DATA – BULGARIA

Land area	110,550 sq km (42,684 sq mi)			

Climate		Temperatures		Annual
	Altitude m (ft)	January °C(°F)	July °C(°F)	precipitation mm (in)
Sofia	550 (1,804)	−1 (28)	20 (68)	571 (39)

Major physical features highest point: Musala 2,925 m (9,596 ft); longest river: Danube (part) 2,850 km (1,771 mi)

Population (2006 est.) 7,385,367

Form of government multiparty republic with two legislative houses

Armed forces army 25,000; navy 4,370; air force 13,100

Largest cities Sofia (capital - 1,138,950); Plovdiv (341,464); Varna (312,026); Burgas (189,529)

Official language Bulgarian

Ethnic composition Bulgarian 83.9%; Turk 9.4%; Roma 4.7%; other 2% (including Macedonian, Armenian, Tatar, Circassian)

Religious affiliations Bulgarian Orthodox 82.6%; Muslim 12.2%; other Christian 1.2%; other 4%

Currency 1 lev (BGL) = 100 stotinka

Gross domestic product (2006) U.S. $77.13 billion

Gross domestic product per capita (2006) U.S. $10,400

Life expectancy at birth male 68.68 yr; female 76.13 yr

Major resources bauxite, copper, lead, zinc, coal, timber, apples, barley, cattle, grapes, lignite, limestone, maize (corn), manganese, oil and natural gas, pigs, poultry, sheep, sunflowers, tobacco, uranium, wheat, tourism

Yugoslavia

FEDERAL REPUBLIC OF YUGOSLAVIA
(SERBIA) AND MONTENEGRO

IN 1990 THE FEDERATION OF THE SIX REPUBLICS comprising Yugoslavia (Slovenia, Croatia, Bosnia Herzegovina, Serbia, Montenegro and Macedonia) began to dissolve with the secession of Slovenia and Croatia. This prompted Serbia to declare war on Croatia. Meanwhile Bosnia Herzegovina and Macedonia also declared independence, leaving Serbia and Montenegro in the new rump state of Yugoslavia.

SERBIA

Serbia, the largest of the former Yugoslav republics, covers some 88,361 sq km (34,116 sq mi) and incorporates the previously independent regions of

Vojvodina in the north and Kosovo in the south. It has international borders with eight states: clockwise from Hungary in the north these are Romania, Bulgaria, the Former Yugoslav Republic of Macedonia, Albania, Montenegro, Bosnia Herzegovina and Croatia.

Geography

Excluding the two provinces of Vojvodina and Kosovo, Serbia is almost completely mountainous. The most fertile land is found on the Pannonian plain, which runs from Hungary and Croatia across into Vojvodina and northern Serbia. A number of rivers crisscross this plain, including the Sava, which eventually flows into the Danube. In the southwestern corner of the country bordering Albania and Montenegro, the province of Kosovo rises out of the Sar and Kopaonik Mountains. Serbia's highest peak, Daravica, at 2,656 m (8,712 ft), is at the meeting point of these three states, amid the most rugged mountain terrain on the Balkan peninsula.

Eastern Serbia is also rugged and wild. Here the Balkan Mountains, which form the border between Serbia, Bulgaria and Romania, rise to 1,800 m (6,000 ft). The climate is almost totally continental. Temperatures are generally mild, but can range from very hot in summer to very cold in winter. Precipitation is steady throughout the year, mostly taking the form of snow on higher ground.

Society

Slavonic in origin, the Serbs first settled in the Balkans during the 6th century AD. Divided into warring clans, the Serbs were dominated by the neighboring Bulgarian and Byzantine empires until the 13th century. The Byzantine influence is still evident: most Serbs belong to the Greek Orthodox church, setting them apart from the Bosnian Muslims and Catholic Croats.

As Byzantium declined the Serbs briefly came to dominate the Balkans under their king Stephen Dušan (c. 1308–55). The rise of the Ottoman Turks, however, signaled an end to Serbian expansionism. After its crushing defeat at the Battle of Kosovo ("the field of black birds") in 1389, Serbia became an Ottoman vassal state. In 1537 the Ottoman empire took complete control. The Serbs continued to wage guerrilla attacks against the Turks, allying themselves by turn with Hungary, Austria, the Holy Roman Empire and Russia. In 1817 Serbia eventually gained

Concrete apartment blocks and offices dominate the skyline of modern Belgrade, capital of Eastern Orthodox Serbia. Built as cheap accommodation for workers, they are a legacy of the rapid process of industrialization that took place after World War II.

recognition as an autonomous principality with its capital at Belgrade. Total independence came in 1878, granted under the Treaty of Berlin concluding the Russo-Turko War. However, Serbia's aggressive forays into neighboring territory continued. By the outbreak of World War I, the Serbian state still lacked stable borders. After the collapse of the Habsburg monarchy in 1918 Serbia was instrumental in setting up the forerunner to the Yugoslav federation, the Kingdom of Serbs, Croats and Slovenes. Serbian domination of the new state became complete in 1921 with the adoption of the Serbian regent as King Alexander I (1888–1934).

After World War II Serbia became part

NATIONAL DATA – SERBIA

Land area	88,361 sq km (34,116 sq mi)			
Climate		Temperatures		Annual
	Altitude m (ft)	January °C(°F)	July °C(°F)	precipitation mm (in)
Belgrade	132 (433)	1 (45)	22 (63)	1,145 (45)

Major physical features	highest point: Daravica 2,656 m (8,714 ft) in Prokletija Range on Albanian border
Population	9,396,411
Form of government	republic
Armed forces	no figures
Largest cities	Belgrade (capital – 1,120,092); Novi Sad (191,405); Nis (173,724)
Official language	Serbian
Ethnic composition	Serb 66%; Albanian 17%; Hungarian 3.5%; other 13.5%
Religious affiliations	Serbian Orthodox, Muslim, Roman Catholic, Protestant
Currency	1 new Yugoslav dinar (YUM) = 100 euro cents
Gross domestic product	(2006) U.S. $44.83 billion
Gross domestic product per capita	(2005) U.S. $4,400
Life expectancy at birth	male 71 yr; female 76 yr
Major resources	oil, gas, coal, iron ore, copper, lead, zinc, antimony, chromite, nickel, gold, silver, magnesium, pyrite, limestone, marble, salt, agricultural produce

Tito, Marshal of Yugoslavia

Marshal Tito held together the disparate peoples of the region, with their cultural and religious differences, from 1945 until his death in 1980. His brand of communism allowed a higher standard of living than other communist countries.

When Josip Broz - better known as Marshal Tito - died on 4 May 1980, Yugoslavia lost its most powerful symbol of unity. Born in Croatia on 7 May 1892, Broz was the seventh of 15 children. During World War I, while serving in the Austro-Hungarian army, he was captured on the Russian front. After his return to Croatia in 1920 (then part of the Kingdom of the Serbs, Croats and Slovenes) he joined the illegal Communist Party. He was arrested three times for communist activities, and in 1928 was sentenced to five years' imprisonment for conspiring against the regime in the state that was about to be renamed Yugoslavia. Following his release in 1934 he rose to prominence in the Communist Party of Yugoslavia (CPY), becoming its secretary-general in 1940.

Tito's leadership was quickly put to the test. In April 1941 German forces invaded and occupied Yugoslavia. Two months later Serbian peasants in Herzegovina rose against savage massacres perpetrated by the Ustase – fascist Croatian separatists who supported German rule. With the German invasion of the Soviet Union, Tito called a general uprising and soon half the country was in the hands of the Yugoslav resistance. Wherever his partisans were in control, Tito created revolutionary councils in preparation for the united Yugoslavia of his dreams – a policy that antagonized both the government in exile and the Soviet Union. The partisans survived a series of increasingly powerful German offensives and civil conflict with the royalist chetniks. In 1943 Tito established a new government with himself as marshal, and the king was deposed.

Despite a growing rift between Tito and the allies – Great Britain, the United States and the Soviet Union – Yugoslavia was given military assistance by them in 1944. By the time the war had ended in 1945 Yugoslavia was in ruins and more than one-tenth of its population was dead. Tito's new 1946 constitution was modeled on that of the Soviet Union, but important differences swiftly became apparent. In 1948 Tito broke with his Stalinist masters and the Soviet Union effectively isolated Yugoslavia in the hope of bringing Tito down. They succeeded only in unifying the country behind its leader as he built his own independent style of communism (popularly known as "Titoism"). Five years later, in 1953, Tito became Yugoslavia's first president, supervising the enactment of a new federal constitution.

The same spirit of independence also showed itself in Tito's determinedly nonaligned foreign policy. While denouncing Western colonialism on the one hand, he was swift to condemn the 1968 Soviet invasion of Czechoslovakia, and responded by strongly reinforcing his own country's defenses. The 1974 constitution included the idea of a collective presidency – an idea that became reality on Tito's death in 1980. In the event, government by committee proved a poor substitute for his effective dictatorship. In 1990 the peoples of Yugoslavia had a brief tantalizing glimpse of a freer, more open society until rising ethnic tensions finally plunged the country into civil war.

Tito's dream of a federal Yugoslavia, where all its nationalities would be equal, had occupied much of his energy and skill for more than 40 years. Within 11 years of his death it had become apparent that his vision of harmony and mutual cooperation between the disparate peoples of Yugoslavia had not been realized.

of the federal Yugoslav state. Its role in the federation began to change significantly, however, when Slobodan Milosevic (b.1941) was elected president of Serbia in 1989 and again, under a new constitution, in 1990. This removed the term "socialist" from the official name of the Republic of Serbia, and revoked the special rights and autonomous privileges of Kosovo and Vojvodina.

During this period Serbia continually vetoed federal assembly motions by Croatia and Slovenia to devolve to a looser federation of Yugoslav republics. As the federation began to fragment, Serbia was able to hold onto majority control of the Yugoslav People's Army (YPA). The YPA failed to prevent the independence of Slovenia in 1991, but in Croatia and in Bosnia Herzegovina it intervened in support of ethnic Serbian minorities who were opposed to independence, plunging both republics

Macedonia

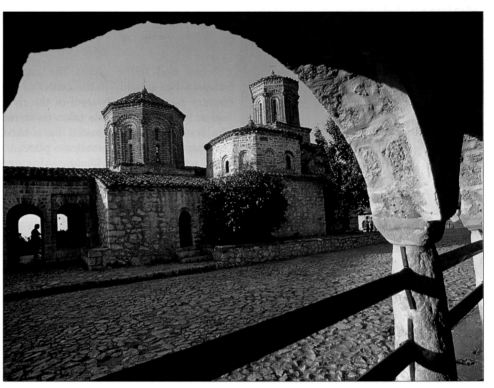

THE FORMER YUGOSLAV REPUBLIC OF Macedonia (FYROM) is a small state in the middle of the Balkan peninsula. It shares borders with Greece to the south, Bulgaria to the east, Serbia to the north and Albania to the west. It gained United Nations membership in 1993 following a compromise with Greece by which it is known as FYROM, whereupon Greece reopened its border to Macedonian trade. It reimposed the economic blockade in 1994 after the majority of European Union states recognized FYROM.

GEOGRAPHY

A large part of Macedonia is a high plateau 600–900 m (2,000–3,000 ft) above sea level; the rest of the roughly rectangular country is mountainous. The Kozuf and Nidze ranges separate it from the

Greek region of Macedonia. The highest point at 2,764 m (9,066 ft) is in the Korab Mountains in the far northeast of the country on the border with Serbia.

The climate is continental; rain is heaviest in the summer and there are significant snow falls in winter. Less traveled than the other more accessible republics of former Yugoslavia, it has large areas around its three great inland lakes – Ohrid, Prespa and Doiran – that remain unspoilt by human activity.

SOCIETY

Modern Macedonia was a part of the ancient Empire of Macedonia, which partly covered territory now in southwestern Bulgaria and northern Greece. Founded in the 6th century BC as the Kingdom of Macedon, its most famous ruler was Alexander the Great (356–323 BC). After his death, Macedon, like Alexander's wider empire, was split between warring factions. It was eventually defeated by the Romans and became a Roman province in 148 BC.

The Romans were quick to exploit the strategic location of this region, which controlled the fastest land route from the Adriatic to the Bosporus (the narrow strait linking the Black Sea to the Aegean) and Asia Minor. Following the decline of Roman power, in the 6th century AD the region was settled by the Slav peoples from whom the majority of modern Macedonians are descended. Macedonia was fiercely fought over by the Byzantines, Bulgars and Serbs until it was conquered by the Ottomans in 1371. With the demise of the Ottoman empire in the

The medieval monastery of St Naum near Lake Ohrid is typical of many such churches throughout the country. Most Macedonians belong to the Eastern Orthodox church, which has been independent from the Serbian Orthodox church since 1967.

19th century, each of these three countries claimed that the Macedonians were their ethnic descendants and the region was therefore their legitimate territory. The Balkan Wars of 1912–13 resulted in the division of the country between all three claimants. The largest Serbian part of the region (Vardar Macedonia) was brought by Serbia into the newly formed Kingdom of the Serbs, Croats and Slovenes after World War I.

Macedonia declared independence in 1992 following a referendum overwhelmingly in favor. In 1993 Macedonia joined the UN but Greece blocked formal recognition and imposed a trade embargo when the European Commission began legal proceedings against it. In 1995, after Macedonia agreed to redesign its flag and renounce any present or future claims to the Greek region of Macedonia, Greece capitulated. UN peacekeeping troops remained in Macedonia through 2000.

ECONOMY

Civil war in the Balkans has destabilized Macedonia since its independence. The Greek trade embargo in 1994 caused energy and food shortages, and in 1999 large numbers of refugees from Kosovo arrived. The economy nevertheless showed signs of recovery. The country is rich in minerals. Agriculture is a key sector, with livestock, cereals and tobacco. Manufacturing includes steel, chemicals and machinery.

NATIONAL DATA - MACEDONIA

Land area	24,856 sq km (9,597 sq mi)
Climate	continental with cold winters and hot summers
Major physical features	highest point: within Korab Mountains 4,634 m (15,203 ft); longest river: Po 620 km (380 mi); largest lake: Lake Garda 370 sq km (140 sq mi)
Population	(2006 est.) 2,050,554
Form of government	multiparty republic with one legislative house
Armed forces	army 9,760; air force 1,130
Capital city	Skopje (467,257)
Official language	Macedonian
Ethnic composition	Macedonian 64.2%; Albanian 25.2%; Turkish 3.9%; Roma 2.7%; Serb 1.8%; other 2.2%
Religious affiliations	Macedonian Orthodox 64.7%; other Christian 0.37%; Muslim 33.3%; other and unspecified 1.63%
Currency	1 Macedonian denar (MKD) = 100 deni
Gross domestic product	(2006) U.S. $16.91 billion
Gross domestic product per capita	(2006) U.S. $8,200
Life expectancy at birth	male 71.51 yr; female 76.62 yr
Major resources	iron ore, copper, lead, zinc, chromite, manganese, nickel, tungsten, gold, silver, asbestos, gypsum, timber, antimony, arsenic, barley, maize (corn), cotton

Albania

REPUBLIC OF ALBANIA

A LBANIA IS A MOUNTAINOUS COUNTRY ON the western (Adriatic) coast of the Balkan peninsula, bordered by Montenegro to the northwest and Serbia and the Former Yugoslav Republic of Macedonia to the east. To the southeast is Greece.

GEOGRAPHY

Albania's mountainous topography has contributed to its isolation from Europe, and much of the broad coastal lowlands are swampy and inaccessible. The climate is largely Mediterranean, though winters in the mountains are often severe.

SOCIETY

The Albanians are descended from the Indo-European Illyrians. In addition to strong regional differences between the northern and southern dialect groups, there are numerous ethnic minorities, including Greeks and Macedonians.

Conquest by the Romans in the 1st century BC was followed by incursions by Germanic Goths, Greek Byzantines and Slavic Serbs, until in the late 14th century Albania succumbed to Ottoman rule.

Independence, won from the Turks in 1912, ended with invasion by fascist Italy in 1939. In 1946, after a bitter civil war, Albania reemerged as a single-party communist republic under the isolationist rule of General Enver Hoxha (1908–85).

Albania's hardline government was the last communist regime in Europe to give way to popular pressure for reform in the early 1990s. The collapse of the economy that followed was exacerbated by the arrival of ethnic Albanian Muslims fleeing former Yugoslavia, and Kosovan refugees in 1999. Civil violence erupted in 1997, after the collapse of investment schemes.

ECONOMY

The government faces massive problems in its attempts both to modernize Albania's underdeveloped economy and to provide for a population that is still primarily rural. The limited export market relies heavily on minerals such as chromite, iron ore, petroleum and natural gas. However, Albania still lacks the infrastructure necessary to promote the growth of a successful market economy.

NATIONAL DATA – ALBANIA				
Land area 27,398 sq km (10,578 sq mi)				
Climate		Temperatures		Annual
	Altitude m (ft)	January °C(°F)	July °C(°F)	precipitation mm (in)
Tirana	89 (292)	7 (45)	24 (75)	1,353 (53.2)
Major physical features highest point: Mount Korabit 2,751 m (9,026 ft); longest river: Drin (part) 282 km (175 mi); largest lake: Lake Scutari 370 sq km (143 sq mi)				
Population (2006 est.) 3,581,655				
Form of government emergent democracy				
Armed forces army 16,000; navy 2,000; air force 3,500				
Capital city Tiranë (376,642); Elbasan (123,270); Durrës (116,275); Shkodër (109,604); Vlorë (104,118)				
Official language Albanian				
Ethnic composition Albanian 95%; Greek 3%; other 2% (Vlach, Roma, Serb, Macedonian, Bulgarian)				
Religious affiliations Muslim 70%; Albanian Orthodox 20%; Roman Catholic 10%				
Currency 1 lek (ALL) = 100 qintars				
Gross domestic product (2006) U.S. $20.21 billion				
Gross domestic product per capita (2006) U.S. $5,600				
Life expectancy at birth male 74.78 yr; female 80.34 yr				
Major resources petroleum, natural gas, coal, bauxite, chromite, copper, iron ore, nickel, salt, timber, hydropower, asphalt, bitumen, cereals, cotton, grapes, olives, potatoes, tobacco				

The Vijosë river in southern Albania cuts its way through rugged mountain peaks typical of 70 percent of the country's landscape. Albania's geographical isolation was reinforced by the policies of the communist leader Enver Hoxha (1908–85).

authority was broken and the great monolith of the Soviet Union began to collapse. It was formally dissolved in December 1991, and 11 of its 15 constituent republics, including the Russian Federation, formed the Commonwealth of Independent States (CIS). Gorbachev had no option but to resign.

Yeltsin found the Russian parliament reluctant to embrace sweeping changes he proposed, and in 1993 armed conflict erupted when demonstrators seized the parliament building. Forces loyal to the president successfully aborted the attempted coup and a new parliament was elected later in the year under a new constitution, which gave the president greater powers. In the late 1990s, a state of continuing economic crisis was accompanied by a number of government corruption scandals. In 1999 Yeltsin, who had been suffering ill health, retired, naming Vladimir Putin as his successor. Putin was elected president in 2000.

Government

The Russian Federation has taken over many of the enterprises, resources and responsibilities of the Soviet Union. However, its political structure is still undergoing major adjustment. Perhaps the most dramatic change was the introduction of multiparty democracy at all levels of government, a process begun in 1989.

There are 10 autonomous areas, 49 provinces, 6 territories and 21 officially independent republics. The republics are highly autonomous and each elects its own president. Provinces and territories also elect their own governors.

People

The Russian Federation contains within its boundaries an enormous diversity of peoples. There are more than 60 different nationalities, who can be broadly divided into four main linguistic groups. The Indo-European group – those languages that developed in Europe and southern Asia – includes some 150 million Russian speakers, and accounts for the great majority of the population from the Baltic to the Pacific. The other three language groups are the Finno-Ugric group, widely dispersed across the forest and tundra of the north and in the Balkans; the Turkic languages spoken in southeastern Europe and Asia; and the Caucasian group encompassing the languages spoken in the west of the Caucasus Mountains, as well as Chechen, Ingush and Dagestanian languages spoken in the east of the Caucasus.

The majority of Russians differ marked-

Waiting in line for food (*above*) at a mobile stall in the St Petersburg suburbs, where housewives gather to buy potatoes trucked in from a collective farm. In the early 1990s a fall in agricultural output, plus huge price rises, made daily living extremely hard.

ly from many of the non-Slav groups, in culture and religion as well as in language. Since the early 1990s there is now a tremendous resurgence in religious practice – considered reactionary and actively suppressed during most of the Soviet period – as the Russian people rediscover their spirituality. A strong Orthodox Christian tradition has survived within the Slavic communities, Baptists are numerous, and many ethnic and national groups have retained their own traditional religions; Islam among some of the Turkic peoples, and Buddhism among many of the Kalmyks and Buryats.

In wintry Siberia (*above*) a herd of domesticated reindeer are rounded up. Under Soviet rule the distribution of reindeer meat was collectivized, but herders were allowed to keep to the ancient custom of following the seasonal migration routes of the animals.

Cooperation in space (*below*)The American and Russian shuttles *Atlantis* and *Mir* docked together in 1995; here, Russian cosmonauts Solovyev and Budarin smile at *Atlantis'* departing crew before the hatch closes it off again from *Mir*.

Transportation and communications

For the Russian economy to function at all, massive quantities of food and raw materials have to be carried over enormous distances. Since the abolition of central planning, distribution has been one of the greatest challenges facing economic reformers. Much needed foodstuffs are frequently left to rot in warehouses before they can be transported to the cities.

The heart of Russia's transportation system is the railroad network, which carries some three-quarters of all freight. Most track is in the west, especially around Moscow; the best-known link with the far east is the Trans-Siberian Railroad from Moscow to Vladivostok.

Surfaced roads cover six times the distance of the entire railroad system, but are less used. Most bulk freight is carried by rail or by water, and buses account for a high proportion of passenger traffic. Relatively few people own a car.

Foreign trade is chiefly conducted by a sophisticated merchant fleet. In the far north and east, shipping provides a lifeline between communities scattered along some 50,000 km (30,000 mi) of coastline. Although most harbors become icebound in winter, some, such as Murmansk, keep open all year. Airline services ensure year-round contact with the most isolated regions. An international airline flies to many major destinations.

Radio, television and the press are still adapting to the recent relaxation of censorship. Even before the breakup of the Soviet Union though, there was a more open approach to news reporting.

Welfare and education

Under Soviet rule, medical care was free throughout the Russian Federation. Although housing was inadequate and the standard of living relatively low, unemployment was practically nonexistent. The collapse of the Soviet system has thrown the old social welfare network into chaos. Russia now faces the same kind of social problems that have plagued the capitalist West: inflation, unemployment and a soaring crime rate.

Hyperinflation in the early 1990s brought a dramatic decline in living standards, and more and more Russians found themselves living below the poverty line. The welfare system is being reformed into an insurance-based system of benefits for sickness and unemployment and retirement pensions. Health and average life expectancy have declined and infant mortality has increased.

ECONOMY

Russia was the main driving force behind the economy of the former Soviet Union. Its rich deposits of minerals supplied heavy industry with raw materials, and its energy resources supplied power to the other republics and Eastern Europe.

In the 1990s Russia struggled to establish a modern market economy hampered by an entrenched bureaucracy and a dilapidated industrial base. Falling international oil prices and a failing currency led to a financial crisis in 1998 as a result of which the ruble was devalued and Russia decided to default on its international debt.

Agriculture

Wheat, barley, oats and rye form the major part of the agricultural sector, but the country cannot grow enough for its own needs. Only about a tenth of the land area is under cultivation, and most of this lies to the west of the Urals. The most suitable land for intensive livestock breeding, including beef and dairy cattle also lies in the west. Russia also has the world's largest reserves of timber.

The Russian fishing fleet, operating from the Arctic waters to the Pacific, is one of the biggest in the world.

Industry

Heavy industry is the backbone of the Russian economy and draws on the country's enormous resources of raw materials. There are rich deposits of iron ore on the Kola Peninsula, in the Kursk Magnetic Anomaly near the Ukrainian border, and in Siberia. There are also major deposits of nonferrous metals in the Urals and Siberia including gold, copper, zinc and various rare metals.

In addition, Russia possesses vast energy reserves. There are huge coal fields in eastern Siberia, the Kuznetsk basin north of the Altai, and the Urals. Northwestern Siberia has the world's greatest reserves of oil and natural gas. The Volga and Yenisei rivers supply hydroelectric power to large stations nearby, and more hydroelectric plants are being built. The nuclear power program suffered a setback in 1986, when a major accident at Chernobyl in the Ukraine (then part of the Soviet Union) exposed safety problems in many installations.

Under the former Soviet system, Russia met most of the country's demands for industrial, agricultural and transportation equipment; few of these items were ever imported. But Russian-built machinery is only just now beginning to benefit from the microprocessor revolution.

Georgia

REPUBLIC OF GEORGIA

THE REPUBLIC OF GEORGIA LIES AT THE eastern end of the Black Sea. To the north is the Russian Federation, to the southeast Azerbaijan and Armenia, and Turkey lies to the south. Its decision not to join the Commonwealth of Independent States after the breakup of the Soviet Union in 1991 was largely due to political and interethnic conflicts, but it also reflected the independent spirit of the nation. Membership was ratified in 1994.

GEOGRAPHY

Georgia is a country of mountains. The northern frontier with Russia runs along the massive mountain barrier of the Great Caucasus. These snowy peaks overlook the Black Sea in the northwest and have numerous spurs and secondary ranges cut by deep gorges. A central plateau called the Kartalinian Plain divides these mountains from a parallel range known as the Little Caucasus, whose lofty peaks and plateaus extend south into Turkey and Armenia. Farther east, beyond the capital, Tbilisi, the Kura river and its tributaries drain southeast into Azerbaijan through a series of upland valleys. The swampy plains of Kolkhida fringe the Black Sea coast.

The mountainous terrain produces a varied climate. Generally it is cold in the mountains in winter, with heavy snows, but the Great Caucasus also block still colder air coming from Russia to the north. Balmier winds from the Black Sea bring a damp, almost subtropical climate to the Kolkhida lowlands, whereas the eastern areas have much drier weather.

Much of Georgia is wooded, from sparse birch cover on the mountain slopes to dense broadleaf forests of beech, oak and chestnut at lower levels. In the drier east the forest gives way to grassland. Wildlife flourishes too, especially forest species such as roe deer, Caucasian deer, brown bears and wolves. Wild goats and their close relatives, tahrs, live in more mountainous areas.

SOCIETY

The Georgians pride themselves on a history that goes back to some of the earliest known records, and on a unique language with its own distinctive script. To the ancient Greeks, Georgia's western plain of Kolkhida was known as Colchis – the land whose fabled wealth inspired the legend of the Golden Fleece. Indeed, until recently sheepskins were still used to

Golden fields of corn lie within clear sight of the foothills and snow-capped peaks of the Caucasus Mountains in western Georgia. This region, with its almost subtropical climate, produces many agricultural products including tea, tobacco, flowers and wine.

include sugar beet and essential oils from roses and other flowers. Sheep are the main livestock. The collectivization of agriculture had less impact here than in other former Soviet republics, so privately owned plots have flourished.

Georgia's rich mineral and energy resources encourage local industrial activity. There are deposits of coal, natural gas and peat, as well as fast-flowing rivers, ideal for generating hydroelectric power. Large manganese-ore deposits support most of the mining industry, but many other metals and minerals are also extracted. Traditional industries have included steelworking, the construction of locomotives, engineering and chemicals. The richness of Georgian arts and culture can be seen in a long tradition of icon-painting and metalworking, embroidered textiles, and wood and stone carving.

Despite the rugged terrain, transportation is well developed, with links between the busy ports of Batumi and Poti and the densely populated hinterland. Healthcare and education in Georgia enjoy a high reputation among the former Soviet states. In particular, the widespread teaching of the ancient Georgian language helped the republic to retain a strong national identity and separate culture throughout the Soviet period. Tourism on the Black Sea coast has declined.

catch the gold dust in its rivers. After coming under the domination first of the Greeks and later of the Romans the country converted to Christianity in 330 AD. The Georgians have since retained their distinctive Orthodox faith, along with a strong sense of national identity, despite many centuries of domination by their powerful neighbors – the Persians and the Ottoman Turks.

By the 18th century Georgia had been fragmented into several principalities. Its rulers sought Russian help in their struggle for independence, but the Russian response was simply to take over the country piecemeal during the 19th century. At the end of the century Georgia was reunited – but under Russian rule. Despite this, the Georgians remained resistant to Russian influences. Under later more liberal-minded governors a renewed sense of national identity emerged, including an artistic revival and movements for social reform. One such Marxist-inspired group was joined in 1896 by a young Georgian called Joseph Dzhugashvili (1879–1953); in later years he became known to the world by his adopted name "Stalin" – meaning "steel".

Following the 1917 Russian Revolution, Georgia was established as an independent state. But in 1921, after a short occupation by the British, the Red Army invaded and declared a Soviet republic. Many nationalists, and even some communists, were murdered, and several peasant uprisings were brutally put down. During Stalin's purges of the 1930s still more Georgians were killed.

Candles are lit during a service of commemoration. Christianity was adopted in Georgia in the early 4th century; the Georgian Orthodox Church (one of the world's oldest Christian communities) survives – with Christian Armenia – in a region dominated by Islam.

Glasnost (openness) in the 1980s gave Georgian nationalism a new voice that led in 1989 to a peaceful demonstration calling for independence. This was savagely crushed, but in April 1991, following a national referendum, the Georgian Supreme Soviet declared the country independent. Its chairman, the former dissident, Zviad Gamsakhurdia (1939–93) was elected president. A month later, however, his dictatorial methods provoked opposition, and soon afterward the country was plunged into armed conflict.

Early in 1992, after weeks of bloodshed, Gamsakhurdia was forced to flee the country. A new government was installed, under the former Soviet foreign minister Eduard Shevardnadze (b. 1928), himself a Georgian, who promised a return to stability. Gamsakhurdia returned to western Georgia in 1993 to ferment a rebellion which was only quelled with Russian help. Conflict with the self-declared Abkhazian Autonomous Republic in the northwest over its desire to secede from Georgia remains unresolved, Abkhazia again declaring independence in 1994. Shevardnaze was reelected in November 1995.

ECONOMY

Georgia produces tea, citrus fruits, tobacco and fine wines; additional products

NATIONAL DATA – GEORGIA

Land area	69,700 sq km (26,911 sq mi)			
Climate		**Temperatures**		**Annual**
	Altitude m (ft)	January °C(°F)	July °C(°F)	precipitation mm (in)
Tbilisi	490 (1,608)	1 (34)	24 (75)	500 (19.7)

Major physical features highest point: Shkhara 5,201 m (17,063 ft); longest river: Kura (part) 1,514 km (941 mi)

Population (2006 est.) 4,661,473

Form of government multiparty republic with one legislative house

Armed forces army 7,042; navy 1,350; air force 1,350

Largest cities Tbilisi (capital - 1,026,874); Kutaisi (173,000); Batumi (115,806); Rustavi (105,393)

Official language Georgian

Ethnic composition Georgian 83.8%; Azeri 6.5%; Armenian 5.7%; Russian 1.5%; other 2.5%

Religious affiliations Orthodox Christian 83.9%; Muslim 9.9%; Armenian-Gregorian 3.9%; Catholic 0.8%; other 0.8%; none 0.7%

Currency 1 lari (GEL) = 100 tetri

Gross domestic product (2006) U.S. $17.79 billion

Gross domestic product per capita (2006) U.S. $3,800

Life expectancy at birth male 72.8 yr; female 79.87 yr

Major resources coal, natural gas, forests, hydropower, manganese, copper, citrus fruits, grapes/wine, sugar beet, livestock, vegetables

Armenia

REPUBLIC OF ARMENIA

M ODERN ARMENIA IS A TINY, LANDLOCKED republic in the mountains to the south of the Caucasus. Its people are survivors of an ancient civilization that was once fought over by Mongols, Turks and Persians and of a nation that was all but destroyed. Armenia shares borders with Georgia to the north, Azerbaijan to the east, Iran and the Azerbaijani region of Nakhichevan to the south, and Turkey to the west. Armenia also claims the Armenian populated enclave of Nagorno-Karabakh, which lies across its eastern frontier in Azerbaijan.

NATIONAL DATA - ARMENIA

Land area 28,400 sq km (10,965 sq mi)

Climate	Altitude m (ft)	Temperatures January °C(°F)	July °C(°F)	Annual precipitation mm (in)
Yerevan	990 (3,248)	-4 (25)	20 (68)	277 (10.9)

Major physical features highest point: Aragats 4,090 m (13,418 ft); longest river: Aras (part) 914 km (568 mi); largest lake: Lake Sevan 1,360 sq km (525 sq mi)

Population (2006 est.) 2,976,372

Form of government multiparty republic with one legislative house

Armed forces army 45,000; air force 3,160

Largest cities Yerevan (capital - 1,086,174); Gyumri (151,450); Vanadzor (104,242)

Official language Armenian

Ethnic composition Armenian 97.9%; Yezidi (Kurd) 1.3%; Russian 0.5%; other 0.3%

Religious affiliations Armenian Apostolic 94.7%; other Christian 4%; Yezidi (monotheist with elements of nature worship) 1.3%

Currency 1 dram (AMD) = 100 luma

Gross domestic product (2006) U.S. $15.99 billion

Gross domestic product per capita (2006) U.S. $5,400

Life expectancy at birth male 68.25 yr; female 76.02 yr

Major resources gold, copper, molybdenum, zinc, alumina, antimony, arsenic, barley, chromites, citrus fruits, cotton, figs, iron, limestone, livestock, magnesium, mercury, potatoes, pumice, silver, sugar beet, tobacco, wheat, wine/grapes

GEOGRAPHY

Armenia is occupied almost entirely by the mountains of the Little Caucasus, whose name belies their considerable height. The rugged landscape includes extinct volcanoes and high lava plateaus cut by deep ravines; frequent earthquakes show that mountain-building is still going on. The highest peak, Aragats at 4,090 m (13,418 ft), is in the northwest. Farther east is Lake Sevan, nearly 2,000 m (some 6,000 ft) above sea level. The Gegam mountain range to the west of the lake overlooks the capital, Yerevan, and the broad Ararat plain, which is crossed by the river Aras as it follows its southeasterly course along the borders with Turkey and, in the far south, Iran.

The country has a dry, continental climate; short, sharp winters and long, hot summers on the plains. The mountains are cooler, and winter conditions there are much harsher.

Steppe vegetation predominates, with drought-resistant grasses and sagebrush at lower altitudes, where jackals, wildcats and sometimes leopards are found. Oak woodlands in the southeastern forests merge with beechwoods in the more humid northeast. Here the animals include squirrels and brown bears. Alpine grassland on the higher slopes provides good summer grazing for wild goats and mouflons (wild sheep).

SOCIETY

The Armenians are descended from the various Indo-European peoples that inhabited the Ararat plain and surrounding areas from the 12th to the 6th century BC. Here, legend has it, Noah landed his ark after the Flood and founded the ancient city of Yerevan. The kingdom of Urartu (as Armenia was then known) flourished here between the 9th and 6th centuries BC, and became one of the most powerful states in the Middle East. In 520 BC the kingdom was conquered by the Persian empire, and then from 330 BC it was ruled by the Macedonian Alexander the Great (356–323 BC) and the Greek dynasty that succeeded him. Ultimately, Armenia fell under the dominion of Rome in 67 BC, becoming a buffer state between the Roman empire and the empire of Parthia – Rome's major rival for power in the east. When Armenia adopted Christianity in about 300 AD, it became a bulwark of Christianity in Asia.

In the centuries that followed, Armenia

struggled to retain its independence and cultural identity as it became a theater of war between the Persians, the Seljuk Turks and the Mongols. The population became scattered across the world, but nothing could destroy the strong sense of nationhood kept alive by the depth of its faith in the Armenian Church.

In the early 19th century, as the Russian empire advanced its frontiers across the Caucasus, the Armenians looked to the tsar for liberation from Persian and Turkish domination. By 1829 Russia had taken all the land now occupied by Armenia and Azerbaijan, except for western Armenia, which remained under Turkish rule. In subsequent decades there was a resurgence of Armenian culture, but following the Russo-Turkish War (1877–78) fear of Armenian nationalism led to repression in both Russian and Turkish Armenia. In 1915, during World War I, 1.75 million Armenians living in the western region ruled by Turkey were deported from their homes and c. 600,000 were massacred by the Turks. The establishment of the Soviet Republic of Armenia in 1920 prevented a Turkish takeover of the whole region and in 1921,

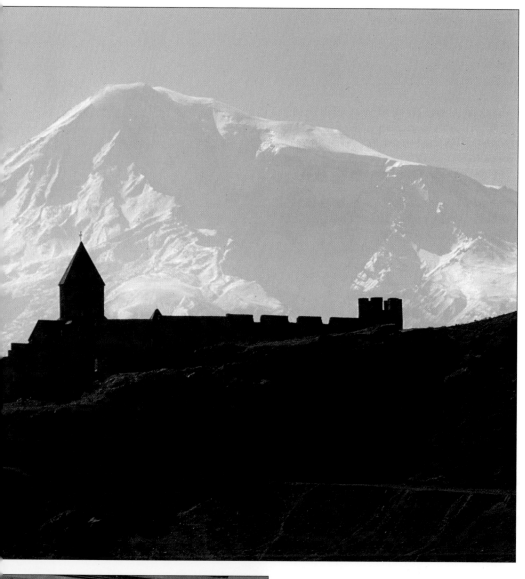

joined the Commonwealth of Independent States in 1991 and became a member of the UN in 1992. Conflict broke out again in Nagorno-Karabakh in 1992, with Armenian forces seizing much of the territory. A ceasefire agreement ended the conflict in 1994, but sporadic outbreaks of violence continued. These troubles exacerbated the already acute problems of supply and distribu-tion during the restructuring of the postcommunist economy and the introduc-tion of multiparty democracy.

The vast majority of the republic's population are Armenians, who despite long periods of Muslim rule have retained their strong Christian tradition. They belong either to the Armenian Orthodox Church or to the much smaller Armenian Catholic Church, which is in communion with Rome.

ECONOMY

Agriculture is the main source of income, especially in irrigated areas of the Aras valley and around Yerevan. Here the crops include grapes, almonds, figs and olives. Apples, pears and cereals are grown on higher ground, while sheep and cattle are raised in the mountains.

In the second half of the 20th century the country industrialized rapidly, exploiting its valuable deposits of copper, zinc and molybdenum, and developing its capacity to generate hydroelectricity. The chief industries are mining, metallurgy, mechanical engineering and chemicals. Armenia also exports precision instruments and textiles, as well as quality brandies and wines.

In December 1988 a devastating earthquake destroyed many towns in northern Armenia and seriously affected its economic infrastructure. Numerous factories and power stations were lost and much of the country's financial resources had to be diverted to meet the immediate demands of humanitarian aid to earthquake victims. The disruption of power services, exacerbated by the continuing conflict with neighboring Azerbaijan, and a subsequent blockade on fuel imports caused much suffering for Armenia's population.

Several years later, in the early 1990s, factories and hospitals were still unheated, power cuts were routine and for increasingly longer periods each day. The government reopened its antiquated nuclear power plant at Metzanor, which had been closed since the 1988 earthquake, in an attempt to solve the country's fuel crisis.

Spiritual home (*above*) The Armenian church of Khor Virap, in the shadow of Turkey's Mount Ararat, marks the site where the evangelist Gregory the Illuminator (257–332) was imprisoned by the pagan king Tiridates III, whom he later converted to Christianity.

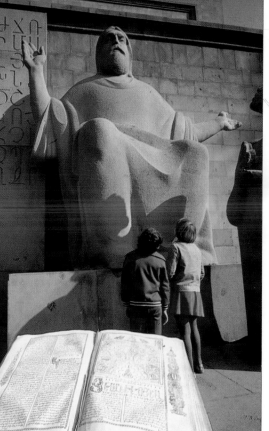

Words in stone (*left*) Youngsters examine a statue of St Mesrob (c. 350–440), founder of the Armenian alphabet. It stands in front of the Matenaradan library, Yerevan, whose archives hold a fine collection of ancient Armenian manuscripts, such as the one shown.

after the Russian Revolution and civil war (1917), the Bolsheviks gave the regions of Nagorno-Karabakh (which was mostly populated by Armenians) and Nakhichevan to Azerbaijan. The following year, Armenia and its neighbors Azerbaijan and Georgia were united in the Transcaucasian Soviet Federative Republic.

Armenia achieved separate status as a Soviet republic in 1936, but its territory did not include either Nagorno-Karabakh or Nakhichevan. This continued separation of ethnic Armenians from their native state created tensions between Armenia and Azerbaijan that the Soviet regime contained for many years until they erupted violently in the late 1980s.

Both republics achieved independence on the break-up of the Soviet Union and

Azerbaijan

REPUBLIC OF AZERBAIJAN

L YING ON THE WEST COAST OF THE CASPIAN Sea, Azerbaijan occupies the southeast corner of Transcaucasia, with Russia to the north and Georgia to the northwest. Armenia is to the west and Iran to the south. Its territories include the autonomous oblast region of Nagorno-Karabakh in the southwest, which is also claimed by Armenia, and the autonomous republic of Nakhichevan, which lies farther southwest, beyond a strip of Armenian territory. It withdrew from the CIS in 1992 and rejoined in 1993, but is also making strong links with Turkey.

GEOGRAPHY

Northern Azerbaijan is dominated by the peaks of the Great Caucasus Mountains that extend southeastward, partly along the Russian border, as far as the capital and chief port, Baku. In the foothills of the Caucasus, the Kura river valley drops southeastward from the Georgian border, opening onto a broad floodplain that lies mainly below sea level. The mountains and plateaus of the Little Caucasus occupy Nakhichevan and western Azerbaijan, while in the far south the Talish Mountains on the Iranian border overlook the Caspian coastline.

In the lowlands the climate is dry and subtropical, with mild winters, long, hot summers and frequent droughts. Mountain areas that have a more continental climate experience cold winters with heavy snow; farther south it is wetter and more humid. The vegetation varies from steppe grassland in the drier lowlands to forests in the mountains and marshlands in the southeast. Wildlife includes leopards, gazelles and ground squirrels.

SOCIETY

The first people to inhabit Azerbaijan were Iranians, and for most of its long history Azerbaijan belonged to the Persian

Trapped innocent (*above*) A young girl from Nagorno-Karabakh – an Armenian Christian enclave in Azerbaijan – with straw brooms for sale. Like many thousands, she is caught by the fighting between Armenians and Azerbaijanis over the enclave's political status.

The waterfront of Baku (*right*) Azerbaijan's capital and major port on the west coast of the Caspian Sea. The city is heavily dependent on the nearby oil fields for its industry.

Olives from reclaimed land (*below*) Azerbaijani men and women clear weeds from a newly planted olive grove near Baku. The land they are tilling has been reclaimed partly from the desert and partly from an oil field that has ceased production.

NATIONAL DATA - AZERBAIJAN

Land area	86,100 sq km (33,243 sq mi)
Climate	humid subtropical to dry continental
Major physical features	highest point: Bazardyuze 4,480 m (14,698 ft); lowest point: Caspian Sea -28 m (-92 ft); longest river: Kura (part) 1,510 km (940 mi)
Population	(2006 est.) 7,961,619
Form of government	multiparty republic with one legislative house
Armed forces	army 56,840; navy 2,000; air force 7,900
Largest cities	Baku (capital – 1,118,904); Gāncā (304,472); Sumgayit (268,336)
Official language	Azerbaijani
Ethnic composition	Azeri 90.6%; Dagestani 2.2%; Russian 1.8%; Armenian 1.5%; other 3.9%
Religious affiliations	Muslim 93.4%; Russian Orthodox 2.5%; Armenian Orthodox 2.3%; other 1.8%
Currency	1 Azerbaijani manat (AZM) = 100 gopiks
Gross domestic product	(2006) U.S. $58.1 billion
Gross domestic product per capita	(2006) U.S. $7,300
Life expectancy at birth	male 59.78 yr; female 68.13 yr
Major resources	petroleum, natural gas, iron ore, nonferrous metals, alumina, cotton, tobacco, fruit, fisheries

empire. Azerbaijanis themselves, many of whom live farther south in the Iranian province of the same name, are a people of Turkic origin who arrived with the 11th-century Seljuk invaders who conquered much of Asia Minor. Persians and Turks fought for control of Azerbaijan throughout the ensuing centuries until finally the Russians gained a foothold during the early 19th century. The Treaty of Turkmenchai in 1828 divided the country between Persia (which gained the south) and Russia (the north).

During the chaos that followed World War I and the Russian Revolutions of 1917, the Azerbaijanis set up an independent anti-Bolshevik republic, but were invaded first by Turkey and then, in 1922, by the Soviet Red Army. In the subsequent resistance to Soviet rule, many religious and political leaders were killed and collectivization of farmland was fiercely resisted by Azerbaijan's peasantry. For a brief period from 1922 it became part of the Soviet Transcaucasian Federation, but it was eventually incorporated into the Soviet Union as a constituent republic in 1936.

Today most of the inhabitants are Muslim Azerbaijanis. However, there is a sizable minority of Armenians, who are mainly Christian, concentrated in the Nagorno-Karabakh enclave. In 1988 a long-standing dispute with Armenia over this area erupted into violence and resulted in the departure of thousands of Armenians from the country. When Azerbaijan and Armenia both achieved political independence on the dissolution of the Soviet Union in 1991, their conflict over Nagorno-Karabakh escalated into a war. A 1994 cease-fire brought temporary peace, but the status of the enclave remains unresolved. Azerbaijan has a 125-member parliament elected by majority vote and a directly elected president. There were several attempted coups in the 1990s and it remains politically unstable.

ECONOMY

Azerbaijan has little arable land, but nevertheless is a major producer of cotton, tobacco, grapes and other fruit. The Caspian Sea is a ready source of fish, especially sturgeon for its highly-prized caviar, but this important industry is now threatened by severe pollution.

Other natural resources include metal ores, natural gas and vast reserves of oil, which is refined mostly in the Baku area. Baku once provided half the world's supply of oil, and the petroleum and chemical industries expanded rapidly around this area until it became the Soviet Union's fifth largest city. The burgeoning industrial sector is now extremely diversified, from chemicals and gasoline to textiles, and electrical and other consumer goods. A market-based economic reform program and increased oil production was encouraging foreign investment and boosting GDP growth rates in the late 1990s.

Transportation includes good railroads and an extensive road network. There are few inland waterways, but shipping lines from the busy port of Baku serve Russian, central Asian and Iranian destinations. Free health, welfare and education services were well developed under the Soviet regime. Azerbaijan's strong intellectual tradition is shown in its prestigious scientific institutions.

The Middle East

COUNTRIES IN THE REGION

TURKEY · SYRIA · LEBANON · ISRAEL

JORDAN · IRAQ · KUWAIT · IRAN · YEMEN

AFGHANISTAN · SAUDI ARABIA · BAHRAIN · QATAR

UNITED ARAB EMIRATES · OMAN

Black Sea

Istanbul
Izmit
Samsun
Bosporus
Ankara
TURKEY
Erzurum
Izmir
Lake Tuz
Kizil Irmak
Elazig
Ararat 5165
Antalya
Anatolia
*Taurus Mt*s
Adana
Lake Van
Tabriz
Aras
Urmia
Lake Urmia
Rasht
Caspian Sea
Concord Peak 5407
Aleppo
Mosul
Elburz Mountains
Mashhad
Baghlan
Hindu Kush
Nowshak 7485
Mediterranean Sea
Tripoli
Homs
SYRIA
Kirkuk
Tehran Damavand 5671
Shah Fuladi 5143
Hari Rud
Nowshak
LEBANON
Beirut **Damascus**
Dasht-e-Kavir
Herat
Kabul
ISRAEL
Tel Aviv
IRAQ
Baghdad
Tigris
Zagros Mountains
Isfahan
IRAN
AFGHANISTAN
Jerusalem
Amman
An Najaf
Kandahar
Dead Sea -400
Syrian Desert
Euphrates
Karun
Kerman
Dasht-e-Lut
Helmand
JORDAN
Basra
Seistan
Aqaba
Shiraz
An Nafud
Kuwait
KUWAIT
Persian Gulf
Bandar Abbas
Hejaz
Nejd
Dammam
BAHRAIN
Strait of Hormuz
Tropic of Cancer
Al Manamah
QATAR
Medina
Riyadh
Doha
Dubai
Abu Dhabi
Muscat
Jedda
Mecca
SAUDI ARABIA
UNITED ARAB EMIRATES
Sur
Cape Hadd
Arabian Peninsula
OMAN
Masirah
Red Sea
Arabian Sea
Rub al Khali
Cape Madrakah
Kuria Muria Islands
Salalah
YEMEN
San'a
Hadhramaut
Al Mukalla
Aden
Gulf of Aden
Socotra

Legend

■ capital city
● major town

height of land (meters)

	5000
	3000
	2000
	1000
	500
	200
	0 (sea level)

▲ mountain peak
▼ depression

Luxury in an arid land (*left*) Water is one of the most precious resources in the Middle East. Here a Bedouin nomad waters his herd of camels at a well in Qatar. But it was the discovery of another liquid – oil – that brought the Arabian Peninsula into the modern world and made it rich beyond measure.

Turkey

REPUBLIC OF TURKEY

TURKEY SPANS TWO CONTINENTS, LINKING mainland Europe with Asia across the narrow straits of the Bosporus and the Dardanelles. European Turkey (Thrace), to the north of the Sea of Marmara, shares borders with Bulgaria and Greece. Asian Turkey, much of which is known as Anatolia or Asia Minor, is larger in area and borders Syria and Iraq to the south and southeast, Iran and Armenia to the east, and Georgia to the northeast. The long coastlines are washed by the Black Sea to the north, the Aegean to the west and the Mediterranean to the south.

NATIONAL DATA – TURKEY

Land area	770,760 sq km (297,592 sq mi)			
Climate		Temperatures		Annual
	Altitude m (ft)	January °C(°F)	July °C(°F)	precipitation mm (in)
Ankara	861 (2,825)	1 (34)	23 (73)	384 (15.1)
Izmir	28 (92)	9 (48)	28 (82)	686 (27)

Major physical features highest point: Ararat 5,165 m (16,945 ft); largest lake: Lake Van 3,675 sq km (1,419 sq mi); longest river: Kizil Irmak 1,150 km (715 mi); (Euphrates and Tigris also rise in Turkey)

Population (2006 est.) 70,413,958

Form of government multiparty republic with one legislative house

Armed forces army 402,000; navy 52,750; air force 60,100

Largest cities Istanbul (10,291,102); Ankara (capital – 3,641,931); Izmir (2,615,568); Bursa (1,504,817); Adana (1,294,460)

Official language Turkish

Ethnic composition Turkish 80%; Kurdish 20%

Religious affiliations Muslim 99.8% (mainly Sunni); other 0.2% (mainly Christians and Jews)

Currency 1 Turkish lira (YTL) = 100 kurus

Gross domestic product (2006) U.S. $6.27.2 billion

Gross domestic product per capita (2006) U.S. $8,900

Life expectancy at birth male 70.18 yr; female 75.18 yr

Major resources coal, iron ore, copper, chromium, antimony, mercury, gold, barite, borate, celestite (strontium), emery, feldspar, limestone, magnesite, marble, perlite, pumice, pyrites (sulfur), clay, cotton, tea, tobacco, wheat, barley, rice, sugar beet, olives, grapes/sultanas, figs, hazelnuts, tourism, hydropower

GEOGRAPHY

Frequent earthquakes, together with geysers and other volcanic phenomena, are evidence that this is a tectonically active region.

The European part of Turkey is characterized by fertile rolling plains surrounded by low mountains. On the Asian side, western Anatolia is crossed by long mountain ridges separated by deep valleys. South of the central Anatolian plateau, the three main ranges of the Taurus Mountains flank the Mediterranean coast. Farther east a complex of even higher mountain ranges culminates in the massive cone of Mount Ararat. To the southwest is Lake Van.

Annual rainfall reaches about 2,500 mm (100 in) along the Black Sea coast, where temperatures are relatively high all year. The central plateaus of the interior are semiarid. In the mountainous east the winter snows may last for several months.

Thick, scrubby undergrowth characterizes the Mediterranean south and west of the country. The Black Sea coast supports dense forest, while the drier interior is covered mainly in steppe grassland. Bears and red deer are still found in the forests and colorful flocks of greater flamingoes thrive on some of the lakes.

SOCIETY

Asia Minor has seen the rise and fall of many of the world's great empires. However, by the early 20th century Turkey was a weak and impoverished country. It was the country's first president, Kemal Atatürk (1881–1938), who laid the foundations for a stronger, modern nation.

The arrival of Indo-European-speaking Hittites from central Asia in about 1900 BC marked the beginning of the first great Anatolian empire. Anatolia was absorbed into the Persian empire in the 6th century BC. Alexander the Great (356–323 BC) then brought the region under Greek control before it fell to the Roman empire in the 1st century BC.

From the 11th century AD Asia Minor was invaded by the Muslim Seljuks – Turkic rulers from the east. Osman I (1258–1326), the ruler in northwestern Anatolia, is traditionally regarded as the founder of the Ottoman state. By the later 15th century the Ottomans had taken the old Roman capital, Constantinople, which they renamed Istanbul, and gained control of the greater part of Asia Minor. Later rulers, notably Suleiman the Mag-

Straddling Europe and Asia (*above*), Istanbul is strategically positioned at the entrance to the Black Sea, controlling a major shipping route. Among its many attractions are the colorful bazaar, the Topkapi Palace, and the great Roman walls and underground cistern.

Colossal ruins atop Mount Nemrut (*below*) in the far east of Turkey. The huge statues of gods were commissioned by Antiochus I of Commagene (62–32 BC) – together with one of himself – "in commemoration of [his] own glory and that of the gods".

The terraces of Pamukkale (*right*) in west-central Turkey descend for some 90 m (300 ft). As the hot, mineral-rich spring water flows down the slope, it evaporates, depositing the minerals in fantastic formations.

nificent (1494/5–1566), pushed the empire's boundaries still farther into the Middle East, Northern Africa and Europe.

By the 19th century the Ottoman empire had begun to disintegrate, and defeat in the Balkan Wars of 1912–13 reduced Turkey's European territories to their present size. In 1918, following its defeat in World War I, the Ottoman empire was dismembered. Meanwhile a nationalist revolutionary movement led by Mustafa Kemal had set up a government in Ankara. As Turkey's first president, Kemal, who took the name Atatürk, or "Father of Turkey", transformed and Westernized Turkish society.

In 1951 Turkey became a member of the North Atlantic Treaty Organization (NATO). A long-running dispute with Greece over the Mediterranean island of Cyprus has been an obstacle to Turkey's aim of joining the European Union (EU). In 1980 there was a military coup but from 1983 political parties were allowed to operate again. During the late 1980s there was growing insurgency among the largely Kurdish population of the eastern provinces. This situation was complicated further by the 1991 Gulf War between Western Allies and Iraq over the invasion of Kuwait. Turkey imposed sanctions in accordance with NATO policy, at consid-

erable cost to its own economy, but was unhappy about having foreign troops stationed on Turkish soil. Moreover, when Kurdish refugees crossed the border from Iraq, Turkey was obliged to help those who had long been regarded as a dangerous minority.

Under the 1982 constitution the 550 members of the Grand National Assembly (GNA) are elected by proportional representation to serve a five-year term. Executive power rests with the president.

Most of the population are Turkish-speaking Sunni Muslims. Kurdish communities in the southeast constitute the only large minority. Since 1984 Turkey has been fighting Kurdish separatists.

ECONOMY

Turkey's economic development has been slowed by the need to import some of its fuel and food. Local coal is available but petroleum has to be imported. Up to half the nation's energy comes from hydroelectric power. Plans to build large dams

and hydroelectric plants will increase Turkey's energy production by 70 percent. Irrigation in the coastal lowlands has enabled the country to produce the largest crops of hazelnuts and sultanas in the world; these and other cash crops such as cotton, tea and tobacco are vital to its export trade.

Iron, aluminum and copper ores supply raw materials for industry. The textile industry, which produces yarn, fabrics, rugs (from locally grown cotton) and carpets (from wool), accounts for about 40 percent of the country's exports. Tourism is growing rapidly. Turkey has the best-developed road and railroad systems in the Middle East. Istanbul is the country's largest port. The state airline, Turkish Airlines, serves most major international and domestic airports.

The national welfare provision includes health insurance and retirement pensions. Healthcare is free for the poor. Housing is in short supply. Free education is provided at primary and secondary levels. There are some 30 universities.

Syria

SYRIAN ARAB REPUBLIC

S YRIA LIES AT THE EASTERN END OF THE
Mediterranean Sea. To the north is
Turkey, to the east Iraq, to the south
Jordan, to the southwest Israel, and to the
west Lebanon and the Mediterranean
coast. Ever since its frontiers were drawn
up at the end of World War I by the Allies,
the country has played a leading role in
Middle Eastern politics; the government's
hardline Arab nationalism has led Syria
into frequent conflicts with its neighbors,
particularly Israel.

GEOGRAPHY

Syria's Mediterranean seaboard extends
for some 180 km (110 mi) between Turkey
and Lebanon and is one of the country's
most fertile and densely populated areas.
Inland are the mountains of the Jebel
Ansariye. Across the mountains, the Asi
(Orontes) river flows through the deep
trench formed by the Ghab Depression.
The mountainous terrain continues along
the Lebanese frontier as far as the Golan
Heights in the southwest, presently
under Israeli military occupation. Extending north and east from the mountains is
a broad, semiarid plateau – crossed by the
Euphrates river in the southeast – that
turns into the Syrian Desert.

The climate is generally dry with hot
summers and mild winters. Rain falls
mostly in the western highlands between
September and May, but varies considerably from year to year, with frequent
summer drought. Hot dry khamsin winds
from the desert blow in summer.

The vegetation varies from thin desert
scrub in the east and dry grassland in the
north to Mediterranean maquis in the
west, and forests on the highest mountains. Gazelles, jerboas and lizards live in
remote desert areas; the high mountain
forests are home to Wild boars, wolves,
badgers and porcupines, as well as
several species of warbler.

SOCIETY

Situated on ancient trade and military
routes between the Mediterranean and
Mesopotamia (present-day Iraq), Syria
from about 2000 BC was part of Akkadia –
one of the earliest urban civilizations.
Over time it became a province of neighboring empires such as those of the
Egyptians, the Hittites in the northwest,
and the Assyrians and Babylonians. The
Persians were driven out by Alexander
the Great (356–323 BC); his successors
introduced Greek culture. In 64 BC Syria
became a Roman province.

From 395 AD Syria was ruled from
Constantinople, the capital of the eastern
Roman empire, until it was overrun by
Muslim Arabs in 634. Most Syrians converted to Islam under the Umayyad dynasty, and Damascus became the political
and cultural center of the Islamic world.
The region fell to the Seljuk Turks in the
11th century. During this period Syria
was the scene of many battles between
Western Christian crusaders and Muslims. The crusaders seized much of Syria
in the 12th century, but were expelled in
1187 by the Kurdish general Saladin (c.
1137–93). Saladin's Ayyubid descendants
and their Mameluke successors ruled until the Ottoman Turkish conquest of 1516.

Syria came under a French mandate in
1920, after Turkey's defeat in World War
I, and its present boundaries were established. During World War II, Syria was
briefly controlled by the pro-German
Vichy French regime, before being occupied by a joint British and Free French
invasion force. The French did not with-

A peaceful mosque courtyard (*above*) in the heart of
Syria's capital, Damascus, a city that has been
continually inhabited since 2500 BC. Its ancient crafts
of weaving and metal working are major industries.

The traditional domed roofs of mud brick (*below*) –
derived from a scarcity of wood – in a rural settlement
near Aleppo. Although many agricultural workers have
moved to the cities in search of work, about half of
Syria's workforce still makes a living from agriculture.

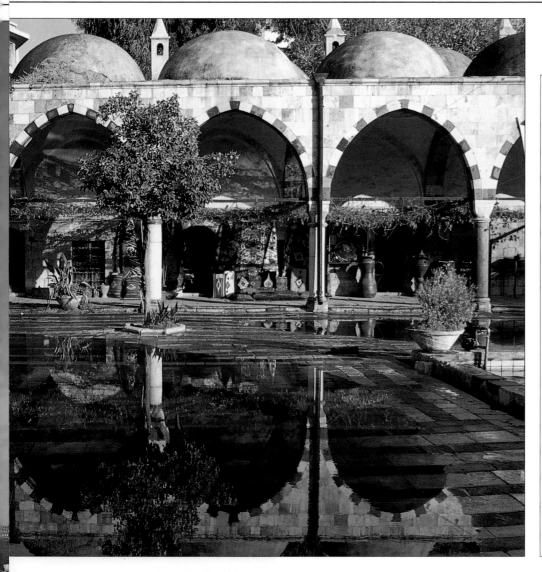

NATIONAL DATA – SYRIA

Land area	184,050sq km (71,062 sq mi)			
Climate		**Temperatures**		**Annual**
	Altitude m (ft)	January °C(°F)	July °C(°F)	precipitation mm (in)
Damascus	720 (2,362)	7 (45)	27 (81)	133 (5.2)

Major physical features highest point: Mount Hermon 2,814 m (9,232 ft); longest river: Euphrates (part) 3,596 km (2,235 mi)

Population (2006 est.) 18,881,361

Form of government multiparty republic with one legislative house

Armed forces army 200,000; navy 7,600; air force 40,000

Largest cities Aleppo (1,649,694); Damascus (capital – 1,592,206); Homs (822,440); Hama (495,390)

Official language Arabic

Ethnic composition Arab 90.3%; Kurds, Armenians, and other 9.7%

Religious affiliations Sunni Muslim 74%; Alawite, Druze, and other Muslim sects 16%; Christian (various sects) 10%; Jewish (communities in Damascus, Al Qamishli, and Aleppo)

Currency 1 Syrian pound (SYP) = 100 piastres

Gross domestic product (2006) U.S. $75.1 billion

Gross domestic product per capita (2006) U.S. $4,00

Life expectancy at birth male 69.01 yr; female 71.7 yr

Major resources petroleum, phosphates, chrome and manganese ores, asphalt, iron ore, rock salt, marble, gypsum, hydropower, barley, cattle, cotton, fruit, goats, natural gas, potatoes, sheep, sugar beet, vegetables, wheat

draw until 1946, when Syria became independent. Three years later Syria's new civilian government was overthrown, and in 1963, after a succession of military coups and a brief union with Egypt (1958–61), the Arab Socialist Baath (Renaissance) Party took power.

The Baathist regime has survived despite internal unrest and conflict with foreign powers. The country suffered major defeats by Israel in the Arab–Israeli War of 1967, when Israel captured the Golan Heights, and in the Yom Kippur War of 1973. Syria became deeply embroiled in the civil war in Lebanon after 1975, leading to more than a decade of further clashes with Israel until a fragile stability was achieved in Lebanon in 1992. During The Gulf War of 1991 Syria joined the coalition forces against its old rival Iraq, and so improved its erstwhile stormy relations with the West.

Hafez al-Assad (1928–2000) was president until he died in 2000. He was succeeded by his son, Bashar al-Assad, who polled 97 percent of a referendum vote. The president, elected for a seven-year term, appoints two vice-presidents and a cabinet. The 250-member People's Assembly is elected every four years; all parties

within it are officially linked as the National Progressive Front, which in turn is dominated by the Baath Party. Although the 1973 constitution defines Syria as a "socialist popular democracy", Baathist domination effectively controls the elections.

Apart from Kurdish communitites in the north, the population consists predominantly of Arabic-speaking Arabs. Sunni Muslims (who include the Kurds) are by far the largest religious group, but political power rests mainly with the 'Alawites – a Shi'ite Muslim sect to which the Baath leaders belong. There are many other religious minorities, among them Druze Muslims and Christians.

ECONOMY

The Syrian economy, burdened by military spending that consumes some two-thirds of the budget, is still predominantly agricultural despite a program of industrialization since World War II. Modest petroleum reserves provide the leading source of export revenue.

Agriculture is concentrated along the coast and beside the Asi and Euphrates rivers, but dry farming is also important.

Cotton is the chief cash crop, and also the most reliable; other crops include sugar beet, lentils, olives, cereals and high-quality tobacco. Sheep, goats and camels feed on pastures irrigated from artificial terraces that collect rainwater: a technology some 4,000 years old.

There are many mineral resources. Hydroelectric stations on the Euphrates, supplemented by thermal stations burning oil, supply electricity. Industries include textile production, food processing and engineering.

Road and rail networks are well developed. Damascus, the capital, and the northwestern city of Aleppo both have an international airport. Syrian Arab Airways serves a small domestic network. The Syrian press is subject to strict censorship, and broadcasting is directly controlled by the state.

Healthcare, concentrated in the cities, is free for the poor. Water quality and sanitation remain inadequate, and infant mortality is fairly high. A limited welfare system provides retirement, injury and invalidity pensions. Education, which is free of charge, is compulsory for six years; some secondary and higher education is also available.

Israel

STATE OF ISRAEL

T HE SMALL, POPULOUS STATE OF ISRAEL WAS established as a Jewish homeland in Palestine in 1948. Its people, who for many centuries have been scattered all over the world, have historical ties with the region that date back more than 3,000 years. However, the Jews' claim to the land conflicts with that of the Palestinian Arabs, whose historical ties are no less ancient, and a state of hostility between Israel and neighboring Arab countries has existed ever since.

NATIONAL DATA - ISRAEL

Land area	20,330 sq km (7,849 sq mi)			

Climate		Temperatures		Annual
	Altitude m (ft)	January °C(°F)	July °C(°F)	precipitation mm (in)
Jerusalem	757 (2,484)	9 (48)	24 (75)	544 (21.4)

Major physical features highest point: Mount Meron 1,125 m (3,962 ft); lowest point: Dead Sea –418 m (–1,371 ft); longest river: Jordan (part) 320 km (200 mi)

Population (2006 est.) 6,352,117

Form of government multiparty republic with one legislative house

Armed forces army 125,000; navy 7,600; air force 35,000

Largest cities Jerusalem (capital – 743,512); Tel Aviv-Jaffa (388,288); Haifa (265,796)

Official languages Hebrew, Arabic

Ethnic composition Jewish 76.4% (Europe/America-born 22.6%; Israel-born 67.1%; Africa-born 5.9%; Asia-born 4.2%); non-Jewish 23.6% (mainly arab)

Religious affiliations Jewish 76.4%; Muslim 16%; Arab Christians 1.7%; other Christian 0.4%; Druze 1.6%; unspecified 3.9%

Currency 1 new Israeli shekel (ILS) = 100 agorot

Gross domestic product (2006) U.S. \$166.3 billion

Gross domestic product per capita (2006) U.S. \$26,200

Life expectancy at birth male 77.33 yr; female 81.7 yr

Major resources timber, potash, copper ore, natural gas, phosphate rock, magnesium bromide, clays, sand, tourism, citrus fruit, vegetables, cotton, groundnuts (peanuts), diamond processing, crude oil, figs, grapes, livestock, olives, sugar beet, vegetables, wheat

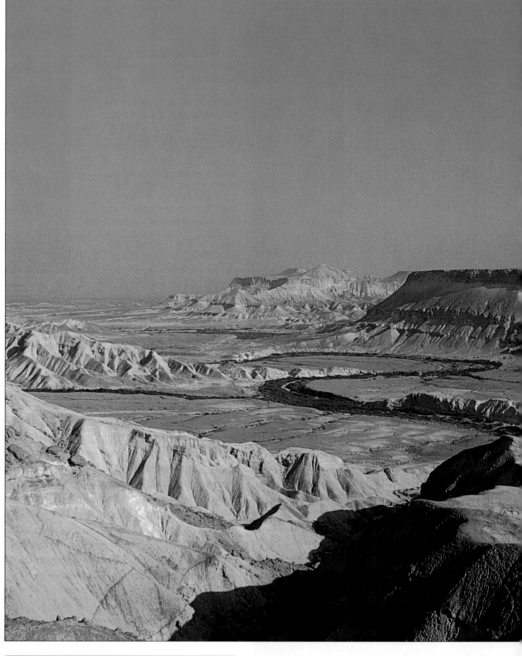

GEOGRAPHY

Israel is bounded by the Mediterranean Sea to the west, Lebanon to the north, Syria to the northeast, Jordan to the east and Egypt to the southwest.

Israel can be divided into four main areas: the fertile densely populated strip of the Mediterranean coastal plains; the central hills extending from Galilee in the north to Judea in the center; the Great Rift Valley running the length of the eastern borders to the Gulf of Aqaba; and the Negev desert in the south. The river Jordan flows south into the Dead Sea.

Winter brings heavy rainfall to the cooler northern highlands, but very little to the south. Summers are hot and dry. Although most of the original evergreen forests have been cleared over the centuries, there has been widespread reforestation, with conifers, eucalyptus and citrus trees. Scrub is the main vegetation, including Mediterranean maquis in the central hills. Several wildlife nature reserves have been established to preserve the enormous variety of wild life.

SOCIETY

The hinterland of Palestine is significant not only for Jews throughout the world, but also for Christians and Muslims; Jerusalem is a holy city for all three faiths.

Modern Israel occupies about three-quarters of the area known as Palestine. The Old Testament of the Bible (the Jewish Torah) describes how nomadic Hebrew peoples (the Jews) escaped Egyptian captivity to settle in the land of the Canaanites (from whom the modern Palestinian Arabs claim descent).

By 40 BC all of Palestine was under Roman domination. Palestine's Jewish population was greatly reduced and scattered. By the time Palestine was conquered by Muslim Arabs in the 7th century, Jews formed only a small minority in a largely Christian population. The Ottoman Turks captured the region in 1516 and ruled for some 400 years.

During the 19th century Jewish nationalists called for a homeland in Palestine. Ever since the 6th century BC, millions of Jews had left Palestine to settle all over the Middle East, the Mediterranean and Europe. In 1917 the British, who occupied Palestine during World War I, declared support for a national Jewish homeland.

Persecution of European Jews by Nazi Germany between 1933 and 1945 accelerated Jewish migration to Palestine, though this was bitterly opposed by its Arab population. In 1947 the United Nations partitioned Palestine and in 1948 the Jewish state of Israel was declared. Since then there has been almost continuous strife between Israel and its Arab neighbors. Several wars (1948, 1967 and 1973) ended with Israel making territorial gains – the Gaza Strip, the West Bank of the river Jordan (with East Jerusalem) and the Golan Heights on the Syrian border. A peace accord in 1993 led to progressive withdrawal of Israeli troops, and expansion of Palestinian self-rule into occupied West Bank and Gaza. However, a "final

The barren Negev desert (*left*) extends across much of southern Israel, but some parts of it have been made to bloom; water from the huge underground aquifers is used to irrigate crops such as pears and tomatoes.

Hasidic Jews in Jerusalem (*below*) during Sukkot, a festival associated with the Jews' exodus from Egypt. Many holy days in the Jewish calendar are observed as religious and public holidays in Israel.

status" agreement has yet to be reached. Violence continues between Israelis and Palestinians with terrorist attacks occurring daily. Assassinations of Israeli ministers have also come from both sides opposing the peace process – in November 1995 Prime Minister Yishtak Rabin was assassinated by an extreme Jew and in 2001 Tourism Minister Rehavam Zeevi was killed by a Palestinian militant.

The Israeli parliament, or Knesset, is elected every four years by proportional representation. Members of parliament choose the president for a five-year term. Power rests with the cabinet, appointed by the president from the parties in government and chaired by the prime minister.

Jews constitute the majority of the population, and more than half of them are Israeli-born. However, a strong ethnic divide exists between Ashkenazi Jews, originally from Central and Eastern Europe, and Sefardic Jews from Asia and the Mediterranean.

ECONOMY

Israel's economic growth and stability has been greatly helped by financial aid from the United States and support from Jewish communities across the world.

Israel won international admiration for its pioneering collective and cooperative agricultural settlements – *kibbutzim* and *moshavim* – and for their achievements in "making the desert bloom". Productivity has benefited from improved efficiency and mechanization, while irrigation has increased the area of cultivable land. Crops include citrus fruits (a major export), cotton and peanuts. Local fishing is supplemented by Atlantic trawling.

The Dead Sea yields valuable bromides and magnesium salts. Some petroleum is also extracted locally. Food processing is the principal manufacturing activity, but textiles, chemicals and electronics are also important along with armaments. Tourism brings valuable foreign currency.

The road and railroad networks are excellent. There are major ports on the Mediterranean and at Elat on the Gulf of Aqaba and an international airport near Tel Aviv. Newspapers and magazines come in Hebrew, Arabic, Yiddish and many European languages.

The welfare system includes a full range of health and social security services. The state provides free schooling at Hebrew and Arab schools. Many students go on to higher education. Literacy rates and standards of health and housing tend to be higher among the Jewish population.

Jordan

HASHEMITE KINGDOM OF JORDAN

THE SMALL ARAB KINGDOM OF JORDAN shares a frontier with Israel to the west, Syria to the north, Iraq to the northeast and Saudi Arabia to the east and south; a short coastline in the far southwest borders the Gulf of Aqaba, giving access to the Red Sea and beyond. Until the Arab–Israeli war of 1967 Jordan held an area to the west of the river Jordan known as the West Bank, which included the Old City of Jerusalem. The king formally renounced his country's claim to this territory in 1988; while it is still under Israeli occupation, Palestinian self-rule was being progressively introduced in 1995/96.

GEOGRAPHY

Jordan's western frontier with Israel runs along the bottom of the Ghor depression, a deep rift valley that contains the Dead Sea. This large saline lake is the lowest point on the Earth's land surface at –400 m (–1,312 ft); it is fed from the north by the river Jordan and from the south by the Wadi Araba. The valley is overlooked by mountains, beyond which lie the barren plateaus and basins of the great North Arabian desert, which merges in the northeast with the rolling lands of the Syrian Desert.

Jordan is an arid country, though the western highlands receive some rainfall during the cooler winter season. In summer it is hot everywhere except on the highest slopes. Between April and October there is little rain anywhere, and water is often in short supply.

Vegetation ranges from Mediterranean plants in the mountains to sagebrush, grasses and scattered shrubs in the dry steppes. Some hardy plants grow in the eastern deserts, especially in basins watered by seasonal streams from the mountains.

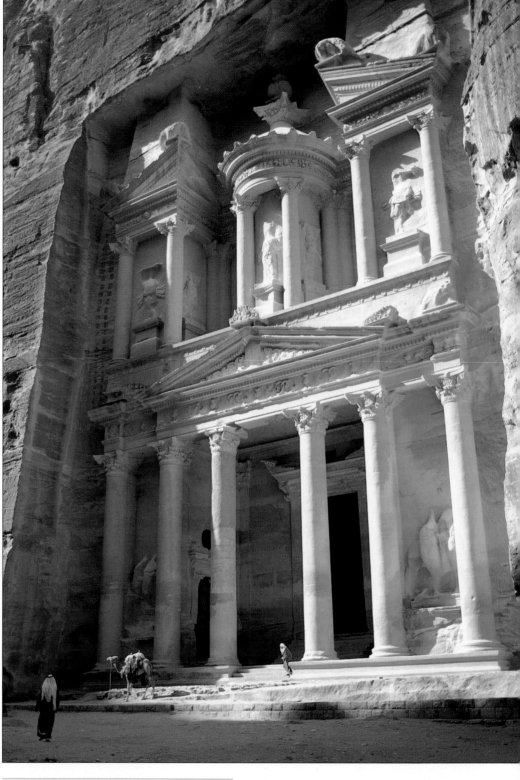

"The rose-red city" of Petra, cut from sandstone rock, was founded about 1000 BC by the Edomites. One of the best-preserved parts is the Street of Façades, which includes the treasury of Khazneh (shown here); in fact it was probably a temple.

SOCIETY

The part of the Middle East now called Jordan was originally claimed by the ancient Moabites and Edomites (told of in the Old Testament of the Bible) – traditional enemies of their western neighbors, the Hebrews. From 800 BC the region was ruled successively by the Assyrians, the Babylonians, the Persians and the Greek Seleucids. It was conquered by the Romans in 64–63 BC, and by Muslim Arabs in the 7th century AD. Muslims clashed with invading Christian crusaders during the 12th century. From the 16th century the region became an imperial province of the Ottoman Turks.

During World War I an Arab rebellion drove out the Ottoman Turks. In 1921 the British persuaded the Hashemite emir Abdullah ibn-Hussein (1882–1951) to establish an Emirate of Transjordan in part of Palestine. This was proclaimed an independent constitutional state in 1923, but it remained a British protectorate. Full independence was not achieved until 1946, when Abdullah became the first king of the Hashemite Kingdom of Transjordan. He joined forces with other Arab leaders to fight the creation of a Jewish state in neighboring Palestine in 1948. In the following year an armistice was signed with Israel, whereby Jordan (as it then became known) retained part of central Palestine (the West Bank), along with the quarter of Jerusalem known as the Old City.

King Hussein ibn-Talal (1935–1999) inherited the throne in 1952 at the age of 17, after his grandfather Abdullah had been assassinated and his mentally ill father deposed. Hussein's own position remained insecure. Raids on Israel by Palestinians living on Jordanian territory provoked sharp reprisals. In 1967 Jordan again joined Arab neighbors in a war against Israel, but this resulted only in the loss of all Palestinian territories west of the river Jordan. Hussein survived a short but bitter civil war with his Palestinian subjects (including refugees from the West Bank). His relations with other Arab countries were also frequently troubled. He supported Iraq during the Iran–Iraq War (1980–88), but was neutral in the 1991 Gulf War against Iraq, when a huge influx of Palestinian refugees from Kuwait strained the already beleaguered Jordanian economy. A peace treaty was signed between Jordan and Israel in 1994, and Israeli forces withdrew from occupied Jordanian land in the Arava Valley in 1995. Hussein died in 1999 and was succeeded by his son Abdullah ibn al-Hussein.

Jordan is a constitutional monarchy. The prime minister, the cabinet and the 30 members of the upper parliamentary chamber are all appointed by the reigning monarch; the 80 members of the lower house are directly elected by the people.

The vast majority of Jordan's inhabitants are Arabic-speaking Arabs, among whom are an ever-increasing number of Palestinian refugees cared for by United Nations' agencies. Although most Jordanians are Sunni Muslims, there is a significant Christian minority.

ECONOMY

The loss of agricultural land and tourist revenue from the West Bank in 1967 had a significant impact on the economy, and the country is now heavily dependent on foreign grants and loans. Natural resources are scarce and Jordan imports half of its food needs.

Irrigation in parts of the Jordan valley and adjoining mountain valleys has made arable farming possible, but dry farming is also important. Although the spread of agriculture has reduced the pastureland available to livestock, some sheep and goats are kept for meat and milk. There is some fishing in the Gulf of Aqaba. Reforestation in the eastern highlands has increased the area devoted to forestry.

The discovery and exploitation of phosphates and potash reserves have boosted the country's industrial development. Manufactures range from foodstuffs and textiles to phosphatic fertilizers, pharmaceuticals, cement and electrical goods. Electricity is generated by thermal power stations burning oil imported mostly from Iraq. Jordan's ancient archeological sites, notably the ruins of Petra in the southern mountains, attract tourists from all over the world.

There is a well-developed network of surfaced roads, and the Hejaz railroad runs north–south between the Syrian and Saudi borders through the capital, Amman, with a branch line to the port of Aqaba. There are international airports at Amman and Aqaba.

Although the welfare system is generally good, government-funded healthcare is available only in the cities. Housing, food and sanitation are often inadequate, with diseases such as typhoid, hepatitis and dysentery not yet fully under control. Education is compulsory up to the age of 14, whether at state, private or refugee schools. However, literacy rates among women have remained low. Further education is available to a limited number.

Sunlight on the hills of Amman Jordan's capital is a new town, built almost entirely since the 1880s over the ruins of the Roman city of Philadelphia. Its modern infrastructure is now severely strained by the presence of thousands of Palestinian refugees.

NATIONAL DATA – JORDAN				
Land area 91,971 sq km (35,510 sq mi)				
Climate		Temperatures		Annual
	Altitude m (ft)	January °C(°F)	July °C(°F)	precipitation mm (in)
Amman	771 (2,529)	8 (46)	25 (77)	273 (10.7)
Major physical features highest point: Jebel Ramm 1,754 m (5,755 ft); lowest point: Dead Sea -418 m (-1,371 ft); longest river: Jordan (part) 320 km (200 mi)				
Population (2006 est.) 5,906,760				
Form of government multiparty constitutional monarchy with two legislative houses				
Armed forces army 85,000; navy 500; air force 15,000				
Largest cities Amman (capital - 1,331,028); Zarqa (494,655); Irbid (327,543)				
Official language Arabic				
Ethnic composition Arab 98%; Circassian 1%; Armenian 1%				
Official religion Islam				
Religious affiliations Sunni Muslim 92%; Christian 6% (mainly Greek Orthodox, but some Greek and Roman Catholics, Syrian Orthodox, Coptic Orthodox, Armenian Orthodox, and Protestants); other 2% (small Shi'a Muslim and Druze populations)				
Currency 1 Jordanian dinar (JOD) = 1,000 fils				
Gross domestic product (2006) U.S. $28.89 billion				
Gross domestic product per capita (2006) U.S. $4,900				
Life expectancy at birth male 75.9 yr; female 81.05 yr				
Major resources phosphates, potash, shale oil, olives, citrus fruits, grapes, tourism, olive oil, vegetables, wheat				

Iraq

REPUBLIC OF IRAQ

I RAQ IS AN ARAB REPUBLIC AT THE NORTH-west end of the Persian Gulf, bordered by Kuwait and Saudi Arabia to the south, Jordan to the west, Syria to the northwest, Turkey to the north and Iran to the east.

GEOGRAPHY

Southwestern Iraq is desert, while the northeast is mountainous. In between is the fertile Tigris–Euphrates river basin; the rivers flow southeast in parallel, meeting to form a vast swamp on their way into the Persian Gulf.

Toward the center of the country the land rises gently through steppeland northwest of Baghdad. In the far north is higher ground, while northeast of the Tigris the land rises to the Hamrin uplands and high mountains on the Turkish and Iranian frontiers. The western deserts slope gently upward toward Saudi Arabia and Jordan.

Summers are hot and humid in the Tigris–Euphrates basin. The winters – from October to May – are cooler but also humid. Northeastern Iraq receives more rainfall, and seasonal temperatures are generally lower, with snow in winter.

The steppe vegetation of the north and east gives way to thorny tamarisks and salt-resistant plants in the south and west. Willow and poplar trees line the river banks, while sedges and reeds flourish in the Tigris–Euphrates marshlands. During the winter migrant birds flock to the swamps and large lakes.

SOCIETY

Mesopotamia, the fertile land between the Tigris and Euphrates, was the cradle of ancient civilization. Agriculture created the wealth that led to the first city-states in about 3450 BC. The Akkadian ruler Sargon (reigned c. 2334–2279 BC) united these city-states in the first of many empires to dominate the region. Babylon, which defeated its thousand-year-old rival Assyria in 612 BC, was itself conquered by the Persians in 538 BC.

The Persians, in turn, were conquered by Alexander the Great (356–323 BC), and Mesopotamia became part of the Greek Seleucid empire. Roman, and later Persian, occupation followed, before the region fell to the Muslim Arabs in the mid 7th century AD. The city of Baghdad quickly became the greatest cultural and

The persecuted Marsh Arabs are seminomadic marsh dwellers who have lived for thousands of years in the marshes of southeastern Iraq, herding water buffalo and hunting wildfowl. Their elaborately structured houses are supported by arches woven from reeds.

A blue and yellow glazed bull, characteristic decoration of the ruined city of Babylon. Dating from before the 18th century BC, the city was rebuilt by Nebuchadnezzar II in the 6th century BC, and is famous for having been the site of the Hanging Gardens, one of the Seven Wonders of the Ancient World.

commercial center of the Muslim world. Following the Mongol invasion of 1258, Baghdad was destroyed and Mesopotamia lost much of its importance as a political and commercial center. In 1638 it was taken by the Ottoman Turks.

Following Turkey's defeat in World War I, Britain established an Iraqi monarchy under British protection. Eleven years later Iraq achieved full independence under King Faisal I (1885–1933), but pro-German activities led to another short British occupation during World War II. In 1958 the monarchy was overthrown in the first of a series of army coups.

In 1967 Iraq fought against Israel in the Arab–Israeli war. The following year the Baath Party took power and, in 1979, its leader Saddam Hussein (b. 1937) became president. In 1980 Iraq invaded Iran following a dispute over control of the Shatt-el-Arab waterway (where the Tigris and Euphrates meet along the Iraq–Iran border), hoping for a quick victory over a country paralyzed by revolution. Instead, the war dragged on until 1989, leaving Iraq nothing to show for heavy casualties and a huge war debt. In 1990 Saddam Hussein ordered the invasion of Kuwait to seize control of its rich oil fields. An international force led by the United States drove the Iraqis out of Kuwait in 1991, crippled Iraq's infrastructure and inflicted a massive defeat on its armed forces. Because of the war and Saddam's continued persecution of the rebellious Kurds in the northeast and Shi'ites in the southeast, the United Nations imposed stringent economic sanctions on Iraq and set up protected zones, including "no-fly-zones". Incursions of these in 1993 led to bombing of strategic targets by the US-led alliance. In 1994 a build-up of Iraqi forces near Kuwait was countered by the

NATIONAL DATA – IRAQ

Land area 432,162 sq km (166,859 sq mi)

Climate	Altitude m (ft)	Temperatures January °C(°F)	July °C(°F)	Annual precipitation mm (in)
Baghdad	34 (112)	9 (48)	35 (95)	155 (6.1)

Major physical features highest point: Huji Ibrahim 3,904 m (12,808 ft); longest river: Euphrates (part) 3,596 km (2,235 mi)

Population (2006 est.) 26,783,383

Form of government parliamentary democracy with one legislative house

Armed forces army 105,700; navy 800; air force 500

Largest cities Baghdad (capital – 5,831,541); Mosul (2,220,624); Basra (2,271,544); Irbil (993,468); Sulaimaniyah (773,496); Kirkuk (610,954); An Najaf (502,400)

Ethnic composition Arab 75%–80%; Kurdish 15%–20%; Turkoman, Assyrian or other 5%

Official religion Islam

Religious affiliations Muslim 97% (Shi'a 60%–65%; Sunni 32%–37%); Christian or other 3%

Currency 1 New Iraqi dinar (NID) = 20 dirhams = 1,000 fils

Gross domestic product (2005) U.S. $94.1 billion

Gross domestic product per capita (2006) U.S. $1,900

Life expectancy at birth male 67.76 yr; female 70.31 yr

Major resources petroleum, natural gas, phosphates, sulphur, camels, cattle, cereals, cotton, goats, sheep, tomatoes

increased presence of American and other forces, and the Iraqis moved away. US troops patrolled the area once more in 1996 following Iraq's installation of anti-aircraft systems in the "No-fly" Zones. In late 1998 American and British planes bombed Iraqi bases after Iraq blocked United Nations weapons inspectors.

Under Iraq's 1968 provisional constitution, power rests with the nine-member Revolutionary Command Council (RCC); one member is elected president. The RCC and a 220-member elected assembly legislate. Saddam Hussein was re-elected president in 1995.

Arab peoples form an overwhelming majority of the population. The largest single minority is the semi-nomadic Kurds. More than 95 percent of the country is Muslim. Shi'ite Muslims (most of them Arabs) now outnumber the Sunnis (including Arabs, and Kurds), but Sunni Muslims continue to wield political power. Some 3.5 percent of the population are Christian.

Baghdad, Iraq's hot, dusty capital, expanded rapidly as soon as the seasonal flooding of the Tigris river was brought under control. Historic mosques and bazaars now stand in the shadow of high-rise apartment blocks.

ECONOMY

Oil is the mainstay of the Iraqi economy although since 1991 sanctions have held back oil sales of around $100 billion. Agriculture has been severely affected by sanctions with output down by up to 50 percent. Cultivated land is concentrated in the Tigris–Euphrates valley, and approximately one-third of the farms are irrigated. Vegetables and cereals are the most important crops, but international sanctions, still in place in 1996, have damaged trade. About one-tenth of the land area provides grazing for livestock. Local waters supply fish for domestic needs.

Apart from petroleum, Iraq's chief mineral resources are phosphates, used in the production of fertilizer. Manufactured goods include building materials, textiles and petrochemical products. Most of the country's power is supplied by hydro-electric stations on the Tigris. A nuclear reactor was destroyed by Israel in 1981 and has not been replaced. Following the Gulf War, the UN placed a total ban on any further nuclear research in the country, although Iraq is suspected abroad of secretly developing nuclear weapons.

The war years inflicted severe damage on the transportation network. Only one-third of roads are surfaced, but the western deserts can support even heavy vehicles. The railroad runs from Baghdad to the southeastern port of Basra, and northwest to Turkey and Syria. Ships operate on the Tigris as far as Baghdad, and some goods are still carried downstream by raft on the Tigris and the Euphrates. Baghdad and Basra both have international airports. Iraq's media are strictly controlled.

War damage and sanctions have adversely affected health. Food and medicine shortages have led to increased mortality, especially among children, and health care facilities are limited. Welfare includes retirement, sickness and unemployment benefits. Children are supposed to receive six years' free compulsory education, but many rural areas lack facilities, and illiteracy is widespread.

Kuwait

STATE OF KUWAIT

KUWAIT IS AN INDEPENDENT ARAB EMIRATE at the head of the Persian Gulf, sandwiched between Saudi Arabia to the south and Iraq to the west and north, with a coast facing east across the Persian Gulf. In 1990 it was occupied by Iraq, regaining its sovereignty only after a war in which both countries suffered considerable economic and environmental damage.

GEOGRAPHY

The landscape consists largely of undulating sandy plains that rise gently to the southwest toward the Saudi frontier. To the northeast there are a few salt marshes. Most of the country is dry, barren desert, except for a few basins that occasionally

NATIONAL DATA - KUWAIT

Land area	17,820 sq km (6,880 sq mi)			
Climate		Temperatures		Annual
	Altitude m (ft)	January °C(°F)	July °C(°F)	precipitation mm (in)
Kuwait City	5 (16)	13 (55)	37 (98)	99 (3.9)

Major physical features	highest point: ash-Shaqaya 290 m (951 ft)

Population (2006 est.) 2,418,393

Form of government nonparty constitutional monarchy with one legislative house

Armed forces army 11,000; navy 2,000; air force 2,500

Capital city Kuwait City (67,103)

Official language Arabic

Ethnic composition Kuwaiti 45%; other Arab 35%; South Asian 9%; Iranian 4%; other 7%

Religious affiliations Muslim 85% (Sunni 70%; Shi'a 30%); Christian, Hindu, Parsi, and other 15%

Currency 1 Kuwaiti dinar (KD) = 1,000 fils

Gross domestic product (2006) U.S. $52.17 billion

Gross domestic product per capita (2006) U.S. $21,600

Life expectancy at birth male 76.13 yr; female 78.31 yr

Major resources petroleum, fisheries, shrimps, natural gas, chemical fertilizers, dates, fruits, goats, sheep, vegetables

catch rainwater. The little rain, which falls mainly in winter, brings a welcome coolness after the extreme summer heat.

Vegetation on the coast is limited to salt-tolerant plants, while in urban areas green oases have been created through the irrigation of imported soil. Wildlife was severely depleted as a result of oil pollution during the 1991 Gulf War.

SOCIETY

Kuwait takes its name from the Arabic word *kut*, meaning fort. Its capital city, also called Kuwait, is in a strategic location on the coast that may have been settled in ancient times. The present Arab inhabitants are probably the descendants of nomads who arrived from the nearby desert in the early 18th century seeking permanent settlement. In 1756 the Sabah family established the emirate that rules today. In the late 19th century the country was threatened by the colonial ambitions of both Germany and the Ottoman Turks, and thus formed an alliance with Britain, becoming a British protectorate in 1914. The discovery of oil in the 1930s brought with it unexpected prosperity.

Kuwait became independent in 1961 and was immediately claimed by Iraq. The claim was soon dropped, but was revived in 1990, when the Iraqi leader Saddam Hussein (b. 1937), following a dispute over oil prices, invaded on the pretext of supporting native insurgents. This invasion met with international condemnation. When Iraq refused to withdraw, a United Nations' coalition sent in a military task force consisting primarily of American, British, French and Saudi troops. The Iraqis were expelled in 1991 with heavy casualties, but the damage done to Kuwait was immense. Vast quantities of oil were spilled into the Persian Gulf and hundreds of wells were deliberately set on fire by the Iraqi troops or accidentally by Allied bombing. However, by late 1993 oil production was back to its quota level. Iraq formally recognized the state of Kuwait in November 1994.

After the Iraqi withdrawal, the emir, an absolute monarch under Islamic law, was pressed to initiate democratic reforms. Before the invasion more than half of the country's population consisted of foreign workers denied the same rights as Kuwaitis. Some were killed in the occupation, but many left at the time or were subsequently encouraged to leave by the Kuwaitis. Kuwait's Palestinian population, many of whom had supported Iraq, were distrusted and harshly treated.

ECONOMY

Kuwait's economy had largely recovered from the effects of the Gulf War by the late 1990s. Livestock farming is of some importance, but intensive farming using hothouses and hydroponics (growing plants without soil) provides large crops of melons, tomatoes and dates. Commercial fishing has been important, especially the export of prawns, but catches have been reduced by pollution in the Persian Gulf following the war. Oil accounts for 93 percent of export revenue; chemical fertilizers and natural gas supply most of the remainder. The modern road system was badly damaged in the war but has since been repaired.

Excellent health benefits and free education, which is compulsory to the age of 14 and extends to university level, are available to native Kuwaitis. Press and broadcasting are privately owned but subject to government controls.

The desert city of Kuwait Until the 1960s the city had only one main street and a surrounding mud wall; since then, the country's vast oil revenue has been used to build a modern city with luxurious air conditioned buildings such as the dramatic Kuwait Towers.

Iran

ISLAMIC REPUBLIC OF IRAN

IRAN, A MOUNTAINOUS COUNTRY IN SOUTH-western Asia, is one of the largest of The Gulf states. It shares frontiers with 10 neighboring countries. A short coastline in the north borders the Caspian Sea, while the much longer southern coast borders the Persian Gulf. The Iranian plateau was the home of the ancient Persian civilizations, and its cultural heritage goes back 5,000 years.

GEOGRAPHY

Much of Iran consists of a massive semi-arid plateau ringed by mountains. It is subject to frequent earthquakes. Fertile lowland is extremely scarce; the only large area is the Karun river basin in the southwestern province of Khuzistan. The river basin is bounded to the northeast by the Zagros Mountains. East of the Zagros range lies a vast plateau at an average height of 1,200 m (4,000 ft) above sea level, characterized by extensive salt deserts and occasional oases. Its southeastern side is fringed by coastal mountain ranges, while to the east are the peaks of the Baluchistan and Khorasan mountains. The Elburz Mountains of the north and northeast include several active volcanoes. North of the Elburz range a narrow lowland fronts the marshy shore of the Caspian Sea, broadening to the east into the Turkoman steppe.

The climate varies from continental to subtropical. In the north the plains and mountains have rain all year round. By contrast, precipitation in the mountainous west and northwest is confined mainly to winter. The rest of the country is much drier. The southern lowlands have mild winters and hot, humid summers, whereas the interior is hot in summer and bitterly cold in winter.

Plant life thrives on the shores of the

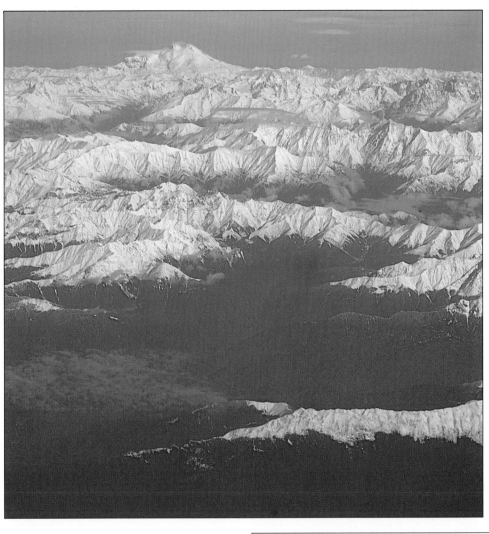

The snowcapped Elburz Mountains of northern Iran run from the capital, Tehran, to the Caspian Sea in Northern Eurasia. The country's highest peak, at 5,671 m (18,606 ft) is the dormant volcano Damavand. Oak and beech trees cover the lower mountain slopes.

Caspian Sea, and the northern slopes of the nearby Elburz Mountains are covered in broadleaf forests; dense forest also covers the western mountains. Elsewhere the vegetation is sparse.

Iran has a rich variety of animal life and many wildlife sanctuaries. The shoreline of the Caspian Sea is a refuge for numerous water birds. Marine life in the Persian Gulf is still recovering from the effects of oil released during The Gulf War in 1991.

SOCIETY

The Iranian plateau has seen the rise and fall of many civilizations: archeological evidence of village life dates back as far as c. 4000 BC. The earliest-known civilization in the area, that of the Elamites in present-day Khuzistan, was founded in the 3rd millennium BC. There followed periods of control by the Medes, Persians, Greeks and Parthians. In 224 AD Ardashir Papakan (d. 241), ruler of the vassal kingdom of Persis (Fars), defeated the Parthians and founded the Sasanian dynasty of Persian rulers.

NATIONAL DATA –IRAN

Land area	1,636,000 sq km (631,663 sq mi)			
Climate		**Temperatures**		**Annual**
	Altitude m (ft)	January °C(°F)	July °C(°F)	precipitation mm (in)
Tehran	1,220 (4,002)	3 (37)	30 (86)	230 (9)
Abadan	2 (7)	12 (54)	36 (97)	193 (7.5)

Major physical features highest point: Damavand 5,771 m (18,934 ft); longest river: Karun 850 km (528 mi)

Population (2006 est.) 68,688,433

Form of government theocratic republic

Armed forces army 350,000; Islamic revolutionary guard corps 125,000; navy 18,000; air force 52,000

Largest cities Tehran (capital - 7,185,831); Mashhad (2,463,393); Esfahan (1,600,554); Karaj (1,602,350); Tabriz (1,496,319); Shiraz (1,307,552)

Official language Farsi (Persian)

Ethnic composition Persian 51%; Azeri 24%; Gilaki and Mazandarani 8%; Kurd 7%; Arab 3%; Lur 2%; Baloch 2%; Turkmen 2%; other 1%

Official religion Islam

Religious affiliations Shi'a Muslim 89%; Sunni Muslim 9%; Zoroastrian, Jewish, Christian, and Baha'i 2%

Currency 10 Iranian rial (IRR) = toman

Gross domestic product (2006) U.S. $610.4 billion

Gross domestic product per capita (2006) U.S. $8,900

Life expectancy at birth male 68.86 yr; female 71.74 yr

Major resources petroleum, natural gas, coal, chromium, copper, iron ore, lead, manganese, zinc, sulfur, building materials, cement, fisheries, textiles, timber, cereals, sugar beet, fruits, nuts, cotton

The Sasanian emperors presided over a golden age of Persian culture, until in the mid 7th century they succumbed to an Arab invasion. In the 13th century the Arabs were overrun by the Seljuk Turks, and later by Mongol invaders from the north. The Iranian Safavids achieved supremacy in 1502, adopting the Shi'ite form of Islam, but fell to an Afghan usurper in 1722. During the reign of the Turkmen warrior Nader Shah (1688–1747) the empire expanded into India.

During the Qajar dynasty (1794–1925) Persia was threatened by Russia to the north, the Ottoman Turks to the west and the British in India to the east. In 1906, following a revolution, Persia's first constitution ended the absolute rule of the Qajars. The discovery of oil in 1908 by a British consortium intensified colonial interest, and from then on the country became a semicolonial state dominated by both Great Britain and Russia.

In the chaotic aftermath of World War I an army officer, Reza Khan (1878–1944), overthrew the ruling Qajar dynasty and had himself elected shah in 1925. He adopted the name Reza Shah Pahlavi, changed the country's name from Persia to Iran, and sought to create a modern state free of foreign domination. However, during World War II British and Soviet occupation forces forced him to

Weavers at work Two girls of the Turkmen minority from northeastern Iran weave a carpet, one of Iran's most ancient crafts. Much in demand by affluent foreigners, the rugs are a valuable export. According to tradition a flaw is deliberately woven into each design, because "only Allah is perfect".

abdicate in favor of his son and heir Mohammad Reza Shah Pahlavi (1919–80).

After the war Iran reasserted its independence under Mohammad Mossadeq (1880–1967). In 1951 he nationalized the petroleum industry, forcing the British Oil Company to withdraw. His attempt to depose the shah led to his downfall. In 1961 the shah initiated the so-called "White Revolution", a program of radical economic and social change that caused major unrest. In 1978 antigovernment demonstrations were put down by force, fueling opposition by Islamic fundamentalists who were followers of the exiled religious leader Ayatollah Ruhollah Khomeini (c. 1900–1989). At the start of 1979 the shah was forced into exile, and Khomeini returned in triumph.

Khomeini's call for an Islamic republic was overwhelmingly supported by referendum. That same year Islamic militants stormed the United States' embassy and seized 66 American hostages, most of whom were not released until 1981. The West, supported by conservative Arab states fearing the spread of the Islamic revolution, responded with diplomatic and economic sanctions. Meanwhile, Iraq had invaded the petroleum-producing southwestern province of Khuzistan, starting a bitter and bloody war that lasted until 1988. Islamic fundamentalism continued to be a powerful force in Iran after Khomeini's death in 1989.

Under the 1979 constitution, supreme authority in all matters rests with the Islamic spiritual leader, or *valiy-e faqih*. The president and the 270-member par-

liament are both freely elected for a four-year term. The president appoints a cabinet, subject to the approval of the Majles and the *valiy-e faqih*. Legislation is vetted by the 12-member Council of Guardians to ensure it conforms to the Islamic constitution.

Speakers of Farsi, the official language, account for less than half the population. A few Armenian Christians remain in the cities; they are the only non-Muslim group apart from some Jews and Zoroastrians.

ECONOMY

Iran's economy is almost totally dependent on petroleum, which for many years brought great prosperity. Economic development was disrupted by the Islamic revolution and the Iran–Iraq war and in the 1990s fluctuating oil prices have prevented stable economic progress.

Despite widespread irrigation, good arable land is in short supply. Yields are low, so that much of the country's food has to be imported. Forestry and fisheries are both well developed.

Iran's mineral resources include coal and various metal ores, though only its vast reserves of oil and natural gas have been fully exploited. Iran is self-sufficient in energy. Apart from traditional textiles – Persian carpets and rugs – most of the manufacturing sector has been developed since 1954. Products range from processed food and clothing to petrochemicals, electrical machinery and armaments.

Railroads cater largely for long-distance travel on major routes; most local passenger and freight traffic relies on poorly maintained roads. Regular shipping services operate on Lake Urmia in the northwest and from ports on the Caspian Sea, but the only navigable river is the Karun in the southwest. Bandar Beheshti in the Gulf of Oman is now the chief trading port.

Radio and television are strictly controlled. Broadcasts are in Farsi, several local languages and also some foreign languages. Press freedom is guaranteed under the constitution, but strict censorship is imposed to ensure adherence to Islamic principles. Healthcare is free to the poor, but many rural areas lack medical supplies and trained staff. Some welfare is available from the state.

Education is compulsory and free at elementary level, but illiteracy remains widespread. There are a number of universities, together with teacher-training colleges and technical schools.

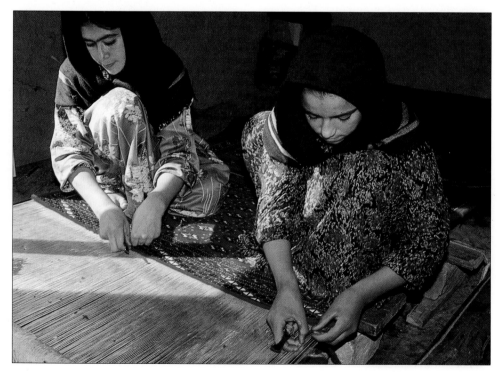

Yemen

REPUBLIC OF YEMEN

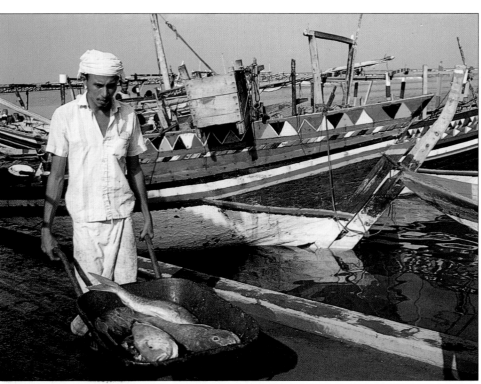

THE REPUBLIC OF YEMEN WAS CREATED IN 1990 by the unification of the former Yemen Arab Republic (Yemen), whose coast faces west across the Red Sea, and the People's Democratic Republic of Yemen (Southern Yemen), which overlooks the Gulf of Aden to the south.

GEOGRAPHY

The Yemeni landscape is one of contrasts, from the fertile highlands in the west to the arid coast and rocky uplands of the east. The narrow sandy plain along the Gulf of Aden is intersected by wadis – seasonal watercourses that flood in summer, providing water for irrigation.

A hot, dry desert climate prevails except in the western highlands, which have high rainfall during the summer. The Red Sea coast sometimes experiences fierce dust storms. Drought-resistant plants are widespread. More fertile areas support fruit trees such as fig and carob.

SOCIETY

Ancient Yemen was ruled by a number of kingdoms – including the Kingdom of Saba (Sheba), which flourished from about 750 to 115 BC. Yemen submitted to the rule of Muhammad in 631 and the inhabitants began to adopt Islam.

In the 16th century Yemen became part of the Turkish Ottoman empire, but in 1635 the Turks were expelled. The British took the port of Aden in 1839 while the Ottomans gradually occupied northwestern parts of Yemen until 1914, when they were again overthrown, and northwestern Yemen fell to a number of despotic *imams* (Islamic leaders).

Aden became a British crown colony in 1937, and the protectorate later extended to neighboring states to form the Federation of South Arabia. Following independence in 1967, a Marxist republic was established, called the People's Democratic Republic of Yemen.

In northwestern Yemen the last of the despotic imams was deposed in 1962 and the Yemen Arab Republic was proclaimed. Disputes with Southern Yemen during the 1970s led to a war in 1979.

When the war ended in the same year, discussions began between the two states that resulted in their unification in May 1990. Civil war broke out between Northern and Southern forces in May 1994 after years of tension. Peace returned in July when Northern forces were victorious. The new Yemeni constitution provides for a multiparty republic with a president as head of state and a government led by a prime minister.

The vast majority of Yemenis are Muslim Arabs. Three-quarters of the population live in rural areas. The culture is based on traditional Islamic values.

ECONOMY

More than half the workforce is engaged in agriculture. Subsistence farming dominates, using traditional methods. Staples include millet, sorghum, fruit and nuts. Coffee and cotton are cash crops.

Petroleum was disovered in the region during the 1980s, but so far industry is confined largely to Aden, which has petroleum and food-processing plants. There are no railroads and few surfaced roads, most of which are in the west. Press and media are limited and few people have access to telephones.

Brightly painted fishing boats are unloaded at Al Hudaydah, a port on the Red Sea coast of western Yemen. Most of the fresh catch is taken directly to the local fish market and is consumed locally, but some is dried for sale to villages in the interior.

Welfare and education are in need of much greater investment. Birthrates are high but so too is infant mortality, while life expectancy is low. Education is free, but is only accessible to a limited number of children in the towns and cities.

NATIONAL DATA – YEMEN

Land area 527,970 sq km (203,850 sq mi)

Climate	Altitude m (ft)	Temperatures January °C(°F)	July °C(°F)	Annual precipitation mm (in)
Aden	7 (23)	25 (77)	31 (88)	47 (1.8)

Major physical features highest point: Jabal an–Nabi Shu'ayb 3,666 m (12,028 ft)

Population (2006 est.) 21,456,188

Form of government multiparty republic with one legislative house

Armed forces army 60,000; navy 1,700; air force 5,000

Largest cities San'a (capital – 2,015,515); Tai'zz (729,583); al-Hudaydah (713,458); Aden (646,180)

Official language Arabic

Ethnic composition Arab 92%; Afro-Arab 3%; Indian 3%; others 2%

Religious affiliations Sunni Muslim 53.0%; Shi'ite Muslim 46.9%; others 0.1%

Currency 1 Yemeni rial (YER) = 100 fils

Gross domestic product (2006) U.S. $20.38 billion

Gross domestic product per capita (2006) U.S. $900

Life expectancy at birth male 60.23 yr; female 64.11 yr

Major resources petroleum, natural gas, rock salt, marble, coal, gold, lead, nickel, copper, cotton, coffee, hides and skins, fisheries, fruit, millet, sesame, sorghum, sugarcane, wheat

Afghanistan

ISLAMIC STATE OF AFGHANISTAN

A FGHANISTAN IS A LANDLOCKED STATE IN the mountains of south-central Asia, sharing borders with Pakistan to the southeast and Iran to the west. Turkmenistan, Uzbekistan and Tajikistan lie across the northern frontier, and China lies to the northeast at the end of the narrow Wakhan corridor. Afghanistan has been plagued by civil war for decades and has endured a harsh Islamic regime since 1996. In 2001, US-led air attacks in response to Afghanistan sheltering the terrorist Osama bin Laden led to the fall of the Islamic Taliban government. A UN-brokered deal led to an interim council and a peacekeeping force was sent to stabilize Kabul.

NATIONAL DATA - AFGHANISTAN

Land area	647,500 sq km (250,001 sq mi)			
Climate	Altitude m (ft)	Temperatures January °C(°F)	July °C(°F)	Annual precipitation mm (in)
Kabul	1,815 (5,955)	-1 (30)	24 (76)	311 (12.2)

Major physical features highest point: Mount Nowshak 7,485 m (24,557 ft); longest river: Helmand (part) 1,400 km (870 mi)

Population (2006 est.) 31,056,997

Form of government Islamic republic

Armed forces army 27,000

Largest cities Kabul (capital – 3,199,091); Kandahar (411,727); Mazar-i-Sharif (326,737)

Official languages Pashto, Dari Persian

Ethnic composition Pashtun 42%; Tajik 27%; Hazara 9%; Uzbek 9%; Aimak 4%; Turkmen 3%; Baloch 2%; other 4%

Religious affiliations Sunni Muslim 80%; Shi'a Muslim 19%; other 1%

Currency 1 afghani (AFA) = 1 puls

Gross domestic product (2004) U.S. $21.5 billion

Gross domestic product per capita (2004) U.S. $800

Life expectancy at birth male 43.16 yr; female 43.53 yr

Major resources natural gas, petroleum, coal, copper, chromite, talc, barites, sulfur, lead, zinc, iron ore, salt, precious and semiprecious stones, barley, cotton, fruit, goats, maize/corn, nuts, rice, sheep, sugar beet, vegetables, wheat

GEOGRAPHY

Afghanistan is at the heart of southern Asia's great mountain belt. The Hindu Kush and neighboring ranges cover eastern and central Afghanistan, a rugged and inhospitable region prone to earthquakes. Most of the population lives in irrigated valleys on the fringes of the mountains, particularly in the fertile valley of the Kabul river in the southeast, after which the capital is named.

The northern plains adjoining the Amu Darya river are fertile and densely populated. South of the mountains, the deserts and saltflats of the broader southwestern plateau are sparsely watered by rivers flowing out of the Hindu Kush.

The extreme climate of cold winters and hot summers is typical of semiarid steppe country, though there is considerable local variation; the winter months are coldest in the northeastern mountains. The southwest is the driest area, and the eastern valleys are humid.

Vegetation is sparse in the semiarid southwest and in some mountain areas. Elsewhere the mountains are forested with pine, cedar and fir, while walnut and oak grow at lower levels. Many large mammals – including tigers – have all but disappeared, but bears, gazelles and snow leopards still survive. Bird life is plentiful, especially birds of prey.

SOCIETY

Afghanistan has long been disputed by foreign powers anxious to control its strategic position astride the great land route to India via the Khyber Pass. The present boundaries are relatively recent.

From the 6th century BC Afghanistan was part of the great Persian empire. After its conquest by Alexander the Great (356–323 BC) in 328 BC, the region was the center of the Greek kingdom of Bactria, which saw a remarkable fusion of Greek, Hindu and Buddhist cultural influences. After 135 BC the region fell to a succession of empires – Parthian, Kushan and Sasanian Persian – before being conquered by the Arabs in the 7th century AD. The Arabs introduced Islam, which by the 9th century had become the dominant cultural force in Afghanistan. The Mongols under Genghis Khan (c. 1162–1227) overran the country in the 13th century, and his descendant Babur (1483–1530) later made Kabul the headquarters for the Mughal invasion and conquest of India.

When the Persian monarch was assas-

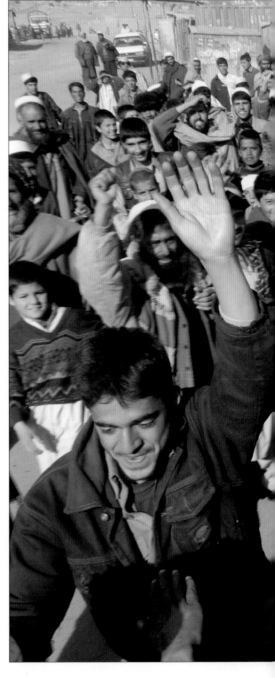

sinated in 1747, the chief of his Afghan bodyguard, Ahmad Khan Abdali (1722–72), was elected shah by Afghan chieftains, and became known as Ahmad Shah Durrani. Afghanistan became an independent state, though it continued to be disturbed by factional rebellions and conflicting British and Russian territorial ambitions. In 1880, following the second Anglo-Afghan war, new boundaries made Afghanistan a buffer state between British India and Russia. After the third Anglo-Afghan war in 1919, Afghanistan gained full independence, and formed a special relationship with the new Soviet regime that was to last until 1979.

Later Afghan rulers attempted to bring stability to the fragile monarchy through economic and constitutional reforms, but with little success. Unrest increased after a severe famine in 1972; the shah was deposed the following year, and the coun-

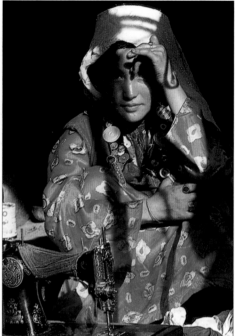

A Kyrghyz seamstress (*above*) with her sewing machine. The Kyrghyz are a small group of Turkic speakers in the extreme northeast of Afghanistan. There is great ethnic diversity among the isolated, often feuding, communities each with their different culture.

migration. The Uzbeks and Turkmens of the far northwest retain their Turkic culture and language.

ECONOMY

Afghanistan's economic development has been severely retarded by war and political instability, and the country remains one of the poorest in the world, with agriculture the mainstay of the economy. Although irrigation projects increased the amount of cultivable land, much of this has been lost through war. The principal crop is wheat, but rice, barley, fruit and cotton are also grown.

The most vital mineral resource is natural gas. There are also sizable deposits of iron and other metal ores, but their potential, like that of hydroelectricity, has not yet been fully exploited. The main industrial activity is the manufacture of woolen and cotton textiles; traditional handicrafts and woven carpets are important exports.

With no railroads and few navigable rivers, transportation is normally by road. There are domestic air services to most provinces, and international airports at both Kabul and Kandahar.

Education, healthcare and welfare provision are all inadequate, especially in rural areas; nearly all healthcare provision depends on foreign aid. Literacy levels are low, particularly among women.

Liberation of Kabul On November 13 2001, the Northern Alliance entered Kabul greeted by cheering crowds. Soon after, the strict, Islamic Taliban regime collapsed and Afghanis were free to sell music and men were able to visit barber shops to shave off their beards.

try was declared a republic. In a further coup in 1978, the president was killed and power passed to the Soviet-backed People's Democratic Party of Afghanistan (PDPA). The new regime was violently opposed by Afghan Islamic groups, and in December 1979 Soviet forces invaded Afghanistan to install a new PDPA government. Fighting between Soviet troops and Afghan rebels continued for nine years.

The last of the Soviet troops were withdrawn in February 1989, and a military council was established under President Mohammad Najibullah (b. 1936). Peace proved elusive as many different rebel factions continued to wage war. During late 1996 the fundamentalist Taliban militia emerged, bringing a measure of stability to the country. However, the Taliban were internationally criticised over their harsh implementation of Islamic law. In 2001, the Taliban were accused of harboring the al-Qaeda terrorist leader Osama bin Laden who was blamed for the horrific September 11 terrorist attacks on the United States. In October the US initiated air attacks on Afghanistan hoping that bin Laden would be flushed out. The attacks allowed opposition groups such as the Northern Alliance to drive the Taliban from power but bin Laden remained elusive. An interim power-sharing council was set up until an elected representative government and new constitution could be adopted. An international peacekeeping force was also set up to guarantee security for the Afghan capital, Kabul.

The ethnic diversity of the Afghan people reflects a history of conquest and

Saudi Arabia

KINGDOM OF SAUDI ARABIA

S AUDI ARABIA, NAMED AFTER THE RULING royal dynasty of Sa'ud, occupies most of the Arabian Peninsula. This desert kingdom, rich in petroleum, borders Jordan, Iraq and Kuwait to the north and Bahrain and Qatar along The Gulf coast. Southeast are the United Arab Emirates and Oman, and to the south is Yemen.

GEOGRAPHY

The Arabian Peninsula is largely made up of a vast plateau. To the west the plateau forms a steep escarpment running parallel to the Red Sea coast; in the northwest there is virtually no coastal plain. The highest mountain peaks are in the Asir highlands in the southwest.

East of the escarpment, the barren plateau drops gently to the shallow waters of The Gulf, whose shores are lined with swamps and salt flats. The plateau is mostly sandy desert, with some areas of bare volcanic rock. A vast swathe of desert known as the Empty Quarter occupies the whole of the south.

Saudi Arabia has no permanent rivers. The Empty Quarter can go without rain for 10 years at a time. Other regions experience the occasional downpour, especially near The Gulf coast in the spring, and in the summer in the Asir highlands. Inland the winters are cool, often with frost at night, but in summer temperatures in excess of 48°C (120°F) are not unusual.

In most parts of the country vegetation is limited to small herbs and shrubs, with date palms in scattered oases. The Arabian oryx has been hunted almost to extinction, and the Arabian tahr (a goat-like antelope) and some birds of prey are now endangered. But wild cats, desert-dwelling sandgrouse, burrowing rodents such as jirds and jerboas, and diverse

reptiles and insects are common. The coastal waters of the Red Sea teem with life, particularly among the coral reefs.

SOCIETY

In 622 AD Muhammad (c. 570–632) – a merchant from the caravan city of Mecca who believed himself to be divinely inspired – traveled north to Medina and founded a new religion: Islam. He returned with an army of converts to capture Mecca, and by the time of his death all Arabia was under the sway of Islam. In 661, however, the center of Islamic power was transferred northward to newly conquered lands. Soon Arab unity was broken, and for many centuries Arabia became a battleground for rival rulers, until the Ottoman Turks added it to their growing empire in 1517.

In the 18th century a Muslim scholar, Muhammad ibn 'Abdal-Wahhab (1703–92), founded the Islamic fundamentalist Wahhabi movement. His ultrapuritanical teachings were spread by the Sa'ud princes of Najd as far as Yemen. During the 19th century the area they controlled was greatly reduced, but in 1904 Ibn Sa'ud (c. 1880–1953) restored Saudi rule in Najd, and in 1924 he invaded the British protectorate of Hejaz in the west, fulfilling his ambition to take Mecca and Medina. Three years later the British

The tall, narrow houses of Al Kana, a small oasis village in southern Saudi Arabia, show a strong Yemeni influence in both their design and painted decoration. On the flat rooftops vegetables are dried, bread is baked and handwoven rugs are cleaned.

The changing role of women (*above*) Formal education for women in Saudi Arabia was introduced as part of the progressive reform program of King Faisal (1906–75), who came to power in 1964. Veils are not worn in the state-run girls' schools.

Riyadh airport (*right*), built at great expense for the burgeoning capital – the seat of the royal family, and until recently a small oasis town. The airport is a symbol of Saudi Arabia's new-found wealth and its entry into the international arena.

acknowledged his kingship over the regions of Najd and Hejaz, which he formally united in 1932 to form the Kingdom of Saudi Arabia.

The country remained neutral in World War II, and played little part in the postwar affairs of other Arab states. But when Iraq invaded neighboring Kuwait in 1990, the Saudi king invited Western and Arab forces to support his army along the northern frontier. Victory in The Gulf War of 1991 strengthened relations between Saudi Arabia and its Arab allies. The war made little impact on the prosperous petroleum-based Saudi economy despite the damage caused by pollution and Iraqi missiles.

The Saudi king, whose official title is Custodian of the Two Holy Mosques (at Mecca and Medina), rules as absolute monarch. His heir and deputy, the crown prince, is selected from among the Sa'ud family by its leading members in consultation with the *ulama*, or supreme religious council. The king governs according to Islamic law, choosing a council of ministers, many of whom are Sa'uds.

The population is predominantly Arab, including many immigrants from other Arab states, especially Yemen. There is a large transient population made up of pilgrims from other Muslim countries visiting Mecca, and also large numbers of foreign industrial workers. Most Saudis are Sunni Muslims – chiefly the Wahabi interpretation – apart from some Shi'ites in areas bordering the Persian Gulf. Islam is strictly enforced (though not, critics say, in some areas concerning human rights), affecting every aspect of Saudi life from its laws to its culture and its visual arts. It is forbidden to practice any non-Muslim religion openly.

ECONOMY

The Saudi economy is based on petroleum exploitation. Revenue from this finances heavy spending elsewhere.

High government subsidies have helped produce a surplus of wheat for export. But two-thirds of all food still has to be imported. Apart from a tiny area of fertile farmland in the mountains of Asir, agriculture is dependent on old and new irrigation techniques. Major crops include cereals, bananas, dates and coffee. Sheep, goats and camels are herded by the country's few remaining nomads.

Saudi Arabia possesses one-quarter of the world's oil reserves as well as significant quantities of natural gas and various metal ores. Manufacturing industry, powered by electricity from oil-fired stations, produces petrochemicals, fertilizers and rolled steel.

The Saudi road network connects all main centers of population, with further links to Yemen, Jordan, Kuwait and Qatar. A railroad runs northeast from the capital, Riyadh, to the major Gulf port of Dammam. Pipelines from the nearby port of Jubail carry oil and natural gas across the country to Yanbu on the Red Sea coast, west of Medina. Jeddah, another major port on the Red Sea, near Mecca, has an international airport, as do Riyadh and Dhahran (next to Dammam).

Wide-ranging welfare services include free medical care for all citizens and visitors. Healthcare is excellent in the cities, but in rural areas diseases such as malaria, trachoma and bilharzia are all common. Education is free but not compulsory, and literacy levels are generally very low. However, students graduating from secondary school have a wide choice of university education, or can opt for technical or vocational training.

NATIONAL DATA – SAUDI ARABIA

Land area	2,149,690 sq km (830,000 sq mi)			

Climate		Temperatures		Annual
	Altitude m (ft)	January °C(°F)	July °C(°F)	precipitation mm (in)
Riyadh	609 (1,998)	14 (58)	35 (95)	113 (4.4)

Major physical features highest point: Jabal Sawda 3,133 m (10,279 ft)

Population (2006 est.) 27,019,731

Form of government absolute monarchy

Armed forces army 75,000; navy 15,500; air force 18,000

Largest cities Riyadh (capital – 4,453,447); Jedda (3,002,839); Mecca (1,383,707)

Official language Arabic

Ethnic composition Saudi 82.0%; Yemeni 9.6%; other Arabs 3.4%; others 5.0%

Religious affiliations Muslim 96.7%; Christian 2.9%; others 0.4%

Currency 1 Saudi riyal (SAR) = 100 hallalah

Gross domestic product (2006) U.S. $374 billion

Gross domestic product per capita (2006) U.S. $13,800

Life expectancy at birth male 73.66 yr; female 77.78 yr

Major resources petroleum, natural gas, iron ore, gold, copper, wheat, dates, bananas, coffee, alfalfa, cattle, goats, camels, grapes, poultry, sheep, sorghum, watermelons

Bahrain

STATE OF BAHRAIN

THE TINY INDEPENDENT MONARCHY OF Bahrain consists of a group of some 30 islands in the Gulf of Bahrain, a small branch of the Persian Gulf between Saudi Arabia and the peninsula of Qatar.

GEOGRAPHY

The main island of Bahrain is joined by causeways to the neighboring islands of Muharraq and Sitrah, and to Saudi Arabia. Its low-lying landscape ranges from barren rock to sandy plains and salt marshes. A little rain falls mainly during the warm winters; the hot summers are also humid. Underground springs and

The King Fahd causeway, 25 km (16 mi) long, links the tiny island nation of Bahrain to Saudi Arabia, with which it maintains close relations. To offset its dependence on oil revenue, it also has a major industrial complex and shipbuilding and repair yard.

desalination plants provide the water needed for irrigation. Tough desert and saltmarsh plants thrive on the island.

SOCIETY

Bahrain has been a flourishing trading center for over 4,000 years. Since the 7th century AD it has been an Arab Muslim country. From 1521 until 1783 it was ruled by the Portuguese and then the Persians. Bahrain became a British protectorate in the mid 19th century, and achieved full independence in 1971. The emir presides as head of state and prime minister over an appointed cabinet. Before 1975 there was an elected National Assembly, but this was dissolved following tension between Shi'ite and Sunni Muslims. There were Sh'ite protests in 1994/95 demanding its reinstatement.

There are large numbers of foreign workers, mainly from Iran and southern Asia. Shi'ite Muslims are in the majority, but the ruling family and many of the richest Bahrainis are Sunnis. The culture remains traditionally Arab, despite the predominantly urban lifestyle and widespread Western television and cinema.

ECONOMY

Bahrain's prosperity is based on natural gas and oil, but with reserves dwindling, aluminum processing and financial services have been growing in importance. Agricultural production meets domestic

demand for both fruit and vegetables.

Transportation and telecommunications are highly developed. The principal port is close to the capital, Al Manamah, and there is an international airport on Muharraq. Oil revenue has enabled Bahrain to offer free medical care and schooling up to secondary level. Tertiary-level education is also available.

NATIONAL DATA – BAHRAIN

Land area 665 sq km (257 sq mi)

Climate		Temperatures		Annual
	Altitude m (ft)	January °C(°F)	July °C(°F)	precipitation mm (in)
Al Manamah	6 (18)	17 (63)	34 (93)	70 (2.7)

Major physical features largest island: Bahrain 583 sq km (225 sq mi); highest point: Durkan Hill 134 m (440 ft)

Population (2006 est.) 698,585

Form of government constitutional hereditary monarchy with one appointed council

Armed forces army 8,500; navy 1,200; air force 1,500

Capital city Al Manamah (148,622)

Official language Arabic

Ethnic composition Bahraini 62.4%; non-Bahraini 37.6%

Official religion Islam

Religious affiliations Muslim (Shi'a and Sunni) 81.2%; Christian 9%; other 9.8%

Currency 1 Bahraini dinar (BHD) = 1,000 fils

Gross domestic product (2006) U.S. $17.7 billion

Gross domestic product per capita (2006) U.S. $25,300

Life expectancy at birth male 71.97 yr; female 77 yr

Major resources oil, natural gas, aluminum processing, fisheries, pearls, dates, eggs, livestock, lucerne, vegetables

---

Qatar

STATE OF QATAR

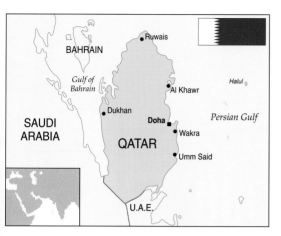

THE SMALL EMIRATE OF QATAR OCCUPIES THE peninsula of the same name on the southwestern coast of the Persian Gulf. It shares borders with Saudi Arabia to the southwest and the United Arab Emirates to the southeast. The island state of Bahrain lies immediately to the west.

GEOGRAPHY

The long Qatar peninsula is mainly flat except for a few hills in the west and some low cliffs in the far northeast. Rainfall is scarce, apart from rare showers during the winter. Summers are generally hot and humid, but winter nights are often chilly. Vegetation, restricted to plants that can survive prolonged drought, is most abundant in the north. Farther south salt-flats and sand dunes predominate. The wildlife includes desert creatures such as gerbils and other sand-burrowing rodents, reptiles and insects.

SOCIETY

Qatar, one of many emirates established in Arabia during the spread of Islam in the 7th century AD, was ruled for many centuries by the Khalifah dynasty. However, the transfer of the emir's headquarters to nearby Bahrain in 1783 led to a rift with his Qatari subjects, and eventually to war in 1867. The British intervened, establishing a separate emirate under a member of the leading Qatari family, al-Thani. In 1916 Qatar became a British protectorate under the terms of a treaty. The discovery of oil in 1939 changed the emirate's fortunes. Qatar chose not to join the United Arab Emirates after the withdrawal of British troops in 1971, and opted for complete independence. In 1995 the emir was deposed by his son, Sheikh Hamed ibn Kalifa ath-Thani, in a bloodless coup.

The emir of Qatar rules as absolute monarch, though his authority depends on the cooperation of the ruling family, whose members make up two-fifths of the indigenous Qatari Arab population. Islam is the official religion though religious law is enforced only within the indigenous community.

ECONOMY

Revenue from the limited oil reserves is increasingly being supplemented by exploitation of natural gas. The country's industries, which are highly dependent on these primary resources, include oil and gas refining and petrochemical, steel and fertilizer production. The use of irrigation and high-technology allow fruit and vegetables to be grown, and the local fishing industry is also being modernized. However, most foodstuffs still have to be imported. Qatar is also a major banking center.

The road network in Qatar is excellent, and there are many sea and air services. Welfare and education are provided for all residents, and university education is also available to a limited number.

Flames from Qatar's oil wells light up the sky The small peninsula has both onshore and – increasingly important – offshore reserves. Since the discovery of oil in 1939, Qatar, formerly one of the poorest countries in the world, has become one of the richest.

NATIONAL DATA – QATAR

Land area	11,437 sq km (4,416 sq mi)			
Climate		Temperatures	Annual	
	Altitude m (ft)	January °C(°F)	July °C(°F)	precipitation mm (in)
Doha	11 (36)	17 (63)	35 (94)	81 (3.2)

Population	(2006 est.) 885,359
Form of government	absolute monarchy
Armed forces	army 8,500; navy 1,800; air force 2,100
Capital city	Doha (358,098)
Official language	Arabic
Ethnic composition	Arab 40%; Indian 18%; Pakistani 18%; Iranian 10%; other 14%
Official religion	Islam
Religious affiliations	Muslim 92.4%; Christian 5.9%; Hindu 1.1%; others 0.6%
Currency	1 Qatari rial (QAR) = 100 dirhams
Gross domestic product	(2006) U.S. $26.05 billion
Gross domestic product per capita	(2006) U.S. $29,400
Life expectancy at birth	male 71.37 yr; female 76.57 yr
Major resources	petroleum, natural gas, fish, steel and cement, petrochemicals, fertilizer, fodder, fruit, livestock, vegetables

United Arab Emirates

THE UNITED ARAB EMITATES (UAE), FORMERLY the Trucial States, are a federation of seven states ruled by emirs on the southern coast of the Persian Gulf. Qatar is to the northwest of the UAE, Saudi Arabia to the southwest and Oman to the east. From west to east the emirates are Abu Dhabi (by far the largest), Dubai, Sharjah, Ajman, Umm al-Qaiwain, Ras al-Khaimah and Fujairah. They all face north across the Persian Gulf except for Fujairah, which fronts the Gulf of Oman on the east coast of the Musandam peninsula.

NATIONAL DATA - UNITED ARAB EMIRATES

Land area 83,600 sq km (32,280 sq mi)

Climate		Temperatures		Annual
	Altitude m (ft)	January °C(°F)	July °C(°F)	precipitation mm (in)
Abu Dhabi	5 (16)	18 (64)	35 (95)	798 (88.9)

Major physical features highest point: Jabal Hafib 1,189 m (3,901 ft)

Population (2006 est.) 2,602,713

Form of government multiparty republic with one appointed council

Armed forces army 44,000; navy 2,500; air force 4,800

Largest cities Dubai (1,225,137); Abu Dhabi (capital - 633,136); Sharjah (584,286)

Official language Arabic

Ethnic composition Emirati 19%; other Arab and Iranian 23%; South Asian 50%; others 8%

Official religion Islam

Religious affiliations Sunni Muslim 80%; Shi'ite Muslim 16%; Christian 3.8%; others 0.2%

Currency 1 Emirati dirham (AED) = 100 fils

Gross domestic product (2006) U.S. $129.4 billion

Gross domestic product per capita (2006) U.S. $49,700

Life expectancy at birth male 72.92 yr; female 78.08 yr

Major resources petroleum, natural gas, camels, cattle, fish, fruit, vegetables, goats, sheep, tourism

GEOGRAPHY

The landscape is generally sandy and low-lying except for coastal salt flats and a range of steep mountains in the east – an extension of Oman's Hajar Mountains that runs northward along the Musandam peninsula. The climate is hot and dry all year round, though more humid near the coast; strong seasonal winds bring fierce dust storms that can cause considerable damage. Apart from date palms in oases in the northeastern emirates, vegetation is generally sparse and scrubby; the wildlife includes many sand-dwelling animals such as jerboas.

SOCIETY

The Muslim emirates that make up the federation were probably founded between the 7th and 8th centuries AD, reflecting local divisions among the different Arab peoples. From the early 17th century onward the British were the most influential Europeans in the area. They established a military base to suppress piracy in the early 19th century. A treaty in 1892 gave Britain control over foreign policy in what became known as the Trucial States, but each emirate maintained its traditional local rule.

In the 20th century the tapping of vast petroleum reserves made the emirates extremely rich. They began to cooperate more closely and in 1971, when British forces left the Persian Gulf area, most chose to join the UAE, though Bahrain and Qatar did not. The union maintained close links with both Britain and neighboring Arab states, with whom they sided in the 1991 Gulf War against Iraq.

The indigenous Arab population is today outnumbered by immigrant workers, mainly from southern Asia and neighboring Arab states. Islam, including both Sunni and Shi'ite sects, is the official religion, and forms the basis of law. Each emirate is an absolute monarchy, whose ruler is a member of the Supreme Council. The emirs elect a federal president and vice-president from among themselves, in addition to appointing a 40-member advisory council.

ECONOMY

The wealth from petroleum reserves has made this otherwise unproductive land one of the most prosperous countries in the world. Only a tiny proportion of the land area is suitable for agriculture: either

Abu Dhabi, largest of the 7 emirates and capital of the UAE, is a thriving modern city of wide boulevards, high-rise buildings and extensive public services. Its wealth is founded on oil, but other traditional industries include fishing and pearl harvesting.

around the oases or where irrigation has been possible. Crops include dates and alfalfa, along with other fruit varieties and cereals. Some crops are grown using high-technology methods such as hydroponics (without soil). Fish are plentiful in neighboring waters, but livestock has been limited mainly to goats and camels because of widespread overgrazing.

Most food has to be imported, but the cost is small in comparison to huge revenues from the oil and gas industries. Aluminum smelting, trade and financial services are also important. Economic development has been further assisted by good road and shipping facilities.

Social services, generously funded by petroleum revenues, are the main source of employment for local people. Healthcare and education to university level are both provided free of charge. However, these developments are relatively recent, and many foreign teachers and doctors are employed.

Oman

SULTANATE OF OMAN

OCCUPYING THE SOUTHEASTERN PORTION OF the Arabian Peninsula, Oman looks southeast across the Arabian Sea and northeast across the Gulf of Oman. It shares borders with Yemen, Saudi Arabia and the United Arab Emirates. The small Omani enclave of Ru'usal-Jibal, at the tip of the Musandam peninsula, guards the entrance to the Persian Gulf.

GEOGRAPHY

In ancient times this part of the Arabian Peninsula was green and fertile, but major climatic changes then turned the land to desert. The only highland areas are the limestone Hajar Mountains in the north, overlooking a fertile coastal plain and the Gulf of Oman, and a lower range in the southern province of Dhofar. In the mountains, especially in Dhofar, summer monsoons bring cooling mist, and in the desert rare but torrential rainstorms fill the dry riverbeds or wadis, causing occasional flash floods.

SOCIETY

Muscat, on the Gulf of Oman, was a flourishing trade center as early as the 5th century BC. From the 2nd century AD the land was overrun by warring Arab peoples. With the spread of Islam in the 7th century, power passed to the leaders of the Islamic faith, the imams. After a royal dynasty was established in 1154, Oman grew into a maritime power. In 1507 the Portuguese sacked the capital, Muscat, but in 1650 the imams regained power, spreading their maritime empire. A sultanate established in 1744 ruled from Zanzibar (an Omani conquest) until 1861, when the monarchy split.

From 1798 onwards Oman established close treaty ties with the British, and it retained strong links with Britain even after it ceased to be a British protectorate in 1951. In 1970 Sultan Qabus Ibn Sa'id (b. 1940) deposed his father. The country gained full independence in 1971, and Qabus began to introduce liberal reforms. In 1991 a new consultative assembly was formed consisting of a nominated president and 59 nominated representatives.

Omani Arabs – mostly Sunni Muslims – form the majority of the population. There are also large numbers of foreign workers from Africa and Asia.

ECONOMY

Oman's petroleum reserves have been very important in building up the economy. The many associated manufacturing industries burgeoned during the 1970s.

Fishing and subsistence agriculture (supported by irrigation wells) provide the main source of revenue for local people. Crops include wheat and alfalfa, along with fruits such as dates, limes and coconuts. Goats are the main livestock, with sheep and cattle in irrigated areas.

Since 1970 transportation has been improved and welfare and healthcare provision extended. Education has been expanded at all levels and remains strongly bound to Islamic scholarship.

NATIONAL DATA – OMAN				
Land area 212,460 sq km (82,031 sq mi)				
Climate		Temperatures		Annual
	Altitude m (ft)	January °C(°F)	July °C(°F)	precipitation mm (in)
Muscat	5 (16)	22 (72)	34 (93)	98 (3.8)
Major physical features highest point: Jabal Akhdar 3,074 m (10,086 ft)				
Population (2006 est.) 3,102,229				
Form of government monarchy with one appointed council				
Armed forces army 25,000; navy 4,200; air force 4,100				
Capital city Muscat (252,864)				
Ethnic composition Omani Arab 73%; Indian 13%; Pakistani 7%; Egyptian 2%; others 5%				
Official religion Islam				
Religious affiliations Muslim 88%; Hindu 7%; Christian 4%; others 1%				
Currency 1 Omani rial (OMR) = 1,000 baiza				
Gross domestic product (2006) U.S. $43.88 billion				
Gross domestic product per capita (2006) U.S. $14,100				
Life expectancy at birth male 71.14 yr; female 75.72 yr				
Major resources petroleum, copper, asbestos, marble, limestone, chromium, gypsum, natural gas, fisheries, alfalfa, bananas, coconuts, dates, fruit, vegetables, wheat, tourism				

Goats for sale in the livestock market at Nazwa in the northern interior of Oman. Monsoon rains in the summer create plentiful pasture for livestock, and most families in rural areas keep goats and cows. Half the population lives by subsistence agriculture.

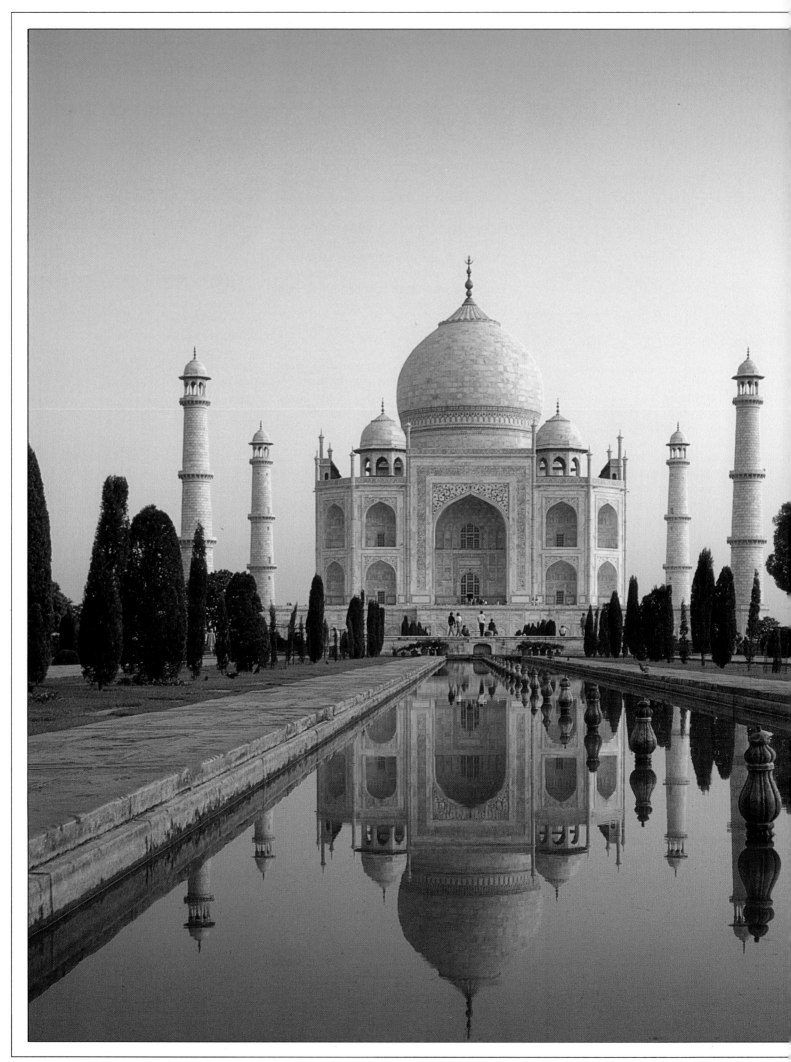

The Indian Subcontinent

COUNTRIES IN THE REGION

INDIA · BHUTAN · PAKISTAN · BANGLADESH · NEPAL
MALDIVES · SRI LANKA

DEPENDENCIES IN THE REGION

BRITISH INDIAN OCEAN TERRITORY

■ capital city
● major town

height of land (meters)

5000
3000
2000
1000
500
200
0

▲ mountain peak

The graceful symmetry of the Taj Mahal (*left*) The marble Taj – one of the finest examples of Moghul architecture – was built in Agra, India, in the 17th century by Emperor Shah Jahan as a mausoleum for his favorite wife Mumtaz Mahal. It is just one of thousands of splendid monuments reflecting India's rich cultural heritage.

409

India

REPUBLIC OF INDIA

THE INDIAN SUBCONTINENT, A LARGE PENIN-
sula in southern Asia, covers an area
about one-third the size of Europe. To the
north the massive mountain ranges of the
Himalayas form a barrier to the rest of
Asia. To the west India is bounded by the
Arabian Sea; the Indian Ocean is to the
south and the Bay of Bengal to the east.
India boasts one of the world's oldest
civilizations (the Aryans and the Dravi-
dians), whose history has been recorded
as far back as the 15th century BC.

The modern republic of India occupies
the greater part of the Indian subcontinent, bordering Pakistan to the west,
Tibet (now occupied by China), Nepal
and Bhutan to the north, Bangladesh and
Burma to the east.

GEOGRAPHY

The varied landscapes of India are a
prime example of plate tectonics (the
movements of continental land masses).
The Himalayas – the highest mountains
in the world – and the great trench to the
south of them were created by the col-
lision of two ancient continental "plates".

The land
The Himalayas descend southward to a
fertile alluvial plain that crosses the country in a broad band from east to west. The
most northerly band of the Himalayas
runs across northern Kashmir and deep
into neighboring Tibet. It is the source of
the great Indus and Brahmaputra rivers.
Farther south lie the Greater Himalayas,
whose mountains are so high that they
are permanently covered in snow and ice.

Still farther south the Lesser Himalayas
give way to the lower sandstone ranges of
the Siwalik Hills, fringing India's northern plains. These were formed from a
trench, originally up to 2,000 m (6,500 ft)
deep, that is now largely filled by stones
and alluvium washed downstream from
the mountains by rivers. Of the main
rivers, the Indus runs southwest into
Pakistan, while the Ganges flows
southeast along the plains toward the Bay
of Bengal. India's other great river, the
Brahmaputra, flows southwest to meet

Dwarfed by the mountains of Ladakh, part of the
mighty Himalayan chain that extends across northern
India, Thikse monastery perches on a barren hill. The
high fertile plains and deep mountain valleys of Ladakh
provide some of the most dramatic scenery in India.

the Ganges, creating a massive delta in
Bangladesh and West Bengal. In the south
lies the massive Deccan plateau, which
contains some of the world's oldest rocks.

Climate
From December to March, cold winds
blowing southward out of the Himalayas

produce cool, dry winter conditions in all regions except the far south and west, which remain hot and humid. Spring, in April and May, brings a change in the prevailing winds, drawing in warmer air from the south. Much of India has little or no rain during this, the hottest time of the year. The summer monsoon rains come as a welcome relief. Most of the country receives at least some rain, but if the monsoon (usually between June and September) fails, drought and famine are the result. The highest rainfall levels are recorded in the northeast and along the southwest coast. In October and November the winds turn once again, sometimes bringing with them violent tropical cyclones.

Plants and animals

Today less than one-fifth of India is forested. Most river deltas have mangrove forests along their seaward edges. The Sundarbans, at the mouth of the Ganges, is one of the world's largest mangrove swamps, covering 4,000 sq km (1,540 sq mi). Dry tropical forest is found on higher ground with low rainfall. Assam in the northeast and the Western Ghats along the southwest coast have moist tropical rainforest. In the wetter parts of the eastern Himalayas, evergreen oak and chestnut predominate, and on higher ground rhododendrons.

India has a remarkably rich variety of wildlife. But as forests are reduced, animal habitats are destroyed. Protected animals include the Asian elephant, the Indian rhinoceros, the gaur (a kind of wild ox), the Asiatic lion, tigers, snow leopards, musk deer, the great Indian bustard and the gharial, a large crocodile.

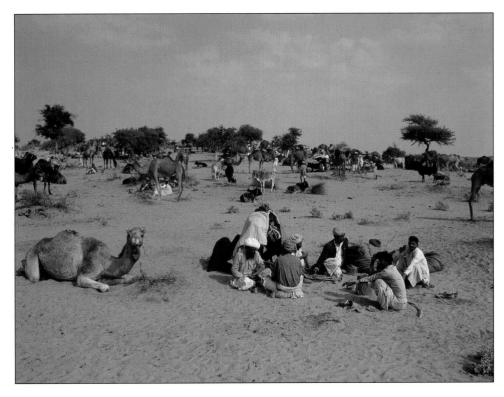

A caravan at rest in the Thar Desert, northwestern India, the eastern extension of the world's largest arid region that encompasses the Sahara and the Arabian desert. Covering some 259,000 sq km (100,000 sq mi), the Thar is home to nomadic cattle herders.

Lakshadweep – land of the giants

Some 300 km (200 mi) off the Malabar coast of southwestern India are the Lakshadweep, the "hundred thousand islands" (also known as the Laccadives). In fact there are only 14 islands, of which 10 are inhabited, the rest of the chain comprising coral reefs. The islands do have a mystery though.

A meter or so (a few feet) beneath the light coral sand there should be a layer of coral some 40 cm (16 in) thick, and yet on all the main islands this coral stratum, at some point in the remote past, was completely dug away. This was a massive task for an area measuring as much as 30 sq km (12 sq mi). As a result, the underlying sands were exposed, creating the ideal growing medium for crops. Local tradition says that the excavation work was done by giants; the truth may never be known.

The people of Lakshadweep are Moplahs, of mixed Indian and Arab descent who are Muslims but speak the Malayalam language of neighboring southwest India. From 1498 the islands were in Portuguese hands, but the Moplahs shook off their rule in 1545 and gave their allegiance to the raja of Cannanore on the Malabar coast. After 1792 the southern group of islands won self-government in return for paying tribute. However all the islands were sequestrated by the British in 1877, and came under Indian control following the withdrawal of the British in 1947. In 1956 they were designated an Indian "union territory".

The islanders are very skilled boatbuilders and sailors. Only local craft make the dangerous journey through the encircling reefs to the Indian mainland. Fishing is the chief occupation, and grain, pulses, bananas, vegetables and coconut palms are farmed. The Moplah women make coir from the fibers of coconut husks.

NATIONAL DATA – INDIA

Land area 2,973,190 sq km (1,147,955 sq mi)

Climate	Altitude m (ft)	Temperatures January °C(°F)	July °C(°F)	Annual precipitation mm (in)
New Delhi	216 (708)	14 (57)	30 (86)	797 (31.3)
Cherrapunji (near Shillong)	1,313 (4,309)	12 (54)	20 (68)	10,799 (425.1)
Chennai (Madras)	16 (51)	25 (77)	30 (86)	1,266 (49.8)

Major physical features highest point: Kanchenjunga 8,586 m (28,169 ft); longest rivers: Brahmaputra (part) 2,900 km (1,800 mi), Ganges 2,505 km (1,557 mi)

Population (2006 est.) 1,095,351,995

Form of government multiparty republic with two legislative houses

Armed forces army 1,100,000; navy 55,000; air force 170,000

Largest cities Mumbai/Bombay (13,073,926); New Delhi (capital – 11,505,196); Bangalore (5,281,927); Kolkata/Calcutta (4,643,011); Chennai/Madras (4,376,400); Ahmadabad (3,819,497); Hyderabad (3,665,106)

Official languages National: Hindi; English State: Bengali; Telugu; Marathi; Tamil; Urdu; Gujarati; Malayalani; Kannada; Oriya; Punjabi; Assamese; Kashmiri; Sindhi; Sanskrit

Ethnic composition Indo-Aryan 72%; Dravidian 25%; Mongoloid and other 3%

Religious affiliations Hindu 80.5%; Muslim 13.4%; Christian 2.3%; Sikh 1.9%; other 1.8%; unspecified 0.1%

Currency 1 Indian rupee (INR) = 100 paise

Gross domestic product (2006) U.S. $4.042 trillion

Gross domestic product per capita (2006) U.S. $3,700

Life expectancy at birth male 63.9 yr; female 65.57 yr

Major resources coal, iron ore, manganese, mica, bauxite, titanium ore, chromite, natural gas, diamonds, petroleum, limestone, tea, coffee, rubber, pepper, coconuts, jute, cotton, barley, fruit, gemstones, ground nuts, millet, potatoes, pulses, rice, sorghum, sugarcane, tea, vegetables, wheat, livestock

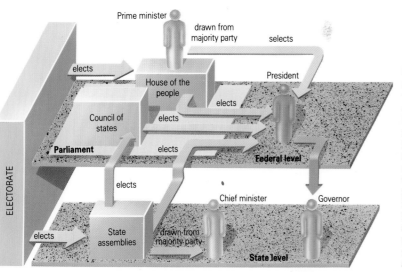

Prime minister
drawn from majority party
selects
elects
President
House of the people
elects
Council of states
elects
Parliament
elects
Federal level
elects
Chief minister
Governor
elects
State assemblies
drawn from majority party
State level
elects
ELECTORATE

India's federal constitution India is a federal republic: its 25 states are self-governing. Power is exercised at national level by the prime minister, who controls the Lok Sabha (House of the People) in the two-chamber federal parliament, elected every five years. The president is chosen by the federal and state parliaments: in practice the prime minister's nominee is always elected.

SOCIETY

The long and eventful history of the Indian subcontinent has left a rich legacy of art and culture. It has also left a very divided society; the gap between rich and poor is very marked.

History

The Indian subcontinent's first trace of a major civilization – in the Indus valley – dates back to 2500 BC. Many Hindu beliefs and customs derive from this Indus culture. The Indus civilization was probably ousted by an invasion of Aryan peoples from the Iranian plateau after about a thousand years. They built towns and cities along the Ganges plains, spreading their culture and their Sanskrit language far and wide. Their concept of social classes, or castes, were vital to the development of later Indian societies.

Two new religious movements arose in the 6th century BC: Jainism, which still claims many adherents among India's merchant community, and Buddhism, which later spread across much of southern and eastern Asia. During the following century the great Kingdom of Magadha, south of the lower Ganges, was established. King Chandragupta Maurya (reigned c. 321–297 BC) expanded its boundaries west of the Indus and south into the Deccan plateau. Under his grandson Ashoka (reigned c. 273–232 BC) the Mauryan empire grew till it dominated the subcontinent.

In the 4th century AD the Gupta dynasty, based at Magadha, ushered in a "Golden Age" of Hindu civilization but Hunnish invasions in the 5th century caused its collapse. In 712 the northwest was occupied by Arab Muslim invaders and in 1001 a Turkish sultan, Mahmud (971–1030) took Islam eastward into the Punjab and beyond. After the 12th century, the Ghurids (from Afghanistan) captured what remained of Mahmud's

Indian empire, and built their own empire. This became the powerful Sultanate of Delhi, until a Mongol invasion from the northwest in 1398 greatly reduced its powers. The 14th century also saw the foundation of the Sikh religion.

From 1519 Babur "the lion" (1483–1530), a Muslim of Mongol extraction, crushed what was left of the Delhi sultanate, and in 1526 founded the Mughal empire. His grandson Akbar (1542–1605) extended it throughout most of northern India and part of the Deccan.

In the 16th century the Portuguese established trading posts and bases in India, and retained a monopoly on trade with India until the 17th century, when the Dutch, English and French East India Companies set up trading stations of their own. By 1799 the British East India Company had become the dominant power in India, and by 1849 virtually all of the subcontinent was under British control.

In 1857 an uprising among Indian soldiers, known as the Indian Mutiny, gained the support of peasants, landlords and princes, but the Sikhs stayed loyal to Britain. After this, the British transferred power in India from the East India Company to the British Crown.

India was now part of the British empire, and in 1877 Queen Victoria was crowned Empress of India. Later, in 1885, the Indian National Congress was formed, the first political organization to claim members from all over the country. By the early 20th century its more radical members had begun to question Britain's right to rule India.

Indian hopes for independence were crushed by the Rowlatt Acts of 1919, which authorized imprisonment without trial. Mohandas Gandhi (1869–1948), a leading member of the Congress Party, resisted the new law by organizing massive displays of civil disobedience and noncooperation. Britain passed the Government of India Act in 1935, intro-

Colorful prayer flags (*above*) flutter above the walls of a Tibetan Buddhist monastery in Leh, capital of India's northern Ladakh district. Every flag has a prayer printed on it, and it is believed that each flap of the cloth is equivalent to the prayer being recited.

A high-caste Hindu (*below*), identified by the sacred yellow thread around his neck, performs ritual ablutions in the river Ganges at Varanasi. According to Hindu myth the river once flowed through heaven, and is worshiped as a goddess able to wash away sins.

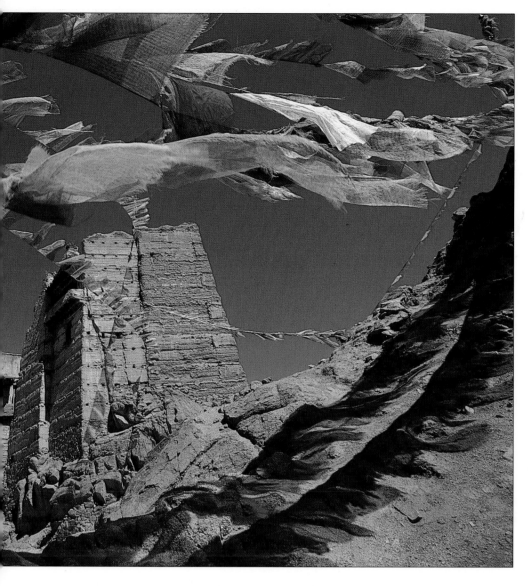

ated by Sikh members of her own bodyguard. Her elder son Rajiv Gandhi (1944–91) was appointed prime minister and leader of the Congress Party, but in 1991 when his Congress Party was returned with a fragile majority, Gandhi himself was assassinated during the elections.

Government and people

In 1950 India became the first republic to achieve independence within the British Commonwealth, and the world's longest written constitution came into force.

The lower house of the two houses of parliament is called the House of the People (Lok Sabha). Members are elected for a five-year term and there may be up to 550 members. The Council of States (Rajya Sabha), the upper house, can have up to 250 members. The head of state is the president, who is chosen by elected members of both houses, as is the vice-president. Real power, however, rests with the prime minister and council of ministers, who are both answerable to the House of the People.

India has a complex racial mix and a rich diversity of languages. Linguists list more than 700 different languages and dialects; 15 major languages have official state recognition. The largest group is the Indo-Aryan languages, which are all related to Sanskrit and are spoken throughout northern and western India. Among them Hindi is used as a first or second language by more than four-fifths of the population, and after independence it became the official language of India alongside English, which

ducing a new federal constitution. But this satisfied no one. Congress won the elections of 1936 and in 1939, when the British viceroy declared India to be at war with Nazi Germany, without having consulted local leaders, all the Congress ministers resigned in protest. A long and troubled series of negotiations, punctuated by rioting and bloodshed, finally brought full independence in 1947. The price of Hindu–Muslim enmity was the partitioning of the country. India retained the states with a Hindu majority, and the mainly Muslim areas to the east and northwest became the separate state of Pakistan. An estimated 8.5 million people were uprooted by partition, of whom some 400,000 died.

India's first prime minister was Jawaharlal Nehru. He tried to avoid alignment with either the Soviet Union or the West, but a border dispute with China in 1962 forced him to appeal for help to Britain and the United States. In 1965, after Nehru's death, a dispute with Pakistan flared into open war, and after Nehru's daughter Indira Gandhi (1917–84) became prime minister in 1967, fighting broke out for a second time in 1971. The

war led to the birth of a new nation – Bangladesh – formerly East Pakistan.

Indira Gandhi became increasingly unpopular, and in 1977 was voted out of office, but in 1980 she was re-elected in triumph. In 1984 Indian troops attacked Sikh militants who had occupied the Golden Temple at Amritsar, the Sikhs' holiest shrine, killing 400 people. Shortly afterward, Indira Gandhi was assassin-

Fishermen carry in the nets from their boat in southern India. The rich waters off India's coasts offer a bountiful harvest of fish, such as sardines, anchovies and bummalo (also known as Bombay duck). In 1995 the country's total fish catch was 4.9 million tons.

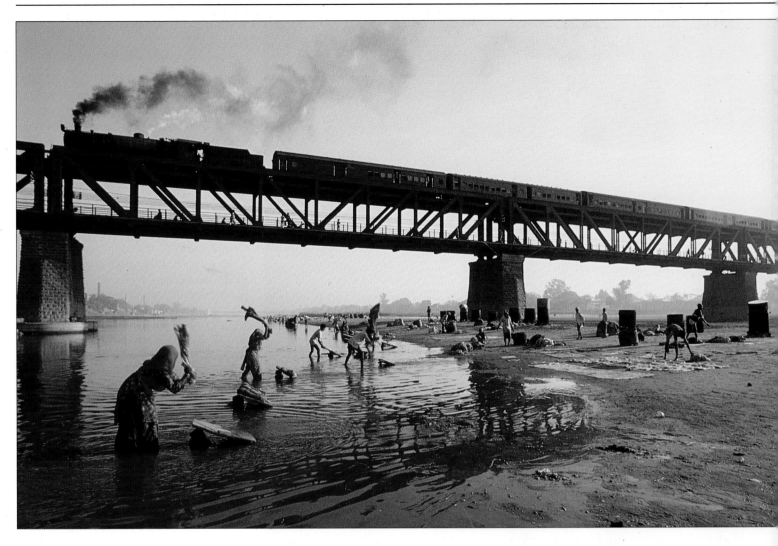

A steam train, en route for Agra, hurtles across a bridge in northern India while below women and men wash clothing in the shallow Yamuna river. India's well-developed railroad system, created by the British in the mid 19th century, is the world's fifth largest.

is preferred in the south. Other major Indo-Aryan languages include Bengali, Marathi, Gujarati and Assamese.

India's dominant religion is Hinduism, but more than one-tenth of the people are Muslim, and there are also significant Christian, Sikh, Buddhist and Jain minorities. The caste system divides Hindus into some 3,000 different groups according to their birthplace and occupation. Members of different castes cannot intermarry without one of them "losing caste". The lowest caste are the "untouchables". Untouchability has been unconstitutional since 1950. However, in 1997 Kocheril Raman Narayanan was elected the first "untouchable" president.

ECONOMY

In recent years India's economy has picked up with exports increasing and an improvement in the balance of payments situation. Investment by developed countries has rocketed and a significant decline has been achieved in foreign debt. There is less need to rely on foreign aid.

Agriculture and industry

More than two-thirds of the total labor force work on the land. Many farms still consist of smallholdings worked by peasants. Floods and droughts are a constant threat but the Green Revolution of the 1970s and 1980s produced better harvests, and irrigation is increasing. India is now self-sufficient in food production.

Rice is the main food crop; over 4,000 varieties are grown. Wheat is cultivated in drier areas. Major cash crops such as tea, coffee, rubber and pepper are grown on large plantations. Others include coconuts, jute and cotton.

India has the largest livestock population in the world, but milk yield per cow is among the lowest in the world because of a shortage of fodder. India's fishing industry has grown rapidly.

Massive deposits of iron ore are being exploited, but iron production has not yet reached its peak. India also has bauxite, manganese, copper, sulfur and salt, and gemstones remain a major export.

There are also huge reserves of coal and some oil. Petroleum production satisfies about one-half of the country's fairly modest requirements. Around three-quarters of India's electricity is supplied by thermal stations, some by hydroelectric power and a small amount by nuclear and wind power. Manufacturing is the fastest-growing sector of the economy. Consumer products such as textiles, bicycles and electrical appliances are made mostly by private companies. State-owned ventures concentrate on heavy industries, but are increasingly open to private and foreign investors.

Transportation and communications

India's rail network is the country's largest state enterprise. Steam locomotives are still widely used, but new equipment is being introduced. Some areas still rely on river transportation. India's major seaports include Kolkata (Calcutta), Chennai (Madras), Mumbai (Bombay) and Cochin. The press functions with few restrictions, but because of the huge number of languages many newspapers have small circulations.

Health and welfare

Poverty and malnutrition are widespread, and medical resources are overstretched. Despite efforts to promote birth control, the population continues to rise. Education is generally free. Literacy has more than doubled since independence. India has a large reservoir of highly educated workers many of whom are employed overseas.

Bhutan

KINGDOM OF BHUTAN

I N THE EASTERN HIMALAYAS, SANDWICHED between Tibet (now occupied by China) to the north and India to the south, is the tiny kingdom of Bhutan. Like Tibet, Bhutan has long been isolated from international politics and commerce, and is regarded by the rest of the world as a secret and mysterious kingdom.

GEOGRAPHY

Bhutan is a country of high mountains and deep valleys that divides naturally into three main geographic regions. The lowlands along the southern border are known as the Duars Plain. From here the land rises sharply to the rugged mountains and broad valleys of the Lesser Himalayas. Above these tower the Great Himalayas, topped by massive peaks along the Tibetan border.

The climate ranges from subtropical in the south to Alpine in the north; the summer rains are heavy in places, falling as snow on the higher peaks. Most of the people live in the temperate central valleys, where there is seasonal pasture-land. The southern lowlands and slopes are a mixture of dense bamboo forests and savanna, which support wildlife such as elephants and tigers. The mountains are home to hardy mountain animals such as yaks and tahrs (a species of wild goat).

SOCIETY

Bhutan's origins are shrouded in legend, but historians believe that the land was settled by migrants from Tibet in the 9th century AD. In about the 16th century a Tibetan monk is thought to have assumed both temporal and spiritual authority over a group of mountain and hill peoples. After him, power was divided between a spiritual leader, the *Dharma Raja*, who was chosen as the reincarnation of his predecessor, and a secular leader, the *Deb Raja*. The last *Dharma Raja* died in the early 20th century, and no new spiritual leader was identified. In theory the *Deb Raja* was elected by a council of governors, but from the 19th century military commanders fought each other for political power. In 1907 the British, the colonial power in the subcontinent, encouraged Bhutan to put an end to this situation by installing a hereditary monarchy – *Druk Gyalpo*, the Dragon King. Bhutan's external affairs were managed by the British until they withdrew from the subcontinent in 1947.

The first Dragon King's grandson, Jigme Dorje Wangchuk (1928–72), introduced limited social reform aimed at improving living standards without sacrificing traditional culture. Slavery was not abolished until the 1960s, when the first real towns were developed. The reign of his son, Jigme Sangye Wangchuk (b. 1954), has been troubled by unrest among refugees from Tibet, and later among the Nepalese minority, who wanted democratic reforms like those being implemented in Nepal. Few concessions were granted. The Nepalese-speaking Hindus of southern Bhutan won little support for change among the majority Bhutia, who are Buddhists and speak Tibetan dialects.

In the early 1990s hundreds of thousands of Nepali-speaking Hindus were expelled and hundreds killed in pro-democracy demonstrations. In 1998 the king gave the National Assembly the right to dismiss him. Tourism is tightly controlled by the government and limited to package tourists with a pre-planned itinerary. Nearly 5,500 foreign tourists visited Bhutan in 1997.

ECONOMY

Agriculture, at a little above subsistence level, is the mainstay of the economy. Most Bhutanese are farmers, for whom rice, potatoes and buckwheat are among the major crops. Livestock farming and forestry are also significant.

The chief mineral and energy reserve is coal, but dolomite, limestone and gypsum are also mined. Industrial output consists largely of food, timber and pulp products. Along with hydroelectricity these are the main exports. Almost all trade is with India, which supplies the country with manufactured goods.

There are few roads in Bhutan, and air travel is largely confined to government helicopters. Welfare, healthcare and education are free but unavailable to most people, and there is widespread illiteracy.

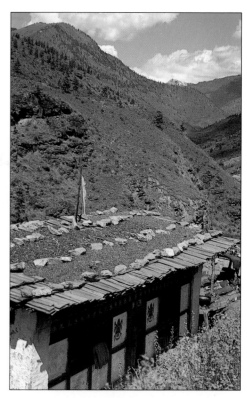

Hot peppers or chilies lie spread out to dry on a roof in a mountain valley in Bhutan. Only 9 percent of the country is cultivated, the rest being either forested or too mountainous. Yet the Bhutanese are largely self-sufficient in food, having to import mainly rice. Together with staple crops they grow oranges, apples and cardamom.

NATIONAL DATA – BHUTAN

Land area	47,000 sq km (18,147 sq mi)			
Climate		Temperatures		Annual
	Altitude m (ft)	January °C(°F)	July °C(°F)	precipitation mm (in)
Thimphu	2,265 (7,431)	4 (39)	17 (63)	3,000 (118)

Major physical features	highest point: Kula Kangri 7,554 m (24,784 ft)

Population	(2006 est.) 2,279,723

Form of government	constitutional monarchy

Armed forces	army 6,000

Capital city	Thimphu (79,334)

Official language	Dzongkha (a form of Tibetan)

Ethnic composition Bhote 50%; ethnic Nepalese 35% (includes Lhotsampas-one of many Nepalese ethnic groups); indigenous or migrant tribes 15%

Religious affiliations Lamaistic Buddhist 75%; Indian- and Nepalese-influenced Hinduism 25%

Currency	1 ngultrum (BTN) = 100 chetrum

Gross domestic product	(2003) U.S. $2.9 billion

Gross domestic product per capita	(2003) U.S. $1,400

Life expectancy at birth	male 55.02 yr; female 54.53 yr

Major resources timber, hydropower, gypsum, calcium carbonate, apples, barley, cardamom, coal, corn, dolomite, maize/corn, millet, oranges, potatoes, rice, buckwheat, yaks

Pakistan

ISLAMIC REPUBLIC OF PAKISTAN

PAKISTAN OCCUPIES THE NORTHWESTERN corner of the Indian subcontinent. To the west is Iran, to the northwest Afghanistan, to the northeast China and to the east India.

GEOGRAPHY

Rising in the Himalayas to the northeast, the Indus river drains the whole of Pakistan from the high northern plateaus to the southern plains. Along the frontier with Afghanistan, the Khyber Pass opens out into an irregular plateau along the foot of the Himalayas. Farther south are the plains of the Punjab – the fertile heartland of Pakistan – watered by the Indus and its tributaries. In the southwest is the arid plateau of Baluchistan.

Pakistan has a continental climate, characterized by very hot summers and cool winters with large daily variations in temperature. Most of Pakistan's rain falls between July and September, when monsoon winds blow from the southwest.

Over much of the country vegetation is sparse. In the northwest, there are forests of pine, deodar and holly, while deciduous trees such as plane and poplar grow in the far north. The mountains are rich in wildlife, including bears, leopards and wild sheep. But with forest cover reduced to 3.5 percent of the land area, animal habitats are threatened.

SOCIETY

Pakistan was created in 1947, when the Muslim-dominated northeastern and northwestern states of India were given autonomy at the end of British rule. Part of Bengal in the east, and part of Punjab in the west were partitioned off from India, creating the new divided nation of East Pakistan and West Pakistan. In 1956

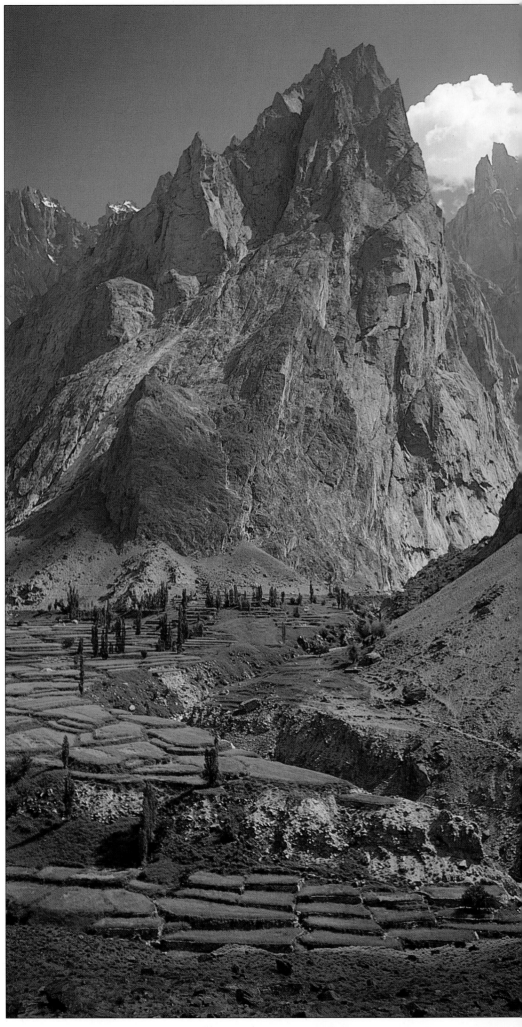

an Islamic republic was proclaimed. Two years later, following a military coup, the country was placed under martial law. In 1965 disputes with India over the Kashmiri border erupted into war. During the elections of 1971 East Pakistan demanded its independence. A civil war followed, in which India supported East Pakistan, and in 1972 that territory became the independent state of Bangladesh.

Zulfikar Ali Bhutto (1928–79) became Pakistan's new head of state, but his election victory in 1977 roused a storm of protest that ended with the renewal of martial law under General Zia-ul-Haq (1924–88). Bhutto was arrested, convicted of attempted murder, and executed.

After Zia-ul-Haq's death in 1988 Bhutto's Pakistan People's Party (PPP) was returned to power under the leadership of his daughter Benazir Bhutto (b. 1953). During the 1990s Bhutto was twice elected and dismissed amid corruption allegations. In 1999 she and her husband were convicted and sentenced to prison. A coup in October 1999 returned Pakistan to military rule under General Musharraf.

Pakistan has two legislative bodies; the national assembly and the senate. The national assembly is headed by the prime minister, and its members are directly elected by popular vote. Members of the senate are chosen by the four provincial assemblies. The president, who is head of state and commander of the armed forces, is chosen by an electoral college.

The people of Pakistan are racially mixed, owing to centuries of invasions and migrations. The main linguistic division is between Iranian languages such as Baluchi and Pashto, and the Indo-Aryan group, which includes Punjabi, Sindhi and Urdu, the official language. The official religion is Islam, which is practiced by most people.

ECONOMY

More than half of Pakistan's population is employed in agriculture, though this now accounts for less than a quarter of national income. The new varieties, techniques and machinery introduced in the "green revolution" of the 1970s have helped to provide a surplus of wheat, rice and sugar, all available for export. Livestock are numerous, but the yields of milk and meat are relatively low because of the shortage of fodder. Fishing is a significant growth industry.

Pakistan has some coal, oil and high-grade iron ore as well as copper and bauxite. Limestone is mined in quantity and is the basis of a growing cement industry. Hydroelectric plants meet about three-fifths of the country's energy requirement, but large reserves of natural gas are now being exploited.

The cotton industry developed rapidly after independence, and is still very important today. Other products including leather, wool and sugar are now processed for the home market and for export. Heavy industry has been government-controlled since 1971.

Pakistan's main road and rail routes link the principal seaport, Karachi, to Lahore, Islamabad (the capital), and ultimately to Peshawar. The national carrier, Pakistan International Airlines, offers an internal and international service. Press censorship in one form or another is generally the rule. The government-controlled radio network is the main source of news and opinion.

Healthcare in Pakistan suffers from a shortage of trained staff. There is some basic welfare provision. Small pensions are paid to the old and disabled. Primary education is free and widely available, but only a minority receive secondary or higher education. Literacy rates remain comparatively low.

Verdant terraces (left) in the Hushe river valley near Skardu in northern Pakistan, where barley is grown in irrigated plots beneath the barren scree slopes of the mountains. Snowmelt feeds the Hushe river, a tributary of the Shigar, which itself feeds the mighty Indus.

Muslim faithful face toward Mecca (below) as they gather in the courtyard of the Mahabuk Khan mosque in Peshawar, northwestern Pakistan; some have already prostrated themselves in prayer. Islam is the dominant and official religion of Pakistan.

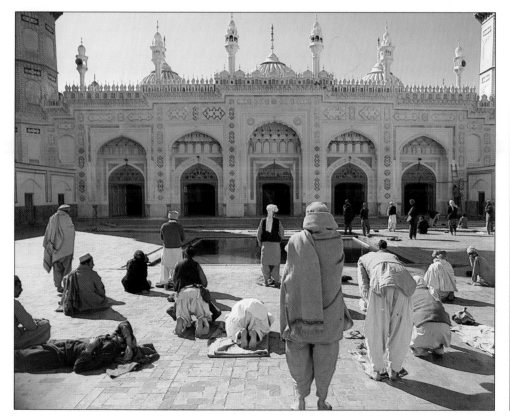

NATIONAL DATA - PAKISTAN

Land area 778,720 sq km (300,665 sq mi)

Climate	Altitude m (ft)	Temperatures January °C(°F)	July °C(°F)	Annual precipitation mm (in)
Islamabad	511 (1,644)	10 (50)	29 (84)	1,142 (44.9)
Karachi	4 (13)	20 (68)	30 (86)	168 (6.6)

Major physical features highest point: K2 8,611 m (28,251 ft); longest river: Indus 2,896 km (1,800 mi)

Population (2006 est.) 165,803,560

Form of government federal multiparty republic with two legislative houses

Armed forces army 550,000; navy 24,000; air force 45,000

Largest cities Karachi (12,315,843); Lahore (6,659,543); Faisalabad (2,658,181); Rawalpindi (1,843,418); Multan (1,506,424); Gujranwala (1,459,139); Hyderabad (1,447,275); Islamabad (capital - 955,629)

Official languages Urdu, English

Ethnic composition Punjabi 48.2%; Pashto 13.1%; Sindhi 11.8%; Saraiki 9.8%; Urdu 7.6%; others 9.5%

Official religion Islam

Religious affiliations Muslim 97% (Sunni 77%, Shi'a 20%); Christian, Hindu, and other 3%

Currency 1 Pakistani rupee (PKR) = 100 paisa

Gross domestic product (2006) U.S. $427.3 billion

Gross domestic product per capita (2006) U.S. $2,600

Life expectancy at birth male 62.4 yr; female 64.44 yr

Major resources coal, cement, cotton, fertilizers, food processing, livestock, fisheries, petroleum, natural gas, sugarcane, yarn, fabric, iron ore, copper, salt, limestone

Bangladesh

PEOPLE'S REPUBLIC OF BANGLADESH

THE SMALL DENSELY POPULATED COUNTRY OF Bangladesh has India to the west, north and east. The southern coastal area fronts the Bay of Bengal.

GEOGRAPHY

Bangladesh occupies the eastern two-thirds of the Ganges–Brahmaputra delta. The land is extremely fertile but highly vulnerable to flooding from both the rivers and the sea especially in the monsoon season. In recent years soil erosion from deforestation upstream has made this flooding even more devastating. Bangladesh is mostly a country of flat plains criss-crossed by rivers. The Sundarbans, a huge mangrove forest in the south, is one of the largest swamplands in the world. The flatness of the land is broken only in the far northeast and in the southeast, where the higher Chittagong Hills run parallel to the Myanmanese border. The Chittagong region in the southeast includes offshore islands and the world's longest beach, at Cox's Bazaar.

Most of the annual rainfall occurs in the summer monsoon season, from June to October. In winter the northeasterly winds bring dry, warm days and cool nights. The winds change direction late in March, when there are torrential thunderstorms. In April and May severe cyclones create massive storm surges causing flooding in coastal areas.

Bamboo and rattan grow in the eastern forests, and central and southern areas are rich in plants adapted to swampy conditions. Fruit trees, including date and coconut palms and mangoes, are cultivated in the densely populated inland areas of Bangladesh. The country's wildlife includes several endangered species, notably the Bengal tiger, the clouded leopard and the Asiatic elephant.

SOCIETY

The name Bangladesh means "land of the Bengalis". Throughout their history the Bengali people have played an important part in the history and culture of the Indian subcontinent.

In the 4th century Bengal was part of the Mauryan Indian empire, ruled by the Buddhist and later Hindu kingdoms and empires of northern India. After the 13th century Islam became the majority religion of Bengal. A British colony from 1760, Bengal was still administered by the Hindu ruling classes, increasing tension between Hindus and Muslims. In 1905 the British partitioned Bengal into a Hindu-dominated western province and a Muslim-dominated eastern province. Western Bengal objected strongly, and in 1912 the two provinces were reunited.

When British rule in India ended in 1947, the country was partitioned along religious lines by creating the state of Pakistan. East Bengal became East Pakistan, despite being separated from West Pakistan by some 1,500 km (1,000 mi). In elections in East Pakistan in 1970, the Awami League, a political party campaigning for Bengali autonomy, won virtually every seat. But the Pakistan government, instead of convening the Assembly, sent troops into East Pakistan. A war broke out, in which India intervened. West Pakistan was defeated, and in December 1971 the new state of Bangladesh was created.

Shaikh Mujib ur-Rahman (1920–75), leader of the Awami League, became the first prime minister of Bangladesh, but he was assassinated in a coup in 1975. Several more coups followed and martial law continued until 1991, when Begum Khaleda Zia (b. 1945), leader of the Bangladesh National Party, was elected prime minister.

Although the original 1972 constitution established a parliamentary government under an elected prime minister and president, years of martial law disrupted this system. In the 1990s the country reverted to an elected parliamentary system where 30 of the 330 seats in the national assembly are now reserved for women.

The great majority of Bangladeshis are Muslim ethnic Bengalis. However, there is a significant Hindu minority, and Buddhists in the Chittagong region. The official language is Bengali. Classical Indian culture, particularly music and dance, exists alongside Bengali culture and Western cinema and sport.

NATIONAL DATA – BANGLADESH

Land area	133,910 sq km (51,703 sq mi)			
Climate		Temperatures		Annual
	Altitude m (ft)	January °C(°F)	July °C(°F)	precipitation mm (in)
Dhaka	8 (26)	19 (66)	29 (78)	1,920 (75.5)

Major physical features longest river: Brahmaputra (part) 2,900 km (1,800 mi)

Population (2006 est.) 147,365,352

Form of government multiparty republic with one legislative house

Armed forces army 110,000; navy 9,000; air force 6,500

Largest cities Dhaka (capital - 6,969,458); Chittagong (3,920,222); Khulna (1,400,689); Rajshahi (727,083); Komilla (419,623); Tungi (368,914)

Official language Bangla (Bengali)

Ethnic composition Bengali 97.7%; Bihari 1.3%; others 1.0%

Religious affiliations Muslim 83%; Hindu 16%; Buddhist 0.6%; Christian 0.3%; others 0.4%

Currency 1 taka (BDT) = 100 paisa

Gross domestic product (2006) U.S. $330.8 billion

Gross domestic product per capita (2006) U.S. $2,200

Life expectancy at birth male 62.47 yr; female 62.45 yr

Major resources natural gas, sugarcane, tea, rice, cotton, jute, fisheries, timber, coal, cattle, glass sand, limestone, peat, salt, tobacco, wheat, white clay

Fishermen near Kurigram (*above*) in northern Bangladesh wait patiently for fish to enter their nets. The fish breed in the rivers, and when the waters flood the neighboring paddyfields – as the Brahmaputra river has done here – the protein-rich fish are caught.

Ox-power (*below*) A Bangladeshi farmer urges his two bullocks on as a youngster feeds sugar cane into a primitive, but effective, crushing machine that is being powered by the animals. Sugar cane – a source of sugar – is one of the country's principal crops.

ECONOMY

Bangladesh is one of the world's poorest countries. Its essentially agrarian economy is vulnerable to the vagaries of the climate; ruined harvests, drought and floods have brought frequent famines. The principal crop is rice, followed by jute, tea and sugar cane. Fishing is also important. Less than 10 percent of the land is forested.

As Bangladesh has few mineral resources, the steel, chemical and textile industries depend on importing raw materials (bought with aid). The industries using local raw materials include jute processing (making rope and other products) and paper-making from bamboo.

An extensive railroad network was built during the years of colonial rule, but today the inland waterways are more important for transporting cargo. Cycle rickshaws are a major form of transport in towns. The main seaport is at Chittagong on the southeast coast.

Comparatively few people own a television or radio; all broadcasting is controlled by the state. By contrast, the press is privately owned.

Healthcare, education and family planning programs are provided by government and private agencies; facilities are based in Dhaka and Chittagong in the east of the country. Even so, malnutrition is rife everywhere. Many Bangladeshis cannot read or write. The government provides five years of free, but not compulsory, primary education. Secondary and higher education are available only to a minority.

Nepal

KINGDOM OF NEPAL

NEPAL IS A SMALL LANDLOCKED KINGDOM in southern Asia lying between India to the south and Tibet (now occupied by China) to the north.

GEOGRAPHY

The Tarai lowlands along Nepal's southern border give way to the Churia foothills farther north, and then to the higher Mahabharat ranges. Still farther north, across the Kathmandu Valley, are the Himalayas, including Mount Everest, the world's highest peak at 8,850 m (29,035 ft). The climate ranges from subtropical in the south to Alpine in the north. The summer brings heavy mon-

A cornucopia of vegetables at a market in Kathmandu, Nepal's capital city. Most Nepalese live as subsistence farmers in scattered communities, often producing crops on heavily fertilized, thin-soiled terraces that have been carved out of the plains and lower hillslopes.

soon rains, and snow in the higher mountains. The western half of the country tends to be drier than other areas.

More than half of Nepal is still forested, ranging from acacia and sissoo forests in the Churia, to pines, oaks and rhododendrons in the Mahabharat and lower Himalayas. The Terai is rich in wildlife, with tigers, leopards and even a few rhinoceroses. Above the tree line are wild sheep and goats, snow leopards and yaks. The brilliantly colored Himalayan monal – a species of pheasant – is the national symbol of Nepal.

SOCIETY

Situated between the Indo-Aryan culture of northern India and the Mongoloid culture of Tibet, Nepal has been greatly influenced by both throughout its history. From the 10th to the 18th century, Nepal was ruled by the Malla dynasty. Then in 1769 it was conquered by the expanding Gurkha principality, which established the capital at Kathmandu. In 1860 Nepal faced the prospect of British rule, and agreed, as a price for continued self-government, to allow Gurkhas to be recruited into the British army. In 1959 King Mahendra (1920–72) introduced a constitution giving him greater power

but in 1990 his son Birendra (b. 1945) was forced by popular unrest to approve a new, more democratic constitution. In June 2001 King Birendra and most of the royal family were shot dead by Crown Prince Dipendra in a family quarrel. Birendra's brother Gyanendra assumed the throne when the prince himself died of gunshot wounds.

Most Nepalese are of Indian descent. Southern Nepal is the birthplace of the Buddha (563–438 BC). Today, the majority of the Nepalese population is Hindu.

ECONOMY

Most of the population live by subsistence farming. Crop yields are relatively low, though Nepal is self-sufficient in staple foods. Commercial forestry generates products for export, but deforestation and subsequent erosion is a problem.

Nepal's mineral resources include coal, iron ore and copper, though reserves are small. Manufacturing is mostly local and on a small scale. The largest factories are processing plants for agricultural produce such as jute, sugar, meat and rice. Tourism is expanding, especially around Kathmandu. Many new roads have been built, and the Royal Nepal Airline Corporation operates both domestic and international flights.

Outside the main cities healthcare is rudimentary. The government's school-building program and its drive to increase literacy are having some impact.

NATIONAL DATA – NEPAL

Land area	143,181 sq km (55,208 sq mi)			
Climate		**Temperatures**		**Annual**
	Altitude m (ft)	January °C(°F)	July °C(°F)	precipitation mm (in)
Kathmandu	1,334 (4,376)	26 (79)	25 (77)	1,461 (57.5)

Major physical features highest point: Mount Everest 8,850 m (29,035 ft); longest river: Ghaghara (part) 915 km (570 mi)

Population (2006 est.) 28,287,147

Form of government multiparty constitutional monarchy with two legislative houses

Armed forces army 69,000

Capital city Kathmandu (856,340)

Official language Nepali

Ethnic composition Chhettri 15.5%; Brahman-Hill 12.5%; Magar 7%; Tharu 6.6%; Tamang 5.5%; Newar 5.4%; Muslim 4.2%; Kami 3.9%; Yadav 3.9%; other 32.7%; unspecified 2.8%

Religious affiliations Hindu 80.6%; Buddhist 10.7%; Muslim 4.2%; Kirant 3.6%; other 0.9%

Currency 1 Nepalese rupee (NPR) = 100 paisa

Gross domestic product (2006) U.S. $41.92 billion

Gross domestic product per capita (2006) U.S. $1,500

Life expectancy at birth male 60.43 yr; female 59.91 yr

Major resources quartz, water, timber, hydropower, lignite, copper, cobalt, iron ore, barley, cattle, fruit, jute, medicinal herbs, pepper, potatoes, rice, sugarcane, tobacco, wheat

Maldives

REPUBLIC OF THE MALDIVES

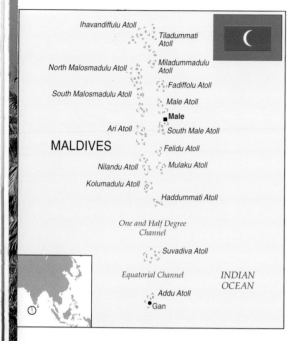

Ihavandiffulu Atoll
Tiladummati Atoll
Miladummadulu Atoll
North Malosmadulu Atoll
Fadiffolu Atoll
South Malosmadulu Atoll
Male Atoll
Male
Ari Atoll
South Male Atoll
MALDIVES
Felidu Atoll
Nilandu Atoll
Mulaku Atoll
Kolumadulu Atoll
Haddummati Atoll
One and Half Degree Channel
Suvadiva Atoll
Equatorial Channel
INDIAN OCEAN
Addu Atoll
Gan

T HE MALDIVES ARE A CHAIN OF SOME 19 island groups in the Indian Ocean to the southwest of India and Sri Lanka, extending over 764 km (475 mi) south to just beyond the Equator.

The islands are atolls – coral reefs surrounding the peaks of a submerged volcanic mountain range. In all there are almost 2,000 islands, but only about 200 are inhabited. The climate is mainly hot and humid. Nearby barrier reefs provide the only protection from occasional violent monsoon-related cyclones.

Parts of the Maldives resemble an idyllic island paradise. The islands are crowned with rich forests of breadfruit and palm, and are ringed with sandy beaches and clear lagoons.

The Maldives were settled some time between 500 BC and 500 AD, probably by fishing people from India and Sri Lanka. The Maldivian language, Divehi, resembles Sinhalese, the majority language of Sri Lanka. Later Arab settlers introduced Islam, which is now the state religion.

The islands fell to the Portuguese in the 16th century. They then became a sultanate under Dutch protection, and in 1796 they became a British protectorate which was formalized in 1887. In 1953 a republic was declared, but the president was soon deposed and the monarch restored. The Maldives attained full independence in 1965. President Maumoon Abdul Gayoom, who was first elected in 1978, survived an attempted coup in 1988 and secured a fifth term in 1998, polling over 80 percent of the vote.

Tourism accounts for 20 percent of GDP. Nearly half a million tourists visited the Maldives in 1998. Fishing provides most other national income. Agriculture is mainly at subsistence level. Crops include

The idyllic atolls of the Maldives lie scattered about the Indian Ocean. At the center of many of them lies a small island – sometimes inhabited, sometimes not – surrounded by a pellucid blue lagoon and enclosed by a reef. For many travelers they are paradise incarnate.

NATIONAL DATA – MALDIVES

Land area	300 sq km (116 sq mi)			
Climate		Temperatures		Annual
	Altitude m (ft)	January °C(°F)	July °C(°F)	precipitation mm (in)
Male	1 (3)	23 (73)	23 (73)	1,911 (75.2)

Population (2006 est.) 359,008	
Form of government nonparty republic with one legislative house	
Armed forces army 900	
Capital city Male (89,287)	
Official language Divehi	
Ethnic composition Sinhalese, Dravidian, Arab, African	
Official religion Islam	
Religious affiliations Sunni Muslim 100%	
Currency 1 rufiyaa (MVR) = 100 laari	
Gross domestic product (2002) U.S. $1.25 billion	
Gross domestic product per capita (2002) U.S. $3,900	
Life expectancy at birth male 63.08 yr; female 65.8 yr	
Major resources fisheries, tourism, coconuts/copra, fruit, vegetables	

millet, cassava and yams, and coconuts are gathered for food and copra.

Vital imports, chiefly from India, include food (especially rice), manufactured goods and fuel for energy. Transportation is mainly by boat, but air travel is increasing. Religious and secular schools provide elementary education, and literacy levels are high. Healthcare still relies largely on traditional medicine.

The last emperor

Emperor Puyi (1906–67), formally known as Xuan Tong, was the last of his line, the Manchu dynasty that had ruled China for 268 years; and the last emperor in a 2,000-year-old tradition. He was crowned emperor in 1908 at the age of two, with his father as regent. But his precarious position mocked his traditional title of "Son of Heaven."

China by the late 19th century was in serious decline. Its government was massive, cumbersome and corrupt; and its army was undisciplined and lacked modern weapons. Conservative and liberal factions at court were bitterly divided. The child emperor Puyi and his immediate family were subordinate to the powerful princes of the Manchu Clan Council, who increasingly depended on the protection of a viceroy and warlord, Li Hongzhang (1823–1901), the richest man in China; and later, on the ambitious general Yuan Shikai (1859–1916).

In 1912 a succession of uprisings threatened the dynasty. The Council called in General Yuan Shikai to quell the rebellion. Yuan instead opened up negotiations that rapidly whittled away the power of the imperial court and ended with his own installation as president of the new Chinese republic after Sun Yat-sen resigned. Puyi was forced to abdicate at the age of six. He continued to live in the Forbidden City – the emperor's traditional residence in a luxurious walled compound within

Royal but vulnerable China's last imperial rulers in a photograph taken in 1909. The emperor Puyi, age 3, stands holding the hand of his father, the regent Prince Chun. Three years later the dynasty collapsed and China became a republic.

Beijing – but despite his new title of Manchu Emperor, he was effectively powerless.

Puyi later became the puppet emperor of the Japanese-controlled state of Manzhouguo (Manchuria; 1931–45). Captured at the end of World War II, he served a 9-year sentence as a war criminal in a Chinese prison before being released to live in Beijing as a private citizen. In the last years of his life he worked in a botanical garden and wrote his autobiography, entitled *From Emperor to Citizen*.

length of 6,400 km (4,000 mi). The emperors built many palaces, including the Imperial Palace at Beijing, and dispatched great sea-going junks to explore as far as the Red Sea. In 1557 the Portuguese were able to establish a settlement in Macao, and they were followed by Jesuit missionaries. Beijing fell to the Manchu, descendants of the Ruzhen, in 1644 and Ming resistance was finally crushed in 1683.

The Manchu, or Qing, dynasty pursued an active policy of rapid expansion. They captured the southeastern island of Taiwan (1683), annexed large areas to the northwest, and posted a garrison in Tibet to put a firm check on Mongol power. The Qing won the loyalty of their subjects but they created growing hostility abroad because of their trading restrictions.

In 1779 the British East India Company acquired a monopoly on opium. The drug's popularity in China depleted imperial reserves and the trade was banned. In 1839 the Chinese authorities seized a consignment of more than 20,000 chests of opium and destroyed it. This triggered the first Opium War, which ended with the Treaty of Nanjing in 1842 granting new privileges to foreign merchants. A second Opium War (1856–60) led to temporary British and French occupation of Beijing in 1860.

The Qing had domestic problems, too. Rural poverty and an exploding population – from 100 million in 1650 to 420 million in 1850 – caused the Taiping Rebellion of 1851–64 in which at least 25

The Boxer Rebellion Between 1898 and 1900 groups of ardent Chinese nationalists known as the Boxers attacked European missionaries and killed thousands of Chinese converts in northern China. The port city of Tianjin, near Beijing, was captured briefly before the rebellion was crushed by an international force.

its golden age: cities grew, and trade with the outside world expanded as junks carried cargoes of tea, silk and porcelain to the East Indies, Africa and India.

In 1126 the Ruzhen (Juchen), a seminomadic people from Manchuria, conquered northern China and set up their own dynasty there. The Song retreated and established a new capital at the port of Hangzhou, south of Shanghai. Early in the 13th century the Ruzhen were attacked by the Mongol leader Genghis Khan (c. 1155–1227), and were finally routed in 1234. Hangzhou also fell to the Mongols in 1276.

Under the Ming (or "Brilliant") dynasty (1368–1644), which followed the overthrow of the Mongols, the Great Wall was repaired and extended to its present

STATE

President

PARTY

General secretary

Standing committee

Standing committee

National people's congress

State council

Central military commission

Politburo

Central committee

National party congress

appoints
appoints

appoints

appoints

responsible to

Provincial people's congress

Provincial people's government

Provincial party committee

Provincial party congress

appoints

appoints

responsible to

County people's congress

County people's government

County party committee

County party congress

appoints

appoints

ELECTORATE

elects

responsible to

Township people's congress

Township people's government

appoints

elects

responsible to

Primary level party congress

China's government (*right*) State and Party organizations mirror each other at all levels. Legislative power is in the hands of the National People's Congress (NPC), which is elected via the provincial and county assemblies for a five-year term. However, as the NPC meets only two or three weeks a year, its day-to-day business is carried out by a 155-member standing committee which it appoints. The NPC also chooses the president, the prime minister and, on the prime minister's advice, the members of the State Council. Its function is to execute the laws enacted by the NPC. The State Council also has its own standing committee, made up of the prime minister, 12 or more vice-premiers and a secretary-general. A parallel organizational structure operates within the Communist Party. The 348 members of its central committee elect six of their number to a politburo, which lays down guidelines for every government action.

Yihe Quan (Righteous and Harmonious Fists – hence their Western name "Boxers") began murdering Chinese Christians. In 1900, they laid siege to foreign legations in Beijing. The affected governments responded by sending a massive international army.

Disillusionment with the Qing was growing. In 1911 a minor army mutiny in Wuchang (Hubei province) triggered widespread opposition and on 1 January 1912, Sun Yat-sen (1866–1925), a Western-educated leader of the budding Nationalist Party, was declared president of the new Republic of China, with its capital at Nanjing.

The republic did not endure, however, and China slid into civil war between regional warlords, whose armies caused immense damage and millions of casualties. In 1919, following World War I, there was a great upsurge of nationalist feeling against the presence of foreign powers within China. Marxism gained in popularity, and the Chinese Communist Party (CCP) was founded in 1921.

The Nationalist Party allied with the communists to confront the warlords in the north and reunite the country. But in 1927 Chiang Kai-shek (1887–1975), leader of the nationalists, turned against his communist allies and massacred many of them. In October 1934, some 100,000 communists began their epic 9,500 km (5,903 mi) retreat northward that has become known as the "Long March". The survivors, numbering fewer than 8,000, established their new Red Army headquarters in the north-central province of Shaanxi under the leadership of Mao

Zedong (1893–1976). Meanwhile the Japanese had invaded Manchuria in 1931, and in 1937 they overran northern and central China. The nationalists were compelled to form a united front with the communists against Japan.

With the end of World War II and the defeat of Japan in 1945, the struggle for power soon developed into an open civil war. The nationalists were completely defeated and the People's Republic of China was proclaimed on 1 October 1949 under Mao Zedong.

China was crippled by decades of conflict, and lacked modern industrial resources. Mao Zedong planned to develop the country's economy by increasing agricultural production and using the surplus for industrial growth. New political structures were modeled on those of the Soviet Union, and Soviet loans boosted initial growth. Tibet – which had expelled Manchu pretenders to Suzerainty in 1912 – was occupied in 1950 by the People's Liberation Army.

In 1958, Mao launched the "Great Leap Forward" – an attempt to speed up economic growth by mobilizing the rural workforce on a massive scale. But the program was a disastrous failure and had to be abandoned. Then in 1960 the Soviet Union cut off military and economic aid and openly criticized Mao.

To regain the ideological initiative, Mao, in 1966, launched the "Great Proletarian Cultural Revolution". It led to anarchy: intellectuals and artists became victims of inquisitions by youthful Red Guards; many were tortured and humiliated. A wave of terror led to the deaths of at least

400,000 Chinese. This, along with fears of a Soviet invasion in 1968, caused the leadership to call in the army to restore order.

China was admitted to the United Nations (UN) in 1971, and was visited by the United States' president Richard Nixon the following year. Mao's death, at the age of 82, on 9 September 1976 was followed by a bitter leadership struggle, which was won by the moderates. After 1982, a new constitution was drawn up under Deng Xiaoping (1904–97).

Deng Xiaoping's resignation from the Party's central committee in 1987 was followed by an upsurge of agitation for democracy. When, in May 1989, the army was called in to disperse a peaceful sit-in in Beijing's Tiananmen Square, 200 were killed according to official figures. Since then, China has made progress on economic liberalization while imposing further political repression, including a campaign against religious sects and the arrest and imprisonment of opposition leaders.

Government

Since 1949 China has been a single-party state, in which the Chinese Communist Party controls most aspects of daily life through mass-membership organizations such as trade unions, youth leagues and professional associations. The 1982 constitution vested legislative power in the 3,000-member National People's Congress (NPC), but in effect more power resides with the politburo of the Central Committee of the Communist Party.

China has a highly complex regional and local government structure. The primary divisions are the 22 provinces,

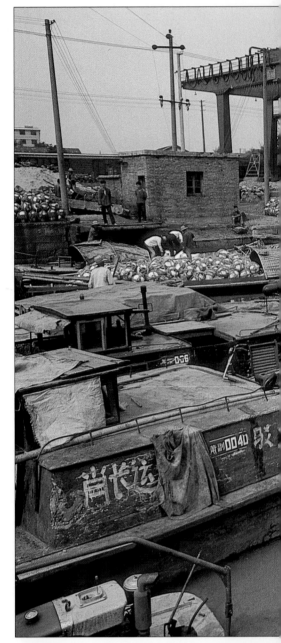

Geese by the barrowload (*left*) in Chengdu, the capital of Sichuan province. Customers buy the geese live and take them home either to cook or to keep for their eggs. Sichuan, with fertile soil and a mild climate, is almost self-sufficient in food production.

The world's longest canal (*right*) China's Grand Canal system extends from Beijing to Hangzhou, through five major river basins and the most populated of China's eastern cities. It was begun in the 5th century BC, but the main sections were built in the 7th century AD and it was completed in the late 13th century. It was built to connect the north with the rice-growing Yangtze valley. Renovated in the 1960s, it is now a major transportation route.

Elbow grease (*below right*) Since the economic reforms of 1978, light industry – including the manufacture of bicycles (shown here), radios, sewing machines, cotton yarn and cameras – has grown rapidly. But despite attempts at modernization, many factories such as this one at Changshu on the Grand Canal in Jiangsu province retain old-fashioned production methods.

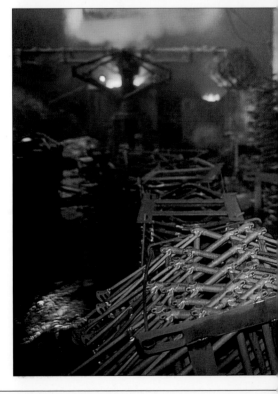

and the four city municipalities of Beijing, Tianjin, Shanghai and Chongqing. These areas are divided in turn into prefectures, then into counties and various smaller units. There are also five nominally "autonomous regions" (Tibet and Xinjiang in the west, Inner Mongolia and Ningxia in the north, and Guangxi in the south).

People and culture

Although Han Chinese form some 92 percent of the population, 55 different minority groups and nationalities are recognized in the country. The largest minority is the 13-million strong Chuang (Zhuang) people of Guangxi (bordering Vietnam). Others include the Tibetans of Tibet, the Turkic Uighurs and Kazakhs of Xinjiang, the Mongols of Inner Mongolia, and the Hui people of Ningxia in the upper Huang valley.

Given the large number of minorities, there is a bewildering array of languages and dialects. Even Chinese is spoken in many "dialects", which constitute major languages in their own right. Mandarin is the dominant tongue, and is understood throughout China. All forms of Chinese share a single written language.

Since the creation of the People's Republic, religious practice has been discouraged by the Communist Party. However, Buddhism is still very strong in Tibet, and Islam in Xinjiang and Ningxia

has also resisted communist secularism.

Traditional Chinese religions or philosophies, notably Taoism and Confucianism, are still openly practiced by about one-fifth of the population, and the small Christian minority includes Protestants and some 3 million Chinese Catholics.

Chinese writing – an art in itself in the form of calligraphy – has provided the primary vehicle for a rich and diverse heritage of painting, poetry, drama and music. Chinese porcelain has for many centuries been the envy of the world, and Chinese cuisine is also much admired and imitated.

ECONOMY

Since the late 1970s the Chinese economy has moved steadily away from the Soviet command economy model toward a market economy, where privately owned produce can be sold for private profit (a percentage tax is paid to the state). Most medium to large industries, however, are still state-owned or cooperative.

Agriculture

Only one-tenth of China's land area is suitable for cultivation, and half of this is given over to paddy fields or irrigated for other crops. Rice, the main crop, is grown mostly in the southern provinces, and China is among the world's leading prowest and around Hainan Island, and plans to build a distribution grid by 2010.

Most of China's electricity is generated from coal. Most hydroelectric production is in the southwest; however, China, is building the world's largest dam on the Yangtze Jiang (Yangtze River). There are two nuclear power plants and three more are being built. There is also heavy investment – some foreign – in renewable sources such as wind and solar energy.

By 2000 China had moved out of heavy industry and become a leading producer of fertilizers, textiles and clothing as well as electronics and transport equipment.

China's most important trading partners are Japan and the United States. Manufactured goods account for around 70 percent of exports, while imports consist mainly of industrial supplies and capital goods.

Transportation and communications

Every province and "autonomous" region has railroad connections, and the total road system now amounts to about 1 million km (600,000 mi). However, most roads outside the cities are unsurfaced and most Chinese still rely on bicycles.

Waterways carry more than one-third of all internal freight. The nine principal ports and other depots around the coast serve an ocean-going merchant. fleet of more than 400 vessels. There are many international airports, the most important being Beijing, Shanghai, Guangzhou and Hong Kong.

The press and broadcasting media are entirely under state control. There are hundreds of radio and television stations. The 1990s saw a boom in telecommunications and by 2000 there was an estimated 10 million Internet users.

Health and education

China has made significant progress in improving healthcare. Medical teams are regularly sent to rural areas, where the so-called "barefoot doctors" are given short but intensive training for work in the country. As a result, serious infectious diseases such as cholera and typhoid have been largely eliminated. A variety of contributory welfare plans, including medical insurance, unemployment insurance and a national pension scheme, have been introduced by the government to support an ageing population.

All children are officially required to attend six years of primary schooling and three years of secondary schooling, though secondary education is sometimes unavailable outside the cities. All university fees are paid by the state.

ducers. Almost one-third of China is pasture, supporting a variety of livestock raised mainly for meat, though milk is taken from sheep, goats and water buffalo as well as from cattle.

Fish are an important part of the Chinese diet. About two-thirds of the catch is obtained from the neighboring seas. The rest comes from freshwater fisheries, mostly in the Yangtze and Zhu river deltas. There are still extensive forests in the Greater and Lesser Hinggan ranges, but nevertheless China has to import timber to meet its needs.

Industry

China's vast mineral wealth includes the largest reserves of antimony in the world. China is both the largest producer (1.2 billion tons in 1998) and consumer of coal in the world. Petroleum reserves are considerable, especially in the north and west, but demand outstrips supply. China has large reserves of natural gas in the north,

Singapore

REPUBLIC OF SINGAPORE

S INGAPORE IS ONE OF THE SMALLEST, MOST densely populated and most prosperous countries in the world. Most of the population of over 4 million people live on Singapore Island; the rest inhabit 54 neighboring islets.

Singapore Island has a low but undulating landscape with a hillier district in the center. The climate is uniformly hot and humid throughout the year, with high rainfall. Indigenous mammals include the crab-eating macaque (a type of monkey).

Modern Singapore was founded in 1819 by Sir Thomas Stamford Raffles (1781–1826) of the British East India Company, and later became a British crown colony.

In 1963 it was incorporated into the Federation of Malaysia, but two years later left the federation to become an independent republic.

The head of state is a directly-elected president, and most legislative power rests with the 87-member parliament, 81 elected every five years. However, the People's Action Party, is effectively the only political party. The three main ethnic groups – Chinese, Malay and Indian – speak a variety of languages, including English, and practice various religions.

Agriculture and fisheries are relatively unimportant; many staple foods, including fish, have to be imported. The prosperous and expanding manufacturing sector is based on imported raw materials. Major industries include petrochemicals, textiles and shipbuilding. Financial services, banking and tourism are also major money earners.

There are road and rail links to neighboring parts of Malaysia, and a busy international airport. Healthcare, welfare and education are all of a high standard.

Singapore's gracious colonial buildings are dwarfed by huge international hotels, tax-free shopping complexes and office blocks in the commercial heart of the city. Already a busy trading center in the 14th century, modern Singapore is an economic colossus.

NATIONAL DATA – SINGAPORE

Land area	683 sq km (264 sq mi)		
Climate		Temperatures	Annual
	Altitude m (ft)	January °C(°F)	July °C(°F)	precipitation mm (in)
Singapore	10 (33)	27 (81)	28 (82)	2,315 (91.1)

Major physical features	largest island: Singapore 541 sq km (209 sq mi); highest point: Timah Hill 162 m (531 ft)
Population	(2006 est.) 4,492,150
Form of government	multiparty republic with one legislative house
Armed forces	army 50,000; navy 4,000; air force 13,500
Capital city	Singapore (3,654,103)
Official languages	Chinese, Malay, Tamil, English
Ethnic composition	Chinese 76.8%; Malay 13.9%; Indian 7.9%; other 1.4%
Religious affiliations	Buddhist 42.5%; Muslim 14.9%; Taoist 8.5%; Hindu 4%; Catholic 4.8%; other Christian 9.8%; other 0.7%; none 14.8%
Currency	1 Singapore dollar (SGD) = 100 cents
Gross domestic product	(2006) U.S. $138.6 billion
Gross domestic product per capita	(2006) U.S. $30,900
Life expectancy at birth	male 79.13 yr; female 84.49 yr
Major resources	fisheries, deepwater ports, shipbuilding, financial services, tourism

Brunei

SULTANATE OF BRUNEI, ABODE OF PEACE

B RUNEI – AN "ABODE OF PEACE" ACCORDING to its unique official title – is an independent sultanate on the northwest coast of Borneo, surrounded and divided in two by the Malaysian state of Sarawak.

GEOGRAPHY

The two parts of Brunei are separated by a few kilometers of coastline where the Limbang river enters Brunei Bay. Both share a landscape of hills and valleys bordering a narrow and often swampy coastal plain. The smaller eastern enclave is more rugged. The humid tropical climate brings heavy monsoon rains between November and March.

Most of Brunei is covered in dense tropical forest, rich in hardwoods. Inaccessibility makes it a haven for wildlife such as monkeys, apes and birds.

SOCIETY

Brunei is known to have had trading links with China since before the 6th century AD. The Portuguese navigator Ferdinand Magellan (1480–1521) was the first European visitor in 1521. At that time the Sultanate of Brunei effectively controlled the whole of Borneo and part of the Sulu Islands and the Philippines to the northeast. In 1839 Sarawak was ceded to the "White Raja" James Brooke (1803–68) and his descendants, while other areas were lost to the British North Borneo Company; Brunei finally became a British protectorate in 1888. In 1929 the first petroleum reserves were discovered.

Following Japanese occupation during World War II Brunei reverted to British rule. In 1959 it adopted a constitutional government and in 1984 achieved full independence within the Commonwealth. Since then, all effective political power has been vested in the sultan.

With over 340,000 people occupying 5,765 sq km (2,226 sq mi), Brunei has the lowest and least dense population of any Southeast Asian country. Malay-descended Muslims make up the majority of the population, followed by Chinese and Indians. The small indigenous population includes Ibans, Dyaks and others.

ECONOMY

Petroleum is the mainstay of the Brunei economy, giving the country one of the highest per-capita incomes in Southeast Asia. There are also natural gas reserves. Other income derives chiefly from small-scale agriculture (pepper and rice), and forestry (cork and rubber).

Main transportation routes are along the coast and rivers, apart from a few roads along the coast. Excellent healthcare is available free of charge, even in isolated locations by means of a "flying doctor" service. Free education also extends to remote areas.

NATIONAL DATA – BRUNEI				
Land area 5,270 sq km (2,035 sq mi)				
Climate		Temperatures		Annual
	Altitude m (ft)	January °C(°F)	July °C(°F)	precipitation mm (in)
Bandar Seri Begawan	27 (80)	27 (81)	27 (81)	2,913 (114.6)
Major physical features highest point: Pagon Peak 1,850 km (6,070 ft)				
Population (2006 est.) 379,444				
Form of government nonparty constitutional monarchy with one advisory body				
Armed forces army 4,900; navy 1,000; air force 1,100				
Capital city Bandar Seri Begawan (69,984)				
Official language Malay				
Ethnic composition Malay 67%; Chinese 15%; indigenous 6%; other 12%				
Official religion Islam				
Religious affiliations Muslim 67%; Buddhist 13%; Christian 10%; traditional beliefs and other 10%				
Currency 1 Bruneian dollar (BND) = 100 cents				
Gross domestic product (2003) U.S. $6.842 billion				
Gross domestic product per capita (2003) U.S. $23,600				
Life expectancy at birth male 72.57 yr; female 77.59 yr				
Major resources petroleum, natural gas, rubber, timber, bananas, cassava, coffee, cork, fisheries, rice, pepper				

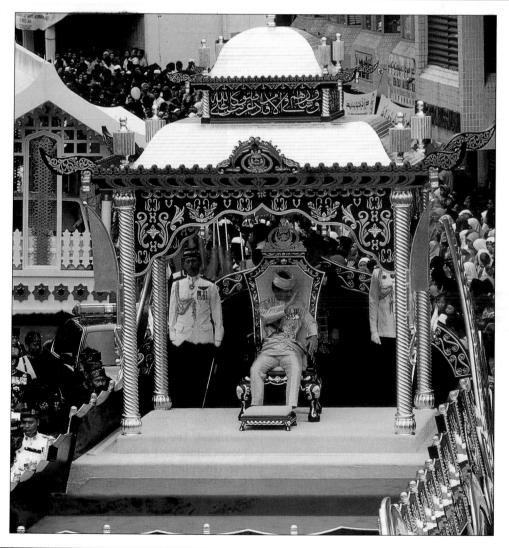

Carried in splendor (*right*) Brunei's absolute monarch Hassanal Bolkiah waves to onlookers from his ornamental float during a procession through the streets of the capital. The sultan's wealth is derived from Brunei's petroleum and natural gas resources.

(5 ft) long. Japan is also rich in insect life – much of it a seasonal problem to the population – such as mosquitoes, cockroaches and fireflies.

Bird life too is abundant, with some 150 different species of songbird. The many sea birds include albatrosses, shearwaters and cormorants. The waters to the east of Cape Inubo in the southeast support some of the richest fishing grounds in the world, with cod, salmon and numerous crustaceans.

SOCIETY

Until the mid 19th century foreigners risked death if they tried to enter Japan, and any Japanese who left were forbidden to return. Following the enforced end to isolationism in 1853 the Japanese embarked on rapid social and economic change, freely adapting Western technologies. Since World War II Japanese society has achieved overwhelming industrial success within one generation.

History

The legendary history of Japan begins in 660 BC, with the accession of the first emperor, Jimmu. He was probably the ruler of the Yamato, a people that gave their name to a province in southwest-central Honshu.

By the 5th century AD, the Japanese had occupied the southern part of Korea and made their first contact with China. From this highly developed civilization came Japan's earliest forms of writing.

By the mid 6th century the Yamato court had lost its grip on Korea, and from there Buddhism, a new cultural influence, reached Japan. The traditional national religion was Shinto, or the Way of the

The first Shogun Minamoto Yoritomo (1147–99) founded the system of military dictatorship by which Japan was ruled for 700 years. He is remembered less as a general than as a politician and administrator.

Gods. Buddhism flourished toward the end of the 6th century under the government of the statesman and scholar Prince Shotoku (574–622).

In 794, Heian-kyo (modern Kyoto in west-central Honshu) became the imperial capital, and continued in that role until 1868. As Japan's isolation grew, the emperor became increasingly subordinate to powerful regent families, especially the Fujiwara, who in effect held power throughout the 11th century. This period was also significant for the rise of the *samurai* – a warrior class that became the only effective focus of local government. In 1192 Minamoto Yoritomo (1147–99) became Japan's first shogun, or military dictator, with his power base at Kamakura (south of present-day Tokyo).

The system of military feudalism in Japan was challenged by two attempted Mongol invasions from Korea, in 1274 and 1281. Both invasion fleets were scattered by typhoons – the so-called *Kamikaze*, or "Divine Wind" – but the shogunate's elaborate preparations for war led to financial collapse. In 1333 the shogunate, or *bakufu*, was overthrown in favor of a new emperor. A new *bakufu*

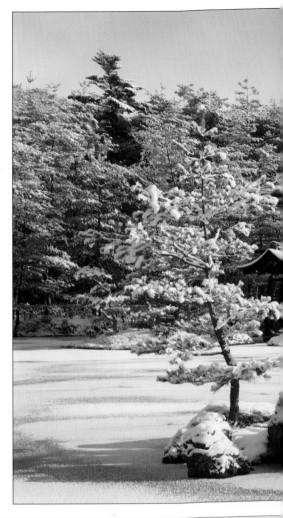

The code of the Samurai

In films such as "Throne of Blood" (1957), the internationally acclaimed director Akiro Kurasawa (b. 1910) gave a vivid picture of Japan's ancient warrior class – the *samurai*. His images combined Hollywood-style spectacle with the emotional intensity of traditional Japanese theater. But the role of the *samurai* in Japanese culture is not restricted to the theater of heroic warfare. Their code of conduct embodies a way of life admired by all walks of Japanese society.

Originally the term *samurai* referred to a small group of aristocratic provincial warriors; from the 12th century it included the entire warrior class then emerging within the Japanese feudal system. In the chaos of frequent power struggles among ruling aristocrats, the *samurai* quickly became a powerful political force. They developed a highly disciplined culture of unconditional loyalty based on an unwritten code of conduct, gradually enshrined as

The fearsome warrior class A 19th-century woodblock print shows a *samurai* in full costume, with the sword that *samurai* alone were privileged to carry. *Samurai* as a social class included women, though they rarely fought in battle.

The arrival of the Portuguese
(*above*) in the mid 16th century.
Jesuit missionaries from Iberia
converted many powerful feudal
lords interested in both Catholicism
and making trade and cultural links
with the West.

**The Kinkakuji or Temple of the
Golden Pavilion** (*left*) at Kyoto,
capital of Japan from 794 to 1868
and one of Japan's most beautiful
cities. Built in the late 14th century
as part of the shogun's villa, the
temple was destroyed by fire in 1950
and rebuilt in 1955.

was established in the Muromachi district
of Kyoto.

In 1542 Portuguese traders from Macao
became the first Europeans to make contact with Japan, and their arrival brought
the feudal lords a powerful new weapon,
the musket. Still more influential was
the introduction of Christianity by the
Spanish Jesuit Francis Xavier (1506–52),
followed by a stream of missionaries.
There was savage fighting as the struggle
for supremacy intensified, until Oda
Nobunaga (1534–82) entered the capital,
Kyoto, in 1548 and set the country on
the path toward unity. On Oda's death,
his able general Toyotomi Hideyoshi
(1536–98) took control, and in 1587 he
made Nagasaki – a port on the far western
side of Kyushu – an imperial city. Soon
Nagasaki attracted regular Spanish, Dutch
and Portuguese trading ships.

Hideyoshi's successor Tokugawa Ieyasu (1543–1616) continued to unite the
country and went on to establish the
Tokugawa shogunate, which lasted from
1603 until 1867. At first Ieyasu favored
foreign trade and was open to Western
influence, but later he became afraid of
the spread of Christianity. After 1636 all
contact with European traders and missionaries was forbidden, except at
Nagasaki, and Japanese Christians were
persecuted.

As the *bakufu* became more repressive
and more corrupt, foreign pressure and
internal discontent eventually combined
to break Japan's isolation. Two major
uprisings in 1836 and 1837 were put
down, while European and American
ships continued to press for a renewal of
foreign trade. Finally, in July 1853, a
squadron of United States' warships
under Commodore Matthew Calbraith
Perry (1794–1858) arrived and anchored in
Tokyo Bay. There they stayed until the
Japanese agreed to a trade and diplomatic
treaty with the United States. Further

bushido – "the way of the warrior." Zen
Buddhism, imported from China in
the 15th and 16th centuries, influenced
bushido, taming some of the more
violent expressions of the code such as
a warrior killing himself after his lord's
death, or carrying out vendettas. In the
17th century, under the Tokugawa
shogunate (1603–1867), *samurai* became the highest of four castes within
a rigid class system.

The true *samurai* was most renowned
for awe-inspiring courage that stem-
med from deep stoicism and fierce
loyalty to a leader. The *samurai* valued
his honor above his life; *samurai* who
neglected to commit ritual suicide (*seppuku*) after defeat or dishonor were cast
out as *ronin* – masterless *samurai* who
became mercenaries or bandits and
were objects of contempt and loathing.
Family devotion also played an important part in the *samurai*'s austere life,
but as warriors their chief duty was to
their overlord.

In the Tokugawa era, *samurai* still
wore two swords to signify their rank,
but their duties shifted from waging
war to serving as civil administrators

and setting a good moral example to
the lower castes. With the growth of
cities and the rise of the merchant
class, the *samurai* became impoverished. In 1868 the feudal system was
abolished and *samurai* lost their
privileged status. *Bushido* was adopted
as a code of behavior for the whole of
Japanese society, and it formed the
basis of the cult of emperor worship
that was taught until 1945. In World
War II *bushido* inspired Japanese soldiers to fight with often deliberately
suicidal ferocity and to prefer death to
the shame of surrender. However,
these same values led the Japanese to
show a callous disregard for the lives of
Allied servicemen taken prisoner by
them during the war.

Although ritual suicide is now rare
(the writer Yukio Mishima committed
seppuku in 1970 after urging the re-
armament of Japan, banned by the
post-World-War-II constitution), the
old virtues of stoicism, loyalty and
obedience have flourished. They can
be observed daily in the classrooms,
workplaces, and households of contemporary Japan.

Government

Australia has a three-tier system of government, with a federal parliament, six state parliaments and about 900 local authorities.

The federal parliament in Canberra is divided into an upper house (the Senate) and a lower houe (the House of Representatives). The cabinet, headed by the prime minister, includes members drawn from both chambers. There are 77 senators, who serve a six-year term and 148 seats in the House of Representatives which are elected for a three-year term.

Every adult aged 18 or over is required to vote in parliamentary elections. The Northern Territory has been self-governing since 1978; the Australian Capital Territory, which includes Canberra, became self-governing in 1989. Australia is a fully independent nation, but retains strong traditional links with the Commonwealth and the UK. In 1999, a referendum to change Australia's status, from a commonwealth headed by the British monarch to a republic, was defeated.

People

Most of the 19 million Australians are of European origin, and many are the descendants of British settlers. The Aboriginals account for 1 percent of the population; 7 percent are either Asian-born or of Asian descent.

After World War II migrants from mainland Europe – mainly whites – were encouraged to settle in Australia. In 1973 the Australian government formally declared an end to the discriminatory "White Australia" immigration policy.

Among Aboriginals continuing poverty, poor health and lack of education have reinforced their mistrust of a white-dominated society. Since the late 1970s, federal governments have given some Aborigines freehold to their land which is of great cultural importance to them. In 1997 a report into the removal of 100,000 Aboriginal children from their families was published, after which the government expressed its regret at the practice of forcing children into white foster homes (discontinued in the late 1960s) and held a "National Sorry Day". English is almost universal in Australia; all but a few of the 260 Aboriginal languages are dead. Christianity is the dominant religion.

ECONOMY

As a new nation Australia owed its early wealth to its extensive agricultural and mineral resources. Since World War II

The Camel train (*above*), nicknamed the "Ghan", is named after the Afghan camels and their drivers who originally explored the Australian desert. Today the federally administered Trans-Australian railway runs from Port Augusta, north of Adelaide, through the Simpson Desert to Alice Springs in the Northern Territory. In all, Australia has more than 40,000 km (24,000 mi) of railroad track.

Herefords at auction in Queensland (*right*) Australia is a major exporter of beef (principally to the United States), and the state of Queensland alone provides grazing for almost half of the country's 26 million cattle. Herefords are the most popular breed in the central and northern part of the state, but agricultural research continues to find new crossbreeds suited to Australian conditions.

Sheep shearing (*below*) is usually a contract job. A gang of about 18 men travel from station to station during the season, averaging 95 sheep per shearer per day. Among others, the gang includes a cook, a wool-classer and numerous "roustabouts" (all-purpose workers) to clean up.

Australia has tried to build up a manufacturing economy to process its minerals.

Agriculture

Much of the farmland is used as pasture and is unsuitable for crops. Water is often in short supply, and soils are generally low in fertility. Production has risen by more than 150 percent since 1950 because of the introduction of new farming methods and mechanization.

A relatively small area of Australia is under permanent cultivation. Sugar cane, mainly from coastal Queensland, and wheat from the southeast and southwest are the major crops. The wide variations in climate allow the cultivation of many kinds of fruit, and wine exports, mostly from vineyards in South Australia and New South Wales, are on the increase.

Almost one-third of the continent is

given over to sheep farming, making Australia the world's leading wool producer. Cattle farming needs the taller pastures found in higher-rainfall areas. Beef cattle are mostly found in Queensland and New South Wales; dairy farming is concentrated in coastal New South Wales and southern Victoria.

Australia has extensive native commercial forests, mostly in the eastern highlands, Tasmania and Western Australia. Japan is the largest single customer for timber. Western Australia and Tasmania are the leading fishing states.

Industry

Australia has some of the world's biggest deposits of iron ore and bauxite and is also a major producer of uranium. New South Wales and Queensland have large reserves of coal, and petroleum and natural gas are found off the coast of Victoria and in northeastern South Australia. Iron ore and coal are the chief mineral exports worldwide. Australia is also a major supplier of energy to Japan.

Most of Australia's own energy is from coal or natural gas. Tasmania is almost totally supplied by hydroelectric power. Australia has one nuclear reactor plant, but no nuclear power stations.

Since the late 1970s manufacturing industries have contracted in the face of competition from, in particular, Southeast Asia. Today the most important industries include iron and steel; motor vehicle production; petroleum refining and petrochemical products; textiles and clothing manufacturing; electrical and electronic equipment; and agricultural machinery. Tourism is also a thriving industry, particularly on the Pacific coast.

Transportation and communications

Some four-fifths of domestic freight is carried on the roads, and there is heavy investment in road building. However, infrastructure in cities has not kept pace with the increase in traffic. Around 85 percent of people live in or near the country's six major cities (Sydney, Melbourne, Brisbane, Adelaide, Perth and Canberra).

The main function of Australia's railroads, apart from passenger services for city commuters, is the bulk haulage of minerals, grain and long-distance freight. There are about 70 commercially significant ports, most of them serving the state capitals and industrial and mining centers. Australia's scheduled domestic airlines operate from some 450 airfields; isolated areas rely on the use of small aircraft. The national carrier Qantas offers

Sailing in Sydney Harbor Yachting is a favored pastime in Australia, with over 150 different types of sailboats registered in local waters. Entrepreneur Alan Bond's victory in the 1983 Americas Cup with his yacht *Australia II* caused nationwide celebrations

a passenger service to 24 countries.

Australia has a vigorous and long-established free press, but the concept of a national newspaper is relatively new. The Australian Broadcasting Corporation is independent but financed by government grants; other radio and television stations are run commercially. An ethnic radio and television network is funded by government grants and advertising.

Health and welfare

Healthcare services are financed principally by the government. Australia's universal health insurance scheme, Medicare, entitles patients to reclaim 85 percent of the schedule fee of most medical expenses. There are separate community health programs for Aboriginals, funded by federal government but run and controlled by Aboriginals themselves. The famous Royal Flying Doctor Service, begun in 1928, serves isolated areas.

Australia, like New Zealand, was a pioneer of social security, introducing state retirement pensions as early as 1909. Today's benefits are much wider-ranging.

Education

Education is free up to and including secondary level. School attendance is compulsory up to the age of 15 (16 in Tasmania). About two-thirds of children are educated in state schools. Many younger children attend preschool centers organized at local level. Most secondary schools are coeducational high schools.

New Zealand

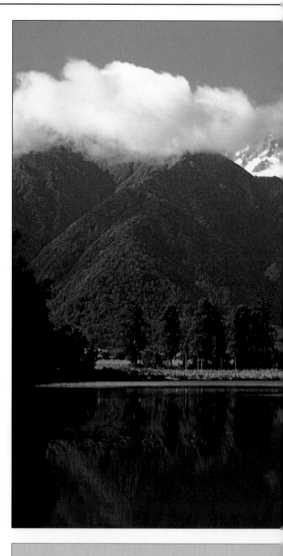

Mt Cook and Mt Tasman, New Zealand's highest mountains, dominate the Southern Alps which were named by Captain Cook in 1770. This spectacular mix of snow-capped mountains and deep glassy lakes is a hallmark of South Island scenery.

N EW ZEALAND IS AN ISLAND STATE IN THE southwestern Pacific Ocean about 1,600 km (1,000 mi) southeast of Australia. It also has a number of small overseas territories in the Pacific, notably Tokelau, Niue and the Cook Islands.

GEOGRAPHY

The country consists mainly of two islands – the North Island and the South Island – divided by the narrow Cook Strait.

The land

The South Island is dominated by the great fold-mountain belt of the Southern Alps. The highest peaks are in the center, reaching 3,754 m (12,349 ft) on Mount Cook; the Tasman Glacier, the largest of New Zealand's many glaciers, flows down its eastern slopes. In the southwest, where the mountains meet the sea, the flooded valleys have formed steep fjords, to create Fiordland, New Zealand's largest national park. The only lowland areas are a few alluvial plains in the east.

The mountains of the North Island are generally much lower and much more fragmented. The central and eastern uplands are an area of considerable geothermal activity, with geysers, hot springs and several volcanoes, two of which are constantly active. The hills slopes, river terraces and scattered lowlands of the North Island are some of the country's best farmland. The capital, Wellington, and the northern city of Auckland are both built around large natural harbors.

Climate

New Zealand generally enjoys a temperate maritime climate with regular rainfall all year round. Summers are warm except in the southern mountains, with plenty of sunshine; winters are mild, especially in the North Island, and snow is common only in mountainous areas. In the South Island the western mountains, which face the prevailing winds, receive heavy rainfall while sheltered eastern areas such as the Canterbury Plains are sometimes unusually dry.

Plants and wildlife

Much of the country is still covered by evergreen rainforest. Nine-tenths of the indigenous plants are unique to the islands, including kauri pines, southern beeches, and certain varieties of ferns. Some other tree species have been introduced. At one time the only medium to large indigenous wildlife, apart from birds, comprised two species of bats and some 30 reptiles. Among the few native reptiles today is the tuatara, sole survivor of a group that died out elsewhere 60 million years ago.

In the absence of mammals and other threatening native predators, some remarkable flightless birds developed on the islands such as the kiwi and the kakapo – the world's largest parrot. Another large parrot is the kea which lives in the snow-covered Southern Alps. Today the islands are populated by a range of species brought by Maori and European settlers. They include cats, dogs, rats and rabbits. Introduced predators have threatened the survival of native species. Conservationists have ensured that the native saddleback bird is established on small predator-free islands.

SOCIETY

In the 19th and 20th centuries British dominion had a powerful influence on New Zealand society. Today, however, New Zealanders are keen to demonstrate their separate identity.

History

New Zealand was first discovered and settled between 800 and 1000 AD by the people now known as the Maori who came from the Polynesian islands. The Maori were farmers and settled mainly on the warmer North Island. The first European to arrive was the Dutch explorer Abel Janszoon Tasman (c. 1603–59),

Tokelau

At first sight the Tokelau island group, some 500 km (300 mi) north of Western Samoa, looks green and lush. Dense vegetation covers the three coral atolls – Fakaofo, Nukunonu and Atafu – and each atoll is made up of many islets, or *motu*. These curved strips of coral sand are never more than 5 km (3 mi) long, or higher than 5m (16 ft) above sea level. Yet, despite the growth of abundant coconut palms, pandanus and other trees and shrubs, there is no fresh water on the islands; the soil is too porous to retain it. However, that has not prevented people from living here since the first settlers arrived from Samoa, probably on drift voyages.

Like many Pacific islanders, the people of Tokelau have led a precarious existence. A raid by Peruvian slavers in 1863 was followed by an outbreak of disease that reduced the population to about 200. In 1889 a British protectorate was established and in 1948 the group became an island territory of New Zealand.

In the past, many islanders emigrated to Samoa, and some more recently to New Zealand, but the present population of Tokelau is relatively stable. Tokelau has a few exports – copra and traditional craft goods – and some revenue comes from stamps and coins. Most people practice subsistence agriculture and the islands receive aid from New Zealand. Even

Tokelau dancers The tiny island territory of Tokelau is made up of three coral atolls covering a mere 10 sq km (3.9 sq mi). Its 1,600 inhabitants are Polynesian with close linguistic and cultural ties to nearby Western Samoa.

today, the only water comes from heavy seasonal rains, which the people collect in tanks or in traditional rain catchments – hollowed-out trunks of coconut palms.

NATIONAL DATA – NEW ZEALAND

Land area 268,021 sq km (103,483 sq mi)

Climate	Altitude m (ft)	Temperatures January °C(°F)	July °C(°F)	Annual precipitation mm (in)
Wellington	126 (413)	18 (64)	9 (48)	1,018 (40)

Major physical features highest point: Mount Cook 3,764 m (12,349 ft); largest lake: Lake Taupo 606 sq km (234 sq mi)

Population (2006 est.) 4,177,937

Form of government multiparty constitutional monarchy with one legislative house

Armed forces army 4,430; navy 1,980; air force 2,250

Largest cities Auckland (1,442,101); Wellington (capital – 448,959); Christchurch (375,145); Hamilton (158,402); Napier (53,658); Dunedin (116,240); Tauranga (117,594)

Official languages English, Maori

Ethnic composition European 69.8%; Maori 7.9%; Asian 5.7%; Pacific islander 4.4%; other 0.5%; mixed 7.8%; unspecified 3.8%

Religious affiliations Anglican 14.9%; Roman Catholic 12.4%; Presbyterian 10.9%; Methodist 2.9%; Pentecostal 1.7%; Baptist 1.3%; other Christian 9.4%; other 3.3%; unspecified 17.2%; none 26%

Currency 1 New Zealand dollar (NZD) = 100 cents

Gross domestic product (2006) U.S. $106 billion

Gross domestic product per capita (2006) U.S. $26,000

Life expectancy at birth male 75.82 yr; female 81.93 yr

Major resources natural gas, iron ore, sand, coal, timber, hydropower, gold, limestone, cattle, cereals, fish, fruit, vegetables, sheep, tourism followed in 1769–70 by the British Captain James Cook (1728–79).

Later whalers and traders from Australia established settlements on the islands, and then came British and French missionaries. In 1840 the Maori chiefs ceded sovereignty to the British crown in return for protection and land rights. The following year New Zealand became a crown colony. The discovery of gold in the 1860s attracted many thousands more immigrants. In the North Island, farmer settlers came into increasing conflict with the Maori over land rights. Rebellion escalated into war, until Maori resistance was finally broken in 1870.

Europeans born in New Zealand were gaining a greater sense of nationhood, and in 1907 the colony was granted dominion status within the British Empire. New Zealand fought alongside Britain in World War I, and did not escape the worldwide depressions of the 1920s and 1930s. During World War II New Zealanders served in the Pacific, where their presence forged closer links with other Pacific nations.

Although a member of the Commonwealth in the postwar period, New Zealand sought new alliances with the United States and some Southeast Asian countries. During the 1970s and 1980s, an energy crisis and loss of trade with Britain caused economic recession. In 1984 a Labour government pursued a monetarist policy of market deregulation but lost the 1990 elections to the National Party.

Government

The members of New Zealand's single-chamber parliament, since a referendum in 1992, are elected under a mixed first-past-the-post and proportional representation system. Every voter has two votes, one for their local candidate and the other for their favored party. The system allows support for minor parties – the old first-past-the-post only system favored the two dominant parties, Labour and the National Party. The head of state is the British monarch, represented in New Zealand by the governor-general.

People

Most of the 3.86 million New Zealanders are of European origin – mainly British – but the Maori account for a significant minority (8.9 percent). In recent decades there has been some improvement in the legal and social status of the Maori, enabling them to recover some ancestral lands from state-owned enterprises. The Maori language has official status along-

Pitcairn Island

In the year 1790, nine British sailors, along with twelve women and six men from Tahiti, arrived on Pitcairn – a small volcanic island in the Pacific Ocean midway between New Zealand and Peru. The sailors were mutineers from the Bounty, and had left most of their companions on Tahiti, over 2,000 km (1,200 mi) to the northeast.

On arrival at Pitcairn, the settlers burned their ship – and for the next 18 years this tiny desert island, some 5 sq km (2 sq mi) in area, was the scene of violence, murder and disease. When a party of American whalers stumbled across the survivors in 1808, there was only one European along with several Tahitian women and a few children.

In 1856 the island's by then 194 inhabitants were moved to Norfolk Island off the Australian coast, but 43 of them later returned to Pitcairn. Today their descendants live in Adamstown, the island's only town with just over 50 inhabitants. They live on what they can grow, and what they can buy from passing ships in return for stamps and curios. Pitcairn is a British dependency. The governor is a British High Commissioner in Auckland, New Zealand, advised by the 10-member island council.

side English. There are also a considerable number of small ethnic groups of non-Maori Polynesians, as well as Chinese and Indians.

The majority of New Zealanders are nominally Christian. Among the small minority sects are two Maori churches that have incorporated aspects of traditional Maori beliefs. Cultural and leisure activities are European-influenced, and sport, particularly rugby football, has a

Sheep farming (*below left*), is ideally suited to New Zealand's mild climate. Refrigerated shipping revolutionized the industry in 1882, opening up the lucrative British market, and there are now some 70 million sheep – more than 20 per capita.

Maori tattooing (*below*) Today the facial tattoo, or *moko*, is applied with paint, but it was originally chiseled into the skin using a *uhi* knife made of human or albatross bone struck with a mallet. *Moko* were also used to adorn thighs, buttocks and waists.

strong following. However, Maori arts and crafts have been revived along with traditional music and dance.

ECONOMY

During the first half of the 20th century, New Zealanders enjoyed one of the highest standards of living in the world. However, in the 1970s the country suffered from the loss of export trade with European Union countries. In the late 1980s and 1990s New Zealand successfully expanded its industrial and service sectors.

Agriculture and fisheries
Agriculture employs 9 percent of the workforce and supplies 8 percent of GDP. Sheep farming is still the chief activity, and lamb and wool are the principal exports. Cattle provide exports of beef and butter, along with cereals, vegetables and fruit.

Plantation forestry and fishing are also important. Forestry, based principally on the introduced Californian Monterey pine, contributes to large paper, pulp, timber, and woodchip exports worldwide.

Industry and commerce
The emphasis has been on food processing and on the manufacture of agricultural machinery and consumer goods for domestic use. Other industries such as chemical manufacture and motor vehicle assembly

The thriving lumber industry (*below*) makes heavy use of the versatile Monterey pine, *Pinus radiata*. It matures in 26 years in New Zealand's favorable climate, making it a lucrative renewable resource. Introduced species account for nearly a million hectares of forest.

rely heavily on imports. Coal is mined for generating electricity and for exporting to Japan. Natural gas has also become an important fuel; converting it into gasoline has led to a reduction in petroleum imports. Hydroelectricity supplies most of the other energy needs.

Today Australia is New Zealand's principal trading partner, followed by Japan and the United States. Since the mid 1980s banking and financial services have become a more important economic sector

Transportation and communications
New Zealand has well maintained roads, an extensive railroad network and regular ferry services between the islands. Local air travel is also available, with international flights from Auckland, Wellington and Christchurch.

Radio and television broadcasting are run by the Broadcasting Corporation of New Zealand. Newspapers are published regionally rather than nationally, but give good news coverage.

Welfare and education
In 1938 the government brought in a system of health and welfare benefits that were then the most extensive in the world. State-provided healthcare remains at a high level, though private-sector medicine is on the increase.

Education is free and compulsory between the ages of 6 and 15, and standards of literacy are high. Subsidized preschool education is also widely available. Higher education establishments include seven universities.

Papua New Guinea

INDEPENDENT STATE OF PAPUA NEW GUINEA

P APUA NEW GUINEA, IN THE WESTERN Pacific, comprises the Eastern portion of the island of New Guinea and numerous smaller islands to the north and east.

GEOGRAPHY

More than four-fifths of the land area is on the New Guinea mainland. The island territories vary from the tiny Trobriand group in the southeast to the long ridges of New Britain and New Ireland.

Papua New Guinea is dominated by mountain ranges subject to earthquakes and volcanic activity. The only extensive lowlands are on New Guinea island, around the massive Central Range.

Papua New Guinea has a tropical monsoon climate – hot and humid, with heavy seasonal rainfall. Vegetation varies from coastal mangroves and lowland swamp forests to oak, beech and pine in the mountains. Wildlife includes many unique species and among the remarkable birds are the large, flightless cassowaries and colorful birds of paradise.

SOCIETY

Although New Guinea has probably been inhabited for about 50,000 years, it was unknown to Europeans until the 16th century. In the late 19th century, the British claimed the island's southeastern quadrant, known as Papua, while Germany took over the northeastern quadrant and the islands.

In 1906 the British gave Papua to Australia. Following World War I the German sector was mandated to Australia by the League of Nations in 1921. Papua New Guinea gained full independence in 1975 within the British Commonwealth.

The head of state is the governor-general, appointed by the British sovereign, but executive power lies with the prime minister and cabinet. Parliamentary elections are held every five years.

The people fall into two main categories – Papuan and Melanesian – but Micronesian, Polynesian and Negrito peoples also live on the island. They speak over 700 separate languages – more than in any other country in the world.

ECONOMY

Papua New Guinea is rich in mineral resources, including copper and gold. The chief economic activity is mining, and minerals account for most exports. Natural gas, oil and hydroelectric power supply the country's energy.

Agricultural activity tends to be either subsistence farming or intensive production of coffee, tea, cocoa and copra for export. Important forestry and fisheries exports include prawns and tuna, palm oil, logs and lumber.

Because of the islands' mountainous and swampy terrain, transportation is mainly by air. Shipping services operate between the coastal ports and Australia.

Telecommunications is a growing industry, but few people have telephones or television. Health, education and welfare facilities are not widely available.

Beautiful but endangered (*above*) This Raggiana bird of paradise is one of over 40 birds of paradise surviving in the tropical forests and mangroves of Australasia. Like most males of the species this one is brilliantly colored; the females are very plain by comparison.

Loading cargo at Port Moresby (*below*) Discovered in 1873 by the British captain John Moresby, this large and sheltered natural harbor has grown to be the commercial center and capital of Papua New Guinea housing over 125,000 people.

NATIONAL DATA – PAPUA NEW GUINEA

Land area   452,860 sq km (174,850 sq mi)

Climate		Temperatures		Annual
	Altitude m (ft)	January °C(°F)	July °C(°F)	precipitation mm (in)
Port Moresby	38 (125)	28 (82)	26 (79)	1,112 (43.7)

Major physical features   highest point: Mount Wilhelm 4,508 m (14,790 ft); longest river: Sepik about 1,000 km (600 mi)

Population   (2006 est.) 5,670,544

Form of government   multiparty constitutional monarchy with one legislative house

Armed forces   army 2,500; air force 200

Capital city   Port Moresby (295,892)

Official language   English

Ethnic composition   Papuan 84%; Melanesian 15%; others 1%

Religious affiliations   Roman Catholic 22%; Lutheran 16%; Presbyterian/Methodist/London Missionary Society 8%; Anglican 5%; Evangelical Alliance 4%; Seventh-Day Adventist 1%; other Protestant 10%; traditional beliefs 34%

Currency   1 kina (PGK) = 100 toea

Gross domestic product   (2006) U.S. $15.13 billion

Gross domestic product per capita   (2006) U.S. $2,700

Life expectancy at birth   male 63.08 yr; female 67.58 yr

Major resources   gold, copper, silver, natural gas, timber, oil, hydropower, fisheries, bananas, cassava, cocoa, coconuts, coffee, rubber, sago, sugarcane, sweet potatoes, tea, vegetables, yams

Solomon Islands

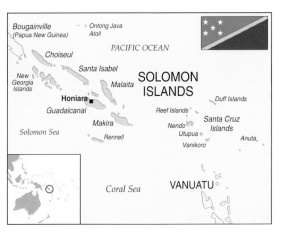

THE SOLOMON ISLANDS LIE IN THE WESTERN Pacific to the northeast of Australia. They were named in 1568 by a Spanish explorer who thought they might contain gold and other riches worthy of the biblical king, Solomon. Bougainville and Buka Island, part of the original group and sometimes known as the Northern Solomons, have been part of the territory of New Guinea for over a century.

GEOGRAPHY

The Solomons consist primarily of two parallel chains of volcanic islands running southeast of Bougainville. Farther east is the smaller Santa Cruz group including Nendo and Vanikoro. Several of the volcanic islands are still active.

The hot, humid climate is relieved by sea breezes, but temperature and rainfall are both very high in the cyclone season (November–April). Most of the land is rainforest. The wildlife includes large numbers of birds, and a unique lizard, the giant Solomon Island skink, with a prehensile tail.

SOCIETY

Melanesian peoples first settled the islands about 2000 BC. After Spanish settlements in the 16th and 17th centuries failed, British missionaries established a settlement in the 1870s. Soon after, Solomon Islanders were put to work on colonial plantations in Queensland, Australia, and on Fiji and other Pacific island colonies. In 1893 the British government declared the Solomons a protectorate.

During World War II the Solomons were the scene of bitter fighting between United States and Japanese forces, and a decisive battle was fought on the island of Guadalcanal. During the occupation nationalist feeling began to grow among the islanders, but full independence from Britain was not achieved until 1978.

Today the Solomon Islands is a member of the British Commonwealth. They have a single-chamber parliament whose 47 members are elected for a four-year term. Most Solomon Islanders are native Melanesians; over 60 indigenous languages are spoken, as well as English and an English-based pidgin.

ECONOMY

Agriculture, forestry and fisheries (the tuna trade in particular) are the largest export earners. Farmers produce copra, cacao, coconuts and palm oil as cash crops, and potatoes, yams, vegetables and rice for subsistence; pigs and cattle are raised for local consumption.

The islands have some gold and bauxite deposits, but no local energy resources. Light manufacturing industries (textiles, food-processing, batteries and fiberglass) flourish alongside traditional handicrafts, including shell-inlay and basketmaking. The islands have poorly developed roads,

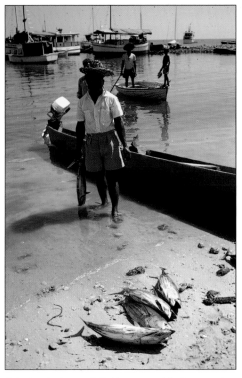

Bringing in the catch Fishing – especially for tuna – is a mainstay of the Solomon Islands economy. Despite the introduction of boats with outboard motors, fishing techniques on the islands have hardly changed over the centuries.

but good shipping and air connections to Australia and other islands.

Healthcare is provided in clinics run by the government or by the church. The incidence of malaria is very high and increasing. Education is provided through church- and government-funded schools.

NATIONAL DATA - SOLOMON ISLANDS

Land area 27,540 sq km (10,633 sq mi)

Climate	Altitude m (ft)	Temperatures January °C(°F)	July °C(°F)	Annual precipitation mm (in)
Honiara	2 (7)	27 (80)	26 (79)	2,094 (82.4)

Major physical features largest island: Guadalcanal 5,336 sq km (2,060 sq mi); highest point: Makarakomburu 2,447 m (8,028 ft)

Population (2006 est.) 552,438

Form of government multiparty constitutional monarchy with one legislative house

Armed forces no armed forces

Capital city Honiara (Guadalcanal - 58,492)

Official language English

Ethnic composition Melanesian 94.5%; Polynesian 3%; Micronesian 1.2%; other 1.1%; unspecified 0.2%

Religious affiliations Church of Melanesia 32.8%; Roman Catholic 19%; South Seas Evangelical 17%; Seventh-Day Adventist 11.2%; United Church 10.3%; Christian Fellowship Church 2.4%; other Christian 4.4%; other 2.4%; unspecified 0.3%; none 0.2%

Currency 1 Solomon Islands dollar (SBD) = 100 cents

Gross domestic product (2005) U.S. $9.413 billion

Gross domestic product per capita (2005) U.S. $2,500

Life expectancy at birth male 70.4 yr; female 75.55 yr

Major resources fish, gold, bauxite, phosphates, lead, zinc, nickel, coconuts, livestock, palm kernels, rice, sweet potatoes, taro, timber

Tuvalu

Dancing the night away Ritual and entertainment are exuberantly combined in the traditional dances of the tiny coral islands of Tuvalu. Pacific arts festivals and tourist interest have helped to keep many ancient customs alive.

TUVALU IS A SCATTERED ISLAND CHAIN IN THE southwestern Pacific between the Gilbert Islands of Kiribati to the north and Fiji to the south. Formerly the Ellice Islands, the group was part of the British colony of the Gilbert and Ellice Islands until 1978.

Tuvalu consists of nine islands, the tiny peaks of an undersea mountain range. Five of them are coral atolls; four are reef islands. They range in size from 4.9 sq km (1.89 sq mi) to 0.4 sq km (0.16 sq mi) and rise only a few meters (yards) above sea level. The islands' humid, tropical climate is cooled by the prevailing southeast trade winds. Annual rainfall is often high, but there are frequent severe droughts. The poor coralline soil supports only coconut palms, breadfruit and cassowary trees.

Wildlife includes mostly rats, lizards and sea birds, but the surrounding sea is rich in fish and other marine life.

It is likely that the islands were first settled by seafaring Polynesians, probably from Samoa, about the 14th century. In 1892 the British persuaded the islanders to join the Gilbert Islands protectorate, later the Gilbert and Ellice Islands colony. During World War II the United States set up bases on the Ellice Islands to combat the Japanese invasion of the Gilberts.

By the 1960s the mainly Polynesian Ellice islanders were becoming increasingly dissatisfied with the Gilbertese (Micronesian) domination of the colony. They eventually seceded in 1976, and the islands became independent as Tuvalu in 1978. Today Tuvalu remains in the British Commonwealth, with the British monarch as head of state.

Traditionally Tuvalu's economy rests on agriculture, mainly subsistence farming, and fishing. Copra, the only cash crop, is marketed by a cooperative body. Fishing is mostly for local consumption except for *beche de mer* (sea cucumbers) which are caught and processed for export. Many islanders go abroad to find work. However Tuvalu's fortunes altered in 2000 following the lease of its Internet domain name "tv." for $50 million.

NATIONAL DATA - TUVALU

Land area	26 sq km (10 sq mi)			
Climate		Temperatures		Annual
	Altitude m (ft)	January °C(°F)	July °C(°F)	precipitation mm (in)
Funafuti atoll	2 (7)	28 (82)	28 (82)	3,544 (139.5)
Major physical features	largest island: Vaitupu 5 sq km (2 sq mi)			
Population	(2006 est.) 11,810			
Form of government	nonparty constitutional monarchy with one legislative house			
Armed forces	no armed forces			
Capital city	Fongafale (Funafuti, 4,921)			
Official languages	Tuvaluan, English			
Ethnic composition	Polynesian 96%; Micronesian 4%			
Religious affiliations	Church of Tuvalu (Congregationalist) 97%; Seventh-Day Adventist 1.4%; Baha'i 1%; other 0.6%			
Currency	1 Australian dollar (AUD) or 1 Tuvaluan dollar = 100 cents			
Gross domestic product	(2002) U.S. $14.94 million			
Gross domestic product per capita	(2002) U.S. $1,600			
Life expectancy at birth	male 66.08 yr; female 70.66 yr			
Major resources	fish, bananas, breadfruit, coconuts, copra, fish, papaya, root crops			

Nauru

NAURU IS A TINY SINGLE-ISLAND STATE IN the western Pacific Ocean, immediately south of the Equator and about 700 km (450 mi) west of Kiribati. It is the world's smallest republic and in the mid-1980s had one of the highest per capita incomes in the world, thanks to a simple precious natural resource – phosphate.

The island is a raised coral reef with a central plateau, where rich phosphate beds have been created by bird droppings deposited over the centuries. The climate is hot and humid, though moderated by sea breezes. During the monsoon season (November to February) rainfall is heavy, but severe droughts can also occur. Vegetation consists mostly of coconut palms, breadfruit trees and scrub.

NATIONAL DATA - NAURU

Land area	21 sq km (8 sq mi)			
Climate		Temperatures		Annual
	Altitude m (ft)	January °C(°F)	July °C(°F)	precipitation mm (in)
Nauru	27 (89)	28 (82)	28 (82)	2,051 (80.8)
Major physical features	highest point: 69 m (225 ft)			
Population	(2006 est.) 13,287			
Form of government	nonparty republic with one legislative house			
Armed forces	no armed forces			
Capital city	No official capital but Yaren is the government seat (4,616)			
Official languages	Nauruan, English			
Ethnic composition	Nauruan 58%; other Pacific Islander 26%; Chinese 8%; European 8%			
Religious affiliations	Christian 100% (two-thirds Protestant, one-third Roman Catholic)			
Currency	1 Australian dollar (AUD) = 100 cents			
Gross domestic product	(2005) U.S. $60 million			
Gross domestic product per capita	(2005) U.S. $5,000			
Life expectancy at birth	male 59.5 yr; female 66.84 yr			
Major resources	phosphates, fish, bananas, coconuts, pineapples			

Kiribati

REPUBLIC OF KIRIBATI

Nauru was probably settled at various times by people from neighboring islands. In 1888 it was annexed by Germany. Some 10 years later phosphate was discovered and mining began in 1906, controlled by a British and a German company. After World War II, the island was made a United Nations' trust territory, and gained full independence in 1968. In 1970 the islanders took over control of all phosphate production.

Nauru's 18-member parliament is freely elected, and chooses a president as head of state. Native islanders are of mixed Pacific ancestry. Naurean and English are spoken.

Phosphate extraction remains the chief export earner – Japan, South Korea, New Zealand and Australia are major customers for phosphate rock which they use as a fertilizer. Although deposits will run out by 2008, government investment overseas should cushion the population from the sudden loss of revenue. In the 1990s, Australia, New Zealand and the UK agreed to compensate Nauru for environmental damage caused by phosphate mining. Much food is imported. Coconuts, bananas and pineapples are the main export crops. Welfare services are excellent on Nauru. Education and healthcare are free and the islanders pay no taxes toward funding these facilities.

KIRIBATI (PRONOUNCED KIREE-BASS) IS MADE up of three widely separated archipelagoes spread across a vast area of the central Pacific. The largest groups are the Gilbert Islands, the Phoenix group and the Line Islands.

The islands are coral atolls, founded on the summits of undersea volcanoes and encircled by reefs. They rise only a meter or so (a few feet) above sea level.

Kiribati's climate is hot and humid. Rainfall is frequent in the Gilberts, but is sparser elsewhere, and is generally unreliable. Only the Gilberts and three of the Line Islands are now inhabited. Vegetation is mostly confined to salt-resistant plants and palms, chiefly coconut and bananas. The reefs teem with marine life.

The Gilbert Islands were settled several thousand years ago by Micronesian peoples, followed later by Polynesian seafarers from the south. In 1900 the British annexed Banaba island for its phosphates and forcibly resettled its population. In 1916 the Gilbert and Ellice Islands colony was founded, and later extended to include the Phoenix and Line Islands.

Earthly paradise A lagoon beach on Tabuaeran (one of the Line Islands). The turquoise waters and palm-fringed beaches have captivated visitors ever since the first Western mariners started to explore the South Pacific seas in the 17th century.

In 1975 the Polynesian-inhabited Ellice Islands seceded and became independent as Tuvalu in 1978. The remaining colony became independent as Kiribati in 1979. The vast majority of today's population are Gilbertese (I-Kiribati) speaking a distinct Micronesian language.

Phosphate deposits on Banaba were exhausted in the late 1970s, but the profits had created a substantial reserve fund. Otherwise, Kiribati's economy is based largely on subsistence agriculture and fishing, with copra the chief cash crop.

NATIONAL DATA - KIRIBATI

Land area	811 sq km (313 sq mi)			
Climate		Temperatures		Annual
	Altitude m (ft)	January °C(°F)	July °C(°F)	precipitation mm (in)
Tarawa	3 (10)	28 (82)	28 (82)	1,914 (75.4)

Major physical features	largest island: Kiritmati (Line Is) 388 sq km (150 sq mi); highest point: Banaba 81 m (266 ft)
Population	(2006 est.) 105,432
Form of government	multiparty republic with one legislative house
Armed forces	no armed forces
Capital city	Tarawa (47,437)
Official language	English
Ethnic composition	Micronesian 98.8%; other 1.2%
Religious affiliations	Roman Catholic 52%; Protestant (Congregational) 40%; Seventh-Day Adventist, Muslim, Baha'i, Latter-day Saints, Church of God 8%
Currency	1 Australian dollar (AUD) = 100 cents
Gross domestic product	(2004) U.S. $206.4 million
Gross domestic product per capita	(2004) U.S. $2,700
Life expectancy at birth	male 59.06 yr; female 65.24 yr
Major resources	bananas, breadfruit, coconuts, copra, fish, papaya

Fiji

REPUBLIC OF FIJI

FIJI IS AN ISOLATED ARCHIPELAGO IN THE southwestern Pacific, some 2,100 km (1,300 mi) north of Auckland, New Zealand. Two of its islands – Viti Levu and Vanua Levu – account for nearly nine-tenths of the land area.

Like most of the islands of Fiji, Viti Levu and Vanua Levu are volcanic in origin with central mountain ranges rising steeply from narrow coastal plains and offshore coral reefs. From November to April the tropical climate brings hot, humid conditions and there is threatened danger from cyclones. The densely forested southeastern coastline receives heavy rainfall, brought by the trade winds. The western slopes, where there are open forests and grasslands, are much drier. More than two-thirds of the land area of Fiji is forested and only a small percentage of land is meadow or pasture.

When the Dutch navigator Abel Janszoon Tasman (c. 1603–59) reached Fiji in 1643, its inhabitants were a mixture of Melanesian and Polynesian peoples. In 1874 Fiji became a British crown colony, and from 1879 the British imported thousands of Indian laborers to work their sugar plantations.

After Fiji became independent in 1970, the political parties divided along racial lines creating considerable political instability. Indian parties won the election of 1987, but three subsequent military coups restored Melanesian dominance which was confirmed in the 1990 constitution.

Agriculture, tourism and light industry are striving to support an economy that suffered badly in the unrest and unfavorable conditions of the late 1980s – bad weather, low sugar prices, labor strikes and political unrest.

Major cash crops include sugar cane, copra, ginger and coconuts. Gold is mined on Viti Levu, and is another major export.

Gaily painted canoes moored against a jetty on the Wainibuki river on Fiji's main island, Viti Levu. Canoes are used for transport, fishing and provide endless photo-opportunities for the country's many tourists.

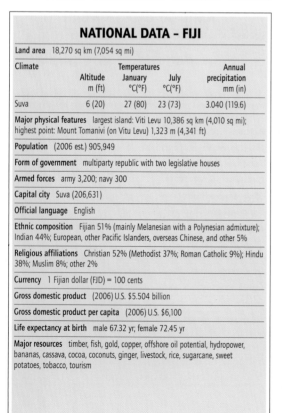

NATIONAL DATA – FIJI				
Land area	18,270 sq km (7,054 sq mi)			
Climate		Temperatures		Annual
	Altitude m (ft)	January °C(°F)	July °C(°F)	precipitation mm (in)
Suva	6 (20)	27 (80)	23 (73)	3.040 (119.6)
Major physical features	largest island: Viti Levu 10,386 sq km (4,010 sq mi); highest point: Mount Tomanivi (on Vitu Levu) 1,323 m (4,341 ft)			
Population	(2006 est.) 905,949			
Form of government	multiparty republic with two legislative houses			
Armed forces	army 3,200; navy 300			
Capital city	Suva (206,631)			
Official language	English			
Ethnic composition	Fijian 51% (mainly Melanesian with a Polynesian admixture); Indian 44%; European, other Pacific Islanders, overseas Chinese, and other 5%			
Religious affiliations	Christian 52% (Methodist 37%; Roman Catholic 9%); Hindu 38%; Muslim 8%; other 2%			
Currency	1 Fijian dollar (FJD) = 100 cents			
Gross domestic product	(2006) U.S. $5.504 billion			
Gross domestic product per capita	(2006) U.S. $6,100			
Life expectancy at birth	male 67.32 yr; female 72.45 yr			
Major resources	timber, fish, gold, copper, offshore oil potential, hydropower, bananas, cassava, cocoa, coconuts, ginger, livestock, rice, sugarcane, sweet potatoes, tobacco, tourism			

Vanuatu

REPUBLIC OF VANUATU

THE SOUTHWESTERN PACIFIC REPUBLIC OF Vanuatu (formerly the New Hebrides) comprises 13 main islands and dozens of smaller ones, forming a Y-shaped group between the Solomon Islands to the north and New Caledonia to the south.

The main islands are Espiritu Santo and Efaté, site of the capital, Port-Vila. Like most of the country, these islands are mountainous and covered in rainforest, with raised coral terraces on some coastal slopes and offshore coral reefs. Earth tremors are frequent, and there are several active volcanoes. The humid, tropical climate produces twice as much rainfall in the north as in the south, and seasonal storms can rise to hurricane force. In 1987 the worst cyclone in the country's history damaged or destroyed most of the houses on the islands.

Vanuatu's largely Melanesian inhabitants lived here for more than 2,500 years before Europeans discovered the group of islands in the 15th century. From the 1840s sandalwood traders and missionaries made their way to the islands. During the 20th century growing pressure from nationalist movements brought independence in 1980, a new constitution and a new name, Vanuatu, "Our Land Forever". Today, members of Vanuatu's single-chamber parliament are elected for a four-year term. Parliament and the heads of regional councils elect a president to serve a five-year term.

The economy depends largely on exports of copra, fish, beef and veal; most manufactured goods and fuel are imported. About three-quarters of the population work on the land, but most are subsistence farmers raising livestock and crops such as yams and bananas. Other exports include manganese, mined on Efaté, and pine lumber. Tourism is a growing source of hard currency, and Vanuatu's favorable tax laws have also made it a financial center.

Tonga

KINGDOM OF TONGA

Tonga, in the southwestern pacific, is a double chain of 169 islands to the southwest of Samoa and east of Fiji. In 1875 the king of Tonga created a constitutional monarchy, and his system of government continues virtually unchanged today.

The islands are in three main groups from north to south, Vava'u, Ha'apai and Tongatapu. The Vava'u group and one island in the Ha'apai group have mountainous landscapes of volcanic rock, and there are four intermittently active volcanoes. All the other islands are low-lying coral formations.

The semitropical climate supports a rich variety of mainly forest vegetation, including trees such as the paper mulberry. Bird life on the islands is plentiful, and some of the world's largest bats, flying foxes, are also found there.

The settlement of Tonga began in about 1000 BC, and since the 10th century AD kingship has been handed down from the original Tu'i Tonga royal line.

A British Methodist mission established in 1826 converted the majority of the islanders to Christianity. In 1900 the kingdom became a British protectorate;

Friendly faces The welcoming smiles that greet modern visitors reflect Tonga's longstanding tradition of hospitality – the British explorer Captain Cook named the islands "The Friendly Islands" when he landed there in the 1770s.

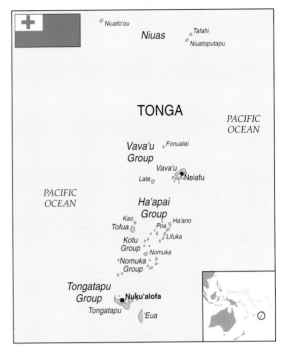

70 years later Tonga regained independence within the British Commonwealth.

Agriculture is central to the economy; cash crops include vanilla, coconuts and bananas, while yams, taro and cassava are among the chief subsistence crops. Traditional handicrafts cater for the growing tourist market, but many Tongans work abroad in New Zealand, Australia or the United States, and send remittances home.

NATIONAL DATA – VANUATU

Land area	12,200 sq km (4,710 sq mi)			
Climate		Temperatures		Annual precipitation mm (in)
	Altitude m (ft)	January °C(°F)	July °C(°F)	
Port-Vila	57 (187)	27 (81)	23 (73)	2,221 (87.4)

Major physical features highest point: Espiritu Santo 3,678 sq km (1,420 sq mi); highest point: Tabwémanasana on Espiritu Santo 1,880 m (6,167 ft)

Population (2006 est.) 208,869

Form of government multiparty republic with one legislative house

Armed forces no armed forces

Capital city Port-Vila (38,438)

Official languages French, English

Ethnic composition Ni-Vanuatu 98.5%; other 1.5%

Religious affiliations Presbyterian 31.4%; Anglican 13.4%; Roman Catholic 13.1%; Seventh-Day Adventist 10.8%; other Christian 13.8%; traditional beliefs 5.6% (including Jon Frum cargo cult); other 9.6%; none 1%; unspecified 1.3%

Currency vatu (VUV)

Gross domestic product (2003) U.S. $276.3 million

Gross domestic product per capita (2003) U.S. $2,900

Life expectancy at birth male 61.34 yr; female 64.44 yr

Major resources manganese, hardwood forests, fish, bananas, cassava, cattle, cocoa, coconuts, coffee, taro, yams, tourism

NATIONAL DATA – TONGA

Land area	718 sq km (277 sq mi)			
Climate		Temperatures		Annual precipitation mm (in)
	Altitude m (ft)	January °C(°F)	July °C(°F)	
Nuku'alofa	3 (10)	26 (78)	21 (70)	1,738 (68.4)

Major physical features highest point: Tongatapu 256 sq km (99 sq mi); highest point: on Kao Island 1,030 m (3,380 ft)

Population (2006 est.) 114,689

Form of government nonparty constitutional monarchy with one legislative house

Armed forces no armed forces

Capital city Nuku'alofa (23,854)

Official languages Tongan, English

Ethnic composition Tongan 96%; Part-Tongan 1.7%; European 0.6%; Fijian 0.2%; Samoan 0.2%; Indian, Chinese, Japanese 0.3%; Other 1%

Religious affiliations Christian 100% (Free Wesleyan Church claims over 30,000 adherents)

Currency 1 pa'anga (TOP) = 100 seniti

Gross domestic product (2004) U.S. $178.5 million

Gross domestic product per capita (2005) U.S. $2,200

Life expectancy at birth male 67.32 yr; female 72.45 yr

Major resources fish, bananas, cassava, citrus fruits, coconuts, sweet potatoes, taro, vanilla, tourism

Samoa

INDEPENDENT STATE OF SAMOA

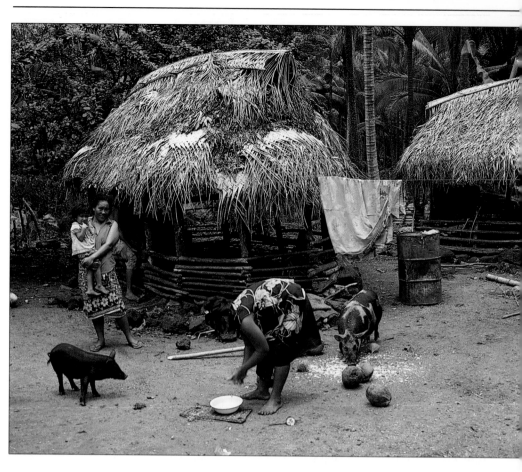

T HE NINE ISLANDS OF SAMOA LIE IN the south-central Pacific Ocean, north of Tonga and northeast of Fiji. Only four islands are inhabited, Upolu, Savai'i, Apolima, and Manono. Today, Samoa's fragile economy depends on agriculture, tourism, and income from Samoans working abroad, especially in New Zealand.

GEOGRAPHY

On the largest islands – Upolu and Savai'i – the interior is mountainous, and coral reefs ring the narrow coastal plain. The islands are all volcanic, and Savai'i suffered some major eruptions in 1902 and 1911. The climate is generally warm and humid, with rainfall especially heavy on southeastern slopes exposed to the frequent trade winds. From November to March there is also a danger of cyclones. Almost half the land is covered with forest, and mangroves thrive in the lowland swamps. Wildlife on the islands is sparse, but includes a unique and rare tooth-billed pigeon.

NATIONAL DATA - SAMOA

Land area	2,934 sq km (1,133 sq mi)			
Climate		Temperatures		Annual
	Altitude m (ft)	January °C(°F)	July °C(°F)	precipitation mm (in)
Apia	2 (7)	27 (72)	26 (79)	2,961 (116.5)

Major physical features largest island: Savai'i 1,813 sq km (700 sq mi); highest point: Silsili (Savai'i) 1,858 m (6,095 ft)

Population (2006 est.) 176,908

Form of government multiparty republic with two legislative houses

Armed forces no armed forces

Capital city Apia (41,204)

Official languages Samoan, English

Ethnic composition Samoan 92.6%; Euronesians 7% (persons of European and Polynesian blood); Europeans 0.4%

Religious affiliations Congregationalist 34.8%; Roman Catholic 19.6%; Methodist 15%; Latter-Day Saints 12.7%; Assembly of God 6.6%; Seventh-Day Adventist 3.5%; other Christian 4.5%; Worship Centre 1.3%; other 1.7%; unspecified 0.1%

Currency 1 tala (SAT) = 100 seniti

Gross domestic product (2002) U.S. $1 billion

Gross domestic product per capita (2005) U.S. $2,100

Life expectancy at birth male 68.2 yr; female 73.94 yr

Major resources hardwood forests, fish, hydropower, bananas, coffee, copra, tropical fruits, tourism

SOCIETY

The Samoan islands were probably settled by Tongans in about 1000 BC. The first European visitor in 1766 was the French explorer Bougainville (1729–1811) who named them the Navigator Islands. Local disputes on the islands in the 19th century were complicated by competing United States', German and British colonial interests. In 1899 the three colonial powers signed an agreement whereby Britain ceded its interests to Germany and the United States retained American Samoa to the east. After World War I (1914–18) the League of Nations mandated German Samoa to New Zealand as Western Samoa, which finally achieved full independence in 1962. The country's name was changed to Samoa in July 1997, but there have been strong objections to this from its neighbor American Samoa.

Samoa is a constitutional monarchy ruled by a paramount chief. He appoints a prime minister on the recommendation of the 47 members of the single-chamber legislative assembly. Two members of the assembly are elected by the communities

Women feeding and tending pigs in a typical village in Samoa. Pigs have a special significance in many South Pacific cultures: they are highly valued not only as a source of food but also as a symbol of wealth. In some parts, pigs are used as dowries.

of European and mixed-descent, while the rest are chosen by the Polynesian clan chieftains.

The great majority of the population are indigenous Samoans, whose language – Samoan – may be the oldest of all the Polynesian languages. English is the second official language.

ECONOMY

The economic mainstays are agriculture, fishing, lumber exports and tourism. Copra and cocoa are grown for export while yams, breadfruit and papaya are grown for local consumption. Fishing has also become an export industry. Timber supplies for local use and for export trade are maintained by reforestation programs. For power, the country relies mainly on hydroelectricity. The range of light industries is growing, while traditional handicrafts meet the demands of a brisk and growing tourist trade.

About one-fifth of Samoa's roads are surfaced, and the islands are linked by sea and air; they have two deepwater ports and one international airport. Healthcare is administered from the main hospital in the capital, Apia, on Upolu. Education, modeled on New Zealand's system, is free though not compulsory.

Palau

REPUBLIC OF PALAU

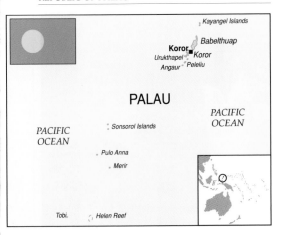

Tʜᴇ ʀᴇᴘᴜʙʟɪᴄ ᴏꜰ ᴘᴀʟᴀᴜ, ᴀʟꜱᴏ ᴋɴᴏᴡɴ ʙʏ its native name Belau, became independent in October 1994. Before then it was a UN trusteeship in free association with the United States. The territory comprises the westernmost portion of the Caroline Islands chain some 850 km (528 mi) southeast of the Philippines. The main islands are part of a coralline reef structure on a volcanic base, with over 300 islets surrounding a huge central lagoon.

The islands vary geologically from the mountainous main island of Babelthuap to low coral islands fringed by barrier reefs. The reefs are home to a variety of marine life, and the hot wet climate supports lush vegetation. Agriculture and fishing remain at subsistence level, though tuna is exported and copra is grown as a cash crop. Coconut oil is exported in small quantities. Transportation is by sea or air, and tourism has not been fully developed.

Marshall Islands

REPUBLIC OF THE MARSHALL ISLANDS

Tʜᴇ ᴍᴀʀꜱʜᴀʟʟ ɪꜱʟᴀɴᴅꜱ ᴀʀᴇ ᴀ ɢʀᴏᴜᴘ ᴏꜰ 5 single islands, 31 widely separated coral atolls (including the largest atoll in the world) and over 1,000 islets and reefs strung out in two chains in the western Pacific Ocean. During World War II the islands were occupied by the Allied Forces, and from 1949 they were part of the Trust Territory of the United States. In 1986 they became self-governing in free association with the US, an arrangement whereby the US retained responsibility for defense in return for assistance. In 1991 the Marshall Islands became politically autonomous.

Local politics are traditionally dominated by the island chiefs, and their high chief holds the post of president, but the economy of the Marshall Islands is still heavily dependent on aid from the United States, and the rent paid by its benefactor for the use of Kwajalein Atoll as a missile base. The tourist industry is an important source of foreign revenue and employment. High-grade phosphate is mined for export on Ailinglapalap Atoll but most other industry is limited to handicrafts, and processing fish or copra. Agriculture is mainly at subsistence level, but some cash crops (coconuts, tomatoes, melons and breadfruit) are grown in small quantities on local farms.

Living with nature (*below*) A typical house in the Marshall Islands is constructed of woven mats and roofed with coconut palm fronds. Coconut palms and breadfruit are the islands' main vegetation; the leaves and fruit serve a variety of purposes.

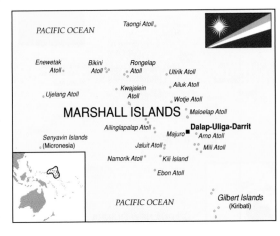

NATIONAL DATA - MARSHALL ISLANDS

Land area 181 sq km (70 sq mi)

Climate		Temperatures		Annual
	Altitude m (ft)	January °C(°F)	July °C(°F)	precipitation mm (in)
Ujelang Atoll	10 (33)	27 (81)	28 (82)	1,976 (77.8)

Population (2006 est.) 60,422

Form of government Constitutional government in free association with the United States of America

Armed forces no armed forces

Capital city Majuro (25,400)

Official language English

Ethnic composition Micronesian 100%

Religious affiliations Protestant 54.8%; Assembly of God 25.8%; Roman Catholic 8.4%; Bukot nan Jesus 2.8%; Mormon 2.1%; other Christian 3.6%; others 2.5%

Currency 1 U.S. dollar (USD) = 100 cents

Gross domestic product (2001) U.S. $115 million

Gross domestic product per capita (2005) U.S. $2,900

Life expectancy at birth male 68.33 yr; female 72.39 yr

Major resources coconuts, copra, fruits, deep-seabed minerals, phosphates, clams, fish, livestock, oysters, tourism

NATIONAL DATA - PALAU

Land area 458 sq km (177 sq mi)

Major physical features 26 islands and over 300 islets; largest island is Babelthuap

Population (2006 est.) 20,579

Form of government constitutional republic with two legislative houses

Armed forces no armed forces

Capital city Melekeok (391) replaced Koror (11,559) in 2006

Official languages Palauan and English are official in all islands except Sonsoral (Sonsoralese and English are official), Tobi (Tobi and English are official), and Angaur (Angaur, Japanese, and English are official)

Ethnic composition Palauan (Micronesian with Malayan and Melanesian admixtures) 69.9%; Filipino 15.3%; Chinese 4.9%; other Asian 2.4%; white 1.9%; Carolinian 1.4%; other Micronesian 1.1%; other or unspecified 3.2%

Religious affiliations Roman Catholic 41.6%; Protestant 23.3%; Modekngei 8.8% (indigenous to Palau); Seventh-Day Adventist 5.3%; Jehovah's Witness 0.9%; Latter-Day Saints 0.6%; other religion 3.1%; unspecified or none 16.4%

Currency 1 U.S. dollar (USD) = 100 cents

Gross domestic product (2004) U.S. $124.5 million

Gross domestic product per capita (2005) U.S. $7,600

Life expectancy at birth male 67.26 yr; female 73.77 yr

Major resources forests, gold, marine products, deep-seabed minerals, fish, root crops, copra, coconuts, tourism

ANTARCTICA
INTERNATIONAL

During the International Geophysical Year (1957–58), scientists from 12 nations spent the winter in more than 50 research stations scattered across the southern polar continent of Antarctica. They had come to study its unique environment, searching for clues to the Earth's past – and its possible future. The following year, despite competing claims for sovereignty over the region, the same 12 nations signed the Antarctic Treaty, reserving the continent for nonpolitical scientific investigation. From the 1970s they imposed restrictions on local seal trapping and whaling, which had already reduced several species to the point of extinction. In 1981 scientists made a discovery that could affect every form of life on the planet: massive damage to the ozone layer above the South Pole.

Antarctica has a bigger land area than Europe or the United States and Mexico combined. About 95 percent is covered by an ice cap that is 4,000 m (13,000 ft) thick in places. This was created by thousands of years of accumulated snowfall, although annual snowfall is minimal, and accounts for some nine-tenths of all the world's glacial ice. The Antarctic landscape is an average of over 2,000 m (7,000 ft) above sea level, making it the world's highest continent, but in many areas as much as four-fifths of the overall height is due to the ice cap. During August and September, in the depths of winter, pack ice up to 3 m (10 ft) deep covers the surrounding Southern Ocean, doubling the area of the continent.

The long Transantarctic Mountain chain runs across the whole of Antarctica, passing close to the South Pole and dividing the continent into two unequal parts: Lesser Antarctica (mostly within western longitudes) and the massive semicircle of Greater Antarctica. From Lesser Antarctica the slender but mountainous Antarctic Peninsula snakes northeast toward the southern tip of South America to form the western boundary of the Weddell Sea. The Ross Sea forms a smaller gulf on the opposite side of Lesser Antarctica, almost directly south of New Zealand. The southern end of both these seas is covered by permanent ice shelves, constantly replenished by new ice forced through gaps in the Transantarctic Mountains by the ice cap beyond.

Several other mountain ranges emerge around the fringes of the ice cap, the highest being the Vinson Massif in Lesser

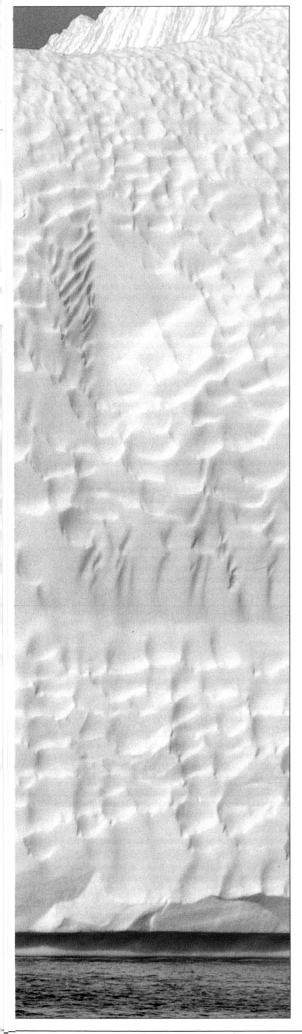

Antarctica – part of the Sentinel Range overlooking the Ronne Ice Shelf on the southern rim of the Weddell Sea. There are some volcanoes, which are largely in Lesser Antarctica, notably the islands north of the Weddell Sea; the active Mount Erebus, on Ross Island at the edge of the Ross Ice Shelf, is closely monitored.

The world's coldest continent suffers its lowest temperatures in August, dropping to -40° F to -95° F (-40° C to -70° C) on the inland ice sheet. On the Antarctic Peninsula, milder winds from neighboring oceans raise summer temperatures to 15° C (59° F), but the rest of Antarctica remains bitterly cold all year round. Rain is virtually unknown, and snowfall is rare except at the coast. Blizzards are frequent, when loose snow is whipped up by winds at the margins of the ice cap.

These hostile conditions, combined with long months of darkness, limit plant life to lichens, mosses, algae, and molds in the few ice-free areas. There are no land mammals in Antarctica, but whales and seals feed on masses of tiny, shrimp-like krill, while porpoises and dolphins are attracted by shoals of fish, predominantly Antarctic perches. With no land predators, the Antarctic coast is a haven for birds; Emperor penguins, Antarctic petrels and South polar skuas breed here and nowhere else, and more than 40 other species live in Antarctica.

A legendary first sighting of the icebound Southern Ocean is claimed by the Maori. The first European to cross the Antarctic Circle was the British Captain James Cook (1728–79) in 1773. In 1820 members of Russian, British and United States' expeditions to the Antarctic first set eyes on the mainland. Later explorers charted the Antarctic Peninsula and the seas around it; the French discovered and claimed Adélie Land in Greater Antarctica, while a British expedition led by James Clark Ross (1800–62) in 1839–43, discovered the sea and the ice shelf that bear his name.

Early this century several sledge expeditions set out to reach the South Pole at the heart of the interior. The Norwegian, Roald Amundsen (1872–1928), was the first to arrive on 14 December 1911. (The southern magnetic pole had been conquered in 1909). He was followed a month later by British explorers led by Robert Falcon Scott (1868–1912) who

Wall of ice (*left*) A tourist dinghy is dwarfed by a massive ice-covered cliff. The awesome scenery of the Antarctic attracts small but growing numbers of tourists, though tourism is strictly controlled to prevent environmental damage.

The leopard seal is perfectly adapted to the harsh conditions of the Antarctic – the coldest continent in the world. It shares its icy habitat with several species of seal, whales, dolphins, porpoises and a wide variety of seabirds.

perished in a blizzard on the return journey. In 1929 the American Richard Evelyn Byrd (1888–1957) became the first man to fly over the South Pole.

By 1942 seven nations – Argentina, Australia, Chile, France, New Zealand, Norway and Britain – had claimed sovereignty over sections of the Antarctic. But since the 1959 Antarctic Treaty those claims are largely ignored, and scientists from many countries work together. Whereas earlier adventurers pitted themselves against the bitter cold of the interior with sledges and dog teams, today's explorers travel in specially adapted vehicles known as snowcats, and use sophisticated seismic and remote-sensing equipment to map the contours of the hidden continent buried deep beneath the ice.

Other research includes planetary geology, hydrology, climatology, and meteoritics. In 1988 teams of scientists in the Antarctic traced the cause of the depletion of the world's ozone layer to the use of chlorofluorocarbons (CFCs) in aerosols, packaging materials, and refrigerants.

Detailed geological study has revealed the presence of many mineral resources. The Transantarctic Mountains include part of what could be the world's largest coalfield, and there are sizable reserves of iron ore in the Prince Charles Mountains near the coast of Greater Antarctica. Antimony, chromium, gold, molybdenum, and uranium have also been discovered, but so far it has been uneconomic to exploit any of these minerals, and potentially harmful to the fragile environment. In 1990 it was agreed to ban mining and mineral exploitation for 50 years.

A small but significant tourist industry has existed since the 1950s. Visits by cruise ships are strictly controlled to minimize environmental damage. Sealing and whaling activity is minimal, but fishing in Antarctic waters has increased greatly, including catching krill, which is the basis of the Antarctic food chain.

INDEX

Numbers in *italics* refer to captions or maps; numbers in **bold** refer to major sections

A

Aalto, Alvar 223
Aare Valley 318
Abacha, General Sanni 169
Abbud, General Ibrahim 148
'Abdal-Wahhab, Muhammad ibn 402
Abdali, Ahmad Khan (Ahmad Shah Durrani) 400
Abdallah, President Ahmed 207
Abdullah ibn-Hussain, King (Jordan) 392–393
Aberdare Range 179
Abidjan 162
Abiola, Chief Moshood 168
Abkhazia 371
Aboriginal people 478–481, *478*
Abu Dhabi 406, *406*
Abu Simbel *141*
Abubakar, General Abdulsam 168
Accra 164, *165*
Aceh (Sumatra) 456
Aconcagua, Mount 124
Aconcagua river 121
Adamawa Highlands 168, 170
Adamstown 484
Addis Ababa 150, 151, 152, 179
Addo Elephant National Park, South Africa 198
Adelaide 481
Adélie Land 497
Aden 399
Aden, Gulf of 151, 152, 153, 399
Adirondack Forest Preserve, USA 65
Aer Lingus 247
Aeroméxico 79
Afghanistan 367, 378, **400–401**, *400, 401*
Agra *408, 414*
Agulhas Current 198
Agulhas Negras, Pico das 116
Ahaggar Mountains *134*, 135
Ahidjo, Ahmadou 170
Air Algérie 135
Air Botswana 195
Air Canada 20
Air France 258
Air Inter 258
Air Lanka 423
Air Malawi 191
Aix-la-Chapelle (Aachen) 253
Ajaccio 252, 255
Ajman 406
Akbar 412
Akosombo Dam 164, *165*
Aksum 150
Akwapim Mountains 164
Al Jabal al Akhdar (Green Mountains) 138
Al Kana *402*
Al Manamah 404
al-Assad, Bashar 387
al-Thani family 405
Alabama 27, **70–71**
Åland Islands 218, 222, 223
Alamo, the 46, 57, *57*
Alaska 13, 14, 26, 30–31, **33**, 35, 42, **50**, *51*
Alaungpaya, King (Burma) 441
Alay Mountains 379
Alayor *282*
Alban Hills 293
Albania **351**, *351*
Albert, Lake 178
Alberta 13, 16, 18, 19
Albufeira *287*
Alderney 234
Aleppo *386*, 387
Aletsch glacier 318
Aleutian Islands 33
Alexander I, King (Yugoslavia) 342, 344
Alexander I, Tsar (Russia) 223, 329, 365
Alexander II, Tsar (Russia)
223, 225, 365
Alexander Nevsky Cathedral, Bulgaria *340*
Alexander the Great 141, 304, 350, 372, 384, 386, 388, 393, 400
Alexandria 140, *141*
Alfonsín, Raúl 125
Alfonso XIII, King (Spain) 281, 283
Algarve 286, *287*, 289
Algeciras 285
Algeria *131*, 132, 133, **134–135**, *134*, 252, 254, 297
Algerian War (1954–62) 135
Algiers 134
Ali, General Zine al-Abidine Ben 137
Alkmaar *268*
Allen, Ethan 63
Allende, Gossens Salvador 123
Allihies *247*
Almagro, Diego de 122
Alo 495
Alps 250, 251, *252*, 255, 292, 293, 294, 296, 311, 318, *319*, 320, 321, 322, 323, 323, 325, 347
Alsace 254, 256
Altai Range 363, 364, 369, 381
Altamira 279
Altiplano 110, 112
Alto Alentejo 286
Altos Cuchumatanes 80
Alumbrera copper and gold mine (Argentina) 126
Alvarado, Pedro de 80
Alzette river 274
Amalfi *293*
Amazon river and basin 104, 111, 112, 116–117, *116*, 119, *119*
Amboina 266
American Broadcasting Corporation 45
American Civil War (1861–65) 35, 39, 53, 62, 63, 64, 65, 66, 67, 68, 69, 70, 71, 72, 73, 88
American Colonization Society 161
American Samoa 26, 492, **495**
American Telephone and Telegraph Corporation (AT&T) 46
American War of Independence (1775–83) 16, 34, 60, 61, 63, 65, 66, 67, 70, 71, 72, 73, 88, 479
Amin, Idi 178
Amman 393, *393*
Amorican Massif 250
Amritsar 413
Amsterdam 266, 269
Amu Darya (Oxus) river 377, 378, 400
Amundsen, Roald 214, 497
Anatolia 304, 384
Anawrahta, King (Burma) 441
Anchorage 50
Ancud, Gulf of 121, 123
Andalusian Mountains 282
Andersen, Hans Christian 227, 228
Andes 104, *105*, 106, 110, 111, 112, *115*, 121, 122, *122*, 124
Andino, General Tiburcio Carlas 83
Andorra **260**, *260*
Andrianampoinimerina, King (Madagascar) 204
Andros 88
Anegada 99
Angkorian dynasty 446
Anglo-Afghan Wars 400
Anglo-Burmese Wars (1824–26, 1852–53, 1885–86) 441–442
Angola **186–187**, *186, 187*, 200
Anguilla 91, **99**
Annaba 134
Annan Highlands 448

Annapolis 67
Antananarivo 205
Antarctic Peninsula 496, 497
Antarctic Treaty 496, 497
Antarctica **496–497**, *497*
Anti-Atlas Range 132
Anti-Lebanon Mountains 388
Antigua and Barbuda **91**, *91*
Antiparos *290*
Antrim Mountains 235
Antwerp 264, 272, 273
Anuradhapura *423*
Aouzou Strip 147
Apennines 292, 293, 294, 295
Apia 492
Apolima 492
Appalachian Mountains 13, 26, 27, 30, 32, 41, 65, 67, 71
Appenzell 320
Appian, Hassan Gouled 152
Apulia 220, 293, 297
Aqaba, Gulf of 143, 390, 391, 393
Aquino, Benigno Simeon 459
Aquino, Corazon 459
Aquitaine 251
Aquitaine 256
Arab–Israeli War (1967) 387, 392, 393, 394
Arab League 138
Arabian Peninsula 402, 407
Aragats 372
Aragón 280, 282
Arakan 442
Ararat, Mount *373*, 384
Ararat Plain 372
Aras river 372, 373
Arava Valley 393
Arc de Triomphe de l'Etoile, Paris *249*
Arctic 14, **15**, *15*, 17, 22, *22*, 23, 212, **215**, 222, 230
Ardashir Papakan 397
Ardennes 270, *271*, 273
Argentina **124–126**, *124, 125, 126*, 128
Aristide, Jean Bertrand 89
Arizona 26, 28, 30, 31, **56**, *57*
Arkansas **68**
Armenia 367, **372–373**, *372, 373*, 375
Armstrong, Neil Alden 43
Arnhem Land 476
Arno river 293, 295
Árpád 336
Arriaga, Manuel José de 288
Arroyo, Gloria 459
Artibonite river 88
Aruba **99**
Ascension Island **208–209**
Ashkhabad 376
Ashoka 412
Asi (Orontes) river 386, 387, 388
Asir Highlands 402, 403
Askaniya Nova Nature Reserve, Ukraine 360
Aspen *54*, 55
Assad, Hafez al- 387
Assam 411, 441, 442
Assyria 394
Astor, John Jacob 51
Asturias 285
Asuncíon 120, *120*, 124
Aswan High Dam 140, 141, *141*, 142, 143
Atacama, Peru 110
Atacama Desert 121, 122
Atafu 482
Atakora Mountains 167
Atakpá, Lake *80*
Atatürk, Kemal 384, 385
ath-Thani, Sheikh Hamed ibn Kalifa 405
Athabasca, Lake 20
Athens 303–304, 305, *305*
Atitlán, Lake *80*
Atlanta 45, 73
Atlantic City 66
Atlas Mountains 132, 134, 136
Auckland 482, 485
Augsburg, Peace of (1555) 313
Augustus, Emperor 295

Australasia 475–495
Australia **476–481**, *476–481*
agriculture *480*, 481; climate 477–478; economy 480–481; education 481; government 479, 480; history 478–480, *478*; industry 481; landscape 476–477, *476, 477*; people *478*, 480; transportation *480*, 481; vegetation and wildlife *476, 478, 478*; welfare 481
Australian Alps 477
Australian Broadcasting Corporation 481
Australian Capital Territory 476, 479, 480
Austria **322–325**, *322, 323, 324, 325*
Austrian Airlines 325
Austrian Broadcasting Company 325
Austro-Prussian War (1866) 313, 323
Auvergne 251
Aviaco 285
Avilés 285
Awash river 150
Ayer's Rock (Uluru) *476*, 477
Ayutthaya dynasty 444
Azerbaijan 373, **374–375**, *374*
Azócar, President Patricio Aylwin 123
Azores 286, **287**, *287*, 289
Azov, Sea of 360, 361
Azov Hills 360
Aztec civilization 77, *79*
Azuero Peninsula 86

B

Bab al Mandab *152*
Babenberg dynasty 323
Babur (the lion) 400, 412
Babylon 394, *394*
Bach, Johann Sebastian 316
Bacon, Francis 239
Baffin Island 14
Baghdad 220, 394, 395, *395*
Bahamas **88**, *88*, 97
Bahia (Salvador) 117
Bahrain **404**, *404*
Bahrain, Gulf of 404
Baikal, Lake 363
Baines, Thomas *478*
Baja California 76
Baku 374, *374*, 375
Bale Plateau 150
Balearic Islands 278, **282**, *282, 283*, 285
Bali 454, *456*, 457, *457*
Balkan Mountains 302, 340, 341, 342, 344, 345, 346, 349
Balkan Wars (1912–13) 344, 346, 350, 385
Ballarat 479
Baltic Heights 328
Baltimore 67
Baltimore, Lord 67
Baluchistan 416
Baluchistan Mountains 397
Banaba Island 489
Banat Lowlands 338, 339
Banda, Hastings 190
Bandama river 162
Bandar Beheshti (Chah Bahar) 398
Bandaranaike, Chandrika 422
Bandaranaike, Sirimavo 422
Bandaranaike, Solomon 422
Bangkok 444, 445
Bangladesh 410, 412, 417, **418–419**, *418, 419*
Bangui 171, *171*
Bangweulu, Lake 188
Bannack 53
Bantry Bay 244
Barbados **94**, *94*
Barcelona *283*, 285

Barisan Mountains 454
Barnacle, Nora 246
Barre, Major-General Siyad 153
Basel 320
Bashir, Lieutenant-General Omar Hassan Ahmad al-148, 149
Basque country 279, 283
Basra 395
Basse-Terre, Guadeloupe 100, 101
Basseterre, St Kitts-Nevis 90, 91
Bastia 252
Bastille *254*
Basutoland see Lesotho
Bata 172
Batavia (Jakarta) 267
Batista, Fulgencio 87
Batlle y Ordóñez, José 127
Baton Rouge 69
Battle of the Boyne (1690) 245
Battle of Culloden (1746) 235
Battle of Guilford Court House (1781) 72
Battle of Kennesaw Mountain (1864) *34*
Battle of Kosovo (1389) 342
Battle of the Nile (1798) 255
Battle of Rancagua (1814) 122
Battle of Trafalgar (1805) 255
Battle of Waterloo (1815) 238
Batumi 371
Baudouin I, King (Belgium) 272
Bavaria 310, *310*, 311
Bay of Bengal 410, 418
Bay of Biscay 278
Bay of Pigs 87
Bear Island 215
Beauharnais, Josephine de 255
Bechuanaland Protectorate 195
Beckett, Samuel 246
Bedie, Henri Konan 162
Beerenberg 215
Beethoven, Ludwig van 316, 324
Beijing 426, 429, 430, 431, 432, *432*, 433
Beira 191, 193
Beirut 388, 389
Bekáa Valley 388, 389, *389*
Belarus **358–359**, *358, 359*
Belau see Palau
Belaúnde Terry, Fernando 113
Belfast 235
Belgian Congo 271, 272
Belgian Radio and Television (BRT) 273
Belgian Television Broadcasting (RTBF) 273
Belgium **270–273**, *270, 271, 273*
Belgrade 342, *342*
Belize 76, **81**, *81*
Bella, Ahmed Ben 135
Belmopan 81
Belovezhskaya Puscha 358
Ben Nevis 234
Bengal 410, 416, 418
Benguela Current 186, 198, 200
Benin **167**, *167*, 168
Benue River 168
Berg, Alban 324
Bergen 215
Berlin *312*
Bern 319
Bernadotte, Jean-Baptiste (Charles John) 218
Bernese Alps 318
Bernini, Gian Lorenzo 300
Bessarabia 338, 357
Betancourt, Rómulo 106
Bette Mountain 138
Bhutan **415**, *415*
Bhutto, Benazir 417
Bhutto, Zulfikar Ali 417

Biafra 169
Bialowieza National Park, Poland 329
Big Sur 27
Bighole Basin, Battle of 54
Bihor Mountains 338
Bijagós Arquipélago 159
Bilbao 285
bin Laden, Osama 400, 401
Binger, Louis-Gustave 162
Bintimani Peak 160
Bío-Bío River 121
Bioko (Fernando Póo) 172
Birendra, King (Nepal) 420
Bishkek (Frunze) *379*
Bishop, Maurice 95
Bismarck, Otto von 254, 315
Bitter Lakes 143
Biwa, Lake 462
Biya, Paul 170
Black Forest 311
Black Hawk War (1832) 35, 60
Black Hills 58, *58*
Black Mountains 346
Black Volta river 163, 164
Blackwater river 244
Bled, Lake *347*
Bloemfontein 198
Blue Mountains, Australia *476, 477*
Blue Mountains, Jamaica 89
Blue Nile River 148, 150, *151*
Blue Ridge Mountains 27
Bo Hai 426
Boali 171
Boccaccio, Giovanni 295
Boer Wars (1899–1902) 199, 203
Boggy Peak 91
Bogotá 104
Bohemia 311, 332–333
Bokassa, Colonel Jean-Bédel 171
Bolilands 160
Bolívar, Pico 106
Bolívar, Simón 104, 106, 110
Bolivia **110**, *110*, 122
Bolkiah, Hassanal (Sultan of Brunei) *453*
Bologna 243
Bolsena, Lake 293
Bonaire 99, 100
Bonaparte, Joseph 281
Bonaparte, Louis-Napoleon 254
Bond, Alan *481*
Bongo Massif 171
Bongo, Omar 173
Bonin Islands 462, 463
Boone, Daniel 69
Borde 310
Bordeaux 257, 258
Borneo 450, 451, 453, 454, 455, 457
Bornholm 226, 229
Boru, Brian 245
Bosna river 349
Bosnia Herzegovina 37, 342–346, *345*, **349**, *349*
Boston 35, 63–64, *64*
Boston Massacre (1770) 34, 63
Botany Bay 479
Botha, Pieter Willem 199
Bothnia, Gulf of 216, 218, 222
Botrange 270
Botswana **195**, *195*
Botticelli, Sandro 295, *295*
Boudiaf, President Muhammad 135
Bougainville 487
Bougainville, Louis Antoine de 492
Boulogne 257
Boumedienne, General Houari 135
Boundary Range 33
Bounty 479, 484
Bourguiba, Habib ibn Ali 136–137
Bouterse, Lieutenant-Colonel Dési 109

Boyacá 104
Boxer Rebellion *430*, 431, 466
Brabant 272
Braga, Téofilo 288
Brahmaputra river 410, 418, *419*
Brasília 118
Bratislava 335
Brazil **116–119**, *116*, *119*, 288
Brazilian Highlands 116
Brazza, Pierre Savorgnan de 174
Brenner Pass 298
Brentano, Clemens 315
Breughel the Elder, Pieter 272
Brezhnev, Leonid 366–367
Bridgetown 94
Brisbane 481
British Airways 243
British Broadcasting Corporation 243
British Cameroon 168, 170
British Columbia 13, 16, 18, 19, 20
British Columbia Railway 20
British East India Company 412, 430, 452
British Guyana *see* Guyana
British Honduras *see* Belize
British Indian Ocean Territory 423
British Isles 233–247
British North Borneo Company 453
British Somaliland 153
British South Africa Company 188, 196
British Virgin Islands **99**
British Virginia Company 97
Brittany 250, *251*, 256
Brno 332
Broadcasting Corporation of New Zealand 485
Brooke, James "White Raja" 453
Brooks Range 33
Broz, Josip *see* Tito, Marshal
Bruce Peninsula 12
Bruckner, Anton 324
Bruges 273, *273*
Brunei **453**, *453*, 458
Brunelleschi, Filippo 295
Brunswijk, Ronnie 109
Brussels 270, 272, 273, *273*
Bucharest 338
Budapest 337
Buddha 420
Buenos Aires 120, 124, *125, 126, 126*
Buganda 178
Bugiri *178*
Bugti 453
Bukhara 377, 378
Bukovina 338
Bulgaria **340–341**, *340*, *341*
Bungo Channel 462
Bunya 203
Bunyoro-Kitara 178
Burgenland 324
Burgundy 251, 257
Burkina Faso 162, **163**, *163*
Burma *see* Myanmar
Burnham, Forbes 108
Burundi **176**, *176*, 177, 272
Bush, George W. 38
Buyoya, Major Pierre 176
Buzâu river 338
Byrd, Richard Evelyn 497
Byrranga Mountains 363

C

Cabinda 174, 186, 187
Cabo de São Vicente 286
Cabora Bassa Dam 193
Cabot, John 16
Cabral, Pedro Álvares 117, 287
Cabrera 282

Cabrillo, Juan Rodríguez 51
Cacheu river 159
Cadiz 278, 285
Cadiz, Gulf of 285
Caetano, Dr Marcelo 288
Cairngorm Mountains 234
Cairns 477
Cairo 140, 143
Calabar (Nigeria) 168
Calabria 292, 293
Caledon river 202
California 27, 28, 29, 30, 31, 32, *39*, 41, *41*, 42, 44, **51–52**, *52*
Callao 114
Calvert, Cecil and Leonard 67
Calvi 252
Camargue 252
Cambodia **446–447**, *446*
Cambrian Mountains 235
Cambridge University **243**, *243*
Cameroon **170**, *170*
Cameroon, Mount 170
Campos basin 119
Canada 10, **12–21**, *12*
agriculture 18–19; climate 14, *14*; economy 18–20; fishing and forestry 18–19, 20; fur trade 19; government 16–17, *17*; health and welfare 20; history 15–16, *17*, *20*; industry and resources 19–20; landscape 12–14; media 20; people 15, 16, 17–18; society 15–16; trade 20; transport and communication 20; vegetation and wildlife 14–15
Canadian Airlines 20
Canadian Broadcasting Commission (CBC) 20
Canadian National Railway 20
Canadian Pacific Railway 20
Canadian Radio-Television and Telecommunications Commission 20
Canadian Shield 12, 13, 28
Canary Islands 278, **280**, *280*, 283, 285
Canberra 476, 479, 480, 481
Cannon Temple, Tokyo *466*
Cantabrian Mountains 278, *278*
Canterbury Plains 482
Cape Canaveral 73
Cape Cod 26
Cape Colony 198–199, 202
Cape of Good Hope 198, 266
Cape Horn 121
Cape Inubo 463, 464
Cape Mount 161
Cape Sao Roque 116
Cape Town 198, *199*, 201, *201*, 209
Cape Verde Islands 159, **183**, *183*
Cape Vert Peninsula 156
Caprivi Strip 194
Carabobo 106
Caracas 106, 107
Cardamom Mountains 444, 446
Caribbean 75–101
Carinthia 324
Carnegie, Andrew 35, 66
Caroline Islands 466, 493
Carpathian Mountains 328, 329, 332, 334, *335*, 338, 360, 361
Carrera, Rafael 80
Cartier, Jacques 16
Casa Rosada, Argentina *125*
Casablanca 133
Casamance River 156
Casbah of Tifoultoute *133*
Cascade Range 28, 30, 51, *51*
Casco Bay *62*
Castile 280, 287
Castilla, General Ramón 113

Castro, Fidel 87, *87*
Catalonia 279, 280, 283
Catherine, Mount 140
Catherine II (the Great), Empress (Russia) 355, 365
Cauca river 104, *105*
Caucasus 362, 363, 364, 365, 372
Causses 251
Cavally river 161, 162
Cavelier, René-Robert 68
Cavour, Conte Camillo Benso di 295
Cayenne 128
Cayman Islands 97
Ceauşescu, Nicolae 338, 339, *339*
Cebu 458, 459
Central Africa 155–183
Central African Republic **171**, *171*
Central America 75–101
Central Europe 309–325
Central Range, New Guinea 486
Central Range, Taiwan 434
Central Siberian Plateau 362, 363
Central Valley, California 28, 30
Central Valley, Chile 121, 122, 123
Central Valley, Mexico 77
Cephalonia 303
Cerro del Aripo 96
Cerro del Punta 97
Cervantes, Miguel de *281*, 284
Cetinje 346
Ceuta 285, 287, 289
Ceylon *see* Sri Lanka
Chacabuco 122
Chaco War (1932–35) 120
Chad **147**, *147*
Chad, Lake 146, 147, 168, 170, 171
Chaillu, Massif du 173
Chakkri dynasty 444
Chalatenango Mountains 82
Champagne **258–259**, *259*
Champlain, Samuel de 16, 63
Yangtze (Chang) River 426, 427, 428, 433
Changshu 432
Channel Islands **234–235**, *235*
Channel Tunnel 243, *244*, 258
Chao Phraya River 444
Chari river 147
Charlemagne 253, 260, 265, 270, 280, 312, 321, 323, 332, 347
Charles, King (Portugal) 288
Charles I, King (England) 67, 237
Charles I, King (Czech Republic) 332
Charles II, King (England) 72, 237
Charles II, King (Spain) 271
Charles IV, Emperor (Luxembourg) 274
Charles V, Emperor (Germany) 312–313
Charles V, King (Spain) 266
Charles VI, King (Austria) 323
Charles VII, King (France) 253
Charles XII, King (Sweden) 217
Charles the Rash (Belgium) 270
Charleston 35, 71, 72, *72*
Charlottenberg Palace *312*
Château-Thierry 258
Chatkal Mountains 377
Cheju 472, 473
Chek Lap Kok (Hong Kong) 436
Chengdu *432*
Chennai 414
Cherangany Hills 179
Chernigov 220
Chernobyl nuclear explosion

(1986) 221, 223, 358, 359, 361, 369
Cherskogo Range 363
Chesapeake Bay 27, *67*
Cheviot Hills 234
Chiang Ching-kuo 435
Chiang Kai-shek 431, 435
Chicago 45, 60, *61*
Chieng-Mai 445
Chile *103*, 110, 113, **121–123**
Chiluba, Frederick 189
China **426–433**, *426–432*, 436–437
agriculture 432–433, *432*; climate 427; economy 432–433, *432*; government 428, 431–432, *431*; health and education 433; history 428–431, *429*, *430*, 466, 470, 473; industry 433; landscape 426–427, *426*; people 432; transportation *432*, 433; vegetation and wildlife 427
Chirac, President Jacques 254
Chittagong 418, *419*
Chobe river 195
Chopin, Frédéric 331
Chott El Jerid 136
Christchurch 485
Christian IV, King (Denmark) 227
Christian, Fletcher 479
Chugoku Mountains 462
Chukot Peninsula 363
Chulalongkorn, King (Thailand) 444
Chun Doo Hwan, President 473
Chun, Prince *430*
Chuquisaca (Sucre) 110
Churia Hills 420
Cibao Valley 90
Cienfuegos 87
Cincinnati 62
Cisco Sytems 44
Citlaltépetl 76
Ciudad Guayana 107
Clare river 244
Clark, William 34
Cleveland 62
Clunies-Ross, George 477
Clunies-Ross, John 477
Cochin 414, 448
Cockpit County 89
Cocos Islands **477**
Codrington family 91
Cold War 366–367
Collor, President Fernando 118
Colôane 437
Cologne 310, 317
Colombia 86, **104–105**, *104*, *105*, 106
Colombo 423
Colón Free Zone 86
Colorado 28, 30, 54, **55**
Colorado river 28, *42*, 49, 53
Columbia river and basin 28, 50, 51
Columbus, Christopher 33, *33*, 75, 87, 93, 95, 97, 98, 101, 104, 114, 280, 281
Comino 306
Commonwealth 89, 96, 157, 199, 206, 240, 245, 413, 422, 453, 480, 484, 486, 487, 488, 491
Commonwealth of Independent States 357, 359, 361, 368, 370, 373, 376–380
Communism Peak *see* Qullai Garmo
Como, Lake 292, 297
Comoros Islands **207**, *207*, 208
Compaoré, Captain Blaiso 163
Concarneau 257
Concepción 121
Confederation of Senegambia 156, 157
Conference of Non-Aligned Countries (1970) *188*
Confucius 428, 432

Congqing 432
Congo (D.R.O.) **174-175**, *174, 175*, 272
Congo (R.O.) 173, **174**, *174*, 271, 272
Congo (Zaire) river and basin 174, 175, 176, 177, 186
Congress of Berlin (1878) 340, 346
Congress of Vienna (1814–15) 313, 323
Connecticut 63, **64**
Constance, Lake 311, 318, 322
Constantine 134, 295
Conté, Brigadier-General Lansana 158
Cook, Frederick Albert 15
Cook, Captain James 52, 238, *478, 479*, 484, *491*, 497
Cook, Mount 482, *482*
Cook Islands 482
Cook Strait 482
Copacabana Beach 119
Copán *83*
Copenhagen 226, *227*, 229, *229*
Copernicus 331
Corantijn river 109
Coras Iompair Eireann (Irish State Transport) 247
Corcovado Peak *116*
Corderilla Central 85, 88, 90, 97, 104
Corderilla de Guanacaste 85
Corderilla de Mérida 106
Corderilla de Talamanca 85
Corderilla Occidental 104, 110
Corderilla Oriental 104, 110
Córdoba 280
Corfu 303
Corinth Canal 302
Cork 247, *247*
Cornwallis, Lord 34
Coronado, Francisco Vásquez de 56
Corsica *252, 252*
Cortés, Hernán 77
Corubal river 159
Corvo 287
Costa del Sol 279
Costa Rica **85**, *85*
Côte d'Azure 250
Côte des Blancs 258
Côte d'Ivoire *162, 162*
Cotentin peninsula 250
Cotonou 167
Cotopaxi, Mount 111
Cotswolds 234
Courantyne River 108
Cox's Bázaar 418
Crete 303, 305
Crimean Mountains 360
Crimean Peninsula 360, 361
Crimean War (1853–56) 338, 365
Cristóbal Colón, Pico 104
Croatia 342–346, **348**, *348*, 349
Crockett, Davy 46
Cromwell, Oliver 237, 245
Cuba 36, 73, **87**, *87*
Cumaná 106
Cumbria 234
Curaçao 99, 100
Cuzcatlán 82
Cuzco 113, 115
Cyclades *290*, 303, *303*, 305
Cyprus 304, **307**, *307*
Cyrenaica 138
Czech Republic **332–333**, *332, 333*, 334–335
Czechoslovakia 332, 333, **334–335**, *334*

D

da Vinci, Leonardo 295
Dacko, David 171
Daddah, President Moktar Ould 144
Dahab, General Abd ar-Rahman Siwar ad- 148
Dahomey *see* Benin
Dakar 156, *156*
Dali 284

Dallas 57
Dalmation coast 348
Damascus *386, 387*
Dammam 403
Danakil Plains 150
Daniel, Prince of Moscow 365
Danmarks Radio 229
Dante Alighieri 295, 296
Danton, Georges 254
Danube (Dunar) river 311, 317, 322, 325, 334, 336, 337, 338, 339, 340, 342
Dar es Salaam 180
Daravica 342
Darling river 477
Darwin, Charles 111
David, Jacques-Louis *255*
Davis, Jefferson 73
Dayton 62
Dayton Peace Accord (1995) 345
De Beers Consolidated Mines 202
de Gaulle, Charles 136, 254
De Haar Castle *266*
De La Warr, Baron 67
de Lesseps, Ferdinand 143
de Rivera, General Prima 281
de Valera, Eamon 245
Dead Sea 390, 391, 392
Deadwood 30
Déby, President Idriss 147
Deccan Plateau 410, 412
Defenestration of Prague (1618) 333
Delagoa Bay 192
Delaware 34, **67**
Delcambre *69*
Delfzijl 269
Della Robbia family 295
Demerara river 108
Denali National Park, USA *51*
Deng Xiaoping 431
Denmark 22–23, 213, 217–218, 219, 221, **226–229**, *226, 227, 229*, 231, 242
Denver 45
Des Moines 59
Detroit 61, *61*
Devil's Island *128*, 129
Devon 235
Dhahran 403
Dhaka 419
Dhofar 407
Diaz, Bartolomeo 198
Díaz, Porfirio 77
Diego Garcia 423
Dinaric Alps 302, 345, 348, 349
Dingle Bay 244
Diori, Hamani 146
Diouf, President Abdou 156
Dipendra, Crown Prince 420
Disney, Walt **46–47**, *47*
Disneyland 45, 46–47, 73
District of Columbia *68, 68*
Dixcove *165*
Djenné *145*
Djibouti 151, **152**, *152*
Djohar, Said Mohamed 207
Dnieper river 220, 358, 360, *361*, 362, 364
Dniester river 357
Dobruja Uplands 338
Dodecanese 303
Doe, Master Sergeant Samuel 161
Doiran, Lake 350
Dolomites 292
Doma Cathedral, Latvia *355*
Dominica *92, 92*
Dominican Republic 88, **90**, *90*
Donets Basin 361
Donets Ridge 360
Dordogne 251, 253
Dorr's Rebellion (1842) 65
Dorsale 136
Douala 170
Douglas, Stephen Arnold 35
Douro River 278, 286
Dover, USA 63, 67
Dow, Lake 195
Doyle, Sir Arthur Conan 108
Draa River 132
Dracula, Count *326*

Drakensburg Range 198, 202
Drava River 342, 347
Drenthe 268
Drimoleague *247*
Drumcliffe *245*
Duars Plain 415
Duarte, Juan Pablo 90
Duarte, Pico 90
Dubai 406
Dubček, Alexander *334*, 335
Dublin 245, 246, *246*, 247
Dublin, USA 63
Dubrovnik 348, *348*, 349
Duisburg *317*
Durango 78
Durban 201
Dürer, Albrecht *312*
Durmitor, Mount 345
Dutch East India Company 198, **266–267**, *267*, 412, 455
Dutch East Indies 267
Dutch Guiana *see* Suriname
Dutch New Guinea 267
Dutch West India Company 100, 109
Duvalier François ("Papa Doc") and Jean Claude ("Baby Doc") 89
Dzaoudzi 208, *209*
Dzhugashvili, Joseph *see* Stalin, Joseph

E

Eanes, General António Ramalho 288
East European (Russian) Plain 362, 364
East India Company 209
East Pakistan *see* Bangladesh
East Slavonica (Croatia) 348
East Timor 456, **457**
Easter Island **122–123**, *123*
Eastern Desert 140
Eastern Europe 327–351
Eastern Sayan Mountains 363
Ebro river 278
Ecuador 106, **111**, *111*, 115
Edéa 170
Edward I, King (England) 237
Edward II, King (England) 237
Edward, Lake 178
Efaté 490, *491*
Egede, Hans 22
Egypt **140–143**, *140, 141, 142*, 148
Egyptian Radio and Television Corporation 143
Eifel Mountains 310
Eiffel Tower *256*
El Greco 284
El Morro *98*
El Niño 112, 113, 114
El Salvador *82, 82*
El Salvador National Railroad 82
Elat 391
Elba 255
Elbe river 310, 317, 332
Elburz Mountains 397, *397*
Elephant Mountains 446
Elfreth's Alley *66*
Elgeyo Escarpment 179
Elgon, Mount 179
Elizabeth I Queen (England) 71, 237, 239
Ellesmere Island 14, 15
Ellice Islands *see* Tuvalu
Empty Quarter 402
Ems river 310
Endla State National Reserve, Estonia *354*
England *see* United Kingdom
Ennedi Plateau 147
Ente Finanziario per l'Industria Meccanica (EFIM) 298
Ente Nazionale Idrocarburi (ENI) 298
Entebbe 178
EPCOT 47
Epernay 258
Epirus 304
Equatorial Guinea (Río Muni)

172, 172
Er Rif 132, 133
Erdeni Dzuu *381*
Erebus, Mount 497
Erie Canal 61
Erie, Lake 12, 21, 61, 62
Erik of Pomerania (Sweden) 217, *218*
Erik the Red 23
Eriksson, Leif 16, 23, 33, 63
Eritrea 150, **151**
Espiritu Santo 490
Estaing 252
Esterel, Massif d' 250
Esterhazy family *337*
Estonia **354**, *354*, 367
Estrada, Joseph 459
Ethiopia 89, **150–151**, *150, 151*, 153
Ethiopian Highlands 140, 148
Etna, Mount 293, *296*
Etosha National Park, Namibia 194
Euboea 302
Euganean Hills 293
Euphrates river 386, 387, 394, 395
European Commission 350
European Community (EC) 240, 254, 325; *see also* European Union
European Union (EU) 41, 157, 214, 221, 225, 228, 231, 242, 245, 256, 268, 269, 272, 274, 283, 288, 297, 298, 304, 325, 354, 355, 356, 385, 485
European Court of Justice 274
European Free Trade Association (EFTA) 213, 221, 325, 355
European Investment Bank 275
European Space Agency 129
Europoort 268, 269, *269*
Everest, Mount 420
Everglades National Park, USA 73
Exeter, USA 63
Eyadéma, Colonel Gnassingbe 166
Eyre Basin 477
Eyre, Lake 477

F

Fabedougou *163*
Faeroe Islands 219, **227**, *227*, 228
Fairbanks 50
Faisal, King (Saudi Arabia) *402*
Faisal I, King (Iraq) 394
Fakaofo Island 482
Falaise de Banfora 163
Falkland Islands (Islas Malvinas) 125, **128**, *128*, 240
Falklands War (1982) 128, *128*, 209
Famagusta 307
Famenne depression 270
Farim *159*
Faro 289
Federation of South Arabia 399
Federation of the West Indies 93, 94, 97, 99
Ferdinand II, Emperor 313
Ferdinand of Aragon 280
Ferdinand of Habsburg 333
Ferdinand of Saxe-Coburg-Gotha 341
Fergana Basin 377, 378, 379
Fernandina 73
Fertöd Palace *337*
Fezzan 138, 139
FIAT 297
Fichtelgebirge 310
Fiji **490**, *490*
Finland 218, **222–225**, *222, 224, 225*
Finland, Gulf of 222, 354
Finnmark Plateau 212, 214
Fiordland National Park, New Zealand 482
Firestone Tire and Rubber Company 161
Flanders 266, 270, 272

Florence 295, **295**, 296
Flores 287, 454
Florida 26, 27, 31, 32, 47, **73**, *73*
Fogo 183
Fontvieille 261
Forestry Commission 242
Formentera 282
Formosa *see* Taiwan
Fort Sumter 72
Fournaise, Piton de la 208
Fox Corporation (USA) 45
Fram 15, *15*
France **249–261**, *249–259*
  agriculture and fishing 256–257; climate 251; dependencies 23, 100–101, 128–129, 208, 494–495, *495*; economy 256–258; education 259; government 254, 256, 257; health and welfare 259; history 253–254, *254, 255*; industry 257–258, *259*; landscape 250–251; media 259; people 256; transportation 258–259; vegetation and wildlife 251–252
Francia, José Gaspar Rodríguez 120
Francis II, Emperor (Austria) 313
Franck, César 272
Franco, General Francisco 280, 283
Franco-Merina Wars (1880) 204
Franco-Prussian War (1870–71) 315
Franconian Forest 310
Frankfurt *309*, 310
Franklin, John 17
Franz Ferdinand, Archduke 315, 323, 344
Fraser river and valley 16
Frederick VI, King (Denmark) 218
Frederick the Great (Germany) 313
Freetown 160
Frei, President Eduardo 123
Frémont, John Charles 53
French and Indian War (1754–63) 63, 73
French Cameroon 170
French Congo (Ubangi-Shari) 171, 174
French East India Company 412
French Equatorial Africa 147
French Guiana **128–129**, *128*
French Guinea 158
French Polynesia *494*, **495**
French Revolution 254, *254*
French Somaliland 152
French Togoland 166
French West Africa 144, 145, 146, 158
Freud, Sigmund 324
Friesland 264, 268
Frisland Islands 310
Fuerteventura 280
Fujairah 406
Fuji, Mount 462, *462*
Fujimori, Alberto 114
Fujiwara family 464
Fukuoka 469
Funchal 288
Fust, Johann 312
Futa Jalon Plateau 158, 159
Fyn Island 226, 229
Fyodor I, Tsar 365

G

Gabon **173**, *173*
Gadsden Purchase (1853) 35, 56
Galana River 179
Galápagos Islands 111
Galicia 278, 284, *284*
Galilee 388, 390
Galleria, Milan *298*
Galtieri, General Leopoldo 125
Galway 247

Gama, Vasco da 192
Gambia 156, **157**, *157*
Gambia river 156, 157, *157*, 158
Gamsakhurdia, Zviad 371
Gandhi, Indira 413
Gandhi, Mohandas 412
Gandhi, Rajiv 413
Ganges river 410, 411, 412, *412*, 418
Gansu *426*
García, Alan 114
Garda, Lake 292, 297
Gargano, Promentory of 293
Garibaldi, Guiseppe 295, 301
Garonne river 251
Gaudi, Antonio *283*
Gaultier de Varennes, Pierre 58
Gayoom, President Maumoon Ábdul 421
Gaza Strip 391
Gdansk 328, 330, 331, *331*
Gdynia 331
Geba river 159
Gediminas dynasty 356
Gediminas Square, Lithuania *356*
Gefara Plain 138
Gegam Range 372
Geneva 320
Geneva, Lake 251, 318
Geneva Convention (1864) 319
Genghis Khan 381, 400, 430
Genoa 295, 299
George II, King (England) 72
Georgia 367, **370–371**, *370, 371*
Georgia, USA 34, *34*, **72–73**, *72*
Georgian Bay 12
German Confederation 313, 321, 323
Germany **309–317**, *309–312, 314, 315, 317*
agriculture 316–317; climate 311; economy 316–317; education 317; government *314*, 316; history 312–316, 334; industry 316–317; landscape 310–311; media 317; people 316; reunification 312, 316; transportation 317; vegetation and wildlife 311; welfare 317
Geysir 230
Ghab Depression 386
Ghana **164–165**, *164, 165*, 166
Ghardaia *135*
Ghat 139
Ghent 266, 272, 273
Ghor Depression 392
Gibraltar **289**
Gilbert Islands 488, 489
Gilbert and Ellice Islands 488, 489
Gimie, Mount 92
Giotto 295
Gironde river 251
Giscard d'Estaing, Valéry 254
Gisozo *176*
Giza 141, *142*
Glåma River 212
Glarus 319
Globe Theatre, London *239*
Gobi Desert 381, *426*
Godunov, Boris 365
Golan Heights 386, 387, 391
Gold Coast 164, *165*, 166
Gold Rushes 30–31, 30, 50, 51–52, 53, 55, 58, 86
Golden Gate Bridge, San Francisco *52*
Golden Temple, Amritsar 413
Goldsmith, Oliver 246
Gomera 280
Goncz, Árpád 337
Good Friday Agreement (1998) 245
Gorbachev, Mikhail 335, 367, 368
Gordon, General Charles 148
Gosnold, Bartholomew 63
Göta River 216
Göteborg 216, *221*
Gotland 219
Gough Island 209
Gowon, Lieutenant-Colonel

Yakubu 168–169
Goya, Francisco de 284
Gozo 306
Gran Canaria 280
Gran Chaco 116, 120, 124
Gran Sasso d'Italia 293
Granada 280, 281
Grand Bahama 88
Grand Canal, China *432*
Grand Canyon 28, *49*
Grand Cayman 97
Grand Mosque, Tunis *137*
Grande-Terre 100, 101
Grand Teton National Park, USA 28
Gray, Robert 51
Graz 324, 325
Great Abaco 88
Great Barrier Reef 477, *477*
Great Bear Lake 12
Great Belt 226
Great Bermuda 97
Great Caucasus 370, *371*, 374
Great Dividing Range 476, 477
Great Escarpment 194, 198
Great Hungarian Plain 336
Great Karoo 198
Great Lakes 12, 14, 16, 18, 20, 21, 42, 45, 61
Great Northern War 217
Great Plains 13, 14, 28, 30, 31, 57
Great Pyramid, Egypt *142*
Great Rift Valley 150, 151, *155*, 175, 176, 177, 178, 179, *179*, 180, 190, 388, 390
Great Ruins 196, *197*
Great Salt Lake 28, 55
Great Slave Lake 12, 20
Great Swartberge 198
Great Usutu River 202
Great Wall of China 428, **429**, *429*, 430
Great Western Plateau, Australia 476
Greater Antarctica 496
Greater Himalayas 410, 415
Greece *291*, **302–305**, *302, 303, 305*, 350
Greenland **22–23**, *22*, 215, 228
Gregory the Illuminator *373*
Grenada **95**, *95*
Grenadines 95
Grieg, Edvard 214
Grimaldi family 261
Groningen 268
Grotto Beach *199*
Guadalajara 78
Guadalcanal 487
Guadalquivir river 278
Guadeloupe **100–101**, *100*
Guam 26, **494**
Guanajuato *77*
Guangdong 437
Guangxi *426*, 432
Guangzhou 433
Guatemala 76, **80**, *80*
Guayaquil 111
Guban 153
Guernsey 234
Guiana Highlands 106, *106*, 116, 158, *158*
Guinea 158, *158*
Guinea, Gulf of 164, 168, 170, 182
Guinea-Bissau **159**, *159*, 183
Guizhou *426*
Gulf War (1991) 37, 141, 385, 387, 388, 393, 394, 395, 396, 397, 402, 406
Gunnbjorn, Mount 22
Gupta dynasty 412
Gusmao, Xanana 457
Gustav I Vasa, King (Sweden) 217
Gustav II Adolf, King (Sweden) 217
Gustav III, King (Sweden) 218
Gutland *263*, 274, 275
Guyana 106, **108**, *108*
Guzmán, Abimael 114
Gyanendra, King 420

H

Ha'apai Islands 491
Haarlem 269
Haarlemmermeer 264
Habibie, Baccharuddin Jusuf 456
Habré, Hissen 147
Habsburg dynasty 266, 270, 271, 274, 312, 313, 319, 323, 333, 342, 348
Hague, The 268, 269, 345
Haile Selassie, Emperor 150
Hainan 433
Hainaut 272
Haiti 75, **88–89**, *88*, 90
Hajar Mountains 406, 407
Half Dome 32
Halla-san 472
Hallstatt 323
Hamburg 317
Hamilton 97
Hammarskjöld, Dag 175
Hamrin Uplands 394
Han river 472
Hanga Roa 111
Hangayn Range 381
Hangzhou 430, *432*
Hanlé Plains 152
Hannibal 279
Hanoi 449
Hanseatic League 217, 227
Harald Bluetooth (Denmark) 227
Harald Fairhair, King (Norway) 212
Harare (Salisbury) 196, 197
Hardangervidda 212
Hare, Alexander 477
Harerge Plain 150
Harold II, King (England) 237
Harz Mountains 310
Hassan II, King (Morocco) 132
Hathaway, Anne 239
Haud Plateau 153
Hautes Fagnes 270
Havana 87
Havel, Vaclav 335
Hawaii 26, *33*, 36, **52–53**, *52*
Haydn, Joseph 324
Hebbel, Friederich 315
Heidelberg 317
Heimaey 230
Heine, Heinrich 315
Hejaz 393, 402
Helgafell 230
Helgoland 310
Helsingor (Elsinore) 226
Helsinki 225, *225*
Henrietta Maria 67
Henriques, Afonso 287
Henry II, King (England) 245
Henry IV, King (France) 253
Henry VI, King (England) 239
Henry VII, Emperor (Luxembourg) 274
Henry VII, King (England) 237
Henry VIII, King (England) 237, *238*, 245
Henry, Duke of Saxony 312
Henry of Navarre 253
Henry the Navigator, Prince 287, 288
Henson, Matthew 15
Hentiyn Range 381
Herzegovina *see* Bosnia Herzegovina
Hermanus *199*
Hermon Mountains 388
Heyerdahl, Thor 214
Hidaka Mountains 462
Hideyoshi, Toyotomi 465
Hierro 280
High Atlas Mountains 132
High Veld 192, 196, 203
Hiiumaa 354
Himalayas 410–411, *410*, 415, 416, 420, 426, 440, 444
Hindu Kush 400
Hinggan Mountains 433
Hirohito, Emperor (Japan) 466
Hiroshima 466
Hispaniola 75, 88, 89, 90, 98
Hitler, Adolf 315, 323, 334

Ho Chi Minh 448–449
Ho Chi Minh City (Saigon) 449
Hodeida 399
Höfn *231*
Hokkaido 462, 463, 467, 468, 469
Holbein, Hans *238*
Hollywood 45, 52, *52*
Holmenkollen *214*
Holy Cross Mountains 328
Honda River 81
Honduras **83**, *83*
Honduras, Gulf of 80
Hong Kong **436**, *436*
Honshu 462, 463, 464, 468, 469
Hoover Dam (Boulder Dam) 42, 53
Horthy, Admiral Miklos 337
Houphouët-Boigny, Félix 162
Houses of Parliament, London *240*
Houston 57
Hoxha, General Enver 351
Huang (Yellow) river 426, 428, 432
Huascarán 112
Huayna Cápac 115
Hudson, Henry 15, 16, 65
Hudson Bay 12, 14, 20
Hudson's Bay Company 16
Huelva 285
Humboldt Current 111
Hundred Years War (1337–1453) 237, 253, 270
Hungary **336–337**, *336, 337*
Hunsrück 310
Huron, Lake 12, 21
Hurricane Mitch 83, 84
Hus, Jan 332
Husak, Gustav 335
Hushe river 417
Hussein ibn-Talal, King (Jordan) 393

I

Ibarra, José Mariá Velasco 111
Iberia 285
Ibiza 282
Ibsen, Henrik 214
Iceland 23, 219, **230–231**, *230, 231*, 242
Icelandair 231
Idaho **53**
Idris I, King 138–139
Ieyasu, Tokugawa 465
Ife 168
IJsselmeer 264, 268
Ikopa river 204
Ile de France 251
Iles des Saintes 100
Ilha das Rôlas 182
Iliescu, Ion 339
Illinois 35, 42, 45, **60**
Inaccessible Island 209
Inca civilization 110, 111, 112, 113, **114–115**, *115*, 122
Inch'on 472, 473
Independent Radio and Television Commission (Republic of Ireland) 246
Independent Television Commission (UK) 243
India **410–414**, *410*
agriculture *413*, 414; climate 410–411; economy 414; government *412*, 413–414; history *408*, 412–413; industry 414; landscape 410, *410*, 411, *411*; people *412*, 413–414; transportation 414, *414*; vegetation and wildlife 411; welfare 414
Indian Mutiny (1857) 412
Indiana **61–62**
Indiana Territory 60
Indonesia 267, **454–457**, *454, 455, 456, 457*
agriculture 457; climate 454–455; economy 457; government 456; history 455–456; landscape 454; people *455*, 456–457;

transportation 457; vegetation and wildlife 455, *455*; welfare 457
Indus river 410, 412, 416
Industrial Revolution 238, 240, 242, 254
Inle, Lake *442*
Inner Mongolia 381, 426, 432
Instituto per la Ricostruzione Industriale (IRI) 298
Intel 44
International Geophysical Year (1957–58) 496
International Red Cross 319
Inyangani, Mount 196
Ionian Islands 303, 304
Iowa **59**
Iran **397–398**, *397, 398*
Iran–Iraq War (1980–88) 143, 393, 394, 398
Iraq 141, **394–395**, *394, 395, 396*
Ireland, Republic of 235, 240, **244–247**, *244, 245, 247*
Irian Jaya (New Guinea) 456
see also West Papua
Irish Free State 245
Irrawaddy river 440, 442
Isabella of Castile (Spain) 280
Ise Bay 462
Iseo, Lake 292
Ishikari River 462
Isla de la Juventad 87
Islamabad 417
Islas de la Bahía *83*
Islas de Revillagigedo 76
Ismail, Mawlay 132
Israel 141, 143, **390–391**, *390, 391*
Issyk-Kul, Lake 379
Istanbul 220, 384, *384*, 385
Isthmus of Tehuantepec 76
Italy **292–299**, *292–299*
Itil 364
Ivan III (the Great), Tsar 364, 365
Ivan IV, Tsar 365
Ivory Coast *see* Côte d'Ivoire
Izborsk 220

J

Jabal Nafusah Mountains 138
Jacca 285
Jackson, President Andrew 69, 73
Jadwiga, Queen (Poland) 329, 361
Jaffna Peninsula 422
Jagiello, Grand Duke (Lithuania) 329, 361
Jakobshavn 22
Jamaica **89**, *89*, 97
James I (England) 237, 239, 245
James II (England) 237–238, 245
Jamestown, St Helena 209
Jamestown, USA 33, 71
Jan Mayen Island 212, 215
Jansson, Tove 223
Jansz, Captain Willem 478
Japan 44, *461*, **462–469**, *462–469*
agriculture 467–468; climate 463; economy 467–469, *469*; fishing and forestry 468; government 467, *467*; history 464–467, *464, 465, 466, 470, 473*; industry 468–469, *468, 469*; landscape 462–463, *462*; people *466*, 467; samurai **464–465**, *464*; transportation 469; vegetation and wildlife 463–464, *463*; welfare 469
Japanese Alps 462
Japanese Broadcasting Corporation (Nippon Hoso Kyokai) 469
Järnefelt, Aino 225
Jaruzelski, General Wojciech 330
Java 266, *267*, 454, 455, 457
Jebel Ansariye Mountains 386
Jebel Marra 148
Jeddah 403
Jefferson, Thomas 34, *58*, 71

Jersey 234–235, *235*
Jerusalem 390, 391, *391*, 392, 393
Jerz de la Frontera *284*
Jilin 470
Jimmu, Emperor (Japan) 464
Joan of Arc 253
Johannesburg 178, 198, 199, 201
John I, King (Portugal) 287
John I of Castile (Spain) 287
John II Casimir Vasa, King (Poland) 329
John III, King (Portugal) 117
John IV, King (Portugal) 288
John VI, King (Portugal) 117
John, Dom 117
John Frederick, Elector of Saxony 313
John Paul II, Pope 330, *330*
John, Prince of Luxembourg 332
Johnston Atoll **494**
Jonathan, Chief Leabua 202
Jordan 392–393, *392, 393*
Jordan river 390, 391, 392
Jos Plateau 168, 169
Jost Van Dyke 99
Jostedalsbreen 212
Jotunheimen Range 212
Joyce, James **246**, *246*
Juan Carlos, King (Spain) 283
Juárez, Benito 77
Juba river 153
Jubail 403
Judea 390
Jujuy *126*
Juli *113*
Jura Mountains, France/Switzerland 250, 311, 318
Jura Mountains, Poland 328
Jutland 226, 229, 310

K

Kabul 400, 401, *401*
Kabul river 400
Kádár, János 337
Kaema Plateau 470
Kafue National Park, Zambia 188
Kafue river 188
Kalahari Desert 194, 195, 198
Kalahari Gemsbok National Park, South Africa 198
Kaleteur Falls *108*
Kalimantan 454
Kaliningrad 356
Kalmar 217, *218*
Kalmar Union (1397) 212, 217, *218*, 227
Kamchatka Peninsula 363, *363*
Kamehameha I 52
Kamehameha III 52
Kampala 178
Kampuchea *see* Cambodia
Kanadevil sect *160*
Kandahar 401
Kandy 422
Kanem-Bornu 147, 168
Kano 169
Kansas **58**
Kanto Plain 462, 469
Kara Kum Desert 376, *376*
Karachi 417
Karajan, Herbert von 324
Karakumsky Canal 376
Karelia 217, 223, 362
Kariba, Lake 188, 196
Karimov, President Islam 377
Kariyoti river 307
Karlsefni, Thorfinn 16
Kartala, Mount 207
Kartalinian Plain 370
Karun river 397, 398
Kashmir 410, 417
Kassala *149*
Kathmandu 420, *420*
Katowice 331
Kaunda, Kenneth 189
Kazakhstan 365, **380**, *380*
Kebnekaise 216
Keeling Islands 477

Keflavík 231
Kemal, Mustapha 385
Kemi River 222
Kennedy, President John F. 36, 43
Kennedy Space Center 73
Kentucky 28, 34, 42, **69**, *69*
Kenya 155, **179**, *179*
Kenya, Mount 179
Kenyatta, Jomo 179
Kerekou, Lieutenant-Colonel Mathieu 167
Khalifa dynasty 405
Khartoum 148, *148*
Khazan 365
Khomeini, Ayatollah Ruhollah 398
Khor Virap Church, Armenia *373*
Khorasan Mountains 397
Khorat Plateau 444
Khrushchev, Nikita 335, 361, 366
Khuzistan 397, 398
Khyber Pass 400, 416
Kicking Horse river *13*
Kiel Canal 317
Kierkegaard, Soren 228
Kiev 220, 358, *360*, 361, 364
Kilimanjaro, Mount 180
Kim Il Sung 470
Kim Dae Jung 473
Kimberley Plateau 476
Kinabalu, Mount 450
Kindley Field *97*
King Charles Land 215
King Fahd causeway, Bahrain *404*
King, Martin Luther 37, 71
King Philip's War (1675–76) *63*
Kinguélé 173
Kinkakuji (Temple of the Golden Pavilion) *465*
Kinyeti, Mount 148
Kiribati 488, **489**, *489*
Kiruna 221
Kitakami Mountains 462
Kitami Mountains 462
Kitchener, Lord 148, *148*
Kithira 303
Kiva 377
Kivu, Lake 177
Kjolen 216
Klagenfurt 325
Klerk, Frederik Willem de 199
Klimt, Gustav 324
Klondike *13*, 16, 50
Knock 247
Knutsson, Karl 217
Kobar Sink 150
Kobe 463, 469
Koblenz 310
Kokoschka, Oscar 324
Kola Peninsula 362, 369
Kolingba, General André 171
Kolkata 414
Kolkhida Plain 370
Kolyma Range 363
Komoé river 162
Kompong Som 447
Kopaonik Mountains 342
Kopet Dag Mountains 376
Korab Mountains 350
Korea 466, 470, 473 *see also* North Korea *and* South Korea
Korea Strait 462
Korean Broadcasting Service 473
Korean War (1950–53) 36, 469
Korhogo *162*
Kornilov, Lavr 365
Kortrijk 273
Koryak Highlands 363
Koryakskayh Sopka *363*
Kosice 335
Kosovo 342, 343, 345, 350
Kossith, Lajos 336
Kostunica, Vojislav 345
Kotor *346*
Kountché, Colonel Seyni 146
Kourou 129
Kowloon 436
Kozuf Mountains 350

Kpémé 166
Krabi Ao Pranang *445*
Kracow 328
Krajina (Croatia) 348
Kraprayoon, General Suchinda 444
Kras Plateau 348
Krasnovodsk Plateau 376
Kroumirie Mountains 136
Kruger National Park, South Africa 198
Kufuor, John 165
Kura river 370, 374
Kurasawa, Akiro 464
Kuril Islands 363, 462, 466
Kurlgram *419*
Kursk Magnetic Anomaly 369
Kuwait **396**, *396*
Kuznetsk Basin 369
Kwajalein Atoll 493
Kwanga, Lake 178
Kykko monastery *307*
Kyoga, Lake 178
Kyoto 464, 465, *465*, 469
Kyrenia Mountains 307
Kyrgyzstan **379**, *379*
Kyushu 462, 463, 465, 469
Kyzyl Kum Desert 377

L

La Boca *125*
La Coruna 284
La Désirade 100
La Digue *181*
La Mancha *281*
La Palma 280
La Paz 110, *110*
La Rochelle 257
Labrador 14
Labrador Current 22
Ladakh *410*, *412*
Lagos 168, 169
Laikipia Escarpment 179
Lake District 234
Lakshadweep (Laccadives) **411**
Lambaréné *173*
Lanca National Park, Chile *103*
Langeland Island 226
Languedoc 257
Lanzarote 280
Laos **443**, *443*
Lapland 222
Las Palmas 280, 285
Las Vegas 53, *53*
Lassen Peak 28
Latium 297
Latvia 220, **355**, *355*, 367
Laxey Wheel *236*
Le Moyne, Jean-Baptiste 68
League of Nations 36, 170, 194, 319, 486, 492
Leane, Lough 244
Lebanon **388–389**, *388, 389*
Lebanon Mountains 388
Lebombo Escarpment 203
Lee Tenghui 435
Leeuwenhoek, Antonie van 266
Leeward Islands 90–92, 99
Leghorn 299
Legoland 229
Leh *412*
Leipzig 310
Lekhanya, Major-General Justin Metsing 202
Lempa river 82
Lena river 362, 363
Lenin, Vladimir Ilyich 365, *365, 366*
Léon, Juan Ponce de 73
Leopold II, King (Belgium) 175, *175*
Leopold of Saxe-Coburg 271
Leptis Magna *139*
Les Landes 257
Lesotho *185*, 198, **202**, *202*
Lesser Antarctica 496
Lesser Antilles 92–93, 95, 99, 100, 101
Lesser Himalayas 410, 415

Lesser Sunda Islands 454
Letsie III, King (Lesotho) 202
Levkas 303
Lewis, Meriwether 34
Leyte 458
Lézarde river 101
Li Hongzhang 430
Li river *426*
Liberia **161**, *161*
Libya 133, **138–139**, *138, 139*, 147, 306
Liechtenstein **321**, *321*
Liège 272
Liffey river 244
Ligurian Alps 292, 293
Lilongwe 191
Lim Fjord 226
Lim river 345
Lima 114
Limbang river 453
Limburg 264, 272
Limousin 251
Limpopo river 192, 195, 196, 198
Lincoln, Abraham **35**, *35, 58*, 60
Line Islands 489, *489*
Lingayen Gulf 458
Linz 325
Lipari Islands 293
Lisbon 286, 289
Lissouba, President Pascal 174
Litani river, Lebanon 388, 389
Litani river, Suriname 109
Lithuania 220, **356**, *356*
Little Belt Strait 226
Little Bighorn, Battle of 54
Little Caucasus 370, 372, 374
Little Cayman 97
Little Hungarian Plain 336
Little Poland 328
Liu Bang 428
Livingstone 188
Livingstone, David 188, *189*
Ljubljana 347
Lloyd's of London *242*
Lobamba 203
Lofoten Islands *211*
Logan, Mount 13
Logone river 147
Loire river 251, 265
Loma Mountains 160
Lombardy 297
Lombok 454
Lomé 166
Lon Nol, General 446
London 235, *239, 240*, *242*
Londonderry, USA 63
Long Island *26*
Lönnrot, Elias 223
Lopez, Carlos Antonio 120
Lopez, Francisco Solano 120
Lorient 257
Lorraine 254, 257, 270
Los Alamos 56
Los Angeles 52, *52*
Lot river 251, *252*
Louis XIII, King (France) 253
Louis XIV, King (France) 68, 254, 281
Louis XV, King (France) 254
Louis XVI, King (France) 254
Louis the Pious 260
Louisiana 34, 42, **68–69**, *69*
Louisiana Purchase (1803) 54, 55, 56, 57, 58, 59, 60, 68
Louisville *69*
Loukos river 132
Low Countries 263–275
Low Veld 196, 203
Lower Austria 324
Lows Lake 65
Luanda 182, 187, *187*
Luangwa National Park, Zambia 188
Luangwa river 188
Luapula river 188
Luba (Equatorial Guinea) 172
Lublin Plateau 328
Lucerne 319
Lucerne, Lake 318, *319*
Lugano, Lake 318
Lugenda river 192
Luleå 221
Lumumba, Patrice 175

Lüneburg Heath 310, *311*
Lungi 160
Lusaka 188, *188*, 189
Luther, Martin 313, 316
Luxair 275
Luxembourg 263, 272, **274–275**, *274*
Luxor *140*
Luzon 458, *458*, 459
Lyon 251, 258

M

Macao 287, 288, 430, **437**, *437*
McBarron, H. Charles *36*
McCarthy, Senator Joseph 36
McClintock, Francis *17*
Macdonnell Range 476
Macedonia, Former Yugoslav Republic of 302, 304, 342–346, **350**, *350*
Macgillicuddy's Reeks 244
Machu Picchu 113, 115, *115*
Mackenzie Range 13
Mackenzie river and valley 14, 20
McKinley, Mount 51
Madagascar **204–205**, *204, 205*
Madeira 286, 288, 289
Madikizela-Mandela, Winnie 199
Madrid 285
Magdalena river 104
Magellan, Ferdinand 453, 458
Magellan, Straits of 266
Maggiore, Lake 292, 297
Magritte, Rene 272
Mahabharat Range 420
Mahabuk Khan Mosque, Peshawar *417*
Mahé 181
Mahendra, King (Nepal) 420
Mahler, Gustav 324
Mahmud, Sultan (Turkey) 412
Mahón 282
Maïnassara, Ibrahim Barre 146
Main Ridge 96
Main river 310
Maine 33, 62, *62*
Mainz 312
Majorca 282
Makgadikgadi Pan 195
Malabar 411
Malabo 172
Malacca 266, 450, 451
Malacca, Strait of 451
Malawi 189, **190–191**, *190, 191*
Malawi, Lake 180, 190, 191, *191*, 192, 196
Malaya 451
Malaysia **450–451**, *450, 451*, 452, 454
Maldives **421**, *421*
Malév 337
Mali **145**, *145*
Malla dynasty 420
Malmo 229
Maloti Mountains 202
Malta **306**, *306*
Man, Isle of *236*, *236*
Managua 84, *84*
Managua, Lake 84
Mancham, James R. 181
Manchuria 426, 430, 431, 466
Manchurian Plain 426
Mandalay 440
Mandara Mountains 168
Mandela, Nelson 199
Manet, Edouard *256*
Mangoky river 204
Manhattan Island 34, 65, *65*
Manila 458, 459
Manitoba 13, 16, 18, 20
Mannar Island 422
Mano river 160, 161
Manono Island 492
Manuel II, King 288
Manzhouguo 466
Mao Zedong 431
Maori people 482, 484, 485, *485*, 497
Maputo (Lourenço Marques) 192, 193
Maracaibo, Lake 106, *106*, 107

Marburg 317
Marcos, Ferdinand Edralin 459
Margai, Sir Milton 160
Margrethe I, Queen *218*
Maria Theresa, Empress
   (Belgium) 271
Mariana Islands 466
Marie Antoinette 254
Marie Galante 100
Maritsa river 302, 340
Marlí, Jose 87
Marmolada 292
Marne Valley 258
Maromokotro Peak 204
Maroon Bells *54*
Marowijne river 109
Marquesas Islands 495
Marrakesh *133*
Marshall Islands 466, *474*, **493**,
   493
Marshall Plan 36, 257, 267
Martel, Charles 253
Martinique *100*, **101**
Mary II, Queen (England) 238
Maryland 33, *66*, **67**
Masaccio, Tommaso 295
Masaryk, Tomáš 333, 334
Mascarene Islands 206, 208
Maseru 202
Massachusetts 26, 33, 62,
   63–64, *64*
Massif Central 251, *251*, 252,
   252
Matabeleland 196
Mateus 277
Matsapa 203
Matterhorn 318, *320*
Mau Escarpment 179
Maui 33
Maures, Massif des 250
Maurice of Nassau, Prince 206
Mauritania 132, **144**, *144*
Mauritius 181, *206*, *206*
Mauritius Broadcasting
   Corporation 206
Maurya, Chandragupta 412
Maya Mountains 81
Mayan civilization 77, 80, *80*,
   83, *83*
Mayombé, Massif du 174
Mayon, Mount *458*
Mayotte 207, **208**, *209*
Maximilian 266, 270
Mayflower 63
Mazarin, Cardinal 254
Mazowiecki, Tadeusz 330–331
Mbabane 203, *203*
Mbasogo, Brigadier General
   Obiang Nuguema 172
Mecca 402, 403, *417*
Mecklenburg 310
Mecsek Hills *336*
Medici family 295
Medina 402, 403
Medjerda river 136
Meiji, Emperor (Japan) 466
Mekong river 426, 443, 446,
   446, 448
Melbourne 479, 481
Melilla 285
Mendoza 126
Menelik II 150
Menem, Carlos Saúl 125
Menéndez de Avilés, Pedro
   73
Mengistu Haile Mariam,
   Lieutenant-Colonel 150
Mengkabong *450*
Mesa Central 76
Mesaoria plain 307
Meseta 279, 284, 286
Mesopotamia 124, 394
Messina, Strait of 293
Metapán Mountains 82
Metternich, Prince (Austria)
   323
Metzanor 373
Meuse river 251, 264, 270, 272
Mexican War (1845) 56
Mexicana 79
Mexico 35, 44, 51, 53, 56, 57,
   76–79, *76*, *77*, *79*, 83
Mexico City 76, 77, 78, 79
Mexico, Gulf of 30, 42, 57, 76
Miami *38*, 73, *73*
Miami Confederation 61

Michelangelo Buonarroti 295,
   300
Michigan **61**, *61*
Michigan, Lake 60, 61
Micronesia **494**
Microsoft 44
Middle Atlas Mountains 132
Middle East 383–407
Middle Veld 196
Midway Islands **494**
Mieszko I (Poland) 329
Milan 296, *298*
Milosevic, Slobodan 343, 344,
   345
Minahassa 454
Mindanao 458
Mindoro 458
Minneapolis 59
Minnesota *30*, 39, **59**
Minorca 282, *282*
Minsk 359, *359*
Miranda, Francisco 106
Miró, Joan 284
Misery, Mount 90
Mishima, Yukio 465
Mississippi 70, *71*
Mississippi river 27, 28, 41, 45,
   59, 60, 68–69, *71*, 72
Missouri 42, **60**
Missouri Compromise 60, 62,
   68
Missouri river 28, 34, 59
Mitchell, Mount 27
Mitsotakis, Constantine 304
Mittelland plateau 318
Mitterand, François-Maurice
   254
Miyajima Island *461*
Mjøsa, Lake 212
Mobutu, Colonel
   Joseph-Désiré 175
Moco, Mount 186
Mogadishu 153
Mohammed V, Sultan 132
Mojave Desert *45*
Moldova 338, **357**, *357*, 367
Mole Game Reserve, Ghana
   164
Molina, Rafael Trujillo 90
Moluccas (Spice Islands) 454,
   455
Mombasa 178, 179
Monaco **261**, *261*
Monaco Grand Prix *261*
*Monet Working on his Boat in
   Argenteuil* 256
Mongkut, King (Thailand) 444
Mongolia 363, **381**, *381*
Monrovia 161, *161*
Mont Blanc 292
Mont Blanc Tunnel 298
Mont Cenis Pass 298
Montagne de Reims 258
Montaña, Peru 112
Montana, USA **53–54**, *54*
Monte Albán *79*
Monte Amiata 283
Monte Carlo 261
Monte Corno 293
Monte Rosa 292, 318
Montenegro 342–345, **346–347**,
   346
Monterey Peninsula *27*, 51
Monterrey 78, 79
Montevideo 127, *127*
Montreal *14*, 16, *19*, 20, 21
Monts de Cristal 172
Monts Karre 171
Montserrat **99**
Monument Valley, USA *57*
Moosonee 20
Mootwingee National Park,
   Australia *478*
Moraca River 345
Moravia 332, 333
Moravian Gate 328
Moresby, John *486*
Morne Diablotin 92
Morne Trois Pitons National
   Park, Dominica 92
Moro river 160, 161
Morocco **132–133**, *132*, *133*
Morón 107
Moroni 207 *207*
Morphou Bay 307
Moscow 255, 353, 365, 367,

369
Moselle river 251, 274, 275,
   310, 317, *317*
Mosquito Coast 84
Mossadeq, Mohammad 398
Mostar 345
Moulouya river 132
Mozambique 190, 191,
   192–193, *192*, *193*
Mozart, Wolfgang 324
Mswati III, King 203
Mshweshwe 202
Mshweshwe II 202
Mubarak, Hosni 141
Muchinga Escarpment 188
Müellerthal 274
Mugabe, Robert 196, *196*
Mugodzhary Hills 380
Muhammad 399, 402
Muhammad Ahmed 148
Muhammad Ali 141, 148
Muharraq 404
Mujib ur-Rahman, Shaikh 418
Muluzi, President Bakili 190
Mumbai 414
Mumtaz Mahal *408*
Munch, Edvard 214
Munich *314*, *315*
Munich Agreement (1938) 334
Munsterland 310
Mures river 338
Murom 220
Murray river and basin 477,
   478
Mururoa Atoll 495
Musala 340
Musandam Peninsula 406, 407
Muscat 407
Museveni, Yoweri 178
Musgrave Range 476
Musharraf, General Pervez
   178
Mussolini, Benito 295, 300
Mutebi II, Ronald Muwenda
   178
Mutesa II, King (Buganda) 178
Mutsuhito, Crown Prince
   (Japan) 466
Muzorewa, Bishop Abel 197
Mwali (Mohéli) 207
Mwene Matapa 192
Mweru, Lake 188
Myanmar *439*, **440–442**, *440*,
   441, *442*
Mykonos *305*
Mývatn, Lake 230
Mzab oasis *135*

N

Nacala 191, 193
Nader Shah 398
Nagasaki 465, 466, *468*
Nagorno–Karabakh 372, 373,
   374, *374*, 375
Nagoya 462, 466
Naguib, General Muhammad
   141
Nagy, Imre 337
Nairobi 179
Nairobi National Park, Kenya
   179
Naivasha, Lake *155*
Najd 402
Najibullah, President
   Mohammad 401
Nakhichevan 373, 374, 375
Naktong river 472
Namib Desert 194, 196
Namibia 186, **194**, *194*, 199, 200
Nampula 193
Namur 272
Nanjing 431
Nansen, Fridtjof 15, *15*, 214
Nantes 258
Naples 295, 299
Naples, Bay of 293
Napoléon I, Emperor (France)
   68, 209, 238, *249*, 254, **255**,
   255, 266, 271, 274, 281, 288,
   306, 313
Napoleon III, Emperor
   (France) 254, 266
Napoleonic Wars (1796–1815)
   108, 181, 218, 228, 252, 255,

301, 323, 329, 365
Narayanan, Kocheril Raman
   414
Narrows, The 90
Narvik 215
Nashville 70, *70*
Nasser, Colonel Gamal Abdel
   139, 141, 143
Nasser, Lake 140, 142
Natal 198, 199
National Broadcasting
   Corporation (USA) 45
Nauru **488–489**, *488*
Navigator Islands see Samoa
Nazarbaev, President
   Nursultan 380
Nazi-Soviet Non-Aggression
   Pact (1940) 355, 356
Ne Win, General 442
Neblina, Pico da 116
Nebraska 39, **59**
Neckar river 310
Negev Desert 390, *391*
Negros 458, 459
Nehru, Jawaharlal 413
Neisse River 310
Nelson, Admiral Horatio 255
Neman River 356
Nemi, Lake 293
Nemrut, Mount *384*
Nemuro Strait 462
Nemy (Mauritania) 144
Nepal 415, **420**, *420*
Neris river 356
Nestor Chronicle 220
Netherlands 99–100, 139,
   264–269, *264*, *266*, *269*, 271
Netherlands Antilles **99–100**
Netherlands Broadcasting
   Corporation 269
Neuchâtel, Lake 318
Neusiedler, Lake 322
Neva river 365
Nevada 30, 31, 41, **53**, *53*
Nevis 90–91
Nevsky, Alexander 365
New Amsterdam 34, 65, 109
New Britain 486
New Brunswick 13, 16, 20
New Caledonia **495**
New England 27, *27*, 33, 62–65
New Guinea see West Papua,
   Irian Jaya
New Hampshire 33, **63**
New Haven 33, 64
New Hebrides see Vanuatu
New Ireland 486
New Jersey 34, **66**
New Mexico 28, 39, **56**
New Orleans 45, 68, 69, *69*
New Providence 88
New South Wales 476, *478*,
   479, 480
New Territories 436
New Waterway 264
New York 34, 35, 37, 39, *39*, 40,
   45, 65, *65*, 66, 109
New York State 26, 27, 63, **65**,
   65
New Zealand **482–485**, *482*,
   483, *485*
Newfane *63*
Newfoundland 13, 14, 16, 19,
   20
Ngoma Zulu tribe *198*
Ngor *156*
Ngorongoro Crater 180
Niagara Falls 12, 65
Niamey 146
Niari Valley 174
Niassa *193*
Nicaragua 83, **84**, *84*
Nicaragua, Lake 84
Nicholas II, Tsar (Russia) 223,
   225, 365
Nicolet, Jean 60
Nicosia 307
Nidwalden 318
Nidze Mountains 350
Niger 139, **146**, *146*
Niger river 145, 146, 158, 167,
   168, *169*
Nigeria **168–169**, *168*, *169*, 170
Nightingale Islands 209
Nile river and valley 140, *140*,

141, 142, 143, 148, 149, 176,
   177, 178
Nimba, Mount 161, 162
Nimeiry, Colonel Gaafar
   Mohamed el- 148
Nimule 148
Ningxia 432
Nissan **469**
Niue Islands 482
Nixon, President Richard 37,
   45, 431
Niyasov, President
   Saparmurad 376
Nizwa *407*
Njazidja (Great Comoro) 207
Nkhotakota *191*
Nkomati Accords 192
Nkomo, Joshua 196, 197
Nkrumah, Kwame 164
Nobi Plain 462
Nobunaga, Oda 465
Nong Het 443
Nord, Massif du 88, 90
Nordic Countries 211–231
Norfolk Island **479**, 484
Noriega, General Manuel 86
Normandy 250, 256
Norrland 216, 220, 221
North America 12–73
North American Free Trade
   Agreement 44
North Arabian Desert 392
North Atlantic Drift 22, *212*,
   215, 216, 226, 227, 230, 236,
   244, 270
North Atlantic Treaty
   Organization (NATO) 213,
   218, 228, 231, 240, 268, 272,
   274, 283, 304, 316, 345, 385
North Carolina 27, 32, 34, 69,
   72
North Dakota **58**
North German Plain 310
North Holland 264, 268
North Korea **470–471**, *470*, *471*
North Pole **15**, *15*, 215
North Yorkshire Moors 234
Northern Africa 131–153
Northern Eurasia 353–381
Northern Ireland *234*, 235,
   236, 237, 240, 241, 245 see
   also United Kingdom
Northern Mariana Islands
   494–495
Northern Range 96
Northern Territory 476, 480,
   480
Northrup, George *30*
Northwest Territories, Canada
   16, *17*, 19, 20
Norway *211*, **212–215**, *212*, *214*,
   215, *217*, 218, 219, 221, 227,
   231, 236
Nosy Komba *205*
Nouâdhibou (Mauritania) 144
Noumea 495
Nova, Joao da 209
Nova Scotia 13, 16, 20
Novgorod 220, 364, 365
Novotny, Antonin 335
Nuba Mountains 148
Nukunonu Island 482
Nunavut, Canada 16
Nyasaland 189, 190, 196
Nyerere, Julius 180
Nyika Plateau 179, *191*
Nzérékoré *158*
Nzwani (Anjouan) 207

O

Ob river 363
Obasanjo, Olusegan 168
Obock 152
Obote, Milton 178
O'Brien, Flann 246
Oder river 310, 317, 328
Odessa 361
Oesling 274
Office de
   Radiodiffusion-Télévision
   Française (ORTF) 259
Ogaden 150, 153
Oglethorpe, James Edward 72

Ogo Highlands 153
Ogooué river 173, *173*
O'Higgins, Bernardo 122
Ohio 34, 42, **62**
Ohio river 28, 61, 62
Ohrid, Lake 350
Ojukwu, Lieutenant-Colonel C. Odumegwu 168
Oka river *363*
Okavango river and basin 195, *195*
Okinawa 467
Oklahoma **56–57**
Oklahoma City 57
Oktoberfest 315
Öland *218*
Olduvai Gorge 180
Olof Skötkonung, King (Sweden) 217
Olt river 338
Olympic Airways 305
Olympic Mountains 29
Olympio, Sylvanus 166
Oman **407**, *407*
Oman, Gulf of 398, 406, 407
Oñate, Juan de 56
Ontario 12, 14, 16, 18, 20
Ontario, Lake 12, 21
Opium Wars (1839–40; 1856–60) 430, 436
Oporto 289
Orange Free State 199
Orange river 198, 199, 200, 202
Ore Mountains *293*, 310, 332, 334
Oregon 28, 29, 30, 31, 33, 45, **50–51**, *51*
Oriente 110
Orinoco river 104, 106, 107
Orkney Islands 234
Øresund Bridge 229
Orlando 73
Osaka 462, 463, 469
Oscar II, King (Sweden) 218
Oslo 212, 215
Osman I (Turkey) 384
Ostende *273*
Othrys Range 302
Otto I (Germany) 312
Ötztaler Alps 322
Ouachita Mountains 68
Ouaddai Mountains 147
Ouarzazate *133*
Ouatchi Plateau 68, 166
Oulu River 222
Our River 274
Ousmane, Mahamane 146
Outer Hebrides 234
Outer Mongolia *see* Mongolia
Overseas Highway, Florida 32
Owen Falls Dam 178
Oxford University **243**
Oyashio Current 463
Öz Beg 377
Ozark Plateau 28, 68

P

Pacific Islands 475–495
Pacific Islands, Trust Territory of 26
Pacification of Ghent (1576) 266
Paektu, Mount 470, *471*
Pagalu (Annobón) 172
Pagan 440, 441, *441*
Pahlavi, Mohammad Reza Shah (Iran) 398
Pahlavi, Reza Shah (Reza Khan) 398
Pakaraima Mountains 108
Pakistan 413, **416–417**, *416*, *417*, 418
Pakistan International Airlines 417
Palau **493**
Palawan 458
Palestine 390–391, *392–393*
Palk Strait 422
Palma 282, 285
Palme, Olaf 218
Pamir Massif 378
Pampas 124, 126

Pamukkale *385*
Pan-American Highway 85, 123
Panama **86**, *86*, 104
Panama Canal 86, *86*
Panay 458, 459
Pannonian Plain 342, 345, 347, 348
Pantanal 116, 117
Papandreou, Andreas 304
Papua New Guinea 454, 457, **486**, *486*, 493
Paraguay 110, **120**, *120*, 124
Paraguay River 120
Paraná River 116, 120, 124
Parinacota *121*
Paris 243, *249*, 251, *252*, *256*, 258
Paris Basin 251, 252, 257
Park Chung Hee, Major-General 473
Parnell, Charles Stewart 245
Patagonia 124, 126
Patasse, President Ange-Felix 171
Peace of Augsburg (1555) 313
Peace of Westphalia (1648) 266, 313, 319
Pearl Harbor 53, 466, 479
Peary, Robert **15**, *15*
Pedias River 307
Pedras Tinhosas 182
Pedro, Dom 117
Pedro II King (Brazil) 118
Peggy's Cove *20*
Peipus, Lake 354, 364
Pelée, Mount 100, 101
Peloponnese 302, 303, 304
Pemba 180
Pemex 79
Penn, William *33*, 34, 66, 67
Pennine Alps 318
Pennines 234
Pennsylvania 27, 34, 42, 44, **66**, *66*, 67
Pensacola 73
Pentagon 68
Pepin 295
Pereira, Pres. Aristides 183
Pérignon, Dom Pierre 258
Perón, Isabel 125
Perón, Juan Domingo 125, 126
Perry, Commodore Matthew Calbraith 465
Perth, Australia 481
Peru 110, 111, **112–114**, *112*, *113*, *114*, *115*, 122, 124
Pescadores 434
Peshawar 417, *417*
Pétain, Marshal Philippe 80
Petén 80
Peter I, the Great Vladika 346
Peter I (the Great), Tsar 217, 364, 365
Petra *392*, 393
Petrograd *see* St Petersburg
Petrovich Njegosh family 346
Peugeot-Citroën 259
Philadelphia 35, 66, *66*
Philip I (the Handsome) (Spain) 266, 270
Philip II, King (Spain) *238*, 266, 281, 288, 459
Philip II of Macedonia 304
Philip V, King (Spain) 281
Philip VI, King (France) 253
Philippine Trench 458
Philippines 36, **458–459**, *458*
Phlegraean Fields 293
Phnom Penh 446, *446*, 447
Phoenix 56
Phoenix Islands 489
Phuket 444
Picasso, Pablo 284
Pico 287
Pidurutalagala 422
Piedmont 297
Piłsudski, Józef 329
Pinang 450
Pindus Mountains 302
Pinochet Ugarte, General Augusto 123
Pirin Mountains 340
Pisa *294*
Pitcairn Island 479, **484**
Pitons, The *93*

Pittsburgh 45, 66
Piva River 345
Pizarro, Francisco 113
Plaisance 206
Plate River 124
Platte Valley 59
Pliny the Elder 280
Plovdiv *341*
Plymouth, Montserrat 99
Plymouth, USA 33, 63
Plzen 333
Po River 298
Pobedy Peak 379
Podgorica (Titograd) 346
Point-Central (Mauritania) 144
Pointe-à-Pitre 101
Pol Pot 446, 447
Poland **328–331**, *328*, *330*, *331*, 356
Polish Radio and Television (PRT) 331
Polotsk 220
Polskie Linie Lotnicze (LOT) 331
Polynesia 482–483, *482*, 488, 489, 492, 495, *495*
Pomerania 328
Pompeii 293
Pompidou, Georges 254
Popocatépetl 76
Port Bouet 162
Port Dauphin 70
Port Harcourt 169
Port Jackson 479
Port Louis 206
Port Moresby *486*
Port Said 140, 143
Port Stanley *128*
Port Sudan 149
Portland *51*
Portland Head, USA *62*
Porto Santo 288
Portsmouth, USA 63
Portugal 277, **286–289**, *286*, *287*, *288*, 437
Port–Vila 490
Post, Telegraph and Telephone (PTT) 269
Potaro River *108*
Poti 371
Potosí 110
Powell, Lake *29*
Prague 332, *332*
Premadasa, President Ranasinghe 422
Prespa, Lake 350
Pretoria 198
Préval, Renee 89
Prevlaka peninsula (Croatia) 348
Primavera, La 295
Primorski 364
Prince Charles Mountains 497
Prince Edward Island 13, 16
Prince Rupert 20
Princetown, Battle of (1777) 66
Príncipe **182**
Pripyat River 358, 360
Professor van Blommestein, Lake *109*
Provence 251
Providence 65
Prut River 357
Puerto Rico 26, 36, **97–98**, *98*
Puertos Barrios 82
Puget Sound 28, 50
Punjab 416
Pusan 473
Putin, Vladimir 368
Puy de Sancy 251
Puy Mary *251*
Puyi, Emperor (Xuan Tong) **430**, *430*
P'yongyang 470, *471*
Pyrenees 250, 251, 252, 256, 260, *260*, 278, *278*, 279

Q

Qabus ibn Sa'id, Sultan (Oman) 407
Qaddafi, Colonel Muammar 138, 139
Qait Bay *141*

Qajar dynasty 398
Qantas 481
Qatar 382, **405**, *405*
Qawz 148
Qin Shi Huangdi 429
Qualibou 92
Quebec 12, 13, 16, *17*, 18, 19, 20
Queen Charlotte Islands 13
Queensland 476, 477, 479, *480*, 481
Quito 111, *111*, 115
Qullai Garmo 378
Qurnet es Sauda 388

R

Rabat 133
Radama I King (Madagascar) 204
Radio Authority (UK) 143
Radio Telefis Eireann 247
Raffles, Sir Thomas Stamford 452
Rainier, Mount 32
Rainier III, Prince 261
Raleigh, Sir Walter 33
Ramanantsoa, Major-General Gabriel 204
Ramos, Fidel 459
Ramses II *141*
Randstad 269
Rangoon 440, 442
Raphael 295, 300
Rapid City 58
Ras al-Khaimah 406
Ras Dashan 150
Ratsiraka, Lieutenant-Commander Didier 205
Ravenna 295
Rawlings, Flight-Lieutenant Jerry 164, 165
Recife (Pernambuco) 117
Red (Sichuan) Basin 426
Red River 448
Red Sea Mountains 140, 148
Red Volta River 163, 164
Redonda 91
Reims 258
Rembrandt van Rijn 266
Renaissance 295, 296, 300, *312*
René, Albert 181
Reno 53
Republic of Gran Colombia 104, 106
Réunion (Bourbon) **208**, *209*
Reykjavik 230, 231
Rharb, the 132
Rhee, President Syngman 473
Rheris River 132
Rhine Confederation 313, 321
Rhine River 250, 251, 264, 265, 268, 310, 311, **314–315**, *315*, 317, 318, 321
Rhode Island 33, **65**
Rhodes 303, 305
Rhodes, Cecil 188, 195, 196
Rhodesia 188, 189, 190, 196–197
Rhodope Mountains 302, 340
Rhön Plateau 310
Rhône River and Delta 250, 251, 252, 256, 318
Richard's Bay 201
Richelieu, Cardinal 253
Richmond 72
Riel, Louis 16
Riga 355, *355*
Riga, Gulf of 354, 355
Rila Massif 340
Rio Bravo del Norte *see* Rio Grande
Rio de Janeiro 116, *116*, 117, 118, 119
Rio Grande 57, 76
Rivière Noire, Piton de la 206
Riyadh *402*, 403
Roanoke 33, 72
Robert the Bruce 237
Robespierre, Maximilien 254
Rocca Tower Fortress 301, *301*
Rockefeller John D. 35
Rocky Mountains 13, *13*, 18,

26, 28, 34, 41, 53, 55
Rodrigues 206
Rodríguez, Pres. José Joaquin 85
Roggeveen, Jacob 122
Roh Tae Woo, President 473
Romania *327*, **338–339**, *338*, *339*
Romanov dynasty 365
Romanov, Michael 365
Romblon 458
Rome 295, 298, 299
Rondonia *119*
Ronne Ice Shelf 497
Roosevelt, Franklin D. 36
Roosevelt, Theodore 58
Roraima, Mount 108
Roseau 92
Ross Ice Shelf 497
Ross Island 497
Ross, James Clark 497
Rostov 220
Rotterdam 264, 269, *269*
Royal Dutch Airlines (KLM) 269
Royal Nepal Airline Corporation 420
Rubens, Pieter Paul 272
Rudolf I (Germany) 312
Ruhr Valley 317, *317*
Rupununi Valley 108
Rushmore, Mount 58
Russia **220**, 222, 223, 224, *353*, **362–369**, *362*, *363*, *364*, *365*, *367*, *368*
agriculture 369; climate 363; economy 369; government 365, *367*, 368; history 364–367, *367*; industry 369; landscape 362–363; people 368; transportation 369; vegetation and wildlife 363–364; welfare 369
Russian Civil War (1918–22) 361, 366, 381
Russian Revolution (1917) 355, 356, 361, **365**, *365*, 366, 371, 376, 379
Russo-Japanese War (1904–05) 365, 466
Russo-Polish War (1919–21) 358–359
Russo-Turkish War (1877–78) 342, 346, 372
Ru'usal-Jibal 407
Ruvubu River 176
Ruwenzori Range 178
Ruzizi River 176
Rwanda 175, **177**, *177*, 272
Ryazan *363*
Ryukyu Islands 462, 463, 466, **467**

S

Saaremaa 354
Saarinen, Eliel and Eero 223, 225
Saba 99
Sabah 450, *450*, 451, 454
Sabah family 396
Sabha Road 139
Sable, Jean-Baptiste-Point du 60
Sacramento 30, 51
Sadat, Anwar 141
Saddam Hussein 394, 395, 396
Sahara Desert *131*, 132, 134, *134*, 135, *135*, 136, 138, 144, 145, 148
Sahel 144, 145, 146, 147, *147*, 163, 168
Sahul Shelf 454
Saigon *see* Ho Chi Minh City
Saimaa, Lake 222
St Barthélemy 100
St Basil's Cathedral *352*
St Bernard Tunnel 298
St Croix 98
St Domingue 89
St Genevieve 60
St George's 95, *95*
St Goarshausen 315, *315*
St Gotthard Massif 318
St Gotthard Pass 298

St Helena **208–209**, 255
St Helens, Mount 29, 50, 51
St Helier 235
St John, Virgin Islands 98
St John's, Antigua 91
St Kitts (St Christopher) 90–91
St Kitts-Nevis **90–91**, *90*, 99
St Lawrence, Gulf of 12
St Lawrence River and Seaway 12, 16, 20, **21**, *21*, 45
St Louis 34
St Lucia **92–93**, *92*, *93*
St Marinus 301
St Martin 100, 101
St Mary's City 67
St Mesrob 373
St Naum Monastery *350*
St Paul 59
St Peter's Basilica and Square, Vatican City 300, *300*
St Petersburg *364*, 365, *365*, 366, 367, *368*
St Pierre and Miquelon 19, **23**
St Sergius, Trinity Monastery of, Sergiyev Posad *367*
St Thomas 98, *98*
St Vincent and the Grenadines **93**, *93*
Saipan Island 495
Sakhalin Island 363, 462, 463, 466
Saladin 386
Salazar, Antonio de Oliveira 288
Salonika 304
Salt Lake City *40*, 55
Salzburg 325
Samar 458
Samarkand 377
Sambre river 272
Samoa (Navigator Islands) 483, 488, **492**, 494 *see also* American Samoa
San Andreas Fault 29, 52
San Antonio 57, *57*
San Francisco 29, 40, 45, 52, *52*
San José 85
San Juan *98*
San Marino 295, **301**, *301*
San Martín, José de 113, 122
San Salvador 82
San'a 399
Sanaga river 170
Sand Island 494
Sandwich Islands 52
Sankanbiriwa 160
Santa Barbara *39*
Santa Cruz, Bolivia 110
Santa Cruz, Canary Islands 280
Santa Cruz Islands 487
Santa Fe 56
Santa Maria 287
Santiago 121, 122, 123
Santo Domingo 90
São Francisco Basin 116
São Miguel 287, *287*
São Paulo 116, 117, 119
São Tomé & Príncipe **182**, *182*
São Vicente 117
Saône River 251
Sapporo 469
Sar Mountains 342
Sarajevo 344, 345, *345*, 349
Sarawak 450, 451, *451*, 453, 454
Saray 365
Sarchi *85*
Sardinia 292, 293, 294, 295, 297
Sargon 394
Sark 234
Saskatchewan 13, 16, 19, 20
Sassandra 162
Sassou-Nguesso, Colonel Denis 146
Sa'ud dynasty 402, 403
Saudi Arabia 149, 151, **402–403**, *402*
Sauerland 310
Sava river 342, 347, 348, 349
Savai'i Island 492
Savannah river 72, *72*
Save river 192
Saxony 310
Saybrook 64

Scandinavia *see* Nordic Countries
Scandinavian Airlines System (SAS) 221, 229
Schelde river 264, 270
Schellenberg 321
Schiele, Egon 324
Schleswig-Holstein 310, 317
Schmalkaldic League (1530) 313
Schoenberg, Arnold 324
Schubert, Franz 324
Schweitzer, Albert *173*
Schwyz 318
Scotland 234, *234*, 236, 237, 240, 241, 242 *see also* United Kingdom
Scott, Robert Falcon 497
Scutari, Lake 345
Seattle 50
Second Continental Congress (1776) 34
Seefeld 324
Seine river 251
Seljalands Foss 230
Senegal 144, 145, **156**, *156*, 157
Senegal river 144, 156, 158
Senghor, Léopold Sédar 156
Sentinal Range 497
Seoul 472, 473
Sequoia National Park 32
Serbia 342–346, *342*, 348, 349, 350
Serengeti National Park, Tanzania 180
Sergius of Radonezh *367*
Sergiyev Posad *367*
Serra, Junípero 51
Serrania de Tabasara 86
Service, Robert William 31
Sevan, Lake 372
Severn river 235
Severnaya Zemlya 363
Seville 285
Seward, William Henry 50
Seybou, General Ali 146
Seychelles **181**, *181*
Shaba 175
Shah Jahan, Emperor *408*
Shakespeare, William **239**, *239*
Sham Chun river 436
Shan Plateau 440
Shandong Peninsula 466
Shanghai 432, 433
Shannon river 244
Sharjah 406
Shaw, George Bernard 246
Shays' Rebellion (1786) 63
Shebelle river 153
Shebshi Mountains 168
Sherlock Holmes, (The), London *233*
Sherman, General William Tecumseh 73
Shetland Islands 234
Shevardnadze, Eduard 371
Shikoku Island 462, 463
Shimonoseki Treaty 462
Shire river 190, 191
Siam *see* Thailand
Sibelius, Jean 223, **225**, *225*
Siberia 362, 363, 366, *368*, 369
Sichuan 427, *432*
Sicily 292, 293, 295, 296, *296*, 297, 306
Sienkiewicz, Henryk 331
Sierra de los Filabres *279*
Sierra de Los Organos 87
Sierra de Perijá 106
Sierra del Cristal 87
Sierra Leone **160**, *160*
Sierra Madre 76, 80
Sierra Madre del Sur 76
Sierra Maestra 87
Sierra Morena 278
Sierra Nevada 28, 29, *29*, 30, 32
Sighisoara *327*
Sigismund, King (Hungary) 336
Sigismund III Vasa, King (Poland) 329
Sihanouk, Prince (Cambodia) 446, 447
Sikhote-Alin 363
Silesia 328, 331, 332
Silicon Valley 52

Simen Mountains 150
Simplon Pass 298
Sinai, Mount 140
Sinai Peninsula 140, 141, 142, 143
Singapore 450, 451, **452**, *452*
Sino-Japanese War (1894–95) 435
Sino-Portuguese Treaty (1887) 437
Sint Eustatius 99, 100
Sint Maarten 99, 100
Sioux City 59
Sistema Central 278
Sistema Ibérico 278
Sistema Penibética Mulhacén 278
Sithole, Ndabaningi 196
Sitrah 404
Siwalik Hills 410
Six-Day War (1967) 141, 143
Sjaelland Island 226, 229
Skagerrak 212
Skåne 216, 220
Skardu *417*
Skaw 226
Slansky, Rudolf 335
Slovakia 332–333, **334–335**, *334*, *335*
Slovenia 342, 343, 344, **347**, *347*
Småland 216
Smith, Ian 196, 197
Smolensk 220
Snaefell 236
Snake River Plain 53
Snowdonia *236*
Soares, Dr Mario 288
Society Islands 495
Sofia 340, *340*
Sogne Fjord 212
Solomon Islands **487**, *487*
Somalia 150, **153**, *153*
Somes river 338
Somoza family 84
Sonsonate 82
Sorrento 299
Soto, Hernando de 70
Soufrière 92, 93, 100
Souk of the Dyars *133*
Sousa, Martim Afonso de 117
South Africa 187, 192, 193, 194, **198–201**, *198*, *199*, *201*, 202, 203
South America 103–129
South Australia 476, 481
South Carolina 34, 35, **72**, *72*
South Dakota 30, **58**, *58*
South Georgia 128
South Holland 264, 268
South Korea 470, **472–473**, *472*
South Pole 496, 497
South Sandwich Islands 128
South West Africa *see* Namibia
Southeast Asia 439–459
Southern Africa 185–209
Southern Alps 482, *482*
Soweto 199, 200
Soya Strait 462
Spain 260, **278–285**, *278–284*, *287–288*, 289
Spanish–American War (1898) 36, 281, 459
Spanish Armada (1588) 237, *238*, 281
Spanish Guinea 172
Spanish Sahara *see* Western Sahara
Spinola, General Ant†onio de 288
Spinoza, Baruch 266
Spitsbergen 215
Sprúngli, Rudolf *320*
Sri Lanka 266, 421, **422–423**, *422*, *423*
Stalin, Joseph 334, 335, 361, 366
Stanley, Henry Morton 175
Stanley, Mount 178
Staraya Ladoga 220
State Theater of Opera and Ballet, Minsk *359*
Statue of Liberty, New York 25
Stavanger 212, 215
Stephen I, King (Hungary) 336

Stephen Dušan, King (Serbia) 342
Stevens, Siaka 160
Stockholm 216
Stonehenge *237*
Straits Settlements 450
Strasbourg 251, 312
Strauss, Johann 324
Stroessner, General Alfredo 120
Stromboli 293
Strømo 227
Sucre, Antonio José de 110
Sudan 145, **148–149**, *148*, *149*
Sudd marshes 148
Sudetic Mountains 328, 329, 332, 334
Sudhuroy 227
Suez Canal 138, 140, 141, **143**, *143*, 306
Suez, Gulf of 140, 142, 143
Suharto, General 456
Sukarno, Ahmed 456
Sukarnoputri, Megawati 456
Sulawesi 454, *455*
Suleiman the Magnificent 385
Sulu Archipelago 458
Sulu Islands 453
Sumatra 454, 455, 456, 457
Sumba 454
Sumbawa 454
Sun Yat-sen 430, 431
Sunda Shelf 454
Sunda Strait 454
Sundarbans 411, 418
Superior, Lake 21, 42
Sûre river 274, *274*, 275
Suriname **109**, *109*, 267
Sutter, John Augustus 30
Svalbard Islands 212, **215**, *215*
Svealand 216
Sveriges Radio 221
Svyatoslav of Kiev, Grand Prince 364
Swaggart, Jimmy *40*
Swaziland **203**, *203*
Sweden 212–213, **216–221**, *216*, *217*, *218*, *220*, 221, 222, 223, 227
agriculture 220; climate 216; economy 220–221; education 221; government 218–219, *218*; history 217–218, *218*, 220; industry 221; landscape 216, *217*; people 219; transportation 221, *221*; vegetation and wildlife 216–217, *217*; welfare 221
Sweyn Forkbeard, King (Denmark) 227
Swift, Jonathan 246
Swiss Broadcasting Corporation 320
Swiss Confederation 319
Swissair 320
Switzerland **318–320**, *318*, *319*, *320*, 321
Sydney 476, 477, 478, 479, 481, *481*
Syria **386–387**, *386*, 389
Syrian Arab Airways 387
Syrian Desert 386, 392
Szczecin 331

T

Table Mountain 198, *201*
Tabuaeran Island 489
Tadjoura, Gulf of 152
Taebaek Mountains 472
Tagus river 278, 286
Tahiti 479, 484, *494*, 495
Taipa 437
Taipei 434, 435
Taiping Rebellion (1851–64) 430
Taiwan 425, 430, **434–435**, *434*, *435*, 466
Taj Mahal *408*
Tajikistan 367, **378**, *378*
Takla Makan Desert 426
Talish Mountains 374
Tamerlane 377

Tana, Lake 150, *151*
Tana river 179
Tanganyika, Lake 175, 176, 180, 188
Tanzania 178, **180**, *180*
Tara river 345
Tarai Lowlands 420
Taranto, Gulf of 293
Tarim basin 426
Tarn river 251
Tartu 354
Tashkent 377
Tasman, Abel Janszoon 479, 482, 490
Tasman Glacier 482
Tasman, Mount 482
Tasmania 476, 477, 478, 479, 481
Tassili plateau *134*
Tatra Mountains 328, 334, *335*
Tauern Mountains 322
Taunus Mountains 310, 314
Taurus Mountains 384
Taya, Sid'Ahmed Ould 144
Taygetus Peninsula 303
Taylor, Charles 161
Taymyr Depression 363
Taymyr Peninsula 363
Tbilisi 370
Teide, Pico de 280, *280*
Tel Aviv 391
Televisión Española 285
Tell 134
Tema 165
Tenasserim 440, 442
Tenerife 280, *280*
Tennessee 28, 32, 34, **69–70**, *70*
Tenochtitlán 77
Terceira 287
Texas 35, 42, **57**, *57*, 77
Thabana Ntlenyana 202
Thailand **444–445**, *444*, *445*
Thailand, Gulf of 444, 446
Thames river 234
Thar Desert *411*
Thessaly 302, 304
Thira *303*
Thirty Years War (1620–48) 217, 271, 313, 323, 333
Thorshavn 227
Thrace 304
Three Mile Island 44
Thuringen Forest 310
Tiahuanaco 110
Tianjin 426, 432
Tiber river 293, 300
Tibesti Mountains 138, 147
Tibet 410, 415, 430, 431, 432
Tibet, Plateau of 426, 427
Tien Shan Mountains 379
Tierra del Fuego 121, 122, 123, 124
Tigre 150
Tigris river 394, 395, *395*
Timor-Timur 454
Timsah, Lake 143
Tingi Hills 160
Tirol 323, *324*
Tiso, Josef 334
Tiss Isat Falls *151*
Tisza river 337
Titano, Mount 301
Titicaca, Lake 110, 112, *113*, 114
Tito, Marshal **343**, *343*, 344, 347
Toba, Lake 454
Togo **166**, *166*
Togo Mountains 164, 166, 167
Tokelau Islands **482–483**, *483*
Tokyo 47, 462, 463, 464, 466, *466*, 469
Tokyo Bay Aqualine 469
Tolbert, President William 161
Toledo, Alejandro 114
Toledo, Fernando Álvarez de 266
Tom, Mount *29*
Tombalbaye, President François 147
Tombstone 56
Tomé, Pico de 182
Tonga **491**, *491*
Tongatapu Islands 491
Tonking 448
Tonle Sap (Great Lake) 446, 447
Topa 115

Toronto *19*, 20
Torres, Luis Vaez de 478
Tortola 99
Toulon, Siege of 255
Touré, President Ahmed
  Sékou 158
Trans-Alaska Pipeline 50
Transantarctic Mountains 496,
  497
Trans-Caucasian Soviet
  Federative Republic 373, 375
Transdanubia 336
Trans-Dniestria 357
Trans-Gabon Railroad 173
Trans-Mongolian Railway 381
Trans-Siberian Railroad 369
Transvaal 198, 199, *201*
Transylvania *327*, 338, 339
Transylvanian Alps 338
Traoré, Lieutenant-Colonel
  Moussa 145
Treaty of Amsterdam 228
Treaty of Berlin (1878) 338, 342
Treaty of Guadalupe Hidalgo
  (1848) 77
Treaty of Kiel (1814) 218
Treaty of Nanjing (1842) 430
Treaty of Paris (1763) 16, 65,
  67, 267
Treaty of San Lorenzo (1795)
  70
Treaty of Turkmenchai (1828)
  375
Treaty of Vienna (1814–15) 456
Trelew 126
Trentino 296
Trenton, Battle of (1776) 66
Trieste 299, 347
Triglav, Mount 347
Trincomalee 423
Trinidad and Tobago **96**, *96*
Tripoli, Lebanon *389*
Tripoli, Libya 138, 139
Tripolitania 138
Tristan da Cunha 208–209
Trobriand 486
Trøndelag 212
Trondheim 215
Troodos Mountains 307, *307*
Tsaratanana, Massif du 204
Tsavo National Park, Kenya
  179
Tshombe, Moise 175
Tsing Ma Bridge (Hong Kong)
  436
Tsiranana, President Philibert
  204
Tsugaru Strait 462
Tsumeb 194
Tua 495
Tübingen 317
Tucson 56, 57
Tumen river 470
Tunis 136, 137, *137*
Tunis, Gulf of 136
Tunisia **136–137**, *136*, *137*, 139
Turfan Depression 426
Turin 296, 297
Turkish Airlines 385
Turkey 307, **384–385**, *384*, *385*
Turkish Bridge, Mostar *349*
Turkmenistan 367, **376**, *376*
Turkoman 397
Turks and Caicos Islands **97**
Turquino, Pico 87
Tutuila Island 495
Tuvalu **488**, *488*, 489

U

Ubangi river 171, *171*, 174, 175
Ubico, Jorge 80
Uganda **178**, *178*
Ukraine 220, 221, 223, **360–361**,
  360, *361*, 366
Ulan Bator 381
Uluru (Ayer's Rock) *476*, 477
Umbeluzi river 203
Umm al-Qaiwain 406
Ungava Peninsula 14
Union of Soviet Socialist
  Republics 43, 334, 354–381,
  366–368 *see also names of*

Soviet Republics
Union of Utrecht (1579) 266,
  271
United Arab Emirates **406**, *406*
United Kingdom 233, **234–243**,
  234–243
  agriculture and fishing 242,
    242; climate 235–236;
    dependencies *97*, 99, 128,
    208–209, 289, 423, 436;
    economy 242; education **243**;
    government 240–241, *240*;
    health and welfare *243*;
    history 237–240, 245;
    industry 242; landscape
    234–235; media 243; people
    241; transport 242–243;
    vegetation and wildlife
    236–237
United Nations Organization
  36, 128, 153, 175, 180, 192,
  194, 197, 218, 240, 307, 319,
  345, 349, 350, 394, 396, 431,
  457, 470, 473, 489, 492
United Provinces 266, 271, 281
United States of America
  25–73, *25*
  agriculture 41; climate 30;
    conservation 31–32;
    dependencies 97–98,
    493–494; economy 41–46, *44*;
    film industry **46–47**, *47*;
    fishing and forestry 41–42;
    government 38–39, *38*; Great
    Depression 36, *37*, 53, 59, 61;
    health and welfare 46;
    history 33–37; industry and
    resources 42, *42*, 44;
    landscape 26–29; media and
    communication 45–46;
    people *25*, 33, 39–41, *39*;
    Prohibition 36, *37*, 60;
    society 33–37; space
    program **43**, *43*; trade 44–45,
    44; transport *44*, 45, *45*;
    vegetation 30–31; wildlife *29*,
    31–33, *31*, *32*
Upolu Island 492
Upper Volta *see* Burkina Faso
Uppsala 216, 220, 221
Ural Mountains 362, 363, 369
Urgel 260
Uri 318
Urmia, Lake 398
Uruguay 117–118, **127**, *127*
Uruguay river 116, 124, 127
Ustyurt Plateau 377
Utah *29*, 36, **55**
Utrecht 269
Uvéa 495
Uzbekistan **377**, *377*

V

Vaal river 198, 199
Vaalserberg 264
Vaduz 321
Vágar 227
Valdai Hills 362
Valdivia, Pedro de 122
Valencia 285
Valira river 260
Valle Central, Costa Rica 85
Valle d'Aosta 296
Valletta 306, *306*
Van, Lake 384
Van Diemen's Land 479
Vancouver 13, 16, 20
Vänern, Lake 216
Vanikoro 487
Vanua Levu 490
Vanuatu **490–491**, *490*
Varanasi *412*
Vargas, Getúlio 118
Vatican City 295, **300**, *300*
Vatnajökull 230
Vava'u Islands 491
Vega, Lope de 284
Velázquez, Diego 284
Velde, Henri van de 272
Velebit Mountains 348
Velika Kapela Mountains 348
Veluwe 265
Venezuela **106–107**, *106*

Venezuela, Gulf of 106, 107
Venice 295, 299, 304
Verkhoyansk Range 363
Vermont 27, *27*, 34, *63*, *63*
Vernon, Mount 72
Verviers 273
Vespucci, Amerigo 33
Vesuvius 293
Vetle Fjord *212*
Viceroyalty of New Granada
  104
Viceroyalty of the Río de la
  Plata 120, 124, 127
Victor Emmanuel II, King
  (Sardinia) 295
Victoria, Australia 476, 477,
  479, 481
Victoria, Canada 16
Victoria Falls 188, *189*, 196
Victoria Harbor 436, *436*
Victoria, Lake 178, 179, 180
Victoria Peak 81
Victoria, Queen 240, 412, 477
Victorville, California *45*
Vienna 322, 324, 325
Vietnam **448–449**, *448*, *449*
Vietnam War (1954–75) 36, 45,
  254, 446, 447, 448
Vigo 284
Vikings **212**, **217**, **220**, *220*, 230,
  231, 237, 244
Vilnius 356, *356*
Vilyui river 363
Vinson Massif 496
Viracocha 115
Virgin Gorda 99
Virgin Islands of the United
  States 26, **98–99**, *98*
Virginia 27, 33, *34*, **71–72**, *71*
Virunga Mountains 177, 178
Visayas Islands 458
Vistula 328
Viterbo 293
Viti Levu 490, *490*
Vitocha Massif 340
Vlad Tepes, Prince (Romania)
  327
Vladimir of Kiev 364, 365
Vladivostok 363, 369
Vlissingen 269
Vljosë river *351*
Vltava river 332, 333
Vogelsberg plateau 310
Vojvodina 342, 343, 345
Volcano Islands 462, 463
Volga river 220, 362, 369
Volta Lake and Basin 164, 165
Volterra *293*
Vorarlberg 324
Vosges 250, 251, 311
Vridi Canal 162
Vyborg 217

W

Waddenzee 264, 265
Wadi Araba 392
Wagner, Richard 315, 316
Wahid, Abdurrahman 456
Wainibuki river *490*
Wajda, Andrzej 331
Wake Island **494**
Walachia 338, 357
Wales **234**, 235, 236, *236*, 237,
  240, 241, 242 *see also* United
  Kingdom
Walesa, Lech 330, 331
Walking Purchase 34
Wall Street, New York *44*
Wallis and Futuna **495**
Wallonia 272
Walvis Bay 194, 198
Wangchuk, Jigme Dorje 415
Wangchuk, Jigme Sangye 415
Wanké, Daouda Mallam 146
War of 1812 (1812–14) 35, 73
War of Austrian Succession
  (1740–48) 323
War of Independence
  (1808–14) 281
War of the Pacific (1879) 110,
  113, 122
War of the Spanish Succession
  (1702–13) 271, 289

War of the Thousand Days
  (1899–1903) 105
War of the Triple Alliance
  (1865–70) 118, 120
War of the Two Brothers
  (1826–34) 288
Warner, Sir Thomas 91
Warri (Nigeria) 168
Wars of Religion (1562–98) 253
Wars of the Roses (1455–85)
  237
Warsaw 328, 329
Warsaw Treaty Organization
  316
Washington 28, 29, 30, 32, **50**,
  51, *51*
Washington D.C. 45, 68, *68*
Washington, George 34, *34*,
  58, 64, 68, 71
Wellington 482, 485
Wenceslas IV, King (Czech
  Republic) 332, *333*
Weser river 310
West Bank 391, 392, 393
West Frisian Islands 264
West Pakistan *see* Pakistan
West Papua 267, 454, *456*, 486,
  487 *see also* Irian Jaya
West Siberian Plain 362, 363,
  364
West, Thomas 67
West Virginia 42, **71**
Western Australia 476, 479
Western Desert 140, 142
Western Dvina river 362
Western Ghats 411
Western Sahara 132–133, 144
Western Samoa *see* Samoa
Westerwald 310
White House, USA *68*
White Island 215
White Nile river 148, 149
White Volta river 163, 164
Whitney, Mount 29
Wicklow Mountains 244
Wieringermeer 264
Wilde, Oscar 246
William I, King (Netherlands)
  274
William III, King (England)
  245
William, Duke of Normandy
  237
William of Orange 238, 266,
  271
William the Silent, Prince 266
Willamette Valley 28, 51
Williams, Roger 65
Williamsburg 71, *71*, 72
Wilmington 67
Wilson, President Woodrow 36
Winchester Cathedral *238*
Windhoek 194
Windward Islands 95
Wisconsin **60**
Witwatersrand 198
Wojtyla, Karol *see* John Paul II,
  Pope
World Trade Center 65, *65*
World War I
  Australasia and Pacific
    Islands 479, 484; Central
    Africa 170, 177, 180; Eastern
    Europe 329, 337, 341; Europe
    240, 254, 267, 271, 315; Japan
    and Korea 466; Middle East
    386, 392; Nordic Countries
    218; North America 36;
    Russia and Northern
    Eurasia 359, 365, 366, 372;
    Southern Africa 199
World War II
  Australasia and Pacific
    Islands 479, 484, 487; Eastern
    Europe 329–330, 337, 338,
    341, *343*, 344, 346, 347;
    Europe 234, 240, 254, 257,
    272, 283, 295, 306, 312, 315,
    323; Japan and Korea 465,
    466; Middle East 386, 391;
    Nordic Countries 213, 218,
    223, 228, 231; North America
    36, 53, 56; Northern Africa
    132, 135, 136, 138; Russia
    and Northern Eurasia 354,
    357, 359, 361, 366; Southeast

Asia 444, 448, 451, 456, 459;
    Southern Africa 199, 204
Wright, Orville and Wilbur 62
Wuchang (Hubei province)
  431
Wyoming 28, 32, 54

X

Xavier, Francis 465
Xieng Khouang Plateau 443
Xinjiang 426, 432

Y

Yaku 463
Yakut 363
Yalla, Koumba 159
Yalu river 470
Yamuna river *414*
Yanbu 403
Yangmingshan National Park,
  Taiwan 434
Yaroslav the Wise 365
Yeats, William Butler 245, 246
Yellowstone National Park 28,
  31, 32
Yeltsin, Boris 367–368, *367*
Yemen **399**, *399*
Yenisei river 362, 363, 369
Yerevan 372, 373, *373*
Yi dynasty 470
Yokohama 469
Yom Kippur War (1973) 387
Yoritomo, Minamoto 464, *464*
Yorktown 34, 72
Yosemite National Park, USA
  32
Young, Brigham 55
Yu Shan 434
Yuan Shikai 430
Yucatán Peninsula 76
Yugoslavia **342-346**, *342*, *343*,
  345, 347 349, 350
Yukon river 30, 33
Yukon Territory 13, 16, 17, 19
Yungas 110
Yunnan 426
Yupanqui (Pachacuti) 115

Z

Zagorsk *see* Sergiyev Posad
Zagreb 348
Zagros Mountains 397
Zaire *see* Congo (D.R.O.)
Zaire river *see* Congo river and
  basin
Zakinthos 303
Zambales Mountains 458
Zambezi river 188, *189*, 192,
  193, 196
Zambia **188-189**, *189*, 196
ZANU PF District Choir *196*
Zanzibar 179, 180
Zarco, João Gonçalves 288
Zeeland 264, 266
Zeevi, Rehavam 391
Zeroual, President Lamine 135
Zeta 346
Zeta river 345
Zhao Zheng 428
Zhdanov 361
Zhelev, Zhelyu 341
Zheng Cheng-gong 435
Zhivkov, Todor 341
Zhu (Pearl) river 426, 432,
  436, 437
Zia, Begum Khaleda 418
Zia-ul-Haq, General 417
Zimbabwe 189, **196-197**, *196*,
  197
Ziyad, Tariq ibn 289
Ziz river 132
Zoetermeer *268*
Zouérate 144
Zug 319
Zugspitze 311
Zurich 319, 320